Applied and Numerical Harmonic Analysis

Introduction to
Partial Differential Equations
with MATLAB

Jeffery Cooper

1998

Birkhäuser

Boston • Basel • Berlin

Jeffery Cooper
Department of Mathematics
University of Maryland
College Park, Maryland 20742

Library of Congress Cataloging-in-Publication Data

Cooper, Jeffery.
    Introduction to partial differential equations with MATLAB /
Jeffery Cooper.
        p. cm. -- (Applied and numerical harmonic analysis)
    Includes bibliographical references and index.
    ISBN 0-8176-3967-5 (alk. paper). -- ISBN 3-7643-3967-5 (alk.
paper)
    1. Differential equations, Partial--Computer-assisted instruction.
2. MATLAB.   I. Title.   II. Series.
QA371.35.C66   1997
515'.353--dc21   97-32251
                                   CIP

Printed on acid-free paper
© 1998 Birkhäuser Boston   *Birkhäuser*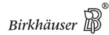

ISBN 0-8176-3967-5
ISBN 3-7643-3967-5
Typeset by T & T Techworks Inc., Coral Springs, FL
Cover design by Dutton & Sherman Design, New Haven, CT
MATLAB® is a registered trademark of The Math Works Incorporated
Printed and bound by Hamilton Printing, Rensselaer, NY
9 8 7 6 5 4 3 2 1

Dedication

For Christina and Rebecca

Contents

Preface xiii

1 Preliminaries 1
  1.1 Elements of analysis 1
    1.1.1 Sets and their boundaries 1
    1.1.2 Integration and differentiation 3
    1.1.3 Sequences and series of functions 5
    1.1.4 Functions of several variables 11
  1.2 Vector spaces and linear operators 14
  1.3 Review of facts about ordinary differential equations 17

2 First-Order Equations 19
  2.1 Generalities 19
  2.2 First-order linear PDE's 21
    2.2.1 Constant coefficients 22
    2.2.2 Spatially dependent velocity of propagation 25
  2.3 Nonlinear conservation laws 30
  2.4 Linearization 39
  2.5 Weak solutions 41
    2.5.1 The notion of a weak solution 41
    2.5.2 Weak solutions of $u_t + F(u)_x = 0$ 43
    2.5.3 The Riemann problem 45
    2.5.4 Formation of shock waves 47
    2.5.5 Nonuniqueness and stability of weak solutions 48
  2.6 Numerical methods 53
    2.6.1 Difference quotients 53
    2.6.2 A finite difference scheme 55
    2.6.3 An upwind scheme and the CFL condition 57
    2.6.4 A scheme for the nonlinear conservation law 60
  2.7 A conservation law for cell dynamics 64
    2.7.1 A nonreproducing model 64

| | | 2.7.2 | The mitosis boundary condition | 67 |
| | 2.8 | Projects | | 70 |

3 Diffusion   **73**
3.1	The diffusion equation	73
3.2	The maximum principle	77
3.3	The heat equation without boundaries	81
	3.3.1 The fundamental solution	81
	3.3.2 Solution of the initial-value problem	85
	3.3.3 Sources and the principle of Duhamel	89
3.4	Boundary value problems on the half-line	95
3.5	Diffusion and nonlinear wave motion	101
3.6	Numerical methods for the heat equation	105
3.7	Projects	110

4 Boundary Value Problems for the Heat Equation   **111**
4.1	Separation of variables	111
4.2	Convergence of the eigenfunction expansions	116
4.3	Symmetric boundary conditions	130
4.4	Inhomogeneous problems and asymptotic behavior	141
4.5	Projects	153

5 Waves Again   **157**
5.1	Acoustics	157
	5.1.1 The equations of gas dynamics	157
	5.1.2 The linearized equations	159
5.2	The vibrating string	160
	5.2.1 The nonlinear model	160
	5.2.2 The linearized equation	163
5.3	The wave equation without boundaries	165
	5.3.1 The initial-value problem and d'Alembert's formula	165
	5.3.2 Domains of influence and dependence	170
	5.3.3 Conservation of energy on the line	170
	5.3.4 An inhomogeneous problem	174
5.4	Boundary value problems on the half-line	181
	5.4.1 d'Alembert's formula extended	181
	5.4.2 A transmission problem	186
	5.4.3 Inhomogeneous problems	187
5.5	Boundary value problems on a finite interval	192
	5.5.1 A geometric construction	192
	5.5.2 Modes of vibration	193
	5.5.3 Conservation of energy for the finite interval	196
	5.5.4 Other boundary conditions	198
	5.5.5 Inhomogeneous equations	199

	5.5.6	Boundary forcing and resonance	201
5.6		Numerical methods	208
5.7		A nonlinear wave equation	211
5.8		Projects	217

6 Fourier Series and Fourier Transform **219**
6.1		Fourier series	219
6.2		Convergence of Fourier series	223
6.3		The Fourier transform	231
6.4		The heat equation again	236
6.5		The discrete Fourier transform	238
	6.5.1	The DFT and Fourier series	238
	6.5.2	The DFT and the Fourier transform	244
6.6		The fast Fourier transform (FFT)	250
6.7		Projects	257

7 Dispersive Waves and the Schrödinger Equation **259**
7.1		Oscillatory integrals and the method of stationary phase	259
7.2		Dispersive equations	263
	7.2.1	The wave equation	263
	7.2.2	Dispersion relations	264
	7.2.3	Group velocity and phase velocity	267
7.3		Quantum mechanics and the uncertainty principle	274
7.4		The Schrödinger equation	278
	7.4.1	The dispersion relation of the Schrödinger equation	278
	7.4.2	The correspondence principle	281
	7.4.3	The initial-value problem for the free Schrödinger equation	282
7.5		The spectrum of the Schrödinger operator	287
	7.5.1	Continuous spectrum	287
	7.5.2	Bound states of the square well potential	290
7.6		Projects	296

8 The Heat and Wave Equations in Higher Dimensions **297**
8.1		Diffusion in higher dimensions	297
	8.1.1	Derivation of the heat equation	297
	8.1.2	The fundamental solution of the heat equation	298
8.2		Boundary value problems for the heat equation	304
8.3		Eigenfunctions for the rectangle	310
8.4		Eigenfunctions for the disk	315
8.5		Asymptotics and steady-state solutions	322
	8.5.1	Approach to the steady state	322
	8.5.2	Compatibility of source and boundary flux	325
8.6		The wave equation	331
	8.6.1	The initial-value problem	331

8.6.2   The method of descent 335

8.7   Energy 339

8.8   Sources 343

8.9   Boundary value problems for the wave equation 347

    8.9.1   Eigenfunction expansions 347

    8.9.2   Nodal curves 349

    8.9.3   Conservation of energy 349

    8.9.4   Inhomogeneous problems 351

8.10 The Maxwell equations 356

    8.10.1   The electric and magnetic fields 356

    8.10.2   The initial-value problem 359

    8.10.3   Plane waves 359

    8.10.4   Electrostatics 360

    8.10.5   Conservation of energy 361

8.11 Projects 365

9   Equilibrium   367

9.1   Harmonic functions 367

    9.1.1   Examples 367

    9.1.2   The mean value property 368

    9.1.3   The maximum principle 372

9.2   The Dirichlet problem 377

    9.2.1   Fourier series solution in the disk 377

    9.2.2   Liouville's theorem 384

9.3   The Dirichlet problem in a rectangle 389

9.4   The Poisson equation 394

    9.4.1   The Poisson equation without boundaries 394

    9.4.2   The Green's function 399

9.5   Variational methods and weak solutions 414

    9.5.1   Problems in variational form 414

    9.5.2   The Rayleigh-Ritz procedure 417

9.6   Projects 423

10 Numerical Methods for Higher Dimensions   425

10.1 Finite differences 425

10.2 Finite elements 433

10.3 Galerkin methods 442

10.4 A reaction-diffusion equation 450

11 Epilogue: Classification   455

Appendices

A Recipes and Formulas   **459**
  A.1 Separation of variables in space-time problems 459
  A.2 Separation of variables in steady-state problems 464
  A.3 Fundamental solutions 471
  A.4 The Laplace operator in polar and spherical coordinates 474

B Elements of MATLAB   **477**
  B.1 Forming vectors and matrices 477
  B.2 Operations on matrices 480
  B.3 Array operations 481
  B.4 Solution of linear systems 482
  B.5 MATLAB functions and mfiles 483
  B.6 Script mfiles and programs 485
  B.7 Vectorizing computations 486
  B.8 Function functions 488
  B.9 Plotting 2-D graphs 490
  B.10 Plotting 3-D graphs 492
  B.11 Movies 496

C References   **497**

D Solutions to Selected Problems   **501**

E List of Computer Programs   **527**

Index   **533**

Preface

Overview

The subject of partial differential equations has an unchanging core of material but is constantly expanding and evolving. The core consists of solution methods, mainly separation of variables, for boundary value problems with constant coefficients in geometrically simple domains. Too often an introductory course focuses exclusively on these core problems and techniques and leaves the student with the impression that there is no more to the subject. Questions of existence, uniqueness, and well-posedness are ignored. In particular there is a lack of connection between the analytical side of the subject and the numerical side. Furthermore nonlinear problems are omitted because they are too hard to deal with analytically.

Now, however, the availability of convenient, powerful computational software has made it possible to enlarge the scope of the introductory course. My goal in this text is to give the student a broader picture of the subject. In addition to the basic core subjects, I have included material on nonlinear problems and brief discussions of numerical methods. I feel that it is important for the student to see nonlinear problems and numerical methods at the beginning of the course, and not at the end when we run usually run out of time. Furthermore, numerical methods should be introduced for each equation as it is studied, not lumped together in a final chapter.

The text is intended for a first course in partial differential equations at the undergraduate level for students in mathematics, science and engineering. It is assumed that the student has had the standard three semester calculus sequence including multivariable calculus, and a course in ordinary differential equations. Some exposure to matrices and vectors is helpful, but not essential. No prior experience with MATLAB is assumed, although many engineering students will have already seen MATLAB.

Organization and features

The material is organized by physical setting and by equation rather than by technique of solution. Generally, I have tried to place each equation in the appropriate physical context with a careful derivation from physical principles. This is followed by a discussion of qualitative properties of the solutions without boundary conditions. Then a nonlinear version of the equation and an appropriate numerical scheme are introduced. Thus Chapter 2 deals with linear first-order equations and the method of characteristics, an example of a scalar nonlinear conservation law, and numerical methods for these equations. Chapter 3 follows the same strategy for the heat equation, while Chapter 4 treats boundary value problems for the heat equation. The wave equation is studied in Chapter 5. Fourier series and the Fourier transform are studied in Chapter 6. Chapter 7 is perhaps novel for this level course. It deals with the method of stationary phase and dispersive equations. The Schrödinger equation is discussed as a dispersive equation. Chapter 8 has fairly standard material on the linear heat and wave equations in higher dimensions. Chapter 9 treats the Laplace and Poisson equations, and includes some examples of nonlinear variational problems. Finally, Chapter 10 provides more numerical methods, suitable for computations in higher dimensions, including an introduction to the finite element method and Galerkin methods.

A substantial amount of time is spent on the standard boundary value problems and the technique of separation of variables. The formulas obtained this way contain much valuable information about the structure of the solutions. Hand computation of the eigenfunctions is done when possible. The usual restriction at this stage in the discussion is that one can only hand compute the Fourier coefficients for a limited number of examples. One could then provide numerical means for evaluating these integrals. However, I have chosen to use finite difference schemes to compute the solutions because they can easily be extended to equations with variable coefficients. Another possibility is the Galerkin method, treated in Chapter 10. In Chapter 6 where Fourier series and Fourier transform are discussed, I have included a brief treatment of the discrete Fourier transform and the fast Fourier transform. If one needs to compute Fourier coefficients in a practical manner, the FFT is the modern, efficient procedure to use.

The nonlinear conservation law discussed in Chapter 2 provides a dramatic example of the difference between linear and nonlinear behavior. It also provides an important motivation for developing numerical methods. However, it is possible to skip the sections on the nonlinear problems in Chapter 2 and go directly to the linear heat equation in Chapter 3. Other examples on nonlinear equations are seen in Chapters 3, 5, 9, and 10.

Computational aspects

I have used the computer extensively in the text with many exercises, although it is possible to use this text without the computer exercises. I feel strongly that students must receive some computational experience as they learn the subject. In addition, current graphics capability provides an intuitive understanding of the properties of the solutions that was not available 20 years ago. Almost every set of exercises has some exercises to be done with a computer. In many cases these computer exercises are paired with analytical exercises, one clarifying the other. In addition, at the end of each chapter there are several computing projects which involve programming. These longer projects provide the most valuable educational experience. Here the student must compute and use analytical examples to check the validity of his results.

My choice of software is MATLAB. I found that MATLAB offers a flexibility and ease of programming that make it preferable to other software packages. In addition MATLAB graphics are excellent and easy to use. There are many good introductions to MATLAB available but I have included a brief appendix containing some basic elements of MATLAB. Within the exercises there are also instructions on how to use MATLAB.

I have compromised between the extremes of having the student do all the programming, or using all prewritten programs. Having the student write all the programs would be too time consuming so I have written a collection of MATLAB mfiles to implement several basic finite difference schemes. In addition there are several mfiles which help with graphical constructions. The mfiles have been tested on both MATLAB4.2 and MATLAB5.0. The only difference may be in the colors that appear in the graphs. The mfiles are grouped by chapter and are available at the following two web sites: `www.Birkhauser.com/book/isbn/0-8176-3967-5` and `www.math.umd.edu/~jec` . This collection of mfiles will be expanded, hopefully with the help of other interested users of the text.

Acknowledgements

I have received many helpful comments and suggestions from my colleagues here at the University of Maryland. I want to extend my warmest thanks to all of them: John Osborn, Stuart Antman, Harland Glaz, Manos Grillakis, Alexander Dragt, Bruce Kellogg, John Benedetto, Robert Pego, and Jerry Sather. I also wish to thank the reviewers, Bob Rogers and Michael Beals; and Thomas Beale who used an earlier version of the text at Duke. Many of their suggestions have been incorporated. Peter Close provided important assistance in checking the MATLAB codes. Finally I wish to thank my students over several semesters who worked through the text, making valuable suggestions.

Chapter 1

Preliminaries

1.1   Elements of analysis

In this section we will introduce some ideas and notations which we will use often in the study of partial differential equations. You will have seen most of this before in calculus. Details and proofs of these results can be found in most calculus books or in an advanced calculus book such as [Bar].

1.1.1   Sets and their boundaries

Definitions

We will often be discussing sets of points in R^n, $n = 1, 2$ or 3. Let G be a subset of R^n. We shall assume that we know what the boundary of the set is, and shall illustrate with examples. We will let the set of boundary points be denoted by ∂G. G need not contain all of its boundary, or even any of its boundary. In fact we define a set G as an *open set* if G contains none of its boundary points. This implies that if G is an open subset of R^2, G contains some two-dimensional area, and if G is an open subset of R^3, G contains some three-dimensional volume. By the *closure* of G, denoted $\bar{G}$, we mean $G \cup \partial G$.

Examples

(i) R^n and the empty set ϕ are open sets of R^n because these sets have no boundary points.

(ii) An interval $\{x \in R : a < x < b\}$ is an open set, called an *open interval*. The boundary of this set consists of the two endpoints a, b. The closure of this set is the *closed interval* $\{x : a \leq x \leq b\}$.

(iii) $G = \{(x, y) \in R^2 : x > 0\}$ is an open subset of R^2. The boundary of G, $\partial G = \{(x, y) : x = 0\}$, is the y-axis. The closure of G, $\bar{G} = \{(x, y) : x \geq 0\}$.

(iv) $G = \{(x, y, z) \in R^3 : x^2 + y^2 + z^2 < \rho^2\}$ is an open set in R^3 called the *open ball* of radius ρ. The boundary $\partial G = \{(x, y, z) : x^2 + y^2 + z^2 = \rho^2\}$ is the sphere of radius ρ. The closure $\bar{G} = \{(x, y, z) : x^2 + y^2 + z^2 \leq \rho^2\}$ is the *closed ball* of radius ρ.

(v) $G = \{(x, y) : 0 < x^2 + y^2 < b^2\}$ is the punctured disk of radius b, and is an open set of R^2. The boundary here consists of two pieces $\partial G = \{(0, 0)\} \cup \{(x, y) : x^2 + y^2 = b^2\}$. $\bar{G}$ is the closed disk of radius b.

Remark   Open sets are often defined by strict inequalities, while the closures are often defined by $\geq$ or $\leq$.

More Definitions

We say that a subset $A \subset R^2$ ($A \subset R^3$) is a *bounded* set if A is contained in some disk of finite radius (some ball of finite radius).

Let A be a subset of R, R^2, or R^3, and let f be a real valued function on A. We say that f is a *bounded* function on A if the range of f is a bounded interval of R. In other words, if $A \subset R^2$, there is a constant M such that $|f(x, y)| \leq M$ for all $(x, y) \in A$.

Finally, we say that a set K contained in R^n is *compact* if it is both closed and bounded.

Examples

(i) $A = \{(x, y) \in R^2 : x \geq 0\}$ is not bounded.

(ii) $B = \{(x, y, z) \in R^3 : x^2 + y^2 \leq \rho^2, 0 \leq z \leq b\}$ is bounded (and closed).

(iii) $f(x, y) = x^2/(1 + x^2 + y^2)$ is bounded on the set A of part (i).

(iv) $f(x) = 1/x$ is unbounded on the interval $(0, 1)$.

(v) The set B above is compact.

(vi) The set $\{(x, y) : x \geq 0\}$ is closed, but not bounded, and so is not compact.

We recall an important result from calculus about continuous functions.

Theorem 1.1
Let K be a compact subset of R^n, and let f be a continuous function $f : K \to R$. Then f is bounded on K and there exist points p_* and p^* such that

$$m \equiv f(p_*) \leq f(p) \leq f(p^*) \equiv M$$

for all $p \in K$.

We write

$$M = \max_K f \qquad and \qquad m = \min_K f.$$

Of course there may be many points in K where f attains its maximum and minimum.

1.1.2 Integration and differentiation

We group three results together under the name of a Null Theorem because they are used to show when a function is zero.

Theorem 1.2

(a) Suppose that $f(x)$ is a continuous function on an interval $I \subset R$ such that $\int_a^b f \, dx = 0$ for all subintervals $[a, b] \subset I$. Then $f(x) = 0$ on I.

(b) Suppose that f is a continuous function on the interval $[a, b]$ with $f(x) \geq 0$ for all $x \in [a, b]$. If $\int_a^b f \, dx = 0$, then $f(x) = 0$ on $[a, b]$.

(c) Suppose that f is continuous on the interval $[a, b]$ and that

$$\int_a^b f(x)\varphi(x)dx = 0 \tag{1.1}$$

for all functions φ which are continuous on $[a, b]$. Then $f(x) = 0$ on $[a, b]$.

We give a brief proof of these assertions. Part (a) is easy to see. By the fundamental theorem of calculus,

$$0 = \frac{d}{db} \int_a^b f(x)dx = f(b)$$

for all $b \in I$.

The proof of part (b) goes as follows. If there is some point $x_0 \in [a, b]$ such that $f(x_0) > 0$, then, since f is assumed continuous, there must be some nonempty interval J containing x_0 such that $f(x) > 0$ for $x \in J$. Hence, because $f(x) \geq 0$,

$$\int_a^b f(x)dx \geq \int_J f(x)dx > 0$$

which contradicts our assumption that the integral is zero.

Now for part (c). In the integral equation (1.1), we can take $\varphi = f$. Hence this condition implies that

$$\int_a^b f^2(x)dx = 0.$$

Then it follows from part (b) that $f(x)^2 = 0$, and hence $f(x) = 0$ for $x \in [a, b]$.

The conclusion still remains true, if instead of requiring that (1.1) hold for all continuous functions φ, we require that it hold for some smaller class of functions ψ, for example, continuously differentiable functions. The key point is that we must be able to approximate the continuous functions φ by functions ψ from the smaller class.

Differentiation of integrals

Often we must differentiate an integral with respect to a parameter which may appear in the limits of integration, or in the integrand.

Let $f(x, t)$ be a continuous function of x, t on the rectangle $\{a \le x \le b\} \times \{c \le t \le d\}$. Assume that the partial derivative $\partial f / \partial t$ is also continuous on this rectangle. Let the function $J(t)$ be defined by an integral:

$$J(t) = \int_{a(t)}^{b(t)} f(x, t)dx,$$

where $a \le a(t) \le b(t) \le b$ are continuously differentiable functions of t.

Theorem 1.3

$$\frac{d}{dt}J(t) = \frac{d}{dt}\int_{a(t)}^{b(t)} f(x, t)dx$$

$$= f(b(t), t)b'(t) - f(a(t), t)a'(t) + \int_{a(t)}^{b(t)} \frac{\partial f}{\partial t}(x, t)dx. \quad (1.2)$$

To understand where this formula comes from, it is best to view $J(t)$ as a composite function,

$$J(t) = F(a(t), b(t), t),$$

where $F(a, b, t) = \int_a^b f(x, t)dx$. Now

$$\frac{\partial F}{\partial t}(a, b, t) = \lim_{k \to 0} \frac{F(a, b, t + k) - F(a, b, t)}{k}$$

$$= \int_a^b \lim_{k \to 0} \frac{f(x, t+k) - f(x, t)}{k} dx = \int_a^b \frac{\partial f}{\partial t}(x, t) dx.$$

For the general case when the arguments a and b are replaced by functions $a(t)$ and $b(t)$, we use the fundamental theorem of calculus and the chain rule.

Improper integrals

The following results can be found in any calculus book.

$$\int_0^1 \frac{dx}{x^p} < \infty \qquad (1.3)$$

for $p < 1$ and diverges for $p \geq 1$.

$$\int_1^\infty \frac{dx}{x^p} < \infty \qquad (1.4)$$

for $p > 1$ and diverges for $p \leq 1$.

1.1.3 Sequences and series of functions

In many situations we must approximate the solution of a partial differential equation by finite sums of simpler, building-block, solutions. We must have some criteria for these approximations, and their integrals and derivatives, to converge.

Definition   *A sequence of functions $f_N(x)$ defined on the interval $[a, b]$ converges* pointwise *to a function $f(x)$ if for each $x \in [a, b]$*

$$\lim_{N \to \infty} f_N(x) = f(x).$$

The sequence f_N converges uniformly *on $[a, b]$ to f if f_N converges pointwise to f and*

$$\lim_{N \to \infty} \max_{x \in [a,b]} |f_N(x) - f(x)| = 0.$$

An example of a sequence of functions which converges pointwise but not uniformly is $f_N(x) = x^N$ which converges pointwise on $[0, 1]$ to the function $f(x) = 0$ for $0 \leq x < 1$ with $f(1) = 1$.

An important result from analysis is

Theorem 1.4
If f_N are continuous on $[a, b]$ and converge uniformly to f on $[a, b]$, then f is continuous on $[a, b]$.

Put another way, if the f_N are continuous and the limit function f is not continuous, then the convergence cannot be uniform. The functions $f_N(x) = x^N$ on $[0, 1]$ illustrate this situation.

Series

First of all, we recall the following results for series of numbers.

(a) The series $\sum_{n=1}^{\infty} \frac{1}{n^p}$ converges if and only if $p > 1$.

(b) The geometric series $\sum_{n=0}^{\infty} z^n$ converges for $|z| < 1$. Here z can be a complex number. Furthermore, $\sum_{n=1}^{\infty} n^p z^n$ also converges for complex $|z| < 1$ and any p.

Let $f_n(x)$, $n = 1, 2, 3, \ldots$ be defined on $[a, b]$ and set $s_N(x) = \sum_{n=1}^{N} f_n(x)$. The s_N are the *partial sums* of the infinite series $\sum_{n=1}^{\infty} f_n(x)$.

Definition   *The series $\sum_{n=1}^{\infty} f_n(x)$ converges* pointwise *to $s(x)$ if the sequence of partial sums $s_N(x)$ converges pointwise to $s(x)$.*
The series converges uniformly *on $[a, b]$ if the partial sums converge uniformly on $[a, b]$ to $s(x)$.*

The Weierstrass test gives a criterion for uniform convergence.

Theorem 1.5
Suppose that the functions $f_n(x)$ are defined on $[a, b]$ and for each $n = 1, 2, 3, \ldots$ satisfy

$$|f_n(x)| \leq M_n \quad \text{for all } x \in [a, b],$$

where M_n is a sequence of (nonnegative) numbers such that $\sum_{1}^{\infty} M_n$ converges. Then the series $\sum_{1}^{\infty} f_n(x)$ converges uniformly on $[a, b]$.

We can draw useful conclusions about series that converge uniformly.

Theorem 1.6
Suppose that the functions $f_n(x)$ are continuous on $[a, b]$ and that the series $\sum_{1}^{\infty} f_n(x)$ converges uniformly. Then
(a) the sum of the series $s(x) = \sum_{1}^{\infty} f_n(x)$ is also continuous on $[a, b]$.

(b) the series may be integrated term by term:

$$\int_a^b \sum_{n=1}^{\infty} f_n(x)dx = \sum_{n=1}^{\infty} \int_a^b f_n(x)dx.$$

(c) *if in addition, each $f_n(x)$ is continuously differentiable on $[a, b]$, and the differentiated series $\sum_1^{\infty} f_n'(x)$ also converges uniformly, then the series may be differentiated term by term:*

$$\frac{d}{dx} \sum_{n=1}^{\infty} f_n(x) = \sum_{n=1}^{\infty} f_n'(x).$$

The delta function

Certain sequences of functions, even though they do not converge in the usual sense, still have a weaker kind of convergence that can be very useful. Here are a couple of important examples.

(a) Let the functions $f_r(x)$ be defined by

$$f_r(x) = \begin{cases} r, & |x| < 1/2r \\ 0, & |x| > 1/2r \end{cases}.$$

(b) Let the functions $\gamma_r(x)$ be defined by

$$\gamma_r(x) = \sqrt{\frac{r}{\pi}} e^{-rx^2}.$$

Functions of the form $A \exp(-rx^2)$ are called *Gaussians*.

There are some properties common to both of these families of functions. It is easy to see that

(i) $f_r(x) \geq 0$;

(ii) $\lim_{r \to \infty} f_r(x) = 0$ for $x \neq 0$, and $\lim_{r \to \infty} f_r(0) = +\infty$; and

(iii) $\int_R f_r(x)dx = 1$ for all $r > 0$.

It is easy to see that the first two properties are true for γ_r, as $r \to \infty$. The third is seen by a change of variable $y = \sqrt{r}x$:

$$\int_R \gamma_r(x)dx = \sqrt{\frac{r}{\pi}} \int_R e^{-rx^2} dx = \frac{1}{\sqrt{\pi}} \int_R e^{-y^2} dy = 1.$$

Theorem 1.7

Let $g(x)$ be a bounded function which is continuous at $x = 0$. Then *(a)* $\lim_{r \to \infty} \int_R f_r(x)g(x)dx = g(0)$, *and*

(b) $\lim_{r \to \infty} \int_R \gamma_r(x)g(x)dx = g(0)$.

There is a simple explanation for both of these assertions. In case (a), the integral $\int f_r g \, dx$ is just the average of the function g over the interval $[-1/2r, 1/2r]$. As the interval shrinks down to 0, the average of g tends to $g(0)$ because g is continuous at $x = 0$. This is also a restatement of the Fundamental Theorem of Calculus.

Case (b) is similar, but we must make an intermediate step first. Let $\varepsilon > 0$ be a very small number. Then

$$\lim_{r \to \infty} \int_R \gamma_r(x)g(x)dx = \lim_{r \to \infty} \int_{|x| \le \varepsilon} \gamma_r(x)g(x)dx$$

because $\gamma_r(x)$ tends to zero for all $x \ne 0$, as $r \to \infty$. Now again because g is continuous at $x = 0$, $g(x) \approx g(0)$ on the short interval $|x| \le \varepsilon$. Thus

$$\int_{|x| \le \varepsilon} \gamma_r(x)g(x)dx \approx g(0) \int_{|x| \le \varepsilon} \gamma_r(x)dx.$$

Making the same change of variable $y = \sqrt{r}x$, we see that, as $r \to \infty$,

$$\int_{|x| \le \varepsilon} \gamma_r(x)dx = \frac{1}{\sqrt{\pi}} \int_{-\varepsilon\sqrt{r}}^{\varepsilon\sqrt{r}} e^{-y^2} dy \to \frac{1}{\sqrt{\pi}} \int_R e^{-y^2} dy = 1.$$

Consequently,

$$\lim_{r \to \infty} \int_{|x| \le \varepsilon} \gamma_r(x)g(x)dx = g(0),$$

which proves assertion (b).

Both of these sequences of functions can be used to define a new mathematical object called the δ function. Rather imprecisely, we can write

$$\lim_{r \to \infty} f_r(x) = \lim_{r \to \infty} \gamma_r(x) = \delta(x),$$

where

$$\delta(x) = \begin{cases} 0, & x \ne 0 \\ +\infty, & x = 0 \end{cases}.$$

This does not properly define $\delta(x)$ as a function because the sequences do not converge at $x = 0$. Furthermore, if we take $g(x) = 1$ in Theorem 1.7, it appears that we should have

$$1 = \lim_{r \to \infty} \int_R f_r(x)dx = \int_R \delta(x)dx.$$

However a function which is zero for all $x \neq 0$ cannot have a nonzero integral.

In fact, this new object, the δ function, is not a function in the usual sense. Rather, it is defined by how it acts on other functions. To each continuous function $g(x)$, the δ function assigns the value $g(0)$. Using an abuse of notation, as if it were a function in the usual sense, we write

$$\int_R \delta(x)g(x)dx = g(0).$$

This definition of $\delta(x)$ is consistent with Theorem 1.7 and says that $f_r(x)$ and $\gamma_r(x)$ converge to $\delta(x)$, as $r \to \infty$, in the sense that the integrals in parts (a) and (b) converge.

We can easily translate the "spike" of the δ function to another position $x = a$ by taking the limits of the functions $f_r(x - a)$ or $\gamma_r(x - a)$. In this case, the definition becomes

$$\int_R \delta(x - a)g(x)dx = g(a)$$

for any continuous function g.

The δ function is an example of a *generalized function* or *distribution*. Parts (a) and (b) of Theorem 1.7 are interpreted to mean that

$$f_r(x) \to \delta(x) \quad \text{and} \quad \gamma_r \to \delta(x)$$

in the sense of distributions. Finally, we remark that there are many other families of functions which also converge to the δ function. They do not have to be nonnegative, and in fact, do not even have to be real valued.

An introductory treatment of distributions is given in [Str1], and a more complete treatment is given in [RR].

Notation: O and o

It is frequently useful to compare the rates at which functions tend to a limit. For example, let $f(x)$ and $g(x)$ be functions defined near $x = 0$. Then we say that

$$f(x) = O(g(x)) \quad \text{as } x \to 0$$

if there is a constant C such that

$$|f(x)| \leq C|g(x)| \quad \text{as } x \to 0.$$

We say that

$$f(x) = o(g(x)) \quad \text{as } x \to 0$$

if $g(x) \neq 0$ for x near 0 and

$$\frac{f(x)}{g(x)} \to 0 \quad \text{as } x \to 0.$$

We can also speak of $f(x) = O(g(x))$ or $f(x) = o(g(x))$ as $x \to \infty$ to compare the behavior of functions, as $x \to \infty$.

Examples

(a) $\sin(x) = O(x)$, as $x \to 0$.

(b) $\cos(x) - 1 = O(x^2)$, as $x \to 0$, which we can also write $\cos(x) = 1 + O(x^2)$.

(c) $x^p = o(e^x)$, as $x \to \infty$, for any power p.

(d) The Taylor expansion of a function $f(x)$ with $(k+1)$ continuous derivatives near $x = a$ is

$$f(x) = f(a) + f'(a)(x-a) + \ldots + \frac{f^{(k)}(a)}{k!}(x-a)^k + \frac{f^{(k+1)}(\xi)}{(k+1)!}(x-a)^{k+1} \quad (1.5)$$

where ξ is some point between x and a. The last term, the remainder term, is sometimes abbreviated as $O((x-a)^{k+1})$ and we may write

$$f(x) = f(a) + f'(a)(x-a) + \ldots + \frac{f^{(k)}(a)}{k!} + O((x-a)^{k+1}).$$

For example,

$$e^{x^2} = 1 + x^2 + O(x^4) \quad \text{as } x \to 0.$$

1.1.4 Functions of several variables

First a bit of notation. We shall be dealing mostly with scalar functions of two or three real variables $u(x, y)$ or $u(x, y, z)$. The first-order partial derivatives are denoted

$$\frac{\partial u}{\partial x} = \partial_x u = u_x, \qquad \frac{\partial u}{\partial y} = \partial_y u = u_y, \qquad \frac{\partial u}{\partial z} = \partial_z u = u_z.$$

Second order partial derivatives are written

$$\frac{\partial^2 u}{\partial x^2}, \quad \frac{\partial^2 u}{\partial y^2}, \quad \frac{\partial^2 u}{\partial z^2}, \quad \frac{\partial^2 u}{\partial x \partial y}, \quad \frac{\partial^2 u}{\partial x \partial z}, \quad \frac{\partial^2 u}{\partial y \partial z}.$$

Although, for the most part, we shall deal only with partial derivatives of first and second order, a more general notation is convenient. A partial derivative of order $m \geq 1$ is an expression

$$\frac{\partial^m u}{\partial x^{m_1} \partial y^{m_2} \partial z^{m_2}}$$

where m_1, m_2, m_3 are nonnegative integers such that $m_1 + m_2 + m_3 = m$. Sometimes we simply write $D^m u$ to represent all the derivatives of order m. Let G be an open set of R^n, $n = 1, 2$ or 3. Let u be defined on G ($u = u(x)$, $u = u(x, y)$, or $u = u(x, y, z)$).

Definition   *We say that u is of class C^k on G if*

$$D^m u \quad \text{is continuous on G for all } m \leq k.$$

We say that u belongs to $C^\infty(G)$ if it has derivatives of all orders.

Examples

(i) $f(x) = |x|$ is continuous everywhere, i.e., $f \in C^0(R)$, but is not differentiable at $x = 0$. In fact, $f \in C^\infty(x > 0)$, and $f \in C^\infty(x < 0)$.

(ii) $g(x) = 0$ for $x \leq 0$, and $g(x) = x^2$ for $x > 0$. $g \in C^1(R)$, but the second derivative does not exist at $x = 0$.

(iii) $u(x, y) = |x| y^2$. u is continuous on R^2. However, the partial derivative with respect to x does not exist on the line $x = 0$ when $y \neq 0$. The partial derivative

$u_x(0, 0)$ does exist, and partial derivatives with respect to y of all orders exist on all of R^2. (The reader should verify this.)

We can speak of u being in $C^k(\bar{G})$ if, for each m such that $m \leq k$, $D^m u$ extends to a continuous function on $\bar{G}$.

The divergence theorem

This theorem is an important tool in the study of partial differential equations. It is a higher dimensional analogue of the fundamental theorem of calculus.

Suppose that G is an open, bounded subset of R^2 (or R^3) and that ∂G is a piecewise-smooth curve (piecewise-smooth surface). On the smooth pieces of the curve or surface, we can define the exterior unit normal vector $\mathbf{n} = (n_1, n_2)$ or $\mathbf{n} = (n_1, n_2, n_3)$. For examples, think of a rectangle in R^2 or hemisphere in R^3.

Theorem 1.8
Let $\mathbf{f}(x, y, z) = (u(x, y, z), v(x, y, z), w(x, y, z))$ be a C^1 vector field on $\bar{G}$ where G is an open, bounded subset of R^3 with piecewise-smooth boundary ∂G. Then

$$\int \int \int_G \nabla \cdot \mathbf{f} \, dxdydz = \int \int_{\partial G} \mathbf{f} \cdot \mathbf{n} \, dS$$

where $\nabla \cdot \mathbf{f} = div\mathbf{f} = u_x + v_y + w_z$ is the divergence of $\mathbf{f}$, $\mathbf{n} \cdot \mathbf{f}$ is the normal component of $\mathbf{f}$ at the boundary ∂G, and dS is the element of surface area on ∂G.

The statement in R^2 for $\mathbf{f}(x, y) = (u(x, y), v(x, y))$ is

$$\int \int_G \nabla \cdot \mathbf{f} \, dxdy = \int_{\partial G} \mathbf{n} \cdot \mathbf{f} \, ds$$

where now $\nabla \cdot \mathbf{f} = u_x + v_y$ and ds is the element of arc length on ∂G.

The divergence theorem can be used to derive an analogue of integration by parts. Let $G \subset R^2$ as above and let $u(x, y)$ and $v(x, y)$ be C^1 functions on $\bar{G}$. Then let $\mathbf{f}$ be the vector field $\mathbf{f}(x, y) = (u(x, y)v(x, y), 0)$, so that

$$div\mathbf{f} = \nabla \cdot \mathbf{f} = ((uv)_x, 0) = (u_x v + u v_x, 0).$$

Then applying the divergence theorem

$$\int \int_G (u_x v + u v_x) \, dxdy = \int \int_G \nabla \cdot \mathbf{f} \, dxdy$$

$$= \int_{\partial G} \mathbf{n} \cdot \mathbf{f} \, ds = \int_{\partial B} n_1 uv \, ds,$$

which can be rewritten,

$$\int \int_G u_x v \, dxdy = \int_{\partial G} n_1 uv \, ds - \int \int_G uv_x \, dxdy. \qquad (1.6)$$

Similarly,

$$\int \int_G u_y v \, dxdy = \int_{\partial G} n_2 uv \, ds - \int \int_G uv_y \, dxdy.$$

These results also hold in higher dimensions.

Change of variables in multiple integrals

Let G be an open, bounded set in R^2, and let $\mathbf{F}(x, y) = (u(x, y), v(x, y))$ be a C^1 function on $\bar{G}$ which maps G in a 1 to 1 fashion onto another set G^*. Define the Jacobian of $\mathbf{F}$ as the 2×2 determinant

$$J(x, y) = det \begin{bmatrix} u_x & u_y \\ v_x & v_y \end{bmatrix}.$$

Then, for any continuous function $f : \bar{G}^* \to R$,

$$\int \int_{G^*} f(u, v) dudv = \int \int_G f(u(x, y), v(x, y)) |J(x, y)| dxdy.$$

A similar formula holds for sets $G \subset R^3$ and mappings

$$\mathbf{F}(x, y, z) = (u(x, y, z), v(x, y, z), w(x, y, z)).$$

In this case the Jacobian is the 3×3 determinant

$$J(x, y, z) = det \begin{bmatrix} u_x & u_y & u_z \\ v_x & v_y & v_z \\ w_x & w_y & w_z \end{bmatrix}.$$

Taylor expansions

Taylor expansions in several variables are difficult to write down without additional notation. We give an expansion up to quadratic terms. Suppose that $f(x, y)$ is of class C^3 for (x, y) near (a, b). Then

$$f(x, y) = f(a, b) + f_x(a, b)(x - a) + f_y(a, b)(y - b) \qquad (1.7)$$

$$+(1/2)f_{xx}(a, b)(x-a)^2 + f_{xy}(a, b)(x-a)(y-b) + (1/2)f_{yy}(a, b)(y-b)^2 + O(r^3)$$

where $r = \sqrt{(x - a)^2 + (y - b)^2)}$ is the distance in R^2 and we have used the O notation introduced at the end of Section 1.1.3.

1.2   Vector spaces and linear operators

You are probably familiar with vectors and vector operations in R^n. We want to extend these concepts to spaces of functions. To make this extension, we extract the essential properties of a vector space and state them as axioms.

Definition   *A set X is a vector space over R if there is an addition between elements $u, v \in X$, $(u, v) \to u + v \in X$, and scalar multiplication between real numbers $\alpha \in R$ and elements $u \in X$, $(\alpha, u) \to \alpha u \in X$ with the following properties:*

(1) $u + v = v + u$ for all $u, v \in X$ (commutative property).

(2) $u + (v + w) = (u + v) + w$ for all $u, v, w \in X$ (associative property).

(3) There exits a zero element $0 \in X$ such that $0 + u = u + 0 = u$ for all $u \in X$.

(4) For each $u \in X$, there is a $v \in X$ such that $u + v = 0$ (existence of an additive inverse).

(5) $\alpha(u + v) = \alpha u + \alpha v$ for all $\alpha \in R, u, v \in X$.

(6) $(\alpha + \beta)u = \alpha u + \beta u$ for all $\alpha, \beta \in R, u \in X$.

(7) $0u = 0$ for all $u \in X$ where the zero on the left is the scalar zero in R and the zero on the right is the zero vector in X.

(8) $1u = u$ for all $u \in X$ where the one on the left is the unit in R.

Examples

(i) $X = R^n$ and $u \in X$ is an n-tuple of real numbers $u = (u_1, \ldots, u_n)$ with addition defined componentwise and scalar multiplication $\alpha u = (\alpha u_1, \ldots, \alpha u_n)$.

(ii) $X = C^k(G)$ where G is an open subset of R^n. If, for example, $G \subset R^2$ and $u(x, y)$ and $v(x, y)$ are functions on G, we define $(u+v)(x, y) = u(x, y)+v(x, y)$ and $(\alpha u)(x, y) = \alpha u(x, y)$. Of course, one must verify that if $u \in C^k$ and $v \in C^k$, then $\alpha u \in C^k$ and $u + v \in C^k$.

We say that a subset $W \subset X$ is a vector *subspace* of X if W also satisfies the axioms (1) - (8) using the vector addition and scalar multiplication inherited from X. One can verify whether or not W is a vector subspace by merely checking the following two conditions:

(1) $u + v \in W$ for all $u, v \in W$.
(2) $\alpha u \in W$ for all $\alpha \in R, u \in W$.

Examples

(i) $C^k(G)$ is a subspace of $C^l(G)$ whenever $l \leq k$.

Vector subspaces often arise in the study of partial differential equations by the imposition of boundary conditions in the spaces of functions.

(ii) Let $X = C^k(\bar{G})$. The reader should verify that

$$W = \{u \in C^k(\bar{G}) : u = 0 \quad \text{on} \quad \partial G\}$$

is a vector subspace of X.

(iii) Assume $k \geq 1$ and let

$$Z = \{u \in C^k(\bar{G}) : \frac{\partial u}{\partial n} = \mathbf{n} \cdot \nabla u = 0 \quad \text{on} \quad \partial G\}.$$

Then Z is also a vector subspace of X.

Linear operators

If A is an $m \times n$ matrix, matrix multiplication with a vector $u \in R^n$ produces a vector $Au \in R^m$. The mapping $u \rightarrow Au$ from R^n to R^m has the property that

$$A(\alpha u + \beta v) = \alpha Au + \beta Av$$

for all $\alpha, \beta \in R$ and all $u, v \in R^n$. We say that $u \rightarrow Au$ is a *linear transformation*.

More generally, if X and Z are vector spaces, a mapping $L : X \rightarrow Z$ with the property

$$L(\alpha u + \beta v) = \alpha Lu + \beta Lv$$

for all $\alpha, \beta \in R$ and all $u, v \in X$ is said to be a *linear operator*. We look at several examples in the function spaces we defined in Section 1.1.4.

Examples

(i) Let $X = C^1[0, 1]$ and $Z = C[0, 1] = C^0[0, 1]$ and define the operator

$$(Lu)(x) = u'(x) + a(x)u(x),$$

where $a(x)$ is a continuous function on $[0, 1]$. Note that here X is a subspace of Z.

(ii) Let $X = C^2(\bar{G})$ where G is an open subset of R^2 and $Z = C^0(\bar{G})$. For functions $u(x, y) \in X$, define

$$(Lu)(x, y) = u_{xx}(x, y) + u_{yy}(x, y) + c(x, y)u(x, y),$$

where $c(x, y)$ is continuous on $\bar{G}$.

(iii) Let $X = C^0[0, 1]$. For functions $u(x) \in X$, define the operator

$$(Lu)(x) = c(x)u(x) + \int_0^1 k(x, y)u(y)dy,$$

where $k(x, y)$ is continuous on the closed square $[0, 1] \times [0, 1]$ and $c(x)$ is continuous on $[0, 1]$. Check that L is a linear operator on X into itself. The linear operators in (i) and (ii) are differential operators, and the linear operator in (iii) is an integral operator.

(iv) Let $X = C[0, 1]$ and for $u \in C[0, 1]$ let the operator be defined by

$$Lu = \int_0^1 h(x)u(x)dx,$$

where $h \in X$. In this case the operator $L : X \to R$. When the linear operator takes on real or complex values, it is usually called a *linear functional*.

(v) With $X = C^1[0, 1]$ and $Z = C^0[0, 1]$ let an operator M be defined by

$$(Mu)(x) = (u')^2(x) + 3u(x).$$

Check that M is not a linear operator.

1.3 Review of facts about ordinary differential equations

Sometimes it is possible to reduce the problem of solving a partial differential equation to that of solving a family of ordinary differental equations. In addition, results from ordinary differential equations can guide us in our study of partial differential equations. Details and proofs can be found in an introductory text on ordinary differential equations such as [BD].

The basic first order ordinary differential equation (ODE) for a single scalar function $y(x)$ is

$$y'(x) = f(x, y),$$

where $f(x, y)$ is a given function on some open set G in the x, y plane (G might be all of R^2). The differential equation assigns a slope $f(x, y)$ at each point $(x, y) \in G$. The vector $(1, f(x_0, y_0))$ is to be tangent to any solution curve $y(x)$ such that $y(x_0) = y_0$.

The basic result about existence and uniqueness of solutions in this general context is given in the following theorem:

Theorem 1.9
Assume that f and f_y are continuous on G, and let $(x_0, y_0) \in G$. Then there is a unique C^1 solution of the initial value problem:

$$y' = f(x, y), \quad y(x_0) = y_0 \tag{1.8}$$

existing on some interval $|x - x_0| < h$. h may depend on the point (x_0, y_0), and may be quite small. The number y_0 is the initial condition.

Examples

(a)

$$y'(x) = \lambda(x)y(x) + g(x), \quad y(x_0) = y_0, \tag{1.9}$$

where $\lambda(x)$ and $g(x)$ are given continuous functions on an interval (a, b) and $x_0 \in (a, b)$. In this case the solution is given explicitly:

$$y(x) = y_0 \exp\left[\int_{x_0}^x \lambda(\xi)d\xi\right] + \int_{x_0}^x \exp\left[\int_s^x \lambda(\xi)d\xi\right]g(s)ds. \tag{1.10}$$

The interval of existence is (a, b), independent of the initial condition y_0.

(b)

$$y' = y^2, \quad y(x_0) = y_0. \tag{1.11}$$

Here again we can solve the equation explicitly using the technique of separation of variables:

$$y(x) = \frac{y_0}{1 - (x - x_0)y_0} \tag{1.12}$$

If $y_0 > 0$, the solutions of the initial problem (b) exist for $x < x_0 + 1/y_0$. If $y_0 < 0$, they exist for $x > x_0 + 1/y_0$.

The difference between examples (a) and (b) is that (a) is a linear differential equation while (b) is nonlinear. In the language of Section 1.2, the operator $Ly = y' - \lambda y$ is a linear operator on the vector space $C^1[a, b]$. Nonlinearity in the equation does not always imply breakdown of the solutions at a finite point as in (b), but explains why the general existence theorem does not specify the interval of existence of the solution.

Exercise

Solve the initial value problem (or look it up in your favorite ODE book)

$$y' = y - y^2, \quad y(0) = y_0. \tag{1.13}$$

This is a nonlinear initial value problem whose solutions exist for all x when $y_0 \geq 0$.

Chapter 2

First-Order Equations

2.1 Generalities

A *partial differential equation* (PDE) for the scalar function $u(x, y, z)$ is an equation

$$F(x, y, z, u, u_x, u_y, u_z, \cdots, D^m u, \cdots) = 0.$$

The *order* of the equation is the order m of the highest order derivative that appears in the equation. We say that u is a (strict) solution of the equation in a region G if $u \in C^m(G)$ and u and its derivatives satisfy the equation at each point $(x, y, z) \in G$.

In many physical situations we must study several scalar quantities simultaneously. For example, in gas dynamics we must determine the density of gas, its pressure, and the velocity of the gas particles. For this reason we often must consider systems of partial differential equations.

We have a system of m PDE's in n unknown functions when $\mathbf{u}$ is a vector of functions

$$\mathbf{u}(x, y, z) = (u_1(x, y, z), u_2(x, y, z), \cdots, u_n(x, y, z))$$

with differential equation

$$\mathbf{F}(x, y, z, \mathbf{u}, \mathbf{u}_x, \mathbf{u}_y, \mathbf{u}_z, \cdots) = 0,$$

where $\mathbf{F}$ is a vector-valued function taking values in R^m.

Well-posed problems

Partial differential equations are used to model a wide variety of phenomena from many areas of science. Usually, it is thought that the solutions of the equations describe a deterministic situation. Thus we expect that if we prescribe the proper data, there should be a unique solution and if small errors are made in the data, these errors should not cause large differences in the solutions. In mathematical terms our study of partial differential equations must address the following three questions.

(1) Find the proper set of conditions (data) under which a solution of the equation exists.

(2) Show that, under these conditions (data), the solution is unique in some class of relevant functions.

(3) Show that the solution depends in a continuous fashion on the data.

When we can show that (1), (2), and (3) are true for some equation, we say that the problem is *well-posed*. In some rare cases we will be able to answer all three questions by finding a solution in closed form. More often we will find solutions in the form of an integral, as the sum of an infinite series, or in terms of a recipe for constructing the solution. In some cases we can show that the solutions of an equation satisfy (1), (2), and (3) by indirect reasoning without finding any formulas for the solution. If we want to see what the solutions look like, we use numerical techniques to calculate solutions. The validity of the numerical results will depend on the abstract reasoning used to establish (1), (2), and (3).

A class of equations that lends itself most readily to solution is the class of linear equations. The equation is said to be *linear* if u and its derivatives appear only to the first power, and the coefficients of u and its derivatives depend only on the independent variables x, y, z. In the language of Section 1.2, a linear equation is written in the form

$$Lu = q,$$

where L is a linear operator on a certain function space.

There is a way to classify second-order linear equations according to the qualititative properties of their solutions. This classification is closely related to the physical context of the equation. We shall study several kinds of second-order equations in later chapters, and finally in Chapter 11 we shall classify them.

We begin here by studying first-order equations for a scalar function $u = u(x, y)$:

$$F(x, y, u, u_x, u_y) = 0.$$

Examples

(a) The *eikonal* equation from optics

$$u_x^2 + u_y^2 = 1.$$

(b) A *transport* equation

$$u_t + a(x, t)u_x = 0.$$

(c) A *conservation law*

$$u_t + F(u)_x = u_t + F'(u)u_x = 0.$$

Note that (a) and (c) are nonlinear equations and (b) is a linear equation. Equation (b) may be expressed $Lu = 0$, where L is the linear operator

$$(Lu)(x, t) = u_t + a(x, t)u_x, \qquad L : X \to Z,$$

and the spaces are $X = C^1(G)$ and $Z = C^0(G)$, provided the coefficient $a(x, t)$ is continuous on some open set G.

Exercises 2.1

1. Verify that, for any C^1 function $f(x)$, $u(x, t) = f(x - ct)$ is a solution of the PDE $u_t + cu_x = 0$, c a constant.

2. Verify that $u(x, y) = \sqrt{x^2 + y^2}$ satisfies the nonlinear PDE $u_x^2 + u_y^2 = 1$ for $(x, y) \neq (0, 0)$.

3. Show that $u(x, t) = f(x/t)$ satisfies

$$u_t + (\frac{x}{t})u_x = 0 \qquad \text{for } t > 0,$$

where f is any C^1 function.

2.2 First-order linear PDE's

The general first-order linear PDE in variables x, y is

$$a(x, y)u_x + b(x, y)u_y + c(x, y)u = g(x, y). \tag{2.1}$$

Equation (2.1) can be written $Lu = g$, where $Lu = au_x + bu_y + cu$ is a linear operator. Linear equations are convenient to work with in the sense that if we can solve partial problems, we can add together the solutions to get another solution. More precisely, if $Lu = g_1$ and $Lv = g_2$, then

$$L(\alpha u + \beta v) = \alpha g_1 + \beta g_2$$

for any scalars α and β. This is known as the *principle of superposition*.

2.2.1 Constant coefficients

We shall begin by studying an especially simple linear equation

$$u_t(x, t) + cu_x(x, t) = 0, \tag{2.2}$$

where $c > 0$ is a constant. We have switched from variables x, y to x, t because this equation is often used to model time-dependent phenomena.

Equation (2.2) says that the directional derivative $(c, 1) \cdot \nabla u = 0$, where the gradient $\nabla u = (u_x, u_t)$. Consequently, u should be constant on the lines $x - ct = x_0$ (see Figure 2.1). x_0 is the point where this line intersects the x axis. The speed of this line is c and the slope is $1/c$. The restriction of u to such a line is given by

$$v(t) \equiv u(ct + x_0, t).$$

If u is a solution of (2.2), then $v(t)$ should be constant. In fact if u solves (2.2), then

$$\frac{d}{dt}v(t) = \frac{d}{dt}u(ct + x_0, t) = u_x(ct + x_0, t)c + u_t(ct + x_0, t) = 0.$$

Thus $v(t)$ is constant so that $v(t) = v(0)$. In terms of u, this says that

$$u(x_0 + ct, t) = v(t) = v(0) = u(x_0, 0)$$

or

$$u(x, t) = u(x - ct, 0)$$

because $x_0 = x - ct$. Thus $u(x, t)$ is simply $u(x, 0)$ translated to the right by the quantity ct. We have found that a C^1 solution of (2.2) is completely determined by its values on the x axis. The 3-D graph of such a solution is shown in Figure 2.2.

In the case of equation (2.2), the analogue to the initial-value problem for an ODE is the initial-value problem (2.3), where the initial condition now consists of

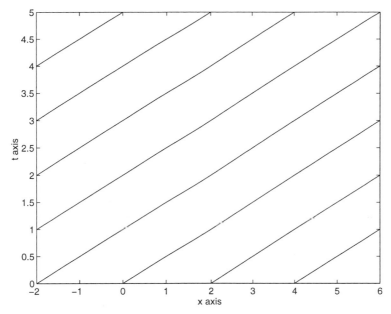

FIGURE 2.1
Characteristics of $u_t + cu_x = 0$ with $c = 2$.

a function rather than a number. We shall see this often in PDE's where time is involved. We shall use the abbreviation IVP from now on for the term initial-value problem.

Theorem 2.1
Let $f(x)$ be a C^1 function on R. Then there is a unique C^1 solution $u(x, t)$ to the IVP

$$u_t + cu_x = 0, \qquad u(x, 0) = f(x). \tag{2.3}$$

Indeed, u is given explicitly by the formula

$$u(x, t) = f(x - ct). \tag{2.4}$$

It is easy to verify that u given by (2.4) satisfies (2.3). The uniqueness follows from the discussion before the theorem because we showed that a solution of (2.2) must be of the form (2.4).

To show that the problem is well-posed, it remains to verify that the solution depends continuously on the initial data. Suppose that $f(x)$ and $g(x)$ are two initial functions, and let $u(x, t)$ and $v(x, t)$ be the corresponding solutions of (2.3).

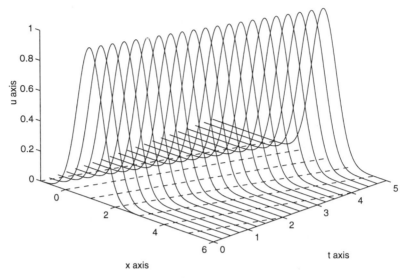

FIGURE 2.2
Solution of (2.3) with initial condition $f(x) = \exp[-2(x-1)^2]$. *Solid curves are sections of the solution surface, dashed lines are characteristics in the x, t plane.*

A key point here is that because equation (2.3) is linear, the function $w = u - v$ solves (2.3) with initial data function $f - g$. Hence (2.4) applied to w yields

$$u(x, t) - v(x, t) = w(x, t) = f(x - ct) - g(x - ct),$$

and this implies that

$$\max_{x,t} |u(x, t) - v(x, t)| = \max_{x,t} |w(x, t)| = \max_{x} |f(x) - g(x)|.$$

Thus if $|f(x) - g(x)| \leq \delta$ for all x, then $u(x, t)$ and $v(x, t)$ never differ by more than δ. Continuous dependence of the solution on the initial data is established in the sense that small changes in the initial data produce small changes in the solution.

No matter what units u may have (e.g., density or velocity), c must have the units of length/time for the terms in (2.2) to have the same dimensions. c is called the *velocity of propagation*. Solutions of (2.2) represent signals or waves propagated to the right (when $c > 0$) with velocity c. When $c < 0$, the signals propagate to the left. The lines $x - ct = $ constant are called *characteristics* for this equation because information (the value of u) is carried along them. In more general circumstances, the restriction of u to a characteristic satisfies an ODE in t. In this special case the ODE was simply $dv/dt = 0$.

This method of constructing the solution by analyzing its behavior along charactertistic lines or curves is called the *method of characteristics*. We can use it for many first-order PDE's.

2.2.2 Spatially dependent velocity of propagation

A constant velocity c indicates a uniform medium where the velocity of propagation does not vary from place to place. However, it is easy to imagine a medium whose properties are not uniform. In this case the velocity of propagation may depend on the spatial coordinate x so that $c = c(x)$. The IVP (2.3) becomes

$$u_t + c(x)u_x = 0, \qquad u(x,0) = f(x). \tag{2.5}$$

Now this equation says that the directional derivative of u at the point (x, t) in the direction $(c(x), 1)$ is zero. Consequently, the vector $(c(x), 1)$ must be tangent to the level curves of u. If these level curves are parametrized by $t \to (x(t), t)$, then it follows that

$$\frac{dx}{dt} = c(x), \tag{2.6}$$

which is a possibly nonlinear ODE. We assume that $c(x)$ is C^1, so that we can apply the basic existence theorem for ODEs. From our experience with the case $c(x) = c$, we expect that if u is a solution of (2.5), then it should be constant along the solution curves of (2.6). This is true. Again let $v(t) = u(x(t), t)$ be the restriction of u to such a curve. Then because u is assumed to be a solution of (2.5),

$$\frac{dv}{dt} = \frac{d}{dt}u(x(t), t) = u_x\frac{dx}{dt} + u_t = c(x)u_x + u_t = 0. \tag{2.7}$$

Thus $v(t)$ is constant along a solution curve $t \to x(t)$ of (2.6). The solution curves of (2.6) are called *characteristic curves* for the equation (2.5) because the PDE (2.5) reduces to an ODE (2.7) along such a curve.

Because $v(t)$ is constant, $v(t) = v(0)$. This means that if (x, t) is on the characteristic curve through $(x_0, 0)$, it follows that $u(x, t) = u(x_0, 0)$. When we had $c(x) \equiv c$, we had $x - ct = x_0$, so that we could solve easily for x_0 in terms of (x, t). In the more general case we are considering now, we follow the characteristic curve through the point (x, t) back in time. We assume that this curve reaches the x axis at the point $(x_0, 0)$. The correspondence $(x, t) \to x_0$ defines a function $x_0 = p(x, t)$, and we can write

$$u(x, t) = u(p(x, t), 0) = f(p(x, t)).$$

This formula generalizes (2.4). The generalization of Theorem 2.1 is

Theorem 2.2
Let $c(x)$ be C^1. Assume that for $t \geq 0$, the characteristic curve through any point (x, t) can be followed back to a point $p(x, t)$ on the x axis.   Then there exists a unique C^1 solution of (2.5) on $x \in R, t \geq 0$.

   Again if $u(x, t)$ is the solution with initial data $f(x)$ and $v(x, t)$ is the solution with initial data $g(x)$, then

$$\max_{x,t} |u(x, t) - v(x, t)| = \max_x |f(x) - g(x)|,$$

so that this problem is well-posed.

Remark   Even though the characteristic through each $(x_0, 0)$ may exist for all $t \geq 0$, the mapping $p(x, t)$ may not be defined for all (x, t), $t \geq 0$. The characteristic through (x, t) may not reach backward to the x axis. To state our result, we had to assume that $p(x, t)$ is defined for each (x, t). See exercise 7 of this section for an equation where this assumption is not satisfied.

Example

$$u_t + x u_x = 0, \qquad u(x, 0) = f(x).$$

The ODE for the characteristics is $dx/dt = x$ with solutions

$$x(t) = x_0 e^t.$$

These characteristics are displayed in Figure 2.3 and the 3-D plot of a typical solution is displayed in Figure 2.4.
   In this example we are lucky that we can solve explicity for x_0 in terms of (x, t). In fact $x_0 = p(x, t) = xe^{-t}$ so that we get an explicit formula for the solution

$$u(x, t) = f(xe^{-t}).$$

   A general treatment of the method of characteristics for first-order partial differential equations is given in [John1].

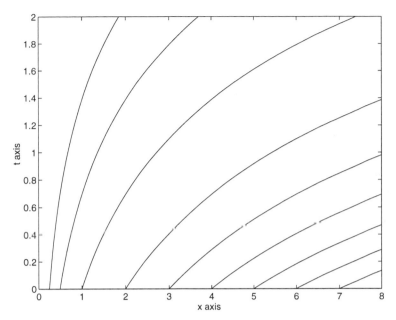

FIGURE 2.3
Characteristics for the equation $u_t + xu_x = 0$.

Exercises 2.2

In the next three exercises, the PDE's all have the same characteristics $x = x_0 + ct$. Let $v(t) = u(x_0 + ct, t)$ be the restriction of u to a characteristic line. Find the ODE solved by v in each case. Solve it, and then restate the answer in terms of u.

1. Solve the IVP

$$u_t + cu_x + u = 0, \quad u(x, 0) = f(x).$$

2. Solve the inhomogeneous equation

$$u_t + cu_x = xt, \quad u(x, 0) = f(x).$$

3. Solve

$$u_t + cu_x = u^2, \quad u(x, 0) = f(x).$$

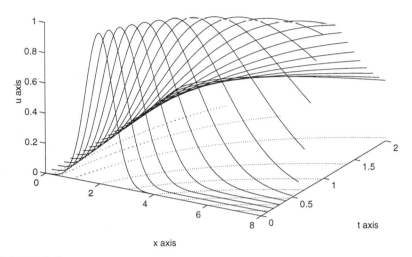

FIGURE 2.4
*Solution of $u_t + xu_x = 0$ with initial data $f(x) = \exp(-2(x-2)^2)$. Solid curves
are sections of solution surface, dotted curves are characteristics in the x, t plane.*

The ODE satisfied by v is nonlinear. The width of the strip containing the
x axis on which u exists depends on $f(x)$.

4. Let l be any line in the x, t plane and consider the IVP

$$u_t + cu_x = 0, \quad u \text{ given on } l.$$

Can this problem always be solved? Pay special attention to the case when
l is a characteristic $x - ct = $ constant.

5. Let $u(x, t) = f(xe^{-t})$ be the solution of $u_t + xu_x = 0$, $u(x, 0) = f(x)$.
    Take $f(x) = \exp[-2(x-2)^2]$.

    (a) Use MATLAB to plot snapshots of the solution on $[0, 15]$ for
         $t = .4, .8, 1.2, 1.6$. Make a fine mesh with $\Delta x = .01$.
         We do this as follows. Set up a function mfile u.m for the func-
         tion $u(x, t)$ (for the definition of a function mfile in MATLAB, see
         Appendix B, Section B.5).

```
function y = u(x,t)
    p = x*exp(-t);
    y = exp(-2*(p-2).^2);
```

Notice that we have made the function "array-smart". Now to plot snapshots of the solution at various times, use the instructions

```
>> x = 0:.01:15;
>> plot(x,u(x,.5))
```

This pair of instructions produces a snapshot at time $t = .5$.

(b) We want to approximate the velocity of the top of the hump at two different times, say $t = .5$ and $t = 1.5$. Plot $u(x, t)$ at times $t = .5, .5 + \Delta t, 1.5, 1.5 + \Delta t$. First use $\Delta t = .1$ Use the zoom feature of MATLAB to determine how far the top of the hump moves over the time interval $[t, t + \Delta t]$. Determine the average velocity of the top of the hump over this time interval. Compute the average velocities of the hump again, this time with $\Delta t = .05$.

(c) Use the formula for the characteristic through $x_0 = 2$ to compute the instantaneous velocity of the top of the hump at times $t = .5$ and $t = 1.5$? How well does the average velocity as you computed it in part b) compare with the instantaneous velocity ?

(d) Why does the hump broaden as the wave moves to the right?

6. Using the example discussed in exercise 5 as a model, solve the IVP

$$u_t - xu_x = 0 \qquad \text{on } x \geq 0, \qquad u(x, 0) = f(x).$$

(a) Solve the ODE's for the characteristics, with initial value x_0. The result should be some function $x = g(x_0, t)$. Plot several of these curves for $0 \leq t \leq 5$ with the following MATLAB commands.

```
>> t = 0:.01:5;
>> plot(g(1,t), t)
>> hold on
>> plot(g(2,t),t)
```

Note that t is the second coordinate in the plots. Plot the curves for values of x_0 ranging from 0 to 10 on the same graph. Do the curves spread apart, or bunch up as t increases?

(b) In the formula for part (a), solve for x_0 in terms of x, t: $x_0 = p(x, t)$, to obtain the solution $u(x, t) = f(p(x, t))$.

(c) Now use the initial data $f(x) = \sin(x)$. Use MATLAB to plot snapshots of $u(x, t)$ on $[.1, 10]$ for several values of t. What happens to the shape of the wave form as t increases? Do the oscillations get broader or narrower? Explain why.

7. We want to find solutions of the IVP

$$u_t + (\frac{1}{x})u_x = 0 \qquad \text{on } x > 0 \qquad u(x,0) = f(x).$$

(a)   Find a formula for the solutions of the characteristic ODE

$$\frac{dx}{dt} = \frac{1}{x}, \qquad x(0) = x_0 > 0$$

(b)   In the formula you find in part (a), solve for x_0 in terms of x and t: $x_0 = p(x,t)$. Note that $p(x,t)$ is not defined for all (x,t). What restriction must you place on x,t ? Now obtain the solution of the IVP $u(x,t) = f(p(x,t))$.

(c)   Put the formula of part (a) in the form $t = h(x, x_0)$. Use MATLAB to plot the characteristic emanating from $(x_0, 0)$ on the interval $0 \le x \le 10$ for several values of x_0. Restrict the curves to lie in the rectangle $[0, 10] \times [0, 5]$ by using the MATLAB command axis([0 10 0 5]). Do the curves spread out or do they come together as t increases?

(d)   Now let the initial data be a hump as in exercise 5 taking $f(x) = \exp[-2(x-2)^2]$. Plot snapshots of the solution on $[0, 10]$ for several values of t using the formula of part (b). You must take into account that $p(x,t)$ is not defined for all x, t. Does the hump broaden, or become narrower as t increases? Explain.

8. Solve the IVP for the linear equation

$$u_t + x^2 u_x = 0, \qquad u(x,0) = f(x).$$

You will need to use the formula of Example (b) of Section 1.3. Over what region in the x, t plane does the solution exist?

2.3   Nonlinear conservation laws

Now we turn our attention to a situation where the velocity of propagation c is not independent of the solution u, as it is in the case of (2.5). We shall consider a nonlinear scalar conservation law.

Suppose that some material is flowing through a pipe so that the density of material is constant across each cross section of the pipe. Then we can assume that the density depends on a single variable x which runs along the length of the pipe. We let $u(x, t)$ denote the linear density at the location x and time t. u has the units of mass/length. The mass of the material in an interval $[a, b]$ is clearly

$$\int_a^b u(x, t)dx.$$

Let $F(x, t)$ denote the rate at which the material is passing the point x. F is the *flux*. We adopt the convention that F is positive when material is passing from left to right and negative when passing from right to left. The balance equation

$$\frac{d}{dt} \int_a^b u(x, t)dx = F(a, t) - F(b, t) \tag{2.8}$$

states that the rate of change of mass in the interval $[u, b]$ is equal to the rate (flux) at which material enters the interval at a minus the rate at which it leaves at b. This equation expresses the conservation law which is *conservation of mass*.

Here we make a strong assumption to simplify the discussion. We assume that the linear density u and the flux F are both C^1 functions. Then according to Theorem 1.3 we can differentiate under the integral sign on the left of (2.8), and we can also write the right hand side of (2.8) as an integral:

$$\int_a^b u_t(x, t)dx = - \int_a^b F_x(x, t)dx,$$

which is the same as

$$\int_a^b [u_t(x, t) + F_x(x, t)] \, dx = 0. \tag{2.9}$$

Since this holds for all intervals $[a, b]$ the integrand must be zero. (see Chapter 1, Theorem 1.2). Thus u and F must satisfy the equation

$$u_t + F_x = 0. \tag{2.10}$$

However this is now a single differential equation in two unknown functions u and F. To reduce the problem to determining a single function, we must make an assumption about the relationship between u and F. We assume that there is a *constitutive relation* between the flux and the density, that is, we assume there is a function $\hat{F}(u)$ such that

$$F(x, t) = \hat{F}(u(x, t)).$$

To avoid additional notation we usually drop the hat. Now assuming that $F(u)$ is a C^1 function of u, the PDE becomes

$$u_t + F'(u)u_x = 0. \tag{2.11}$$

Equation (2.8) is called the integral form of the conservation law and (2.11) is the differential form. Equation (2.8) is more general than (2.11). We shall see later that there are situations where (2.8) is satisfied but where (2.11) is not even defined.

Example A

Suppose that u is the density (cars per mile) of cars on a road moving from left to right. It is observed that the speed at which people drive depends on the density of cars. Let $\beta > 0$ be the maximum density. The speed at which people drive is given by

$$k(\beta - u).$$

Here k is a constant of proportionality with units of (miles)^2/(car hours) to ensure that this expression has units of miles per hour. This law reflects what we have all observed, namely, that as traffic gets heavier, it usually slows down because people do not want to drive at high speeds bumper-to-bumper. Of course, there are always the exciting exceptions. Now the rate at which cars pass a given point (cars per hour) is the product of their speeds and the density:

$$F(u) = ku(\beta - u) \tag{2.12}$$

for $0 \le u \le \beta$. F is the flux. To simplify the computations we shall set $k = 1$. The reader should check that the greatest traffic flow, i.e., the largest flux, occurs when the density $u = \beta/2$.

Back to the general case. Comparing (2.11) with (2.3), we see that the velocity c now depends on u,

$$c(u) = F'(u).$$

Indeed, $F'(u)$ has the units of length/time, no matter what the units of u may be. Following our experience with (2.5), we define the characteristic curves of (2.11) for a given solution $u(x, t)$ as solutions of the differential equation

$$\frac{dx}{dt} = c(u(x, t)). \tag{2.13}$$

Note here that the family of solution curves now depends on the choice of u. In the analysis of (2.5), the characteristic curves were independent of the solution u.

However, the present case is simpler than it looks. As before, we expect u to be constant along the characteristic curves. Indeed, if $x(t)$ is a solution of (2.13), then $u(x(t), t)$ is the restriction of u to this curve and along this curve,

$$\frac{d}{dt} u(x(t), t) = u_x \frac{dx}{dt} + u_t = c(u)u_x + u_t = 0.$$

Let $x(t, x_0)$ be the characteristic curve emanating from the point $(x_0, 0)$ on the x axis. Then $u(x(t, x_0), t) = u(x(0, x_0), 0)$, and the differential equation (2.13) now has a constant right-hand side. This means that $x(t, x_0)$ is in fact a straight line:

$$x(t, x_0) = x_0 + c(u(x_0, 0)).$$

The family of characteristic curves (lines) now depends on the value of the solution at $t = 0$, that is, on the initial data $f(x) = u(x, 0)$. To sketch the characteristic lines emanating from points on the lines $t = 0$, we first assign the velocity $c(f(x_0))$ to each point $(x_0, 0)$. Then the characteristic through $(x_0, 0)$ has the equation

$$x(t) = x_0 + c(f(x_0))t. \tag{2.14}$$

Example A (continued)

When $F(u)$ is given by (2.12) with $k = 1$,

$$c(u) = F'(u) = \beta - 2u. \tag{2.15}$$

Note that $c(u) < 0$ whenever $u > \beta/2$. Thus $c(u)$ is not the velocity of individual cars; they are always moving to the right. Rather $c(u_0)$ is the velocity of a disturbance in the density around the value u_0. For example, when traffic stops at a stop light, there is a wave of increased density (in which the cars are stopped) that moves backward through the line of cars.

Now suppose that we are given an initial density of cars on the line,

$$f(x) = u(x, 0) = \begin{cases} 0 & \text{for } x \leq 0 \\ \beta x^2 (3 - 2x) & \text{for } 0 \leq x \leq 1 \\ \beta & \text{for } x \geq 1. \end{cases}$$

We want to see how the density evolves as a function of x, t. We must solve the IVP

$$u_t + c(u)u_x = 0, \quad u(x, 0) = f(x), \tag{2.16}$$

where $c(u) = F'(u) = \beta - 2u$. The condition $f(x) = \beta$ represents gridlock - the cars are not moving. No cars are present for $x \le 0$, and in the transition region $0 \le x \le 1$, the density of cars increases from 0 to the maximum capacity β. The value $\beta/2$ is attained at $x = 1/2$.

According to (2.14), the characteristic lines starting between $x = 0$ and $x = 1/2$ lean to the right because $c > 0$ there, and those starting between $x = 1/2$ and $x = 1$ lean to the left. In particular, using (2.15), we see that the characteristic line emanating from $x = 0$ (where $f = 0$) has velocity $c = \beta$ with equation $x = \beta t$ while that emanating from $x = 1/2$ has velocity $c = 0$ with equation $x = 1/2$. Finally the line emanating from $x = 1$ (where $f = \beta$) has velocity $c = -\beta$ with equation $x = 1 - \beta t$ (see Figure 2.5).

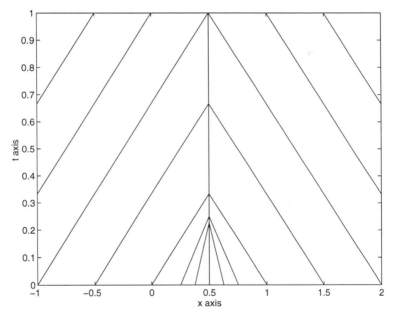

FIGURE 2.5
Characteristics for (2.16) with $\beta = 1.5$.

Now the density $u = 0$ on the line $x = \beta t$, $u = \beta/2$ on the vertical line $x = 1/2$, and $u = \beta$ on the line $x = 1 - \beta t$. At time $t = 1/(2\beta)$, these lines intersect at the point $(1/2, 1/(2\beta))$. Here the density is undefined because it would have to take on three different values. The solution cannot be continued in the strict sense beyond $t = 1/(2\beta)$. In fact the solution actually breaks down at a time $t_* < 1/(2\beta)$ when the graph of the profile has a vertical tangent at $x = 1/2$. From this time on, the profile generated by the method of characteristics no longer defines a function.

To see this we can make a graphical construction which uses the method of characteristics. Draw the initial curve $u(x, 0) = f(x)$. Pick a number of points on the graph of f, say at heights 0, $\beta/4$, $\beta/2$, $3\beta/4$, and β. At each point $(x, f(x))$

on the initial profile, the velocity is $c(f(x)) = \beta - 2f(x)$. Hence the velocities at the heights 0, $\beta/4$, $\beta/2$, $3\beta/4$, and β are $c(0) = \beta$, $c(\beta/4) = \beta/2$, $c(\beta/2) = 0$, $c(3\beta/4) = -\beta/2$, and $c(\beta) = -\beta$. Thus the bottom of the profile moves to the right and the top moves to the left (see Figure 2.6).

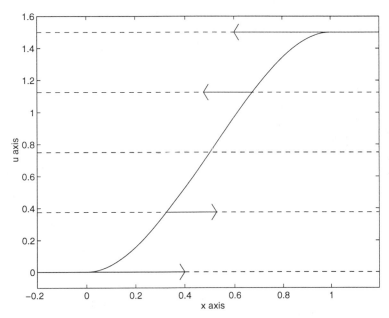

FIGURE 2.6
Initial profile and direction of evolution of the solution profile with $\beta = 1.5$.

Suppose that $f(x)$ takes on the value $\beta/4$ at $x = x_1$. Then because the solution u is constant along characteristics, at time t, the solution will take on the value $\beta/4$ at the point

$$x = x_1 + c(\beta/4)t = x_1 + (\beta/2)t.$$

Successive profiles are shown in Figure 2.7.

Note that although this procedure continues to generate a smooth curve for each t, it is not the graph of a function for $t > t_*$. The breakdown of the strict solution occurs not because the solution $u(x, t)$ itself blows up at $x = 1/2$, but rather because the derivative u_x blows up at $x = 1/2$ as $t \to t_*$. We shall see later that the solution can be continued beyond t_*, but not in the strict sense.

Here is another example of a nonlinear conservation law.

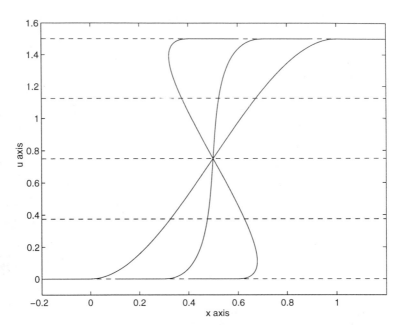

FIGURE 2.7
Snapshots of the solution of (2.16) with $\beta = 1.5$, at times $t = 0, .2, .4$

Example B

In equation (2.11) we now take

$$F(u) = \frac{u^2}{2}, \quad \text{whence } c(u) = F'(u) = u.$$

The IVP is

$$u_t + uu_x = 0, \quad u(x, 0) = f(x). \tag{2.17}$$

This differential equation is called the inviscid Burgers' equation. It is used as a simplified model of the system of equations that governs gas dynamics.

Again, u is constant on the characteristic lines which emanate from the x axis. The characteristic line emanating from x_0 is given by

$$x = x_0 + f(x_0)t$$

because $c(f(x_0)) = f(x_0)$. We shall study (2.17) in the exercises.

We have seen that the nature of the characteristics of a first-order equation depends on the type of velocity function. We summarize these observations here.

- velocity c = constant. Characteristics are parallel straight lines. For solutions of (2.3) all points on the solution profile move at the same speed c.

- velocity $c = c(x)$. Characteristics are curves in the x, t plane which do not intersect. For solutions of (2.5) the speed of a point on the solution profile depends on the horizontal coordinate of the point.

- velocity $c = c(u)$. Characteristics are again straight lines, but they may intersect. For solutions of (2.11) the speed of a point on the solution profile depends on the vertical coordinate of the point.

Exercises 2.3

1. For the equation of Example A, consider the initial data

$$g(x) = \begin{cases} \beta & \text{for } x \leq 0 \\ \beta(1 - x^2(3 - 2x)) & \text{for } 0 \leq x \leq 1 \\ 0 & \text{for } x \geq 1. \end{cases}$$

   (a) In a sketch, indicate the slope of the characteristic lines emanating from the x axis at several points ranging from $x = -1$ to $x = 2$. Will the characteristic lines cross for $t > 0$? What about for $t < 0$?

   (b) Make a graphical construction of the profile of the solution at several times $t > 0$. Does the profile steepen as t increases or does it flatten out? Does the solution exist for all $t > 0$?

   The remaining exercises deal with the IVP of Example B.

2. Show that if the initial data $f(x)$ has $f'(x_0) < 0$ for some x_0, then the C^1 solution of (2.17) must break down at some time $t > 0$.

3. Show that if $f'(x) \geq 0$ for all x, then the characteristic lines emanating from the x axis do not intersect in $t > 0$. What does this say about the existence for $t > 0$?

4. Let initial data

$$f(x) = \begin{cases} u_l & \text{for } x \leq 0 \\ u_l - \frac{(u_l - u_r)}{L}x & \text{for } 0 \leq x \leq L \\ u_r & \text{for } x \geq L, \end{cases}$$

   where $0 < u_r < u_l$. Strictly speaking f is not proper initial data for (2.17) because f is not differentiable everywhere. Nevertheless, the same

graphical construction works to create a kind of generalized solution. Find
the time t_* when the profile of the solution becomes vertical. Show that
$t_* = 1/|f'|$ where f' is the derivative of f in the interval $(0, L)$.

More generally, when the initial data f is not piecewise linear, but has
a point x_0 where $f'(x_0) < 0$, the solution of (2.17) will first break down
(develop a vertical tangent) at the time $t_* = 1/[\max(-f')]$.

The next exercises deal with Example B and should be done with the program
mtc (method of characteristics). Information about the workings of program mtc
can obtained by first invoking MATLAB and then by typing help mtc. There are
four choices of initial data. They are

$$f1(x) = \begin{cases} 1, & x \le 0 \\ 1 - (1/8)x^2(3 - x), & 0 \le x \le 2 \\ 1/2, & x \ge 2; \end{cases}$$

$$f2(x) = \begin{cases} 1, & x \le 0 \\ (x + 1)(x - 2)^2/4, & 0 \le x \le 2 \\ 0, & x \ge 2; \end{cases}$$

$$f3(x) = \begin{cases} 1/2, & x \le -1 \\ 1/2 - (x + 1)^2(x - 2)/8, & -1 \le x \le 1 \\ 1, & x \ge 1; \end{cases}$$

$$f4(x) = \begin{cases} 1.1, & x \le 0 \\ .05x^3 - .15x^2 + 1.1, & 0 \le x \le 2 \\ .9, & x \ge 2. \end{cases}$$

5. Consider the IVP (2.17) with initial data $f3$. Sketch some of the characteristics in the x, t plane. Will they intersect in $t > 0$? Now run mtc with
   this data choice. Pick several times at which you wish to view the profile,
   say $t = 1, 2, 3, 4$. Observe how the wave flattens out.

6. Next consider the IVP (2.17) with initial data $f1$. Sketch some characteristics in the x, t plane. Do they intersect in $t > 0$? Run mtc with this data
   choice at times $t = 1, 2, 3, 4$. The profiles will be plotted together on the
   interval $[-1, 6]$. Note that the top of the wave is at height 1 and so moves
   with velocity $c = 1$, while the bottom is at height $1/2$ and hence moves

with velocity $c = 1/2$. By choosing other times t_1, t_2, t_3, t_4, carefully you can determine the time t_* when the profile first has a vertical tangent. Use the zoom feature to blow up the important parts of the graph. Hint: look at the middle of the wave around height $3/4$ and use the result of exercise 4 as a guide.

7. Finally run program mtc using initial data $f2$. The profile will "roll over." Find the time t_* when the profile first develops a vertical tangent by plotting and by using the formula of exercise 4.

2.4 Linearization

We want to compare the behavior of the solution of the linear equation (2.3) with the solutions of the nonlinear equation (2.11). The solutions of (2.3) retain the shape of the initial profile $f(x)$. The solution is merely a translation by ct. This contrasts sharply with the behavior of the solutions of the nonlinear equation (2.11). In this case the profiles may steepen or flatten out as t increases, and the solution may even fail to exist (in the strict sense) for t beyond a certain time. Nevertheless, the solutions of a linear equation like (2.3) can be used to approximate solutions of (2.11) by a technique known as linearization.

Again we consider the initial-svalue problem for the nonlinear conservation law

$$u_t + c(u)u_x = 0, \quad u(x, 0) = f(x), \tag{2.18}$$

where we have used $c(u)$ to denote $F'(u)$. Note that if the initial data $f(x)$ is a constant function, $f(x) \equiv u_0$, then the unique solution is $u(x, t) \equiv u_0$. Now assume that $f(x)$ deviates by only a small amount from u_0, and write $f(x) = u_0 + g(x)$. We expect that the solution u of (2.18) will differ by only a small amount from u_0, at least for a while. Thus we write $u(x, t) = u_0 + v(x, t)$, where we expect v to be small and we look for what approximate equation v will satisfy. Substituting in (2.18), we see that

$$v_t + c(u_0 + v)v_x = 0, \quad v(x, 0) = g(x). \tag{2.19}$$

Now make a Taylor expansion of $c(u)$ around $u = u_0$.

$$c(u) = c(u_0) + c'(u_0)(u - u_0) + (1/2)c''(u_0)(u - u_0)^2 + \dots$$

$$= c(u_0) + c'(u_0)v + O(v^2).$$

Here we have used the O notation introduced in Chapter 1, Section 1.1.3. Substituting this expression in (2.19), we obtain

$$v_t + c(u_0)v_x + c'(u_0)vv_x + O(v^2 v_x) = 0. \tag{2.20}$$

We assume that v and v_x are "small", and that the products $c'(u_0)vv_x$ and $O(v^2 v_x)$ are much smaller than the other (linear) terms in (2.20). Dropping this term, we see that v approximately solves a linear equation:

$$v_t + c(u_0)v_x \approx 0. \tag{2.21}$$

We let w be the exact solution of

$$w_t + c(u_0)w_x = 0, \quad w(x, 0) = g(x). \tag{2.22}$$

Equation (2.22) is called the *linearized equation* corresponding to (2.18). The linearization is taken about the constant solution $u(x, t) \equiv u_0$. We expect v and w to be close because they have the same initial data g and v solves (2.22) with a small error. The solutions of (2.22) are completely understood:

$$w(x, t) = g(x - c(u_0)t).$$

Thus we expect that the solution u of (2.18) can be approximated as

$$u = u_0 + v \approx u_0 + w = u_0 + g(x - c(u_0)t),$$

at least for short times.

To justify dropping the nonlinear term in (2.20), we must be sure that

$$|c'(u_0)vv_x| \ll |c(u_0)v_x|. \tag{2.23}$$

The symbol $\ll$ means "much smaller than." In practice, it means smaller by at least a factor of $1/10$. Let $\delta = \max|f(x) - u_0| = \max|g(x)|$. Using the exercise at the end of this section, we know that $\max|v(x, t)| = \max|u(x, t) - u_0| = \delta$ for as long as the solution exists in the strict sense. Thus for (2.23) to hold, it suffices that

$$|c'(u_0)|\delta \ll |c(u_0)|.$$

This last inequality simply asserts that, as v runs over the interval $u_0 - \delta$ to $u_0 + \delta$, the change $|c(u_0 + v) - c(u_0)|$ is small relative to $c(u_0)$. A thorough discussion of the validity of linear approximations quickly leads to difficult mathematical questions and is beyond the scope of this text .

Exercises 2.4

1. If u_0 is a constant, and u is a solution of (2.18), show that max $|u(x, t) - u_0| = $ max $|u(x, 0) - u_0|$.

2. Consider the IVP (2.17) with initial data $f4$ (see exercises of section 2.3). Let u be the solution.
    (a) Are conditions satisfied so that we may make a linear approximation to u? What is the linearized equation, and what is the approximate solution?
    (b) Run the computer program `mtc` with the choice of data 4. Profiles of the solution of the nonlinear problem are displayed in yellow, and those of the linear approximation problem are displayed in green. How long can you go before the linear approximation is off by 10%?

3. Go back to the equation of exercise 3 of section 2.2. Linearize the equation about the solution $u \equiv 0$. What conditions should be put on the initial data f to ensure validity of the linearization? If $|f(x)| \le \varepsilon$, how long before the solution of the nonlinear equation differs by more than 10% from the solution of the linear equation?

2.5 Weak solutions

2.5.1 The notion of a weak solution

The usual (strict) sense of solution for a first-order PDE

$$u_t + F(x, u)_x = 0 \qquad (2.24)$$

requires that u have continuous partial derivatives for the equation to make sense. However this notion of solution seems inadequate in at least two situations.

(1) When $F(x, u) = cu, c > 0$ constant, the equation is linear

$$u_t + cu_x = 0,$$

and the solution is of the form $u(x, t) = f(x - ct)$. Notice that this formula makes sense even when f is not differentiable, even discontinuous. What meaning can we give to u in this case? It behaves like a solution in that it is constant on lines $x - ct =$ constant, but it may not be differentiable.

(2) In the case of the nonlinear conservation law $F(x, u) = F(u)$, with $c(u) = F'(u)$, the equation is

$$u_t + c(u)u_x = 0,$$

and the solution can be continued only until the wave profile has a vertical tangent at some point, at which time the profile no longer defines a function. Nevertheless the physical process does not stop just because we cannot continue the solution.

To overcome these difficulties we must introduce a broader notion of solution. This is a common practice in mathematics. We introduce complex numbers when we cannot solve $x^2 = -1$ in the real numbers. What we need is a notion of a generalized solution which incorporates the structure of the solution without using derivatives.

We shall illustrate the notion of a generalized solution of an equation by an example in one dimension. Suppose that $y \in C^1(R)$ and that

$$y'(x) = f(x). \tag{2.25}$$

y is a strict solution of this equation. Now let $\varphi(x)$ be any C^1 function such that $\varphi(x) = 0$ for x outside some finite interval I. φ is called a *test function*. Multiply (2.25) by φ and integrate over R:

$$\int_R y'(x)\varphi(x)dx = \int_R f(x)\varphi(x)dx.$$

Note that we do not have to worry about convergence of the integrals because $\varphi = 0$ outside I. Now integrate by parts on the left. Again because $\varphi = 0$ outside I,

$$\int_R y'(x)\varphi(x)dx = -\int_R y(x)\varphi'(x)dx.$$

Hence we see that, if y is a strict solution of (2.25), then

$$-\int_R y(x)\varphi'(x)dx = \int_R f(x)\varphi(x)dx \tag{2.26}$$

for all test functions φ. The derivative of y does not appear in this integral equation. On the other hand, if y is a C^1 function satisfying (2.26) for all test functions φ, it can be shown that y solves (2.25).

Now consider the function

$$u(x) = \begin{cases} x, & x < 0 \\ 2x, & x > 0 \end{cases}.$$

Let

$$f(x) = \begin{cases} 1, & x < 0 \\ 2, & x > 0 \end{cases}.$$

It is easy to see that $u' = f$ for $x > 0$ and for $x < 0$, but that u is not a strict solution of $u' = f$ in any interval containing $x = 0$. However, u and f do satisfy (2.26) for any test function φ. In fact,

$$-\int_R u(x)\varphi'(x)dx = -\int_{-\infty}^0 x\varphi'(x)dx - \int_0^\infty 2x\varphi'(x)dx$$

$$= -x\varphi(x)\Big|_{-\infty}^0 + \int_{-\infty}^0 \varphi(x)dx - 2x\varphi(x)\Big|_0^\infty + \int_0^\infty 2\varphi(x)dx$$

$$= \int_R f(x)\varphi(x)dx.$$

We see that we may define a more general notion of solution of a differential equation by requiring that the integral equation (2.26) be satisfied for all test functions φ.

2.5.2 Weak solutions of $u_t + F(u)_x = 0$

Now we turn to the two dimensional case. We define the class of test functions as those $\varphi \in C^1(R^2)$ such that $\varphi = 0$ outside a finite rectangle. Next we define the appropriate integral equation corresponding to the equation (2.24). Let $u(x, t)$ be a C^1 solution of (2.24). Multiply (2.24) by a test function φ, and integrate

$$0 = \int \int_{R^2} [u_t + F(x, u)_x]\varphi(x, t)dxdt = \int \int_Q [u_t + F(x, u)_x]\varphi(x, t)dxdt,$$

where Q is any rectangle such that $\varphi = 0$ outside Q.

Now recall that the divergence theorem (see Chapter 1) says that

$$\int \int_Q [u_t\varphi + u\varphi_t]dxdt = \int \int_Q \partial_t(u\varphi)dxdt = \int_{\partial Q} n_t u\varphi dS,$$

where n_t is the t component of the exterior normal to the boundary ∂Q of Q. However, $\varphi = 0$ on ∂Q. Hence,

$$\int\int_{R^2} u_t\varphi dxdt = \int\int_Q u_t\varphi dxdt = -\int\int_Q u\varphi_t dxdt = -\int\int_{R^2} u\varphi_t dxdt.$$

Similarly,

$$\int\int_{R^2} F(x,u)_x\varphi dxt = \int\int_Q F(x,u)_x\varphi dxdt =$$

$$-\int\int_Q F(x,u)\varphi_x dxdt = -\int\int_{R^2} F(x,u)\varphi_x dxdt.$$

Thus for any test function φ,

$$0 = \int\int_{R^2} [u_t + F(x,u)_x]\varphi dxdt = -\int\int_{R^2} [u\varphi_t + F(x,u)\varphi_x]dxdt. \quad (2.27)$$

The integral expression on the right does not involve derivatives of u.

Definition   *A weak solution of (2.24) is a piecewise continuous function $u(x,t)$ such that (2.27) holds for each test function φ which is 0 outside some rectangle Q.*

By reversing the integration by parts, it can be shown that if u is a C^1 function which satisfies (2.27) for all test functions φ, then u solves (2.24) in the strict sense.

To illustrate the use of the definition, suppose that $F(x,u) = cu$, where $c > 0$ is constant. For any piecewise continuous function $f(x)$, we claim that $u(x,t) = f(x - ct)$ is a weak solution of

$$u_t + cu_x = 0.$$

We must show that $u(x,t) = f(x - ct)$ satisfies (2.27) for all test functions φ. This means that we must show that for all test functions φ,

$$0 = \int\int_{R^2} [u\varphi_t + cu\varphi_x]dxdt = \int\int_{R^2} [f(x-ct)\varphi_t + cf(x-ct)\varphi_x]dxdt.$$

To evaluate the integral on the right, we make the change of variable

$$y = x - ct, \quad t = t.$$

Under this change of variable, $dxdt = dydt$, so the integral becomes

$$\int\int_{R^2} f(y)[\varphi_t(y+ct, t)+c\varphi_x(y+ct, t)]dydt = \int_R f(y)\int_R \frac{d}{dt}\varphi(y+ct, t)dtdy = 0.$$

Here we have used the fact that

$$\int_R \frac{d}{dt}\varphi(y+ct, t)dt = \lim_{\tau\to\infty}[\varphi(y+c\tau, \tau) - \varphi(y-c\tau, \tau)] = 0$$

because $\varphi = 0$ outside some finite rectangle.

A discontinuity of f or f' is called a *singularity* of the initial data. The formula $u(x, t) = f(x-ct)$ shows that, in a weak solution of $u_t + cu_x = 0$, the singularities of the initial data are propagated along the characteristic lines $x = ct + const$. This is not true for the nonlinear equation $u_t + uu_x = 0$.

The solution of the conservation law that we studied with the program mtc and the data choice (1) began to roll over when $t = 8/3$. We saw that the profile of the solution became multivalued and no longer defined a function. In fact we can continue the solution beyond $t = 8/3$ as a weak solution with a discontinuity. Before discussing this procedure, we consider a simpler case.

2.5.3   The Riemann problem

To see how a discontinuous function can be a weak solution of the conservation law (2.24), let us consider the IVP

$$u_t + F(u)_x = 0, \qquad u(x, 0) = f(x),$$

where

$$f(x) = \begin{cases} u_l, & x < 0 \\ u_r, & x > 0 \end{cases}.$$

f has only two constant values u_r and u_l and has a step discontinuity at $x = 0$. This IVP is called the *Riemann problem* and is one of the building blocks of the theory of nonlinear conservation laws. As a candidate for a weak solution, we shall try a piecewise constant function

$$u(x, t) = \begin{cases} u_l, & x < st \\ u_r, & x > st \end{cases}.$$

The constant s is the speed of the line of discontinuity and must be determined. For now we note that u clearly solves (2.24) for $x < st$ and for $x > st$ because it has constant values there. Now we state (without proof) that, *if u is a weak solution of (2.24) as defined by (2.27), then u satisfies the integral form (2.8) of the conservation law*, which in this case becomes

$$\frac{d}{dt} \int_a^b u(x,t)dx = F(u(a,t)) - F(u(b,t)).$$

This form of the conservation law can be applied to discontinuous functions, and we shall use it to determine the value of s. Since this form of the law is supposed to hold for all $a < b$, we can suppose that a and b are chosen so that $a < st < b$ for t in some interval $|t - t_0| \leq \delta$. Then for t in this interval,

$$\int_a^b u(x,t)dx = \int_a^{st} u_l dx + \int_{st}^b u_r dx = u_l(st - a) + u_r(b - st).$$

Thus

$$\frac{d}{dt} \int_a^b u(x,t)dx = (u_l - u_r)s.$$

This is the rate of mass flow across the discontinuity and by conservation of mass, it must equal the difference in the fluxes on either side:

$$(u_l - u_r)s = F(u_l) - F(u_r).$$

Thus the conservation law (2.8) determines the speed s to be

$$s = \frac{F(u_l) - F(u_r)}{u_l - u_r}.$$

This is a special case of the *Rankine-Hugoniot* condition. The discontinuous weak solution of (2.24) is called a *shock wave* and s is called the *shock speed*.

Example

Consider the inviscid Burgers' equation $u_t + uu_x = 0$. Here $F(u) = u^2/2$ and the Rankine-Hugoniot condition becomes

$$s = \frac{1}{2} \frac{u_l^2 - u_r^2}{u_l - u_r} = \frac{1}{2}(u_l + u_r).$$

The shock speed is just the average of the values of u on either side of the shock line.

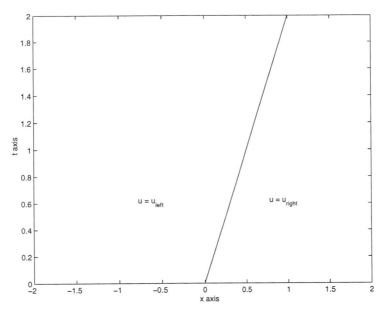

FIGURE 2.8
*Solution of the Riemann problem for (2.24) given by two constant states $u_l \neq u_r$
and the shock line with speed $s = [F(u_l) - F(u_r)]/(u_l - u_r)$.*

More generally the Rankine-Hugoniot condition holds for weak solutions when
there is a curve of discontinuity $t \to \gamma(t)$. In this case Rankine-Hugoniot becomes

$$\gamma'(t) = \frac{F(u_l) - F(u_r)}{u_l - u_r} \tag{2.28}$$

where u_l and u_r are now the limiting values from left and right at the point $(\gamma(t), t)$
on the curve. Now $\gamma'(t)$ is a variable shock speed.

2.5.4 Formation of shock waves

Let us return to the IVP discussed in exercise 4 of Section 2.3. Here the equation
is $u_t + uu_x = 0$, and the initial data is

$$f(x) = \begin{cases} u_l, & x \leq 0 \\ u_l - (\frac{u_l - u_r}{L})x, & 0 \leq x \leq L, \\ u_r, & x \geq I, \end{cases}$$

with $u_l > u_r > 0$. With the top of the wave moving at speed u_l and the bottom at
speed u_r, the solution develops a vertical jump discontinuity at $x_* = Lu_l/(u_l - u_r)$

and time $t_* = L/(u_l - u_r)$. For $t > t_*$, the solution is continued as a shock wave
with shock speed $s = (u_l + u_r)/2$. In the x, t plane the solution has a jump
discontinuity across the line $x = x_* + s(t - t_*)$ for $t > t_*$, with $u = u_l$ to the left
of this line, and $u = u_r$ to the right.

As a generalization of this phenomenon to initial data which is not piecewise
linear, look at the solution of the conservation law $u_t + uu_x = 0$ which we studied
in data option (1) of the program mtc. The solution first began to have a vertical
tangent in its profile at time $t_* = 8/3$. From that time on, we continue u as a
weak solution with a step discontinuity, just enough of a jump to keep the solution
from becoming multivalued. The height of the jump increases until the function
makes a full jump from the value $u_l = 1$ to the value $u_r = 1/2$. From the
Rankine-Hugoniot condition, it is clear now that the shock will propagate with
speed $3/4 = (1/2)(1 + 1/2)$. In fact by applying the more general form (2.28) we
see that the shock line starts at the point $(3, 8/3)$ in the x, t plane with speed $3/4$
for $t \geq 8/3$. We shall see how the shock wave develops in the exercises using the
program shocks. Here is what the characteristics look like for this problem (see
Figure 2.9). Notice here that the characteristics run into the line $x = 1 + 3t/4$ where
the solution becomes discontinuous. This line is a characteristic for $t < 8/3$ but
not for $t > 8/3$. In fact the solution is not even defined on this line for $t > 8/3$. A
3-D display of the snapshots of this solution is shown is Figure 2.10. For $t > 8/3$,
the profiles contain a growing step continuity.

2.5.5   Nonuniqueness and stability of weak solutions

As a way to treat the development of shock waves, we expanded our notion of
solution to include the notion of a "weak solution". However, it happens that we
have enlarged our class of solutions so that we no longer have uniqueness of the
solutions of the IVP. For example the problem $u_t + uu_x = 0$ with initial data

$$f(x) = \begin{cases} 0, & x < 0 \\ 1, & x > 0 \end{cases}$$

has the two weak solutions

$$u(x, t) = \begin{cases} 0, & x < t/2 \\ 1, & x > t/2 \end{cases}$$

and

$$v(x, t) = \begin{cases} 0, & x < 0 \\ x/t, & 0 < x < t \ . \\ 1, & x > t \end{cases}$$

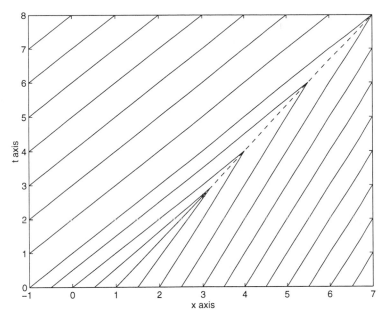

FIGURE 2.9
Characteristics for equation $u_t + uu_x = 0$ with initial data $f1(x)$. The line $x = 1 + 3t/4$ is not a characteristic for $t > 8/3$, indicated by dashed line.

The first solution, $u(x, t)$, has a step discontinuity on the line $x = t/2$. The second, $v(x, t)$, is continuous and is called a *centered rarefaction wave*. Because this IVP has two weak solutions, we see that it is not well posed in the class of weak solutions.

We need an additional criterion to select the physically meaningful solution. To be physically meaningful, a solution of the IVP must be *stable* under small perturbations which might come from numerical or experimental error in the data. If we make a small perturbation of the initial data $f(x)$, the resulting solution should approximate the original solution. We shall see that the discontinuous solution of this initial value problem does not have this property. It is *unstable* with respect to small, smooth perturbations of the initial data f. Let f_ε be a C^1 approximation to f. For example, we might take

$$f_\varepsilon(x) = \begin{cases} 0, & x \le -\varepsilon \\ (3x\varepsilon^2 - x^3)/4\varepsilon^3 + 1/2, & -\varepsilon \le x \le \varepsilon \ . \\ 1, & x \ge \varepsilon \end{cases}$$

Now we know from exercise 3 of Section 2.3 that the solution u_ε of $u_t + uu_x = 0$ with initial data f_ε exists as a strict solution for all $t > 0$. It flattens out as t increases much as the solution discussed in exercise 5 of Section 2.3. Thus as t

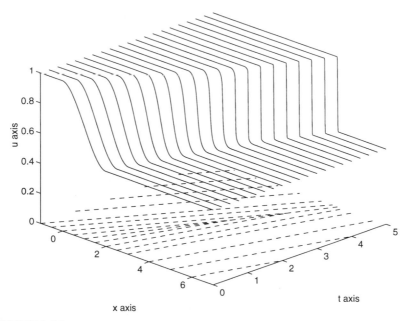

FIGURE 2.10
Solution of $u_t + uu_x = 0$ with initial data $f1(x)$. Solid curves are sections of the
solution surface, dashed lines are characteristics plotted in the x, t plane.

increases, the solution u_ε differs more and more from the discontinuous solution
$u(x, t)$. However, this perturbed solution does resemble the continuous weak solution $v(x, t)$ which is the centered rarefaction wave (see Figure 2.11). Consequently
the stability criterion tells us to reject the discontinuous solution.

In fact, whenever we have two constant states u_l and u_r with $u_l < u_r$, there
will be two weak solutions to the Riemann problem for $u_t + uu_x = 0$, and the
discontinuous solution will be unstable. The stability criterion allows discontinuous solutions of this equation only when $u_l > u_r$. In this case the shock speed
$s = (u_l + u_r)/2$ satisfies $u_r < s < u_l$. Geometrically this means that in the x, t
plane, the characteristics from both sides must run into the line of discontinuity
as t increases. This stability criterion and its geometric expression can also be
formulated for the general equation (2.11).

In the context of gas dynamics, the selection criterion for physically meaningful discontinuous weak solutions is called an *entropy* condition. It takes several
different forms, but one way of stating it is to require that certain families of characteristics run into the line or curve of discontinuity from both sides as t increases.
Generally, when the entropy condition is satisfied, a discontinuous weak solution
is called an *admissible shock wave*. Further discussions of entropy conditions are
found in more advanced texts. The book of Courant and Friedrichs [CF] is the

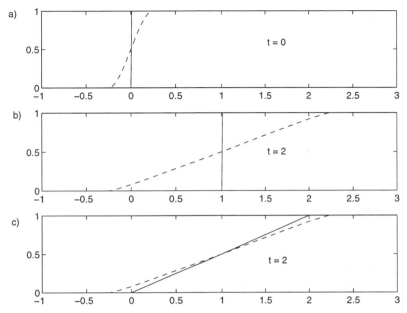

FIGURE 2.11
Frame (a) shows the initial step function $f(x)$ in the solid line, while $f_\varepsilon(x)$ is the dashed curve. Frame (b) shows the solution u in the solid line, and u_ε in the dashed curve, both at time $t = 2$. Frame (c) shows the solution v (rarefaction wave) in the solid line, and u_ε in the dashed curve, both at time $t = 2$.

classical text. More recent treatments can be found in [Wh], [Sm] and [Le]. An excellent introduction to the subject is [La].

Exercises 2.5

1. Let $u(x, t)$ be defined for $(x, t) \in R^2$ by

$$u(x, t) = \begin{cases} 1 & \text{for } x < t/2 \\ 0 & \text{for } x > t/2. \end{cases}$$

   (a) Show that u is a weak solution of $u_t + uu_x = 0$.

   (b) Show that u satisfies the integral form (2.8) of the conservation law when $F(u) = u^2/2$.

The program shocks uses the method of characteristics together with the Rankine-Hugoniot condition to calculate some solutions of $u_t + uu_x = 0$ which evolve into shock waves. The solutions are continued as weak solutions with a discontinuity when they develop vertical tangents. This is in contrast to the program mtc which used only the method of characteristics. In mtc when the profile developed a vertical tangent, it rolled over and no longer defined a solution.

There are three choices of initial data. The first two are the same as choices (1) and (2) of program mtc (see the exercises Section 2.3) The third is the unit step function which is 1 for $x \leq 0$ and zero for $x > 0$. For information on how to run program shocks, invoke MATLAB and enter help shocks .

2. Run mtc with data choice (1), and with $t_1 = 1, t_2 = 2, t_3 = 3$ and $t_4 = 4$. Type the MATLAB command hold on. Then run program shocks with data choice (1) and the same times. You will see the two sets of plots superimposed with the plots of shocks in green.

   (a) We know from exercise 6 of Section 2.3 that, for this choice of data, the time t_*, when the profile first develops a vertical tangent, is $t_* = 8/3$. Verify this graphically by repeating the above procedure with appropriate choices of times. Use the zoom feature of MATLAB.

   (b) Now run the program shocks with times $t1 = 8/3$ and the other times at intervals of .2. The vertical segment of the profile indicates the location of the shock at that time. How fast is the shock moving? The values u_l and u_r are the heights of the top and bottom of the vertical segment. Is the Rankine-Hugoniot condition satisfied?

3. Continue with program shocks and data choice (1). Experiment with various times, and find the time $t_{**} > t_*$ when the weak solution becomes a step wave. You can also calculate this analytically. How fast does the step wave move to the right? Does this agree with the Rankine-Hugoniot condition?

4. Repeat exercises 2 and 3 for data choice (2) of programs mtc and shocks. Use the appropriate time t_* for data choice (2).

5. (a) Verify that, when $u_l < u_r$ and $t > 0$,

$$v(x, t) = \begin{cases} u_l, & x \leq x_0 + tu_l \\ (x - x_0)/t, & x_0 + tu_l < x < x_0 + tu_r \\ u_r, & x \geq x_0 + tu_r \end{cases}$$

   is a continuous weak solution of $v_t + vv_x = 0$. v is a rarefaction wave centered at $x = x_0$.

   (b) Sketch the characteristics of v, i.e., the lines along which v is constant.

   (c) Let $f(x)$ be a piecewise constant function

$$f(x) = \begin{cases} 1, & x < -1 \\ 1/2, & -1 < x < 1 \\ 3/2, & 1 < x < 2 \\ 1, & x > 2 \end{cases}.$$

Construct the weak solution of the IVP, $u_t + u u_x = 0, u(x, 0) = f(x)$ for small values of $t > 0$ using step shock waves and centered rarefaction waves. Sketch the characteristics of the solution.

2.6 Numerical methods

2.6.1 Difference quotients

The method of characteristics is useful for understanding the structure of solutions but produces useful numerical results only in special cases where is it easy to make the analytical calculations. For this reason numerical methods have been developed which can treat a wide variety of problems.

We begin with some basic ideas that we shall use frequently.

If f is a C^2 function, we have the Taylor expansion:

$$f(x + h) = f(x) + f'(x)h + \frac{1}{2} f''(\xi) h^2$$

where ξ is some point between x and $x + h$. This implies that

$$f'(x) = \frac{f(x + h) - f(x)}{h} - \frac{1}{2} f''(\xi) h.$$

We say that the difference quotient approximates f' to within order h, and we write

$$f'(x) = \frac{f(x + h) - f(x)}{h} + O(h).$$

Recall from Section 1.1.3 that $O(h)$ means that the quantity it represents tends to zero as $h \to 0$ at least as fast as h. Taking $h > 0$, we say that

$$\frac{f(x + h) - f(x)}{h} \tag{2.29}$$

is the *forward difference* approximation to $f'(x)$ and

$$\frac{f(x) - f(x-h)}{h} \tag{2.30}$$

is the *backward difference* approximation. Taking the average of these two yields the *centered difference* approximation

$$\frac{f(x+h) - f(x-h)}{2h}. \tag{2.31}$$

As might be expected from a graphical argument, the centered difference approximation is more accurate. In fact, using Taylor expansions, one can show that

$$f'(x) = \frac{f(x+h) - f(x-h)}{2h} + O(h^2).$$

Adding the expansions of $f(x+h)$ and $f(x-h)$, and dividing by h^2, we arrive at the difference approximation for f'':

$$f''(x) = \frac{f(x+h) - 2f(x) + f(x-h)}{h^2} + O(h^2). \tag{2.32}$$

You should recall from your ODE course that numerical methods can be developed by replacing the derivatives in the equation by appropriate difference quotients. Thus, in the ODE

$$y' = f(x, y),$$

if we replace $y'(x)$ by the forward difference quotient (2.29), we obtain the approximate formula

$$y(x+h) \approx y(x) + hf(x, y(x)),$$

which is the Euler method. Using (2.30) to approximate $y'(x)$ yields the backward Euler method:

$$y(x) \approx y(x-h) + hf(x, y(x))$$

usually written

$$y(x+h) \approx y(x) + hf(x+h, y(x+h)).$$

A general reference for numerical methods, including numerical methods for solving ODE's and elementary PDE's, is [KC].

2.6.2 A finite difference scheme

We shall sketch how these finite difference methods can be applied to partial differential equations. Because partial differential equations can model a wide variety of physical phenomena, their solutions can have quite different qualitative properties. To be effective the numerical methods must be tailored to each kind of equation and reflect its structure.

As usual we begin with the simple IVP

$$u_t + cu_x = 0, \qquad u(x, 0) = f(x), \tag{2.33}$$

where $c > 0$ is a constant. First we lay out a lattice (grid) in the (x, t) plane using points $x_j = j\Delta x$ and $t_n = n\Delta t$, where $\Delta x, \Delta t > 0$ are small and j and n are integers. We hope to calculate a two-dimensional array of numbers $u_{j,n}$ such that

$$u_{j,n} \approx u(x_j, t_n),$$

and this approximation should converge in some sense to the exact solution, as Δx and Δt tend to zero.

As a first attempt, let us replace the partial derivatives in (2.33) by forward difference quotients:

$$u_t(x, t) = \frac{u(x, t + \Delta t) - u(x, t)}{\Delta t} + O(\Delta t),$$

$$u_x(x, t) = \frac{u(x + \Delta x, t) - u(x, t)}{\Delta x} + O(\Delta x).$$

Then if u satisfies (2.33),

$$0 = u_t + cu_x = \frac{u(x, t + \Delta t) - u(x, t)}{\Delta t} + c\frac{u(x + \Delta x, t) - u(x, t)}{\Delta x} + \sigma.$$

σ is called the *truncation error*, and it measures the degree to which the difference quotients approximate the PDE. In this case $\sigma = O(\Delta t) + O(\Delta x)$. If we drop the error term, we arrive at the difference scheme

$$\frac{u_{j,n+1} - u_{j,n}}{\Delta t} + c\frac{u_{j+1,n} - u_{j,n}}{\Delta x} = 0.$$

Solving for $u_{j,n+1}$, we obtain (with $\rho = \Delta x/\Delta t$)

$$u_{j,n+1} = u_{j,n} - \frac{c}{\rho}(u_{j+1,n} - u_{j,n}) \tag{2.34}$$

$$= (1 + \frac{c}{\rho})u_{j,n} - \frac{c}{\rho}u_{j+1,n}.$$

The computational diagram (Figure 2.12) is

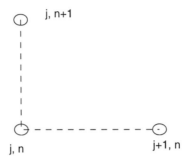

FIGURE 2.12
Computational diagram for scheme (2.34).

Now, because the initial data $f(x)$ is given at $t = 0$, we can take $u_{j,0} = f(x_j)$ for all j, and then calculate $u_{j,1}$ for all j using (2.34). Proceeding forward in this manner, we can calculate $u_{j,n}$ on all the lines $t_n = n\Delta t$.

Unfortunately this is not a satisfactory scheme for this equation. The numbers $u_{j,n}$ produced by this scheme do not always converge to the exact solution. Consider the initial data

$$f(x) = \begin{cases} 0, & x < -1 \\ x + 1, & -1 \le x \le 0 \\ 1, & x > 0 \end{cases}.$$

Now $u_{j,0} = f(x_j) = 1$ for $j \ge 0$. Using (2.33) with $n = 0$ and $j \ge 0$,

$$u_{j,1} = (1 + \frac{c}{\rho})u_{j,0} - \frac{c}{\rho}u_{j+1,0} = 1$$

for all $j \ge 0$. Repeating this argument for each n, we can deduce that

$$u_{j,n} = 1 \quad \text{for} \quad n \ge 0, \ j \ge 0,$$

no matter how small Δx and Δt. However the exact solution for this initial data is $u(x, t) = f(x - ct)$ and $u(x, t) = 0$ for $x \le ct - 1$. Thus $u_{j,n}$ does not converge to the exact solution as $\Delta x, \Delta t \to 0$.

2.6.3 An upwind scheme and the CFL condition

We can remedy this situation if we construct a numerical scheme which respects the analytical structure of solutions of the equation. When $c > 0$, $u(x, t)$ depends on the initial data at the point $x - ct$ on the x axis which lies to the left of x. In the scheme (2.34), we see that $u_{j,n}$ depends on the initial data at points to the right of $x = j\Delta x$. It is not surprising that this scheme is unsuitable for (2.33).

Instead let us replace the forward difference used to approximate u_x with the backward difference

$$u_x = \frac{u(x, t) - u(x - \Delta x, t)}{\Delta x} + O(\Delta x).$$

Replacing u_t by the forward difference and u_x by the backward difference produces an approximation which again has a truncation error $\sigma = O(\Delta x) + O(\Delta t)$. The resulting scheme is

$$\frac{u_{j,n+1} - u_{j,n}}{\Delta t} + c\frac{u_{j,n} - u_{j-1,n}}{\Delta x} = 0.$$

Solving for $u_{j,n+1}$,

$$u_{j,n+1} = (1 - \frac{c}{\rho})u_{j,n} + \frac{c}{\rho}u_{j-1,n}. \tag{2.35}$$

The computational diagram is shown in Figure 2.13.

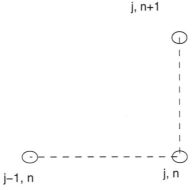

j, n+1

j−1, n   j, n

FIGURE 2.13
Computational diagram for (2.35).

Now $u_{j,n+1}$ depends on the initial data to the left of $x_j = j\Delta x$, which is more consistent with the manner in which information is propagated by the solutions of this equation. However, in addition, we must insist that the ratio $\rho = \Delta x/\Delta t$ is chosen so that

$$\frac{c}{\rho} \le 1. \tag{2.36}$$

In other words, given a spatial step Δx, we must choose the time step Δt so that

$$c\Delta t \le \Delta x.$$

This is known as the CFL (Courant-Friedrichs-Levy) condition. The ratio c/ρ is referred to as the CFL number for the scheme (2.35). The number ρ is the "speed" of propagation of the numerical scheme, and the CFL condition says that this speed must be at least as large as the speed of propagation of the exact solution. The condition (2.36) ensures that the scheme (2.35) uses initial data from an interval on the x axis (the *domain of dependence of the numerical scheme*) that includes the value which determines $u(x_j, t_n)$ exactly (see Figure 2.14).

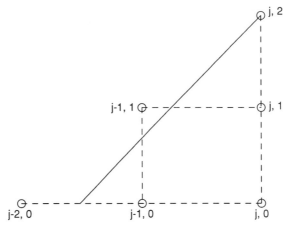

FIGURE 2.14
Domain of dependence for the scheme (2.35). The solid line is the characteristic through the point $(x_j, 2\Delta t)$. In this figure, the CFL condition is satisfied.

To see this, note that $u_{j,n}$ is determined by the values of f at $x_{j-n}, \ldots x_j$, while the exact solution $u(x_j, t_n) = f(x_j - ct_n)$. Because $c/\rho \le 1$, and $\Delta x = \rho \Delta t$,

$$x_{j-n} = (j-n)\Delta x \le (j - n(c/\rho))\Delta x = x_j - ct_n \le x_j.$$

If we think of (2.33) with $c > 0$ as a model for a flow moving from left to right, the scheme (2.35) uses information which comes from upstream in the flow. In the context of gas dynamics, (2.35) is called an *upwind* scheme. In this language, the CFL condition says that the numerical scheme must use information from at least as far upwind as does the exact solution.

When the CFL condition is satisfied, the scheme (2.35) converges as Δt, $\Delta x \to$ 0. However, it has a dissipative effect which makes it lose accuracy as t increases. To see this effect, use the program in the Appendix B, Section B.7 with $\rho = \Delta x / \Delta t > c$ and initial data a Gaussian hump.

Convergence and stability

Another difference between the two schemes (2.34) and (2.35) can be seen when we use the initial data $f(x) = \varepsilon \cos(\pi x / \Delta x)$. $f(x)$ oscillates exactly with the spacing of the mesh points x_j on the x axis. In fact, $u_{0,j} = f(x_j) = \varepsilon(-1)^j$. The exact solution of (2.33) is $u(x, t) = f(x - ct)$, and u is bounded for all (x, t). However, it is easy to calculate (see exercise 1) that, using the scheme (2.34),

$$u_{j,n} = \varepsilon[1 + 2(c/\rho)]^n (-1)^j.$$

This scheme magnifies the amplitude ε of the initial oscillations by the factor of $[1 + 2(c/\rho)] > 1$ at each time step, making it grow without bound. The scheme (2.34) is said to be *unstable*.

On the other hand, for the same initial data, the upwind scheme (2.35) yields

$$u_{j,n} = \varepsilon[1 - 2(c/\rho)]^n (-1)^j.$$

When the CFL condition $c/\rho \le 1$ is satisfied, $u_{j,n}$ remainds bounded because, in this case, $|1 - 2c/\rho| \le 1$. When $c/\rho > 1$, the magnifying factor $1 - 2c/\rho < -1$, and the $u_{j,n}$ grow without bound in an oscillating manner, as n increases. We say that the scheme (2.35) is *stable* when $c/\rho \le 1$ and unstable when $c/\rho > 1$.

Both schemes (2.34) and (2.35) have a truncation error that tends to zero as Δx, $\Delta t \to 0$. These schemes are said to be *consistent*. An important theorem of numerical analysis, the Lax equivalence theorem, states that, for a linear equation such as (2.33), a consistent finite difference scheme is convergent if and only if it is stable.

In addition to knowing that a numerical scheme converges, we also want to be able to estimate how the error decreases as the mesh is refined. In discussing convergence, we assume that $\Delta x = \rho \Delta t$ for some constant ρ, so that it is enough to specify Δt. The *global error* at time T measures the difference between the exact solution $u(x_j, T)$ and the value $u_{j,n}$ generated by the scheme in n time steps Δt where $n \Delta t = T$. Naturally the global error is related to the truncation error. The truncation error measures the error made in approximating the first derivatives of u at each time step. The function values are approximated within order $O(\Delta t^2)$. If the scheme is stable, the error made at each time step is not magnified at later time steps. Therefore in marching forward from $t = 0$ to $t = T$, we accumulate global error $n O(\Delta t^2) = (T/\Delta t) O(\Delta t^2) = T O(\Delta t)$. This intuitive argument can be made rigorous when we carefully define the manner in which we measure the global error.

The subjects of convergence and stability of finite difference schemes are treated in [Le], [RM], and [St].

2.6.4   A scheme for the nonlinear conservation law

Next we turn to a numerical scheme for the nonlinear conservation law

$$u_t + c(u)u_x = 0. \tag{2.37}$$

Now the speed of propagation $c(u)$ depends on u. Let us assume that $c(u) \geq 0$ for all values of u involved in the problem, so that again the flow is moving from left to right. Using our experience with the linear equation with $c > 0$ constant, we approximate u_t by a forward difference, u_x by a backward difference, and evaluate $c(u)$ at $u_{j,n}$. We arrive at the scheme

$$u_{j,n+1} = u_{j,n} - \frac{c(u_{j,n})}{\rho}[u_{j,n} - u_{j-1,n}] \tag{2.38}$$

which is quite similar to (2.35). Here, as in (2.35), $\rho = \Delta x/\Delta t$.

It is a very important feature of the conservation law (2.37) that the solution is constant on the characteristics. This means that, if the initial data $f(x)$ lies in the range $m \leq f(x) \leq M$, then the same is true of the solution u, $m \leq u(x,t) \leq M$. Hence the maximum speed of propagation (maximum speed of the characteristics) is determined by the initial data

$$c_{max} = \max |c(u)| = \max |c(f(x))|,$$

and the CFL condition for (2.38) becomes

$$\frac{c_{max}}{\rho} \leq 1.$$

Example

Let the conservation law have the flux function $F(u) = u^3/3$ so that $c(u) = u^2$. Suppose that the initial data $f(x) = x\exp(-x^2)$. Then

$$c_{max} = \max_{x \in R} c(f(x)) = \max_{x \in R} x^2 e^{-2x^2} = \frac{1}{2e}.$$

The CFL condition will be satisfied if $\Delta t \leq 2e\Delta x$.

However, the evaluation of $c(u)$ at $u_{j,n}$ in (2.38) is somewhat arbitrary. Why not evaluate $c(u)$ at $u_{j-1,n}$, or take an average of the two? A better way to address this question is to use the fact that (2.37) is a conservation law and can be written

$$u_t + F(u)_x = 0, \tag{2.39}$$

where $F'(u) = c(u)$. We continue to assume that $F'(u) = c(u) \geq 0$ for all values of u encountered in the problem. Now it is apparent that we should approximate $F(u)_x$ by a backward difference in the quantity $F(u)$. We no longer must face the question of where to evaluate $c(u)$. We arrive at the scheme

$$u_{j,n+1} = u_{j,n} - \frac{1}{\rho}[F(u_{j,n}) - F(u_{j-1,n})]. \tag{2.40}$$

Specializing to the case $F(u) = u^2/2$ which is equation (2.17),

$$u_{j,n+1} = u_{j,n} - \frac{1}{\rho}[\frac{u_{j,n}^2 - u_{j-1,n}^2}{2}].$$

This scheme is implemented in the program cl which you will use in the exercises. The dissipative tendency of this scheme smears out shocks when they occur. On the other hand, this dissipative feature prevents the numerical solutions from developing oscillations at the discontinuities.

For the upwind scheme (2.40) to work, we require that $c(u) \geq 0$. If $c(u)$ changes sign, we must use a method that takes information from both the left and right of x_j. When we deal with systems of equations, we will have waves propagating in both directions. Again we must have a scheme which takes information from both left and right. The Lax-Friedrichs and Lax-Wendroff methods do this. Another class of methods which captures shocks well and exploits the conservation law form of the equations is the Godunov methods. There is an extensive literature of numerical methods for conservation laws. An excellent introduction is [Le].

Exercises 2.6

1. Let a mesh Δx, Δt be chosen, and set $\rho = \Delta x/\Delta t$. Consider the equation (2.33) with initial data $f(x) = \cos(\pi x/\Delta x)$.

   (a) With $u_{j,0} = f(x_j) = (-1)^j$, show that the scheme (2.34) yields

   $$u_{j,n} = (1 + 2c/\rho)^n(-1)^j.$$

   (b) For the same initial data, show that the scheme (2.35) yields

   $$u_{j,n} = (1 - 2c/\rho)^n(-1)^j.$$

2. Let the flux function $F(u) = u^2/2 + u$, so that $c(u) = u + 1$. Let the initial data be $f(x) = x/(1 + x^2)$. Find c_{max}, and determine the restriction on Δx and Δt which satisfies the CFL condition.

3. (a) Use the flux function $F(u) = u(\beta - u)$ of example A in the scheme (2.40). What is the resultant scheme?

   (b) What restriction must we place on the values of the initial data so that $F'(u) \geq 0$ for all the values of u encountered in the problem?

   (c) Under this restriction on the initial data, how would you state the CFL condition?

The remaining exercises use the computer program cl. The program uses the upwind scheme of Section 2.6 to give a numerical solution to the IVP

$$u_t + (F(u))_x = 0, \qquad u(x, 0) = f(x).$$

The program requires you to give as input the spatial step, Δx, and the time step, Δt. You must choose them properly so that the CFL condition is satisfied. Instead of entering the times t_1, t_2, t_3, t_4 at which you wish to make snapshots, enter the number of time steps *between* snapshots, n1, n2, n3, n4. For example, if $\Delta t = .05$ and you enter [20 30 20 40] , then you will make snapshots at times $t_1 = 20\Delta t = 1$, $t_2 = t_1 + 30\Delta t = 2.5$, $t_3 = t_2 + 20\Delta t = 3.5$, and $t_4 = t_3 + 40\Delta t = 5.5$.

Program cl calls the mfile flux.m to determine the flux function F. Currently F is chosen to be $u^2/2$ so that you may compute the solutions of the IVP of example B, Burgers' equation. All of the remaining exercises deal with this problem.

There are seven choices for the initial data. The first two are the same used in the programs mtc and shocks. Choice (3) is the unit step function, $f(x) = 1$ for $x < 0$, and $f(x) = 0$ for $x > 0$, the same as data choice (3) of program shocks. Choice (4) is another unit step function, this time $f(x) = 0$ for $x < 0$, and $f(x) = 1$ for $x \geq 0$. Choices (5) and (6) are

$$f5(x) = \frac{3}{2}e^{-2(x-1)^2} + \frac{1}{2}$$

and

$$f6(x) = 2(x - 1)e^{-4(x-1)^2} + \frac{1}{2}.$$

Choice (7) can be supplied by the user by writing a formula in the mfile f.m. The formula must be array-smart, and because the scheme is onesided, we require that $c(f(x)) \geq 0$. In the first four cases, $0 \leq f(x) \leq 1$, so that max $c(f(x)) =$ max $f(x) = 1$. For these data choices, the CFL condition will be satisfied with $\Delta x = \Delta t$.

4. Run cl with data choice (1), $\Delta x = \Delta t = .05$, with 20 time steps between snapshots. This will produce plots at times $t = 0, 1, 2, 3, 4$. Then type the

command hold on. Next run program shocks with data choice (1) and the same times.

(a) The plots are not identical. When you do you begin to see significant differences?

(b) Repeat the calculations for shocks and cl, but this time use $\Delta x = \Delta t = .02$ and 50 time steps between snapshots. How do the plots compare now?

5. Run cl with data choices (2) and (3), and with the same parameters as in exercise 4. Compare with the plots generated by program shocks with data choices (2) and (3)

6. Run data choice (1) again, this time with $\Delta x = .04$ and $\Delta t = .05$. Take the number of time steps between snapshots to be $n1 = n2 = n3 = n4 = 6$. Plot the snapshots separately. Explain the results you see.

7. Run data choice (4), with $\Delta x = \Delta t = .02$, and times $t = 1, 2, 3, 4$. You must take $n1 = n2 = n3 = n4 = 50$. The solution here is the centered rarefaction wave.

8. Run cl with data choice (5) with $\Delta x = .02$.

(a) From the initial data, find the maximum speed of propagation of the problem, c_{max}, and determine Δt, so that the CFL condition is satisfied. With this Δt, choose n1, n2, n3, n4 to make snapshots at $t = 1, 2, 3, 4$. The left side of the hump flattens out as we would expect in a rarefaction, while the right side steepens into a shock.

(b) When does the shock first appear? Calculate t_* using the result of exercise 4 of section 2.3. Then run cl with $\Delta x = .0025$ and the appropriate Δt. Make snapshots at times close to t_*, so that you can see when the vertical tangent first appears.

(c) After the shock forms and starts to move to the right, how does the magnitude (strength) of the shock change? Run cl this time with $\Delta x = .01$ (and the appropriate Δt) and make snapshots at several times $t > t_*$.

(d) Does the shock move at a constant speed? Using the plots of part (c) and the Rankine-Hugoniot condition (2.28), estimate the shock speed at several times $t > t_*$.

9. Plot the initial data choice (6) with the MATLAB commands

```
>>  x = -1:.01:6;
>>  plot(x,f6(x))
```

(a) How many shocks and rarefactions should be expected to develop from this data?

(b)   Again set $\Delta x = .02$, and compute the correct Δt for this initial data consistent with the CFL condition.

(c)   Now run cl to make snapshots at times $t = 1, 2, 3, 4$. What do you see? What is the shape of the wave when $t = 6$?

10.   In this exercise we shall investigate the property that the upwind scheme (2.40) "smears out " a shock. The initial data for choice (3) is $f(x) = 1$ for $x \le 0$ and $f(x) = 0$ for $x > 0$.

(a)   What is the exact weak solution of (2.17) with this initial data?

(b)   Calculate (by hand) the numerical solution using (2.40) with $F(u) = u^2/2$ for $n = 1, 2, 3$. Take $\rho = 1$.

The initial data f is scale invariant, that is, $f(\alpha x) = f(x)$ for all constants α. Thus to duplicate the hand calculations you made in part (b), we can use any choice of Δx and Δt as long as $\rho = \Delta x/\Delta t = 1$.

(c)   Run program cl, data choice (3), with $\Delta x = \Delta t = 1$, and with $n1 = n2 = n3 = n4 = 1$. To see the actual numbers that have been computed, use the MATLAB command

$$>> \ [x', snap0', snap1', snap2', snap3', snap4']$$

There should be six columns of numbers with the leftmost column being x. These numbers should agree with the numbers that you calculated in part (b). What is the width of the transition region where the profile drops from 1 to 0 in each snapshot, that is, how many Δx?

Now run again with $\Delta x = \Delta t = 1$, and $n1 = 3, n2 = n3 = n4 = 1$. Again look at the numbers actually computed, this time with the long format, and determine the width of the transition region in terms of Δx. Does the transition region get larger as time increases or does it reach a maximum width?

In both cases, you may wish to plot the snapshots in different colors to better identify them. Use the matlab command

$$>> \ plot(x, snap0, x, snap1, x, snap2, x, snap3, x, snap4)$$

2.7   A conservation law for cell dynamics

2.7.1   A nonreproducing model

In the treatment of cancer with radiation or chemotherapy, one is attempting to kill a certain cell population. The growing cancer cells may be particularly vulnerable to treatment when they reach a certain degree of maturity. The following model attempts to describe the maturity levels in a cell population. This model, along with many other problems in mathematical biology, is discussed in [E].

Let x denote the maturity of a cell. x is a fraction between 0 and 1. Let $w(x, t)$ be the number of cells in the population with maturity x and time t. The total cell population is given by

$$\int_0^1 w(x, t)dx,$$

and the number of cells with maturity x in the range $a \le x \le b$ is given by

$$\int_a^b w(x, t)dx.$$

Let $F(x, t)$ be the cell flux. It is the rate at which cells with maturity x become more mature. The conservation law

$$\frac{d}{dt} \int_a^b w(x, t)dx = F(a, t) - F(b, t),$$

states that no cells die when advancing from maturity level a to maturity level b. As in Section 2.3, we deduce the differential equation

$$w_t + F_x = 0.$$

Rubinow (see [E]) assumed that

$$F(x, t) = v(x)w(x, t),$$

where $v(x)$ is the maturation "velocity." The time needed for a cell to mature from x_1 to x_2 is expressed by

$$\int_{x_1}^{x_2} \frac{dx}{v(x)}.$$

The IVP for w becomes

$$w_t + (v(x)w(x, t))_x = 0, \quad 0 < x < 1, \, t > 0,$$

$$w(x, 0) = f(x).$$

In this case we have a conservation law which yields a linear PDE. We shall consider the special case when $v(x) = kx$ where $k > 0$ has the units of time^{-1}. By changing the time scale, i.e., replacing kt by t, we can state the IVP as

$$w_t + (x\,w(x, t))_x = w_t + xw_x + w = 0, \tag{2.41}$$

$$w(x, 0) = f(x).$$

The characteristic curves for this equation satisfy

$$\frac{dx}{dt} = x,$$

so that the characteristic curve passing through the point (x, t) is

$$\tau \to \xi(\tau; x, t) = xe^{\tau - t}. \tag{2.42}$$

Assume that w solves (2.41). The restriction of w to the characteristic (2.42) is $v(\tau) = w(\xi(\tau), \tau)$, and satisfies the ODE

$$\frac{dv}{d\tau} = w_\xi \frac{d\xi}{d\tau} + w_t = \xi w_\xi + w_t = -w$$

$$= -v.$$

Hence,

$$v(\tau) = v(t)e^{t - \tau},$$

or

$$w(\xi(\tau), \tau) = w(x, t)e^{t - \tau}. \tag{2.43}$$

Now suppose that the characteristic (2.42) crosses the x axis between 0 and 1:

$$0 \le \xi(0; x, t) = xe^{-t} \le 1.$$

Then from (2.43),

$$w(\xi(0; x, t), 0) = w(x, t)e^t,$$

whence

$$w(x, t) = f(\xi(0; x, t))e^{-t} = f(xe^{-t})e^{-t}. \tag{2.44}$$

This model is not very realistic. There is a steady-state solution, $w(x, t) = 1/x$, which has an infinite cell population. But a more important unsatisfactory aspect of this model is that it does not yield a growing population. Indeed, if we take initial data $f(x)$ such that $f = 0$ for $0 \le x \le a$, where a is some number, $0 < a < 1$, then from (2.44) we see that $w = 0$ on all of the characteristics emanating from the interval $[0, a]$. This means that $w(x, t) = 0$ for all (x, t) which lie above and to the left of the characteristic $\xi(t; a, 0) = ae^t$. This characteristic takes time $t_a = -\ln(a)$ to reach the right boundary $x = 1$. All cells reach maturity by time t_a, and no new cells are added.

2.7.2 The mitosis boundary condition

To produce a more realistic model, we must introduce a description of cell reproduction and a minimum cell maturity $x_0 > 0$. Now we assume that, when a cell reaches full maturity at $x = 1$, it splits into two cells (mitosis) with maturity $x = x_0$. Our conservation equation now holds on the interval $(x_0, 1)$:

$$w_t + (xw)_x = 0, \quad x_0 < x < 1, \ t > 0, \tag{2.45}$$

$$w(x, 0) = f(x), \quad x_0 < x < 1,$$

and we add the boundary condition

$$x_0 w(x_0, t) = 2w(1, t). \tag{2.46}$$

For consistency we require that the initial data satisfy the compatibility condition

$$x_0 f(x_0) = 2f(1). \tag{2.47}$$

We solve the problem (2.45), (2.46). Let $\tilde{x}(t)$ denote the characteristic emanating from x_0:

$$\tilde{x}(t) = x_0 e^t,$$

and let $t_1 = -\ln(x_0)$ be the time when this characteristic reaches the boundary $x = 1$. $w(x, t)$ is still given by (2.44) for (x, t) with $\tilde{x}(t) \leq x \leq 1$ (see Figure 2.15). We want to determine the values of $w(x, t)$ for (x, t) lying above and to the left of $\tilde{x}(t)$.

The values of w along the left boundary $x = x_0$ will be determined by the mitosis rule (2.46). For the moment assume that $g(\tau) = w(x_0, \tau)$ is known. Then for (x, t) with $x_0 < x < \tilde{x}(t)$, the characteristic (2.42) meets the boundary $x = x_0$ at some time τ so that

$$x_0 = xe^{\tau - t}. \tag{2.48}$$

Then by (2.43),

$$g(\tau) = w(x_0, \tau) = w(x, t)e^{t - \tau}.$$

Thus

$$w(x, t) = g(\tau)e^{\tau - t}, \tag{2.49}$$

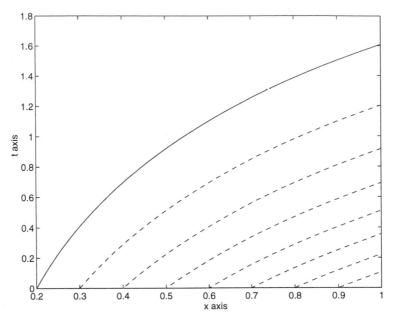

FIGURE 2.15
The solid curve is the characteristic $\tilde{x}(t)$. The dashed curves are other character-
istics. In the region below the curve $\tilde{x}(t)$, the solution is determined by the initial
values. In this graph we have taken $x_0 = .2$.

which shows that in this region w is determined by the values on the left boundary.
From (2.48),

$$e^{\tau - t} = \frac{x_0}{x} \qquad \text{and} \qquad \tau = t - \ln(\frac{x}{x_0}). \tag{2.50}$$

Substituting in (2.49), we see that

$$w(x, t) = (\frac{x_0}{x})g(t - \ln(\frac{x}{x_0})). \tag{2.51}$$

From (2.48) we see that each characteristic starting on the left boundary $x = x_0$
takes the same time $t_1 = -\ln(x_0)$ to cross to the right boundary $x = 1$. We define
the sequence $t_n = nt_1, n = 1, 2, 3, \ldots$. Now $w(1, t), 0 \leq t \leq t_1$, is determined
by the initial data (use (2.44)):

$$w(1, t) = f(e^{-t})e^{-t}.$$

Let $g_1(\tau)$ be the restriction of $g(\tau)$ to the interval $[0, t_1]$. By the mitosis rule (2.46),

$$x_0 w(x_0, \tau) = 2w(1, \tau),$$

or

$$g_1(\tau) = w(x_0, \tau) = \frac{2}{x_0} w(1, \tau), \quad 0 \leq \tau \leq t_1.$$

Now we can use (2.51) to determine $w(x, t)$ in the region between the two characteristics emanating from $(x_0, 0)$ and (x_0, t_1), all the way across to $x = 1$. This yields

$$w(1, t) = x_0 g_1(t - t_1).$$

Again using the mitosis rule, we see that, for $t_1 \leq \tau \leq t_2$,

$$g(\tau) = w(x_0, \tau) = \frac{2}{x_0} w(1, \tau)$$

$$= (\frac{2}{x_0}) x_0 g(\tau - t_1) = 2 g_1(\tau - t_1).$$

If we repeat this process, we deduce that, for $t_n \leq \tau \leq t_{n+1}$,

$$g(\tau) = 2g(\tau - t_1) = 2^2 g(\tau - t_2)$$

$$= 2^n g(\tau - t_n) = 2^n g_1(\tau - t_n).$$

Thus g is determined for all $\tau \geq 0$ in terms of its values on the interval $[0, t_1]$. We see that g grows exponentially, as $\tau \to \infty$. According to (2.51), this means that for each x, $t \to w(x, t)$ grows exponentially, as $t \to \infty$.

This model may be further refined to include some mortality of the cells before they reach full maturity which might be produced by the introduction of a cytotoxic drug.

Exercises 2.7

1. Let $0 < x_0 < 1$. By sketching the characteristics for the equation

$$w_t + (xw)_x = 0,$$

show that the following IBVP has a unique solution in the set $x_0 < x < 1$, $t > 0$:

$$w(x, 0) = f(x), \quad x_0 \leq x \leq 1$$

$$w(x_0, t) = g(t), \quad t \geq 0.$$

Here f and g are given C^1 functions. What compatibility condition must be placed on f and g to ensure a continuous solution?

2. Write down a finite difference scheme for the equation of exercise 1. As in (2.35), use a backward difference to approximate the derivative $(xw)_x$.

3. Write a program in pseudocode to solve the IBVP of exercise 1 using the difference scheme of exercise 2. Pay careful attention how you advance the solution from the line $n = 0$ to the line $n = 1$.

2.8   Projects

1. Write a MATLAB program which uses the upwind finite difference scheme (2.35) to compute numerical solutions to the IBVP (assume $c > 0$)

$$u_t + cu_x = 0, \qquad 0 < x < 10, \quad t > 0,$$

$$u(x, 0) = f(x), \qquad u(0, t) = g(t).$$

Note that we are specifying boundary values for the solution on the left boundary $x = 0$, but not on the right boundary at $x = 10$. Why? Here f and g are given functions to be supplied in function mfiles f.m and g.m. Use the program in section B.7 of Appendix B as a model. You will have to compute a vector of t values

```
t = delt:delt:nsteps*delt;
```

Inside the for loop you must assign the values u(1) = g(t(n)). Try, for example, $g(t) = t \exp(-t)$, or $g(t) = \arctan(t)$ with $f(x) = 0$ or $f(x) = (x - 2) \exp[-2(x - 2)^2]$.
Make enough snapshots so that you can get a good idea what the solution looks like, and check your computed results with the exact solution.

2. Again using the program in Section B.7 as a model, write a MATLAB program as above, but for the equation

$$u_t + (c(x)u)_x = 0.$$

Use a backward difference to approximate $(c(x)u)_x$. You will need a function mfile for $c(x)$. For given $c(x)$ and given Δx, how will you use the CFL condition to determine Δt? Because we are using a one-sided difference scheme, you can only use $c(x)$ such that $c(x) > 0$. Try, for example, $c(x) = 1 + x/10$ on $[0, 10]$. What kind of boundary data will be needed on $x = 0$? Why? Try some of the suggested combinations of data of problem 2.

Make enough snapshots so that you can get a good idea what the solution looks like. When possible, find the exact solution, and check against the computed solution.

3. Further modify the program of problem 2 to solve the equation on the interval $[x_0, 1]$ with $x_0 = .2$, and with $c(x) = x$. Instead of the boundary function $g(t)$, use the mitosis boundary condition (2.46). The initial data $f(x)$ must satisfy the compatibility condition

$$x_0 f(x_0) = 2f(1).$$

For example, use

$$f(x) = \begin{cases} \frac{(x-.3)^2(x-.9)^2}{.3^4}, & |x - .6| \le .3 \\ 0, & |x - .3| \ge .3 \end{cases}.$$

If you have used the mitosis condition correctly, the first element of each snapshot $u(x_0, t)$ should be equal to ten times the last element, which is $u(1, t)$. When you make a 3-D plot of the solution, you should see a surface with hills that grows exponentially.

4. Consider the IVP

$$u_t + (t - x^2/10)u_x = 0, \qquad u(x, 0) = f(x),$$

where $f(x) = \exp[-(x - 3)^2]$. The ODE for the characteristic curves is given by

$$\frac{dx}{dt} = t - x^2/10, \qquad x(0) = x_0. \tag{2.52}$$

Write a MATI AB program which uses the ODE solver `ode23` or `ode45` to numerically integrate (2.52) for a number of values of x_0. Use the vector of initial values $x0 = 0: .1:8$, and integrate the ODE on the intervals $[0, t_{final}]$ for $0 \le t_{final} \le 2$. Plot the curves with the x coordinate first and then restrict the rectangle of viewing by using the command `axis([0 5 0 tfinal])`. Let x_{final} be the vector of terminal t values of the characteristics. It should have the same size as x_0. Use the fact that the solution is constant on characteristics to plot a snapshot of the solution at time t_{final}. Note how the top of the hump moves, first left, then right. Can you explain this by looking at the characteristics?

Chapter 3

Diffusion

In this chapter we investigate equations which model diffusion processes, such as heat flow in a solid, or the spread of dye in water. There is an important difference between a diffusion process and the wave motion studied in Chapter 2. In diffusion, initial data is smeared and smoothed out; there are no sharp fronts. We shall see that this is a result of the constitutive relation (or diffusion law). In the latter part of this chapter we will combine nonlinear wave motion and diffusion.

3.1 The diffusion equation

We shall derive a basic diffusion equation in the context of heat flow. Assume that we have a bar of some material of constant cross section that is surrounded by insulation so that heat can only flow along the bar and not out of the cylindrical surface. We shall eventually assign conditions to the ends of the bar to control the flow of heat in or out of the ends, but for the moment we assume that the bar is so long that we can neglect what happens there, that is, we assume that the bar is infinitely long. Putting the x axis along the bar, we say that there should be a well-defined temperature $u(x, t)$ at each position x in the bar, and at each time $t > 0$. We are assuming that the temperature is uniform across each cross section. Let c be the specific heat of the material: the amount of heat energy, usually in calories, needed to raise the temperature of a unit mass of the material one degree centigrade. We shall assume that c does not depend on the temperature u of the material in the range of temperatures we consider but that c may depend on x. Let ρ be the linear density of the material. ρ may also depend on x. Then

$$\int_a^b u(x, t)c(x)\rho(x)dx$$

is an expression for the amount of heat energy in that portion of the bar in $a \leq x \leq b$. The rate of change of the heat energy in this portion of the bar is given by

$$\frac{d}{dt} \int_a^b u(x, t)c(x)\rho(x)dx.$$

The rate of change of heat energy in $[a, b]$ is the rate at which heat enters and leaves this portion of the bar through its ends, plus the rate at which heat energy is created or absorbed by an internal source (for instance an electric heating element). Assuming for the moment no internal sources, we can write

$$\frac{d}{dt} \int_a^b c(x)\rho(x)u(x, t)dx = F(a, t) - F(b, t), \tag{3.1}$$

where the flux $F(x, t)$ is the rate at which heat energy passes the point x, with the convention that $F(x, t) \geq 0$ if the heat energy is flowing from left to right. Thus if u and F are C^1, then

$$\int_a^b [c(x)\rho(x)u_t(x, t) + \partial_x F(x, t)]dx = 0$$

for all intervals $[a, b]$. We conclude that

$$c(x)\rho(x)\partial_t u(x, t) + \partial_x F(x, t) = 0. \tag{3.2}$$

This is virtually the same as equation (2.10) in Chapter 2 that arose in wave motion. The flux in Chapter 2 depended only on the density u. A moment's reflection, however, shows that the heat flux could not depend only on the temperature, for if the temperature is constant, there should be no heat flow. Rather the heat flux should depend on the spatial rate of change of the temperature. This is expressed in Fourier's law of cooling:

$$F(x, t) = -\kappa u_x(x, t).$$

$\kappa > 0$ is a constant of proportionality determined by the material. It is called the heat conductivity. Thus if the temperature is decreasing from left to right at x, heat flows from left to right, that is, $F(x, t) > 0$. If the material is homogeneous, κ is independent of x. However in a material which varies with x, perhaps because of impurities, κ may depend on x. Allowing κ to depend on x, we substitute $F(x, t) = -\kappa(x)u_x(x, t)$ in (3.2) under the additional assumption that u is C^2 in x. Then

$$c(x)\rho(x)u_t(x, t) = \partial_x(\kappa(x)u_x(x, t)). \tag{3.3}$$

If we add a source term to (3.1), it will be of the form

$$\int_a^b c(x)\rho(x)q(x,t)dx$$

where q has units degrees/time. This yields the inhomogeneous equation

$$u_t(x,t) = \frac{1}{c(x)\rho(x)}\partial_x(\kappa(x)u_x(x,t)) + q(x,t).$$

If c, ρ, κ are constant (a uniform material), and $q = 0$, we arrive at

$$u_t(x,t) = ku_{xx}(x,t), \tag{3.4}$$

where $k = \kappa/c\rho$ is the diffusion constant which has units (length)^2/time. Equation (3.4) is the well known heat equation. When the source term is present, we have the inhomogeneous equation

$$u_t = ku_{xx} + q. \tag{3.5}$$

We can use essentially the same derivation to describe the diffusion of a chemical, say dye, in a liquid. In this case $u(x,t)$ is the concentration of the dye in gm/cm. The analogue of Fourier's law of cooling is known as Fick's law of diffusion which states that the flux is proportional to the spatial rate of change of the concentration, and that dye moves from regions of higher concentration to regions of lower concentration:

$$F(x,t) = -ku(x,t).$$

We are led to the same equation (3.5).

Equation (3.5) also arises in the study of Brownian motion, and we can give a probabilistic interpretation to the solutions of (3.5). See the book [Fe]. For diffusion in the biological context, including the spread of genes in a population, see [E].

Jump conditions

Suppose the bar consists of two different materials, with an interface at $x = a$, so that

$$c(x) = \begin{cases} c_r & \text{for } x > a \\ c_l & \text{for } x < a \end{cases},$$

$$\rho(x) = \begin{cases} \rho_r & \text{for } x > a \\ \rho_l & \text{for } x < a \end{cases},$$

and

$$\kappa(x) = \begin{cases} \kappa_r & \text{for } x > a \\ \kappa_l & \text{for } x < a \end{cases}.$$

Let $k_r = \kappa_r/c_r\rho_r$ and $k_l = \kappa_l/c_l\rho_l$ be the two diffusion constants. Heat flow in this bar is governed by two heat equations and conditions at the interface linking the two solutions. Let u_l solve the first equation and u_r the second:

$$u_t = k_l u_{xx} \qquad \text{for } x < a ,$$

$$u_t = k_r u_{xx} \qquad \text{for } x > a .$$

They are linked across the interface by the conditions

$$u_l(a) = u_r(a),$$

continuity of the temperature, and

$$\kappa_l \partial_x u_l(a) = \kappa_r \partial_x u_r(a), \tag{3.6}$$

which is continuity of the flux. Note that the second of these conditions forces the first derivative u_x to have a jump across the interface if $\kappa_r \neq \kappa_l$.

Equation (3.6), the continuity of the flux, is derived from the integral law of diffusion (3.1) as follows. Apply (3.1) to the short interval $[a - \varepsilon, a + \varepsilon]$. Now, from the left side of (3.1), we see that

$$\int_{a-\varepsilon}^{a+\varepsilon} c(x)\rho(x)u_t(x)dx \to 0, \quad \text{as } \varepsilon \to 0,$$

so from the other side of (3.1), we deduce that

$$F(a - \varepsilon, t) - F(a + \varepsilon, t) \to 0, \quad \text{as } \varepsilon \to 0 .$$

This is exactly the statement that $x \to F(x, t)$ is continuous at $x = a$.

3.2 The maximum principle

Before finding any particular solutions of (3.4), let us look at some qualitative properties of solutions of (3.4). Let Q be the semi-infinite open set

$$Q = \{(x, t) \in R^2 : a < x < b, t > 0\}$$

for some $a < b$, and suppose that $u(x, t)$ is a solution of the heat equation (3.4) in Q. Further suppose that, at some point $(x_0, t_0) \in Q$, the function $u(x, t)$ has a local maximum at (x_0, t_0) with $u_{xx}(x_0, t_0) < 0$. Then at the point (x_0, t_0), (3.4) says that

$$u_t(x_0, t_0) = ku_{xx}(x_0, t_0) < 0,$$

so that the temperature at x_0 must be strictly decreasing. Consequently, for $t = t_0 - \delta < t_0$, it follows that $u(x_0, t) > u(x_0, t_0)$. This is impossible because (x_0, t_0) is supposed to be a local maximum point of the temperature. Thus, as a consequence of Fourier's law of cooling, we see that heat cannot concentrate to produce a local space time maximum at a point $(x_0, t_0) \in Q$. Heat must flow away from hot spots. A precise statement of this idea is the (weak form) of the maximum principle.

Theorem 3.1
Let u be a strict solution of (3.4) in the set Q, which is continuous on $\bar{Q}$. For any T > 0, let

$$Q_T = \{(x, t) \in Q : 0 < t < T\}.$$

Let Γ_T be that part of ∂Q_T described by

$$\Gamma_T = \{(x, t) : x = a, b, \ 0 \le t \le T\} \cup \{(x, t) : t = 0, \ a \le x \le b\}.$$

Then for each T > 0,

$$\max_{\bar{Q}_T} u = \max_{\Gamma_T} u.$$

In words, the temperature in the piece of the bar $a \le x \le b$ can never exceed the larger of the maximum of the initial temperature or the maximum temperature *(over time) at the ends of the interval [a, b] (see Figure 3.1). A similar statement holds for the minimum.*

(a,T)   (b,T)

(a,0)   (b,0)

FIGURE 3.1
The part of the boundary of Q_T which is Γ_T indicated by the solid line.

The intuitive argument preceding the statement of the maximum principle is not quite correct because u can have a maximum in the interior of Q without having $u_{xx} < 0$. To make a tighter argument, we modify u slightly. Let $\varepsilon > 0$ be an arbitrarily small constant. Define

$$v_\varepsilon(x, t) = u(x, t) - \varepsilon t.$$

Both u and v_ε are continuous on $\bar{Q}_T$. Hence both u and v_ε have well-defined maximums on the sets $\bar{Q}_T$ and Γ_T. We see that v_ε satisfies

$$(v_\varepsilon)_t - k(v_\varepsilon)_{xx} = (u - \varepsilon t)_t - ku_{xx} = -\varepsilon.$$

If v_ε has a maximum at (x_0, t_0) with $a < x_0 < b$ and $0 < t_0 \le T$, then $(v_\varepsilon)_{xx}(x_0, t_0) \le 0$, which implies that $(v_\varepsilon)_t(x_0, t_0) = (v_\varepsilon)_{xx}(x_0, t_0) - \varepsilon \le -\varepsilon$. This is impossible since if the maximum occurs at a point with $t_0 < T$, then $(v_\varepsilon)_t(x_0, t_0) = 0$, and if the maximum occurs on the line $t = T$, then $(v_\varepsilon)_t(x_0, t_0) \ge 0$. Thus the maximum of v_ε on $\bar{Q}_T$ cannot occur in Q_T or on the line $t = T$. It must therefore occur on the part of the boundary Γ_T. In symbols,

$$\max_{\bar{Q}_T} v_\varepsilon = \max_{\Gamma_T} v_\varepsilon.$$

Now

$$\max_{\bar{Q}_T} u = \lim_{\varepsilon \to 0} \max_{\bar{Q}_T} v_\varepsilon = \lim_{\varepsilon \to 0} \max_{\Gamma_T} v_\varepsilon = \max_{\Gamma_T} u.$$

The argument is finished. □

Here is another version of the maximum principle, derived from Theorem 3.1. In this version we deal with a solution defined for all $x \in R$ and $t \geq 0$. Since there are no spatial boundaries, we must add another hypothesis to the type of solution we will consider.

Theorem 3.2

Let u be a strict solution of (3.4) in the open set

$$B_T = \{(x, t) : x \in R, \quad 0 < t < T\},$$

and suppose u is continuous on $\bar{B}_T$. Suppose further that there is a constant $M \geq 0$, such that $u(x, t) \leq M$ for all $(x, t) \in \bar{B}_T$. Then

$$u(x, t) \leq M_0 \quad \text{for all} \quad (x, t) \in \bar{B}_T,$$

where M_0 is any number such that $u(x, 0) \leq M_0$ for all $x \in R$.

Example

Suppose that $u(x, t)$ satisfies the hypotheses of Theorem 3.2 with $u(x, 0) = \arctan(x)$. Then we may take $M_0 = \pi/2$ and deduce that $u(x, 0) \leq \pi/2$ on B_T for any $T > 0$.

The proof goes as follows. Let $u(x, t)$ be the given solution of (3.4) and for $\varepsilon > 0$ an arbitrarily small real quantity, set $v_\varepsilon(x, t) = u(x, t) - \varepsilon(kt + x^2/2)$. Then v_ε also solves (3.4). Now we want to apply Theorem 3.1 to v_ε. We define the set

$$Q_{T,a} = B_T \cap \{-a < x < a\}$$

which has Γ boundary

$$\Gamma_{T,a} = \{(x, t) : |x| = a, \ 0 \leq t \leq T\} \cup \{(x, t) : |x| \leq a, \ t = 0\}.$$

Then by Theorem 3.1, for each $\varepsilon > 0$,

$$\max_{Q_{T,a}} v_\varepsilon = \max_{\Gamma_{T,a}} v_\varepsilon.$$

However this last quantity can be estimated using the function u. In fact,

$$\max_{\Gamma_{T,a}} v_\varepsilon = \max_{\Gamma_{T,a}}[u(x,t) - \varepsilon(t + x^2/2)] \le \max\{M - \frac{\varepsilon}{2}a^2, M_0\}.$$

So far we have not made any choice of the constant a. Now we choose $a = a_0$, depending on ε, so that $M - \varepsilon a_0^2/2 = M_0$. Then for all $a \ge a_0$,

$$\max_{Q_{T,a}} v_\varepsilon \le M_0.$$

This implies that

$$v_\varepsilon(x,t) \le M_0 \quad \text{for all} \quad (x,t) \in \bar{B}_T.$$

Thus we have established the conclusion of Theorem 3.1 for the auxilliary function v_ε for each $\varepsilon > 0$. Now take the limit as $\varepsilon \downarrow 0$. We see that

$$u(x,t) = \lim_{\varepsilon \downarrow 0} v_\varepsilon(x,t) \le M_0$$

for all $(x,t) \in B_T$. This concludes the proof of Theorem 3.2. $\square$

The maximum principle is a very important tool in the study of solutions of the heat equation. We shall use it to establish the uniqueness of solutions of initial-value problems and initial-boundary-value problems. The book [PW] provides a thorough treatment of maximum principles in many contexts.

Exercises 3.2

1. Verify that each of the following functions satisfies the heat equation.

    (a) $u(x,t) = kt + \frac{1}{2}x^2 + C$.

    (b) $v(x,t) = \exp(-\gamma^2 kt)\sin(\gamma x)$, for any real γ.

    (c) $w(x,t) = \exp(-\gamma^2 kt)\cos(\gamma x)$, for any real γ.

    (d) $z(x,t) = \exp(kt \pm x)$.

2. (a) For each of the functions in exercise 1, find the maximum and minimum over the rectangle $[-a, a] \times [0, T]$. Verify that the maximum principle of Theorem 3.1 is satisfied in each case. To which of the cases can you apply the maximum principle of theorem 3.2?

   (b) Make 3-D plots of each of these functions using MATLAB, thereby visually verifying your results of part (a). For example, write a function mfile, say u.m, as follows:

```
function y = u(x,t,k)
y = k*t + .5*x.^2;
```

   Then use the MATLAB commands meshgrid and surf to make a 3-D plot.

3. Let $u(x, t)$ and $v(x, t)$ both be solutions of the equation

$$u_t - ku_{xx} = q.$$

   Suppose that $u(x, t) \le v(x, t)$ for $|x| \le L, t = 0$, and for $x = \pm L, 0 \le t \le T$. Show that $u(x, t) \le v(x, t)$ for $|x| \le L, 0 \le t \le T$. Hint: Consider the equation satisfied by $u - v$.

3.3 The heat equation without boundaries

3.3.1 The fundamental solution

It will be useful in our construction of a solution to determine if there is a conserved quantity asssociated with solutions of the heat equation in the absence of any sources. Recall that equation (3.3) is

$$c\rho u_t = \partial_x(\kappa u_x),$$

where c, ρ, κ are allowed to depend on x. Under suitable assumptions on c, ρ, κ and u, we can integrate the equation in x over the whole real line to deduce

$$\frac{d}{dt} \int c\rho u \, dx = \int c\rho u_t \, dx = \int \partial_x(\kappa u_x) \, dx$$
$$= \lim_{a \to \infty} (\kappa u_x)|_{-a}^{a} - \lim_{u \to \infty}(F(-a, t) - F(a, t)).$$

Here, and in the remainder of this subsection, we shall omit the upper and lower limits of integration when the integral is taken from $-\infty$ to ∞. If, for each t, the heat flux $F(\pm a, t)$ converges to the same number as, $a \to \infty$ (for example 0), we can conclude that the total heat energy

$$\int c(x)\rho(x)u(x,t)dx$$

is constant. In other words, if u solves the heat equation written above, $u(x, 0) = f(x)$, and the heat flux $F(-a, t) - F(a, t) \to 0$, as $a \to \infty$, then

$$\int c(x)\rho(x)u(x,t)dx = \int c(x)\rho(x)f(x)dx$$

for all t. In particular, this would be the case if c, ρ, κ are constant and $u_x(x, t) \to 0$, as $x \to \pm\infty$. Then

$$\int u(x,t)dx = \int f(x)dx \quad \text{for all } t \geq 0 \, .$$

From now on we assume that c, ρ, κ are constant.

We want to investigate the situation where a fixed amount of heat energy is concentrated initially in a very short portion of the bar. We can model this situation with a sequence of initial conditions $f_n(x)$ defined by

$$f_n(x) = \begin{cases} n & \text{for } |x| \leq \frac{1}{2n} \\ 0 & \text{for } |x| > \frac{1}{2n} \end{cases} \, .$$

As in Chapter 1, it is easy to verify that $f_n(x) \to 0$, as $n \to \infty$, for each $x \neq 0$, and that the amount of heat energy $\int_R f_n(x)dx = 1$ for all n. Let $u_n(x, t)$ be the solution of the IVP

$$(u_n)_t = k(u_n)_{xx}, \qquad u_n(x, 0) = f_n(x). \tag{3.7}$$

We denote the limiting solution by $S(x, t)$, if it exists. Formally it would satisfy

$$S_t = kS_{xx}. \tag{3.8}$$

What would be the initial condition of S? We saw in Section 1.1.3 that the sequence f_n does not converge to a function in the usual sense. Rather it converges to a generalized function called the δ function, denoted $\delta(x)$. We write the initial condition for S as

$$S(x, 0) = \delta(x). \tag{3.9}$$

Because $\int u_n(x,t)dx = \int f_n(x)dx = 1$ for all $t > 0$ and all n, we expect that the limiting solution S has the same property:

$$\int S(x,t)dx = 1 \quad \text{for all } t > 0. \tag{3.10}$$

For convenience, we set $k = 1$. When we are done, we can replace t by kt.

Next we note that if $v(x,t)$ is any solution of $v_t = v_{xx}$, then, for each $\lambda > 0$, the same is true of a new function v_λ defined by

$$v_\lambda(x,t) = v(\lambda x, \lambda^2 t),$$

because $(v_\lambda)_t - (v_\lambda)_{xx} = \lambda^2(v_t - v_{xx}) = 0$. If $v(x,t)$ has initial condition $v(x,0) = f(x)$, then $v_\lambda(x,0) = f(\lambda x)$. Now applying this invariance property to $S(x,t)$, we see that $S_\lambda(x,t) = S(\lambda x, \lambda^2 t)$ should also solve the heat equation, and since $S(x,0) = 0$ for $x \neq 0$, the same will be true for S_λ. We conjecture that

$$S(\lambda x, \lambda^2 t) = C(\lambda)S(x,t),$$

where $C(\lambda)$ is a constant depending on λ. To find $C(\lambda)$, apply the integral constraint. Combining (3.10) with the previous equation, we see that for any $t > 0$,

$$1 = \int S(x,t)dx = \frac{1}{C(\lambda)}\int S(\lambda x, \lambda^2 t)dx.$$

Now in the last integral, make the change of variable $y = \lambda x$ with $dx = dy/\lambda$. We deduce that

$$1 = \frac{1}{\lambda C(\lambda)}\int S(y, \lambda^2 t)dy = \frac{1}{\lambda C(\lambda)}$$

because by (3.10), $\int S(y,s)dy =$ for any value of $s > 0$. Thus $C(\lambda) = 1/\lambda$, and we find that

$$S(\lambda x, \lambda^2 t) = \frac{1}{\lambda}S(x,t). \tag{3.11}$$

We can use this scaling equation to reduce the problem of finding S to that of solving an ODE for a function of one variable. Since we can make any choice of λ in (3.11), let us choose λ so that the quantity $\lambda^2 t$ is constant. Specifically, let $\lambda = t^{-1/2}$. Then the scaling equation becomes

$$S(x,t) = \frac{1}{\sqrt{t}}S(\frac{x}{\sqrt{t}}, 1).$$

Let $g(s) = S(s, 1)$ so that

$$S(x, t) = \frac{1}{\sqrt{t}} g(\frac{x}{\sqrt{t}}). \tag{3.12}$$

Substituting this candidate into (3.8), we find that g must satisfy

$$-(\frac{1}{2})t^{-3/2}g(xt^{-1/2}) - \frac{1}{2}xt^{-2}g'(xt^{-1/2}) = t^{-3/2}g''(xt^{-1/2}).$$

Multiplying by $t^{3/2}$ and setting $s = xt^{-1/2}$ yields the following ODE for g:

$$g''(s) + \frac{1}{2}sg'(s) + \frac{1}{2}g(s) = 0. \tag{3.13}$$

Standard ODE methods (see [BD]) produce a family of solutions

$$g(s) = A \exp(-s^2/4),$$

where A must be determined by the condition (3.10). Thus

$$S(x, t) = \frac{A}{\sqrt{t}} \exp(-\frac{x^2}{4t}),$$

and, making the change of variable $y = \frac{x}{2\sqrt{t}}$,

$$1 = \int S(x, t)dx = \frac{A}{\sqrt{t}} \int \exp(-\frac{x^2}{4t})dx$$

$$= 2A \int \exp(-y^2)dy = 2A\sqrt{\pi}.$$

Thus we finally obtain (after replacing t by kt)

$$S(x, t) = \frac{1}{\sqrt{4\pi kt}}e^{-\frac{x^2}{4kt}}. \tag{3.14}$$

Referring to Section 1.1.3, we see that for each $t > 0$, $x \to S(x, t)$ is a Gaussian, that is, $S(x, t) = \gamma_r(x) = \sqrt{r/\pi} \exp(-rx^2)$ if we take $r = 1/4kt$. When $t \to 0$, $r \to \infty$. Thus by Theorem 1.7, $S(x, t) \to \delta(x)$, as $t \to 0$. In summary, $S(x, t)$ satisfies (3.8) - (3.10). $S(x, t)$ is called the *fundamental solution* of the heat equation, or the one-dimensional heat kernel.

3.3.2 Solution of the initial-value problem

Now we want to construct solutions of the initial-value problem

$$u_t = k u_{xx} \quad \text{for } t > 0, x \in R, \qquad u(x, 0) = f(x) \text{ for } x \in R. \qquad (3.15)$$

Assume for the moment that the initial data $f = 0$ for x outside some finite interval $I = [a, b]$. We subdivide I into n equal subintervals I_j of length $\Delta y = (b-a)/n$. Finally we sample the initial data f at the midpoints y_j of the I_j. Then the heat flow from the interval I_j is approximated by

$$(x, t) \to S(x - y_j, t) f(y_j) \Delta y,$$

which is the singular heat flow that results from concentrating the heat energy $f(y_j)\Delta y$ at the single point y_j. Because the PDE (3.15) is linear, the sum

$$v_n(x, t) = \sum_{j=1}^{n} S(x - y_j, t) f(y_j) \Delta y$$

is again a solution of the PDE (3.15), and it is an approximation to the solution of the IVP (3.15) which gets better as $\Delta y \to 0$ (i.e., as $n \to \infty$). On the other hand, for each $t > 0$, $v_n(x, t)$ is a Riemann sum approximation of the integral

$$\int S(x - y, t) f(y) dy.$$

Hence our candidate for the solution of the IVP (3.15) is given by

$$u(x, t) = \lim_{n \to \infty} v_n(x, t) = \int S(x - y, t) f(y) dy.$$

Combining with (3.14), this formula becomes explictly

$$u(x, t) = \frac{1}{\sqrt{4\pi kt}} \int_R e^{-\frac{(x-y)^2}{4kt}} f(y) dy. \qquad (3.16)$$

Now that we have derived the formula (3.16), we shall drop the assumption that $f = 0$ outside some finite interval.

Theorem 3.3
Assume that f is continuous on R and that there is an M > 0 such that $|f(x)| \leq M$ for all $x \in R$. Then there is a unique bounded solution $u(x, t)$ of (3.15) which is continuous on $x \in R, t \geq 0$. u is given by (3.16) and satisfies

$$|u(x, t)| \leq M \quad \text{for all } x \in R, \ t \geq 0.$$

In addition, the solution $u \in C^\infty$ for $x \in R$, $t > 0$.

Remark   We conclude from this theorem that the IVP (3.15) is a well-posed problem in the class of continuous, bounded functions in the sense of Chapter 2. In fact, if $u(x, t)$ is the solution of (3.15) with initial data f, and $v(x, t)$ is the solution of (3.15) with initial data g, then $w = u - v$ solves (3.15) with initial data $f - g$. If $|f(x) - g(x)| \leq \delta$ for all $x \in R$, then, by the estimate of the theorem,

$$|u(x, t) - v(x, t)| = |w(x, t)| \leq \delta \quad \text{for all } x \in R, t \geq 0.$$

This shows that u and v never differ by more than the maximum difference in the initial data f and g.

Here is a proof of Theorem 3.3. From (3.14) we see that $y \to S(x - y, t)$ decays very rapidly as $y \to \pm\infty$ for each $x \in R$ and $t > 0$. This means that we will have no problems with the convergence of the integral in (3.16). Now differentiating (3.16) formally, we obtain

$$u_t = -\frac{1}{2\sqrt{4\pi k t^{3/2}}} \int \exp(-\frac{(x - y)^2}{4kt}) f(y) dy$$

$$+ \frac{1}{\sqrt{4\pi kt}} \int \frac{(x - y)^2}{4kt^2} \exp(-\frac{(x - y)^2}{4kt}) f(y) dy.$$

The differentiated kernel still decays very rapidly so that these integrals converge. In fact one can differentiate to any order and still have convergent integrals:

$$\frac{\partial^k}{\partial t^k} u(x, t) = \int \frac{\partial^k}{\partial t^k} S(x - y, t) f(y) dy$$

and

$$\frac{\partial^k}{\partial x^k} u(x, t) = \int \frac{\partial^k}{\partial x^k} S(x - y, t) f(y) dy.$$

The rapid decay of the derivatives of $S(x - y, t)k$, as $y \to \pm\infty$, ensures that the integrals will converge for $t > 0$. In particular,

$$u_t - ku_{xx} = \int [S_t(x - y, t) - kS_{xx}(x - y, t)]f(y)dy = 0.$$

Thus u given by (3.16) is a solution of the differential equation of (3.15).

Next we verify that u takes on the proper initial values. We cannot plug $t = 0$ into the formula (3.16) because $S(0, t)$ becomes infinite as $t \to 0$. Instead we will show that

$$\lim_{t \to 0} u(x, t) = f(x)$$

for each x. As we saw before, we can write

$$S(x - y, t) = \gamma_r(x - y) = \sqrt{r/\pi} \exp[-r(x - y)^2],$$

if $r = 1/4kt$. Hence

$$u(x, t) = \int S(x - y, t)f(y)dy = \int \gamma_r(y - x)f(y)dy$$

because $\gamma_r(x - y) = \gamma_r(y - x)$. Since $r \to \infty$ when $t \to 0$, we can apply part (b) of Theorem 1.7 to deduce that

$$\lim_{t \to 0} u(x, t) = \lim_{r \to \infty} \int \gamma_r(y - x)f(y)dy = f(x).$$

We have shown that the limit of $u(x, t)$, as $t \to 0$, exists and equals $f(x)$. However, this does not completely prove that u is continuous at $(x, 0)$. A more complete argument shows that we can approach $(x, 0)$ from any direction and u has the same limit.

To show that $u(x, t)$ is bounded for all $x \in R, t \geq 0$, we use the fact that $S(x, t) \geq 0$ and (3.10):

$$|u(x, t)| = |\int S(x - y, t)f(y)dy| \leq M \int S(x - y, t)dy = M.$$

Thus the same constant M that bounds the initial data f is also a bound for $|u(x, t)|$.

The uniqueness of the bounded solution follows from the version of the maximum principle of Theorem 3.2. Suppose that u and v are two bounded solutions of (3.15) with the same initial data f. Then $w = u - v$ is also a

bounded solution of (3.15), but with zero initial data. By the maximum principle, $u - v \leq max_{x\in R}(u(x,0) - v(x,0)) = 0$. Reversing the roles of u and v we
see that $v - u \leq 0$ as well. We conclude that $u = v$. The proof of the theorem is
finished. $\Box$

We began our search for the fundamental solution $S(x,t)$ by thinking of S as
the limit, in some sense, of solutions $u_n(x,t)$ of (3.7). In the exercises we will see
that indeed the solutions u_n converge to S, as $n \to \infty$.

The error function

For most choices of initial data, the formula (3.16) is hard to evaluate analytically.
However, when $f(x)$ is piecewise constant, we can express the solution in terms
of the error function.

The error function, $erf(x)$, is defined as

$$erf(x) = \frac{2}{\sqrt{\pi}} \int_0^x e^{-s^2} ds.$$

In most applications, $erf(x)$ is used with $x \geq 0$. However, from the expression
for erf, we see that it can be considered an odd function of x:

$$erf(-x) = -erf(x).$$

We shall refer to this defintion of the error function as the *extended* error function.
You can compute the value of $erf(x)$, for a real number x, or for a vector of real
numbers **x**, with the MATLAB command `erf(x)` .

Now let the initial data $f(x)$ for problem (3.15) be the function

$$f(x) = \begin{cases} \alpha, & x < 0 \\ 0, & x > 0 \end{cases}.$$

Then the solution formula (3.16) becomes

$$u(x,t) = \int_R S(x-y,t)f(y)dy = \frac{\alpha}{\sqrt{4\pi kt}} \int_{-\infty}^0 e^{\frac{-(x-y)^2}{4kt}} dy.$$

To express this integral in terms of the error function we make the change of variable
$z = \frac{(y-x)}{\sqrt{4kt}}$ so that $dy = \sqrt{4kt}\, dz$. Then we find that

$$u(x,t) = \frac{\alpha}{\sqrt{\pi}} \int_{-\infty}^{-x/\sqrt{4kt}} e^{-z^2} dz$$

$$= \frac{\alpha}{\sqrt{\pi}} \int_{x/\sqrt{4kt}}^{\infty} e^{-z^2} dz$$

$$= \frac{\alpha}{\sqrt{\pi}} \left(\int_{0}^{\infty} e^{-z^2} dz - \int_{0}^{x/\sqrt{4kt}} e^{-z^2} dz \right)$$

$$= \frac{\alpha}{2} (1 - erf(x/\sqrt{4kt})).$$

3.3.3 Sources and the principle of Duhamel

Now let us turn our attention to finding solutions of the heat equation with a source,

$$u_t - ku_{xx} = q, \qquad u(x, 0) = f(x). \tag{3.17}$$

First, using the linearity of the equation, we can split the solution $u = v + w$ where

$$v_t - kv_{xx} = 0, \qquad v(x, 0) = f(x), \tag{3.18}$$

and

$$w_t - kw_{xx} = q, \qquad w(x, 0) = 0. \tag{3.19}$$

Since we already know how to solve (3.18), we consider (3.19). We will construct the solution of (3.19) using the *principle of Duhamel*. By way of motivation, recall from Chapter 1 that the solution of the ODE problem (λ constant)

$$\varphi' + \lambda\varphi = q, \qquad \varphi(0) = 0, \tag{3.20}$$

is

$$\varphi(t) = \int_{0}^{t} \exp[-\lambda(t - s)]q(s)ds.$$

Note that for each fixed s, $z(t, s) = \exp[-\lambda(t - s)]q(s)$ is in fact the solution of the IVP

$$z'(t, s) + \lambda z(t, s) = 0, \qquad z(s, s) = q(s),$$

where $'$ is differentiation with respect to t. Thus

$$\varphi(t) = \int_{0}^{t} z(t, s)ds.$$

To verify that φ solves (3.20), we calculate

$$\varphi'(t) = z(t, t) + \int_0^t z'(t, s)ds$$

$$= q(t) - \lambda \int_0^t z(t, s)ds = q(t) - \lambda\varphi(t),$$

where we have used Theorem 1.3.

Now we wish to use the same construction to solve (3.19). Let $z(x, t, s)$ be the solution for $t > s$ of

$$z_t = kz_{xx}, \qquad z(x, s, s) = q(x, s).$$

The solution of this problem is given by (3.16), replacing t by $t - s$ because we are solving for $t \geq s$:

$$z(x, t, s) = \int S(x - y, t - s)q(y, s)dy.$$

Then the solution $w(x, t)$ of (3.19) is

$$w(x, t) = \int_0^t z(x, t, s)ds = \int_0^t \int S(x - y, t - s)q(y, s)dyds.$$

Finally the solution of (3.17) is

$$u(x, t) = \int S(x - y, t)f(y)dy + \int_0^t \int S(x - y, t - s)q(y, s)dyds. \quad (3.21)$$

Exercises 3.3

1. Make an mfile, say S.m, to plot profiles of $S(x, t, k)$ for various values of t and k. Try the following

   ```
   function u = S(x,t,k)
       u = exp(-x.^2./(4*k*t))./sqrt(4*pi*k*t);
   ```

   After choosing an x vector, say x = -4:.05:4 you can plot the profile for $t = 1$, and $k = .5$ with the command

```
>> plot(x,S(x,1,.5) )
```

You can plot several profiles on the same graph (in different colors), say for $k = .5$ and $t = .1, .2, .5$, with the command

```
>> plot(x,S(x,.1,.5),x,S(x,.2,.5), x,S(x,.5,.5) )
```

(a) Plot several profiles with $k = 1$ and several values of t. What happens to the profiles as t increases? What happens as $t \downarrow 0$?

(b) Plot several profiles with $t = 1$ and various values of k. Verify that, if k_1, t_1 and k_2, t_2 are values of k and t such that $k_1 t_1 = k_2 t_2$, then $S(x, t_1, k_1) = S(x, t_2, k_2)$.

(c) Fix $t = .5$ and $k = 1$. Plot $x \rightarrow S(x - y, .5, 1)$ for several values of y. For example, if $y = 2$, you can use the command

```
>> plot(x,S(x-2,.5,1))
```

Now plot the result of adding together heat sources at several points. Plot the profile $x \rightarrow .5S(x - 2, t) + .4S(x - 1, t) - .2S(x + 1), t)$ with the command

```
>> u=.5*S(x-2,.5,1)+.4*S(x-1,.5,1)-.2*S(x+1, .5,1);
>> plot(x,u)
```

2. In this exercise we graphically investigate the way in which the solution of the initial-value problem (3.15) is built up as a superposition of point heat sources. This will be a further development of the technique of part (c) of exercise 1. In our derivation we said that the solution should be given approximately by the discrete sum

$$\sum_{j=1}^{n} S(x - y_j) f(y_j) \Delta y$$

and that the formula for $u(x, t)$ is obtained by taking the limit as $\Delta y \rightarrow 0$. The program heat1 computes this sum for initial data given on the interval $[-1, 1]$. You must enter the choice of k, n (n must be even), and t. The following initial data is "built in " to program heat 1.

$$f(x) = \begin{cases} -1 & -1 \le x < 0 \\ 2 & 0 < x \le 1 \\ 0 & \text{elsewhere} \end{cases}.$$

We subdivide the interval $[-1, 1]$ into n subintervals of length $\Delta y = 2/n$, and we choose y_j as the midpoint of each subinterval. For small $t > 0$, the solution $x \rightarrow u(x, t)$ should resemble f, but with the corners rounded off.

You can also run heat1 with different initial data which must be provided in a supporting mfile, ff.m. For more information enter help heat1. Run heat1 with time $t = .01$, $k = 1$, and with $n = 2, 4, 6, 8, 12$. To get a better view, restrict the plots to the interval $[-5, 5]$ with the command axis([-5 5 -1.5 3]). What happens as n increases?

3. Now run the program heat1 with the same f, $n = 20$, $k = 1$, and $t = .1, .5, 1, 2, 5$. To get the profiles on the same graph, run the program with $t = .1$, then type the command hold on. To get a better view, change the axes with the command axis([-5 5 -1 2]). After you have seen the graphs you want, type the command hold off. Otherwise, all succeeding graphs will be plotted together.

(a) Notice which parts warm up and which parts cool off. How can this behavior be predicted from the differential equation?

(b) Try to estimate the rate at which the solution decays. Look at the value of the profile at $x = 1/2$ at several times (say $t = 10, 11$, or $t = 20, 21$), and see if you can find constants C and γ such that the heights at $x = 1/2$ fit on a curve $Ct^{-\gamma}$. To get the values of the approximate solution (with $n = 20$) at $x = 1/2$, you must first figure out what index of the plotting vector x corresponds to $x = 1/2$. In the program heat1, x = -10:.05:10. Then each time you run the program heat1, save the component of the vector snap1 that has this index.

(c) Explain the rate of decay you found in part b) by analyzing the solution formula (3.16), using, of course, (3.14).

4. Let $f(x) = \alpha$ for $x < 0$, and $f(x) = \beta$ for $x > 0$ and let $u(x, t)$ be the solution of (3.15).

(a) Show that for this piecewise constant data, the formula (3.16) can be rewritten in terms of the (extended) error function. You will need to make the change of variable $z = \frac{(y-x)}{\sqrt{4kt}}$.

(b) Show that for each x, $\lim_{t \to \infty} u(x, t) = (\alpha + \beta)/2$.

(c) Using the error function of MATLAB erf(x), plot the solution on $[-5, 5]$ for $k = 1$ and $t = .01, .1, .5, 2$.

5. A bit harder. Let $f(x) = \alpha$ for $x \le -B$, and $f(x) = \beta$ for $x \ge B$ where B is some positive number. Assume that f is continuous on $[-B, B]$, and hence bounded on all of R. Show that conclusion 4 (b) still holds.

6. Assume that $u(x, t)$ is a solution of $u_t = ku_{xx}$, such that u and u_x tend to zero rapidly, as $x \to \pm\infty$. Let $Q = \int_R u(x, t)dx$. We have already seen that Q is a conserved quantity. Here are two more quantities associated with solutions of the heat equation that have a special behavior. In the following, assume that $Q \neq 0$.

   (a) Show that $m = \frac{1}{Q} \int_R xu(x, t)dx$ is independent of t. (Hint: Differentiate with respect to t under the integral and use the fact that u solves the heat equation.)

   (b) Let

   $$p(t) = \frac{1}{Q} \int_R (x - m)^2 u(x, t)dx.$$

   Show that $p(t) = p(0) + 2kt$. Use the same hint as in part (a).

Remark When $u(x, t) \geq 0$, $x \to u(x, t)/Q$ can be thought of as the probability distribution of a random variable $U(t)$. The probability that $U(t)$ lies in the interval $[a, b]$ is given by

$$\frac{1}{Q} \int_a^b u(x, t)dx.$$

m is the mean of this random variable, which remains constant, and $p(t) = \sigma^2(t)$ where $\sigma(t)$ is the standard deviation which increases as $t^{1/2}$.

   (c) Show that

   $$p(t) = \frac{1}{Q} \int_R x^2 u(x, t)dx - m^2.$$

   (d) Find m and $p(t)$ for the fundmental solution $S(x, t)$.

   (e) Evaluate the integral

   $$\int_{-\sqrt{p(t)}}^{\sqrt{p(t)}} S(x, t)dx$$

   using the error function. The amount of heat contained in the interval $[-\sqrt{p(t)}, \sqrt{p(t)}]$ is a constant fraction of the total amount of heat.

7. We want to construct an approximation $v(x, t)$ to the solution $u(x, t)$ of the heat equation which is easy to evaluate and such that $|u(x, t) - v(x, t)| \to 0$ in a appropriate sense as, $t \to \infty$. This result will extend the analysis of exercise 6.

(a)　Let $Q = \int_R f(y)dy$ (assume again that $Q \neq 0$), $m = \frac{1}{Q}\int_R yf(y)dy$
and $p_0 = \frac{1}{Q}\int(y - m)^2 f(y)dy$ be the quantities associated with the
initial data f.

Expand the function $y \to \exp[-(x - y)^2/4kt]$ in a Taylor series
about the point $y = m$.

(b)　Show that

$$\int_R e^{-\frac{(x-y)^2}{4kt}} f(y)dy = Qe^{-\frac{(x-m)^2}{4kt}}(1 - \frac{p_0}{4kt}) + R$$

where, for each x, $R = O(1/t^2)$, as $t \to \infty$.
Set

$$v(x, t) = Q\frac{(1 - p_0/4kt)}{\sqrt{4\pi kt}}e^{-\frac{(x-m)^2}{4kt}}.$$

We shall refer to v as the *Gaussian approximation*. Note that v is not
a solution of the heat equation, but that

$$\int_R v(x, t)dx = Q(1 - \frac{p_0}{4kt}), \qquad \int_R xv(x, t)dx = mQ(1 - \frac{p_0}{4kt}).$$

(c)　Show that $|u(x, t) - v(x, t)| \leq C/t^{5/2}$ for x in a bounded interval
$[a, b]$. The constant C will depend on the interval $[a, b]$.

(d)　Compute Q, m, and p_0 for the initial data of exercise 2. Note that
$p_0 < 0$, so that in this example $x \to u(x, t)/Q$ is not the probability
distribution of a random variable.

(e)　The program heat1 also computes and plots the Gaussian approximation for the initial data of heat1. To see the graph of the Gaussian approximation, run heat1, and then give the command plot(x,gauss).
Run heat1 this time with $n = 20$, $k = 1$, and $t = .5, 1, 2, 5, 10$.
Compare the graphs of the solution u and the Gaussian approximation
v for each of these values of t by using the command plot(x,snap1,
x,gauss). How large must t be so that the relative error $|u -$
$v|/\max |u|$ is less than 5%?

8.　Show that the solutions $u_n(x, t)$ defined in (3.7) converge to the fundamental solution $S(x, t)$, as $n \to \infty$.

9.　Solve the initial-value problem

$$u_t - ku_{xx} + \gamma u = 0, \qquad u(x, 0) = f(x).$$

Hint: Set $v(x, t) = \exp(\gamma t) \cdot u(x, t)$. Find the equation satisfied by v and solve it.

10. Solve the initial value problem

$$u_t + cu_x - \varepsilon u_{xx} = 0,$$

$$u(x, 0) = f(x) = \begin{cases} \alpha & x < 0 \\ \beta & x > 0 \end{cases}.$$

This equation combines linear diffusion and a linear convection term cu_x. Call the solution $u_\varepsilon(x, t)$. Show that, for each x,

$$\lim_{\varepsilon \to 0} u_\varepsilon(x, t) = f(x - ct).$$

Hint: Change to a coordinate system that moves with speed c. Let $v_\varepsilon(s, t) = u_\varepsilon(s + ct, t)$ so that $u_\varepsilon(x, t) = v_\varepsilon(x - ct, t)$. Use the chain rule to calculate the derivatives of u_ε in terms of those of v_ε, and substitute in the equation for u_ε to find the equation satisfied by v_ε. Solve this equation using the results of exercise 4. Then transform back and take the limit.

3.4 Boundary value problems on the half-line

Up to now, we have assumed that our bar in which heat is flowing is infinitely long, whereas in any real physical situation, we shall have to deal with bars of finite length. Various assumptions can be made about the behavior of the heat flow at the ends of the bar. These are called *boundary conditions*. We start with a bar which occupies the infinite half-line $x \geq 0$. We want to solve a problem

$$u_t - ku_{xx} = 0 \quad \text{in } x > 0, t > 0, \qquad u(x, 0) = f(x) \quad \text{for } x > 0.$$

In addition we must specify boundary conditions at $x = 0$. There are three principal types.

First kind or Dirichlet condition: $\qquad u(0, t) = 0$
The temperature is held fixed at 0, perhaps by putting ice cubes on the end.

Second kind or Neumann condition: $\qquad u_x(0, t) = 0$
The end of the bar is insulated so that no heat can flow (Flux $= 0$).

Third kind or Robin condition: $u_x(0, t) - hu(0, t) = 0$
Here we assume $h > 0$. In this case the heat flux $F(0, t) = -\kappa u_x(0, t) = -\kappa hu(0, t)$ is negative when $u(0, t) > 0$. Heat flows out of the end of the bar when the temperature is positive and flows into the bar when the temperature is negative. In both cases the temperature will tend toward zero. If this condition is imposed with $h < 0$, the temperature may grow when $u(0, t) > 0$ which is an unstable situation.

In all three conditions we can introduce a nonzero function of t on the right-hand side, making the boundary condition inhomogeneous.

We will solve the homogeneous boundary conditions by extending the initial data in an appropriate manner to all of R and then solving the equation on the whole line using (3.16). First we recall the definitions of two classes of functions on R.

$f(x)$ defined on R is an *even* function if $f(-x) = f(x)$ for all $x \in R$.
$f(x)$ is an *odd* function if $f(-x) = -f(x)$ for all $x \in R$.

For example $\cos(x)$ is an even function and $\sin(x)$ is odd. If f is odd, then we must have $f(0) = 0$. If f is even and differentiable at $x = 0$, then $f'(0) = 0$.

Now suppose that f is odd and that u is the solution of the IVP $u_t = ku_{xx}$, $u(x, 0) = f(x)$. Then we claim that for each $t > 0$, the function $x \to u(x, t)$ is odd. In fact the solution is given by (3.16):

$$u(x, t) = \int S(x - y, t) f(y) dy,$$

so that

$$u(-x, t) = \int S(-x - y, t) f(y) dy.$$

Setting $z = -y$ and using the fact that $S(-x + z, t) = S(x - z, t)$ we find that

$$u(-x, t) = \int_{\infty}^{-\infty} S(-x + z, t) f(-z)(-dz)$$

$$= -\int_{-\infty}^{\infty} S(x - z, t) f(z) dz = -u(x, t).$$

It follows that $u(0, t) = 0$ for all $t \geq 0$. A similar calculation shows that when f is even, $x \to u(x, t)$ is even, so that $u_x(0, t) = 0$ for all $t \geq 0$.

Now we solve the initial-boundary-value problem (IBVP)

$$u_t = ku_{xx} \text{ for } x > 0, t > 0, \qquad u(x, 0) = f(x) \text{ for } x > 0 \qquad (3.22)$$

$$u(0, t) = 0 \text{ for } t > 0.$$

We think of the given function $f(x)$ as defined to be zero for $x < 0$. Then we define the *odd extension* of f as

$$\tilde{f}(x) = f(x) - f(-x).$$

$\tilde{f}$ is an odd function on R, and because $f(x) = 0$ for $x < 0$, we see that

$$\tilde{f} = \begin{cases} f(x) & \text{for } x > 0 \\ -f(-x) & \text{for } x < 0 \end{cases}.$$

Then let $\tilde{u}(x, t)$ be the solution given by (3.16) with initial data $\tilde{f}$. Thus

$$\tilde{u}(x, t) = \int S(x - y, t)\tilde{f}(y)dy = \int S(x - y, t)[f(y) - f(-y)]dy$$

$$= \int_0^\infty S(x - y, t)f(y)dy + \int_{-\infty}^0 S(x - y, t)[-f(-y)]dy$$

$$= \int_0^\infty [S(x - y, t) - S(x + y, t)]f(y)dy.$$

The restriction of $\tilde{u}(x, t)$ to $x \geq 0$ is the solution of (3.22).

Next we turn to the solution of the second boundary value problem

$$u_t = ku_{xx} \text{ for } x > 0, t > 0 \qquad u(x, 0) = f(x) \text{ for } x > 0, \qquad (3.23)$$

$$u_x(0, t) = 0 \text{ for } t > 0.$$

To solve this problem, we define the *even extension* of f by

$$\tilde{f}(x) = f(x) + f(-x).$$

Clearly $\tilde{f}$ is an even function. We substitute the even extension of f in (3.16) to produce an even solution $\tilde{u}(x, t)$ of (3.15). Its restriction to $x > 0$ is given by

$$u(x, t) = \int_0^\infty [S(x - y, t) + S(x + y, t)] f(y) dy.$$

This is the solution of the IBVP (3.23).

Uniqueness of solutions to these IBVPs follows from the maximum principle, and continuous dependence of the solution on the initial data follows as in Theorem 3.3.

Exercises 3.4

1.

    (a)   Reformulate the maximum principle as stated in Theorem 3.2 for the quadrant $\{x, t \geq 0\}$, and modify the argument of Theorem 3.2 to prove it.

    (b)   Use the maximum principle as stated in part (a) to prove uniqueness of bounded solutions of (3.22).

2.   Let u be a solution of either (3.22) or (3.23). Assume that $u(x, t) \geq 0$ and that $u, u_x, u_{xx} \to 0$ rapidly as, $x \to \infty$.

    (a)   Compute the rate of change of the total heat energy $\int_0^\infty u(x, t) dx$.

    (b)   For which boundary conditions (Dirichlet or Neumann) is this quantity conserved? Why?

    (c)   Assume the Dirichlet boundary condition. If the temperature in the bar $u(x, t) > 0$ for x near zero, does the total amount of heat increase or decrease? If the temperature outside the bar is greater than the temperature inside, does the total amount of heat increase or decrease?

3.   The program heat2 solves the initial-boundary-value problems (IBVP) (3.22) (Dirichlet) and (3.23) (Neumann) with initial data

$$f(x) = \begin{cases} 0 & 0 < x < 1 \\ 1 & 1 < x < 2 \, . \\ 0 & x > 2 \end{cases}$$

The solution to each of these problems can be expressed in terms of the error function $erf(x)$. Program heat2 evaluates these expressions and plots them together. To get information on program heat 2, invoke MATLAB and type help heat2 . To see the expressions, look at the file heat2.m.

Which solution decays faster? Set $k = 1$, and compare the values at $x = 1$. In the code heat2.m, the vector x = 0:.05:10. Note that x(21) corresponds to $x = 1$. The solutions at time t are stored in vectors dirch and neumann. The values of the solutions at $x = 1$ also have index 21. Find constants C_D and γ_D, and C_N and γ_N, such that the solution of (3.22) decays like $C_D t^{-\gamma_D}$ at $x = 1$, and the solution of (3.23) decays like $C_N t^{-\gamma_N}$ at $x = 1$. Use pairs of values of t, like $t = 10, 11$ or $t = 20, 21$.

4. Now we find an analytic reason for the different rates of decay observed in exercise 3.

   (a) Use the power series for $\exp(-x^2)$ to expand both terms in the solution formula for (3.22). What is the leading power of t in the expansion after subtraction?

   (b) Do the same for the solution formula of (3.23). Now what is the leading power of t? Does this explain your observations in exercise 3?

5. (a) Starting with the solution formula for IBVP (3.22), show that the solution of (3.22) with initial data $f(x) \equiv U$ is given by

$$u(x, t) = U - 2U \int_x^\infty S(y, t)\,dy.$$

   (b) Verify directly that this function u solves the IBVP (3.22).

   (c) Make the change of variable $z = y/\sqrt{4kt}$, and show that $u(x, t) \to 0$ for each x, as $t \to \infty$.

6. (a) Using the solution of exercise 5, show that the solution of the inhomogeneous IBVP

$$v_t - k v_{xx} = 0 \quad x, \ t > 0, \qquad v(x, 0) = 0, \quad x > 0,$$

$$v(0, t) = U, \quad t > 0,$$

   is given by

$$v(x, t) = 2U \int_x^\infty S(y, t)\,dy.$$

   (b) Show that $v(x, t) \to U$ for each x, as $t \to \infty$.

   (c) Write the solution v in terms of the error function erf(x). For the definition of the error function, see the exercises of Section 3.3.

7. A heat source of temperature 100° C is placed at the end of a long metal rod whose cylindrical surface is insulated. If the rod is initially at temperature 0° C, how long does it take for the temperature to reach 50° C at a distance of 10 cm. from the end of the rod? Use the solution of exercise 6, the error function from a table or from a software package, and the following values of the diffusion constants k for iron, aluminum, and copper.

$$
\begin{array}{ll}
\text{iron} & k = .230 \text{ cm}^2/\text{sec.} \\
\text{aluminum} & k = .975 \text{ cm}^2/\text{sec.} \\
\text{copper} & k = 1.156 \text{ cm}^2/\text{sec.}
\end{array}
$$

8. Consider the half-line problem with a source:

$$u_t - ku_{xx} = q, \qquad x, t > 0$$

$$u(0, t) = 0, \quad t \geq 0, \qquad u(x, 0) = f(x), \quad x > 0.$$

Let $H(x, y, t) = S(x - y, t) - S(x + y, t)$.

(a) Using the principle of Duhamel, find the solution of this IBVP (see formula (3.21)).

(b) Assume that $q = 0$ for $x > a$ and for $t > T$. Make a Gaussian approximation to the solution that is valid for large T.

9. Suppose that u solves the IBVP with the Robin condition:

$$u_t - ku_{xx} = 0 \quad x, t \geq 0, \qquad u(x, 0) = f(x), \quad x \geq 0,$$

$$u_x(0, t) = hu(0, t), \qquad t \geq 0,$$

and that $u, u_x, u_{xx} \to 0$ rapidly, as $x \to \infty$. Assume that $u(x, t) \geq 0$. How does the sign of h affect the rate of change of the total heat energy $\int_0^\infty u(x, t)dx$? For which sign of h is the Robin condition a radiating (absorbing) boundary condition?

10. We want to solve the IBVP of exercise 9.

(a) Assume u solves this problem and let $v(x, t) = u_x(x, t) - hu(x, t)$. v is again a solution of the heat equation. What boundary and initial conditions does v satisfy?

(b) Solve the IBVP for v in terms of integrals of the fundamental solution S.

(c) Now solve the first-order ODE for u in terms of v. Assume $h > 0$. The solution which is bounded, as $x \to \infty$, is

$$u(x, t) = - \int_x^\infty e^{h(x-\xi)} v(\xi, t)d\xi.$$

Substitute the formula for v obtained in part (b). This will express u in terms of its initial data.

11. (a) Let $u(x, t)$ be the solution of the IBVP of exercise 9 (as solved in exercise 10) with initial data $u(x, 0) \equiv U$. Now the intermediate solution $v(x, t)$ has initial value $v(x, 0) = u_x(x, 0) - hu(x, 0) = -hU$, and we can use exercise 5 to find v. Finally show that

$$u(x, t) = U + 2U \int_x^\infty S(y, t)[e^{h(x-y)} - 1]dy.$$

Note that the formula works equally well for $h > 0$ and $h < 0$.

(b) What happens to the temperature $u(0, t)$ as t increases (depending on the sign of h)?

3.5 Diffusion and nonlinear wave motion

Now is a good time to compare the properties of wave motion, seen in Chapter 2, and the properties of diffusion. First note that for any $t > 0$, no matter how small, the fundamental solution of the heat equation, $S(x, t)$, is strictly positive for all $x \in R$. Thus the information about a heat source at $x = 0$ spreads instantaneously to all of R, which contrasts with the finite speed of propagation of waves. Next note that the formula (3.16) for the solution of the IVP for the heat equation shows that the value of u at $(x, t), t > 0$, depends on the values of $f(y)$ for all $y \in R$ in the form of an integral. This is quite different from the wave motion we studied where $u(x, t)$ depends only on the initial data at a single point $x - ct$. A third difference which follows from the dependence on initial data is that, in the case of wave motion, $u(x, t) = f(x - ct)$ has the same regularity as f, but in the case of diffusion, $u(x, t)$ given by (3.16) is C^∞, even though the data may be only continuous.

To see what happens when we combine diffusion and nonlinear wave motion we consider

$$u_t + uu_x - \varepsilon u_{xx} = 0, \qquad u(x, 0) = f(x) \text{ for } x \in R. \tag{3.24}$$

Equation (3.24) is Burger's equation (with viscosity). Recall from Chapter 2 that when $f'(x_0) < 0$ for some x_0, solutions of

$$u_t + uu_x = 0, \qquad u(x, 0) = f(x) \tag{3.25}$$

breakdown at a time t_* by developing a vertical piece in the wave profile which corresponds to a discontinuity in the solution. For $t > t_*$, the solution must be continued as a weak solution. An example of a weak solution of (3.25) is the shock wave

$$u(x, t) = f(x - ct), \tag{3.26}$$

where

$$f(x) = \begin{cases} u_l & \text{for } x < 0 \\ u_r & \text{for } x > 0 \end{cases},$$

with $u_l > u_r$. The translation speed (shock speed) c is determined from the Rankine-Hugoniot condition

$$c = \frac{u_l + u_r}{2}. \tag{3.27}$$

The addition of a diffusion term in (3.24) models a viscous effect in gas dynamics which tends to smooth solutions. The small parameter ε is the viscosity. We imagine that the addition of the diffusion term would prevent the breakdown that can happen to the solutions of (3.25). In particular we conjecture that there are solutions of (3.24) which look like (3.26) except that they are smoothed off, without the discontinuity. We call such solutions, when they exist, *travelling waves*.

Let us look for a travelling wave solution of (3.24)

$$u(x, t) = \varphi(x - ct),$$

such that

$$\varphi(s) \to u_l, \quad \text{as } s \to -\infty \tag{3.28}$$

and

$$\varphi(s) \to u_r, \quad \text{as } s \to +\infty.$$

We assume that $u_l > u_r$, and we expect the speed c to be determined by the values u_l and u_r.

We substitute this form in (3.24) and see that φ must satisfy the nonlinear second-order ODE

$$\varepsilon \varphi''(s) = (\varphi(s) - c)\varphi'(s). \tag{3.29}$$

We seek a solution of (3.29) and a value $c > 0$ such that (3.28) is satisfied. Rewrite (3.29) as

$$\varepsilon\varphi'' = [\frac{1}{2}\varphi^2 - c\varphi]',$$

which we can immediately integrate once to get

$$\varepsilon\varphi' = \frac{1}{2}\varphi^2 - c\varphi + A, \qquad (3.30)$$

where A is the constant of integration. Now assuming $\varphi'(s) \to 0$, as $s \to \pm\infty$, we take limits in (3.30) and use (3.28) to deduce that

$$\frac{1}{2}u_l^2 - cu_l + A = 0 = \frac{1}{2}u_r^2 - cu_r + A. \qquad (3.31)$$

The A drops out and we can solve for c:

$$c = \frac{u_l + u_r}{2}.$$

Returning to (3.31), we see that

$$A = \frac{u_r u_l}{2}.$$

With A and c determined this way, substitute in (3.30) to arrive at

$$\varepsilon\varphi' = \frac{1}{2}(\varphi - u_r)(\varphi - u_l).$$

Up to this point we have not used the assumption $u_l > u_r$. If we expect the solution φ to lie between the values u_l and u_r, then the right-hand side of this first-order ODE is negative, which says that φ is decreasing. Thus we must require that $u_l > u_r$ to be consistent with (3.28).

This equation is a separable ODE which can be written

$$(\frac{1}{u_r - u_l})[\frac{1}{\varphi - u_r} + \frac{1}{u_l - \varphi}]d\varphi = \frac{ds}{2\varepsilon}.$$

Now integrating so that $\varphi(0) = c$, we obtain

$$(\frac{1}{u_r - u_l})\log(\frac{\varphi - u_r}{u_l - \varphi}) = \frac{s}{2\varepsilon},$$

whence

$$\varphi(s) = u_l + \frac{u_r - u_l}{1 + \exp[-\frac{s}{2\varepsilon}(u_l - u_r)]}.$$

In the exercises we shall compute an approximate width of the region where the graph of φ makes its drop from u_l to u_r, and shall see that it tends to zero as $\varepsilon \to 0$.

If we think of both (3.24) and (3.25) as models for a fluid or gas, (3.24) is more sophisticated because it takes viscosity into account. The travelling wave solution $u_\varepsilon(x, t) = \varphi_\varepsilon(x - ct)$ of the viscous equation converges to the shock wave (3.26) as the viscosity coefficient $\varepsilon \to 0$. To get a travelling wave solution we had to assume $u_l > u_r$, and the limiting discontinuous solution of the inviscid Burgers' equation is an admissible shock wave. The procedure of approaching the shock wave by solutions of the viscous equation picks out the physically meaningful shock waves which satisfy the entropy condition. The simpler model (3.25) provides a good approximation to the viscous solution for small ε. In particular the shock wave and the viscous travelling wave have the same speed of propagation.

It is a happy accident that the solution of (3.24) can in fact be found in closed form using the Cole-Hopf transformation. We have concentrated on the travelling wave solutions to illustrate a technique that we shall use later with another equation where such a general transformation is not available. For a discussion of the Cole-Hopf transformation, see [Lo].

Exercises 3.5

1. Show that the only steady-state solutions of (3.24) which exist for all x are constant functions.

2. Make an mfile for the function φ, including the dependence on the parameters ε, u_l and u_r. Try

    ```
    function u = phi(s,myeps, uleft, uright)
       denom = 1+exp( -.5*s*(uleft - uright)/myeps );
       u = uleft + (uright -uleft)./denom ;
    ```

    Note that we cannot use eps as a variable, because this is already a system constant. Plot φ for $u_l = 2$, $u_r = 1$, and several values of ε, say $\varepsilon = .02$, .05, .1, .2. Plot the graphs on $[-2, 2]$. In each case estimate the width of the transition region where φ makes 90% of its drop from 2 to 1, that is, from 1.95 to 1.05.

3. Make an analytic estimate of the width of the transition region for general u_l and u_r by solving the defining equation for φ in terms of s. Let s_l

be the value of s such that $\varphi(s_l) = u_l - .05(u_l - u_r)$ and s_r such that $\varphi(s_r) = u_r + .05(u_l - u_r)$. Show that the width of the transition region $\Delta s = s_r - s_l$ tends to zero as the first power of the viscosity coefficient ε. Does this estimate agree with your graphical estimates in exercise 2?

4. Consider the nonlinear equation

$$u_t + u u_x + \gamma u = \varepsilon u_{xx}.$$

What is the ODE that φ must satisfy, so that $u(x, t) = \varphi(x - ct)$ is a travelling wave solution of (3.24)? Can you solve it the same way as we solved (3.29)?

5. The nonlinear equation

$$u_t - k u_{xx} = f(u)$$

is called a reaction-diffusion equation. The nonlinear term $f(u)$ represents the reaction of chemicals while the term $k u_{xx}$ as usual represents diffusion.

(a) Let $f(u) = u(1 - u)$. What constant solutions are there?

(b) Look for a travelling wave solution $u(x, t) = \varphi(x - ct)$. What second-order nonlinear ODE must φ satisfy?

(c) Write this second-order ODE as a first-order system by introducing the new dependent variable $\psi = \varphi'$. Find the critical points of this system. Linearize the system about each of these critical points, and discuss the stability of each (this will depend on the relationship between c and k).

(d) It can be proved that for each $c > 0$, there is a unique trajectory $s \to (\varphi_c(s), \psi_c(s))$ which approaches $(1, 0)$ as $s \to \infty$, and approaches $(0, 0)$ as $s \to -\infty$. Sketch the graph of $\varphi_c(s)$ for different values of c.

3.6 Numerical methods for the heat equation

In this section we shall examine finite difference methods for the heat equation

$$u_t - k u_{xx} = 0, \qquad u(x, 0) = f(x). \tag{3.32}$$

We shall use these methods in the next chapter to solve boundary value problems for the heat equation. These methods are easily generalized to treat situations where the coefficients c, ρ, κ depend on both x and t.

Make a grid with mesh points $x_j = j\Delta x, t_n = n\Delta t$. Let $u_{j,n}$ denote the approximate value at (x_j, t_n). Replace the derivatives by their finite difference approximations. As a first attempt, replace u_t by the forward difference

$$u_t = \frac{u_{j,n+1} - u_{j,n}}{\Delta t} + O(\Delta t)$$

and the second derivative u_{xx} by the centered difference

$$u_{xx} = \frac{u_{j+1,n} - 2u_{j,n} + u_{j-1,n}}{\Delta x^2} + O(\Delta x^2).$$

If we substitute these expressions for the derivatives and solve for $u_{j,n+1}$, we arrive at the scheme

$$u_{j,n+1} = (1 - 2s)u_{j,n} + s(u_{j+1,n} + u_{j-1,n}) \tag{3.33}$$

where $s = k\Delta t / \Delta x^2$. The truncation error for this scheme is $\sigma = O(\Delta t) + O(\Delta x^2)$. The scheme is *explicit*. If we have already computed $u_{j,n}$ at level n, then we can compute $u_{j,n+1}$ from (3.33). The computational diagram is shown in Figure 3.2.

Unfortunately this scheme is unstable (and does not converge) unless the coefficient of $u_{j,n}$ is nonnegative. Suppose we take initial data $f(x) = \varepsilon \cos(\pi x / \Delta x)$. Then

$$f(x_j) = f(j\Delta x) = \varepsilon(-1)^j.$$

This data oscillates exactly with the spatial frequency of the grid. Then it is easy to show that, for this initial data,

$$u_{j,n} = \varepsilon(-1)^j (1 - 4s)^n. \tag{3.34}$$

Now the exact solution of (3.32) with this initial data has $|u(x, t)| \le \varepsilon$ for all x, t. However the computed discrete approximation will be unbounded, as $n \to \infty$, unless

$$|1 - 4s| \le 1,$$

which is true if and only if $0 \le s \le 1/2$, or

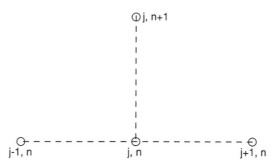

FIGURE 3.2
Computational diagram for (3.33).

$$k\Delta t \le \frac{(\Delta x)^2}{2}.$$

Thus for $\Delta x = .1$, say, this numerical scheme will become unstable if the initial data contains oscillations on the order of Δx, unless $k\Delta t \le (.1)^2/2 = .005$. This is a very small step size, and so, to make this scheme stable, is not practical.

Now let us try an *implicit* scheme. Instead of centering the difference quotient for the second derivative at (x_j, t_n), let us center it at (x_j, t_{n+1}), that is,

$$u_{xx}(x_j, t_{n+1}) = \frac{u_{j+1,n+1} - 2u_{j,n+1} + u_{j-1,n+1}}{(\Delta x)^2} + O(\Delta x^2).$$

This is equivalent to replacing the difference quotient for u_t by a backward difference. We arrive at the scheme

$$(1 + 2s)u_{j,n+1} - s(u_{j+1,n+1} + u_{j-1,n+1}) = u_{j,n}.$$

The computational diagram is shown in Figure 3.3. The truncation error is again $O(\Delta t) + O(\Delta x^2)$. We must solve this system of equations for each time step. To see better what is involved we write out the system for a small number of spatial grid points, say x_j, $j = 0, 1, 2, 3, 4$. From the initial condition $u(x, 0) = f(x)$, we have $u_{j,0} = f(j\Delta x)$ given. Then to advance to the next level $(n = 1)$, we must solve three equations in the five unknowns $u_{j,1}$, $j = 0, \cdots 4$:

$$\begin{array}{rllll}
-su_{0,1} +(1 + 2s)u_{1,1} & -su_{2,1} & & = u_{1,0} \\
-su_{1,1} +(1 + 2s)u_{2,1} & -su_{3,1} & & = u_{2,0} \\
-su_{2,1} +(1 + 2s)u_{3,1} -su_{4,1} & & = u_{3,0}
\end{array}$$

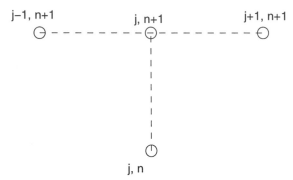

FIGURE 3.3
Computational diagram for the implicit scheme.

Usually such an overdetermined system does not have a solution. We should have a system of three equations in three unkowns. One way to reduce the number of unknowns is to impose the boundary conditions discussed in Section 3.4 at both ends of the bar. We shall discuss this kind of boundary value problem in Chapter 4. Here we use these concepts as a means of getting a solvable system of linear equations. We suppose that the bar has length L and that, for purposes of illustration, $\Delta x = L/4$ with $x_0 = 0$, and $x_4 = L$. Recall that the Dirichlet boundary condition specifies the temperature of the ends of bar in advance for all time. This means that $u_{0,n}$ and $u_{4,n}$ are given for all n. These values are then moved to the right-hand side of the first and third equations, resulting in a system of three equations in three unknowns.

$$\begin{bmatrix} 1+2s & -s & 0 \\ -s & (1+2s) & -s \\ 0 & -s & (1+2s) \end{bmatrix} \begin{bmatrix} u_{1,1} \\ u_{2,1} \\ u_{3,1} \end{bmatrix} = \begin{bmatrix} u_{1,0} + su_{0,1} \\ u_{2,0} \\ u_{3,0} + su_{4,1} \end{bmatrix}.$$

For instance, if the ends of the bar are kept at temperature zero, then $u_{0,n} = u_{4,n} = 0$ for all n. Now to find $u_{1,1}, u_{2,1}, u_{3,1}$, we must solve the system above. This may be done for any $s \geq 0$. In the general case, we use mesh points $x_0, x_1, \dots, x_J$ and the matrix is $J - 1 \times J - 1$ corresponding to the $J - 1$ interior points $x_1, \dots x_{J-1}$.

A second important boundary condition is the (homogeneous) Neumann condition in which we assume that the ends of the bar are insulated, so that the heat flux through the ends of the bar is zero, that is, $u_x(0, t) = u_x(L, t) = 0$ for all t. For the moment, we add "ghost points" x_{-1} and x_5 to our mesh. These points lie outside the bar. We can approximate $u_x(0, t)$ and $u_x(L, t)$ with centered differences

$$0 = u_x(0, t) \approx \frac{u_{-1,n} - u_{1,n}}{2\Delta x} \quad \text{or} \quad u_{-1,n} = u_{1,n},$$

and

$$0 = u_x(L, t) \approx \frac{u_{5,n} - u_{3,n}}{2\Delta x} \quad \text{or} \quad u_{5,n} = u_{3,n}.$$

Using the ghost points, augment the system of equations centered at the interior points by an equation centered at x_0 and another equation centered at x_4.

$$-su_{-1,1} + (1 + 2s)u_{0,1} - su_{1,1} = u_{0,0}$$

$$-su_{3,1} + (1 + 2s)u_{4,1} - su_{5,1} = u_{4,0}.$$

Now we have a system of five equations in the seven unknowns $u_{-1,1}, u_{0,1}, u_{1,1}, u_{2,1}, u_{3,1}, u_{4,1}, u_{5,1}$. However, when we use the Neumann condition, which translates into the discrete framework as $u_{-1,1} = u_{1,1}$ and $u_{5,1} = u_{3,1}$, we can reduce the number of unknowns in these last two equations to arrive at the systems of five equations in five unknowns:

$$\begin{bmatrix} 1+2s & -2s & 0 & 0 & 0 \\ -s & 1+2s & -s & 0 & 0 \\ 0 & -s & 1+2s & -s & 0 \\ 0 & 0 & -s & 1+2s & -s \\ 0 & 0 & 0 & -2s & 1+2s \end{bmatrix} \begin{bmatrix} u_{0,1} \\ u_{1,1} \\ u_{2,1} \\ u_{3,1} \\ u_{4,1} \end{bmatrix} = \begin{bmatrix} u_{0,0} \\ u_{1,0} \\ u_{2,0} \\ u_{3,0} \\ u_{4,0} \end{bmatrix}.$$

Again, we get a tridiagonal matrix. See also exercise 3 of Section 10.1.

In both cases the system is solved by making an LU factorization of the matrix and then using forward and backward substitution. This method is stable for any $s \geq 0$ and converges.

One can get improved accuracy in the approximation of the PDE by using an average of the explicit and the implicit scheme. This is called the Crank-Nicolson method. It has a truncation error $\sigma = (\Delta t^2) + O(\Delta x^2)$. The method is stable for all $s \geq 0$ and converges. The system of equations becomes

$$-su_{j-1,n+1} + (1 + 2s)u_{j,n+1} - su_{j+1,n+1}$$

$$= su_{j-1,n} + (1 - 2s)u_{j,n} + su_{j+1,n},$$

$j = 0, \cdots, J$, where now $s = (1/2)k\Delta t/(\Delta x)^2$. Again, these equations with proper boundary conditions can be solved efficiently with LU factorization. This is the method which is implemented in the programs heat3, heat4, and heat5 which you will use in Chapter 4.

General references for finite difference methods are the books [RM] and [St].

3.7 Projects

1. Write a MATLAB program to implement the explicit finite difference
    scheme (3.33). Set $k = 1$ and use the boundary conditions $u = 0$ at
    $x = 0$ and at $x = 10$. Write an mfile f.m for the initial data. Try out your
    program with initial data $f(x) = \sin(\pi x/10)$.

    Fix $\Delta x = .5$ and then experiment with various values of Δt to see
    when the scheme becomes stable. Your observations should agree with
    the results of Section 3.6.

    Compare your computed results with the exact solution

    $$u(x, t) = \sin(\pi x/10) \exp[-(\pi/10)^2 t]$$

    at various times, with $\Delta t = .125$. Find the maximum error at each time.
    Use help max to see how to have MATLAB do this.
    Now reduce the spatial step size to $\Delta x = .25$, and make $\Delta t = (\Delta x)^2/2$.
    Again compare the computed solution and the exact solution at the same
    times you did before when $\Delta x = .5$. Is the error smaller? How much?

2. Write a MATLAB program which uses Simpson's rule to evaluate the
    convolution integral

    $$u(x, t) = \int S(x - y, t) f(y) dy.$$

    Assume that the initial data $f(y) = 0$ for $|y| > 1$. Compare the plots of
    the solution at some $t > 0$, as $h \to 0$. Do you see convergence?

3. Write a MATLAB program which calculates the solution of exercise 11 of
    Section 3.4 and plots $x \to u(x, t)$ for each t. Let h be an input parameter,
    and compare the plots for different values of h.

4. Use MATLAB to make a phase-plane portrait of the system of ODE's that
    arises in exercise 5 of Section 3.5. Make c and k input parameters and
    comment on how the phase-plane portrait changes as c and k change.

Chapter 4

Boundary Value Problems for the Heat Equation

Up to now we have studied fairly broad qualitative properties of solutions of the heat equation. We have not attempted to solve problems with boundary conditions assigned at both ends of a bar of finite length. In this chapter we study these problems and exploit the theory of linear operators to give a unified treatment of the many possible combinations of boundary conditions.

4.1 Separation of variables

In this section we develop new techniques of solution based on the venerable method of separation of variables. The formulas we obtain for the solutions will contain much valuable information about their structure. However, for computational purposes, we will use the finite difference schemes discussed in Chapter 3. They are easily extended to the case of variable coefficients.

As a model problem, we consider a bar of length L with zero boundary conditions at both ends:

$$u_t = ku_{xx}, \quad \text{for} \quad 0 < x < L, \ t > 0, \qquad u(0, t) = u(L, t) = 0 \ \text{ for } \ t \geq 0, \tag{4.1}$$

$$u(x, 0) = f(x) \ \text{ for } \ 0 < x < L.$$

This is an initial-boundary-value problem. We shall use the abbreviation IBVP from now on.

We begin by looking for certain building-block solutions of the PDE and boundary conditions in the form of a product (ignoring the initial conditions for the moment):

$$u(x, t) = \varphi(x)\psi(t).$$

Substitute this expression into (4.1). There results

$$\varphi(x)\psi'(t) = k\varphi''(x)\psi(t),$$

whence

$$\frac{\psi'(t)}{k\psi(t)} = \frac{\varphi''(x)}{\varphi(x)}. \tag{4.2}$$

Since the left side of (4.2) depends only on t and the right side depends only on x, we conclude that

$$\frac{\psi'(t)}{k\psi(t)} = -\lambda = \frac{\varphi''(x)}{\varphi(x)},$$

where λ is a constant. Thus we are led to two ODE's : the first in the spatial variable

$$-\varphi''(x) = \lambda\varphi(x), \tag{4.3}$$

to which we add the boundary conditions

$$\varphi(0) = \varphi(L) = 0, \tag{4.4}$$

and a second ODE in the t variable,

$$\psi'(t) + \lambda k\psi(t) = 0. \tag{4.5}$$

Now consider (4.3), (4.4) in operator terms. Let $X = C[0, L]$ with W the subspace $W = \{\varphi \in C^2[0, L] : \varphi(0) = \varphi(L) = 0\}$, and let $M : W \to X$ be the linear operator $M\varphi = -\varphi''$. The problem (4.3), (4.4) can be restated as the problem of finding $\varphi \in W$ such that

$$M\varphi = \lambda\varphi \tag{4.6}$$

for some constant λ. In analogy with the terminology used for matrices and vectors, we say that (4.6) is an eigenvalue problem.We must find eigenvalues λ and corresponding nontrivial eigenfunctions φ which satisfy (4.6). We can do this readily for this problem. Assume that $\lambda > 0$. The general solution of (4.3) is

$$\varphi(x) = A\cos(\sqrt{\lambda}x) + B\sin(\sqrt{\lambda}x).$$

The boundary condition $\varphi(0) = 0$ implies that $A = 0$. We lose no generality by setting $B = 1$. Then to satisfy $\varphi(L) = 0$, we must select $\sqrt{\lambda}$ such that

$$\sin(\sqrt{\lambda}L) = 0.$$

This will be true if $\sqrt{\lambda}L = n\pi$, n integer. Thus, there is an infinite sequence of eigenvalues λ_n and eigenfunctions φ_n, $n = 1, 2, 3, \cdots$.

$$\lambda_n = (\frac{n\pi}{L})^2, \qquad \varphi_n(x) = \sin(\frac{n\pi x}{L}). \tag{4.7}$$

Now that we have determined the admissible values λ_n, we can return to equation (4.5) which has solutions

$$\psi_n(t) = \exp(-\lambda_n k t).$$

We must, of course, ask if there are any solutions of the eigenvalue problem (4.6) with $\lambda \leq 0$, or even complex λ. It follows from more general considerations that the eigenvalues of this problem must be real. However, we can see directly that there are no solutions of (4.6) with $\lambda < 0$. In fact, if we look for a solution with $\lambda = -\gamma$, with $\gamma > 0$, the general solution of (4.3) is given by

$$\varphi(x) = Ae^{x\sqrt{\gamma}} + Be^{-x\sqrt{\gamma}}.$$

If we apply the boundary conditions (4.4), we obtain a 2×2 system of linear equations for A and B:

$$A + B = 0$$

$$Ae^{L\sqrt{\gamma}} + Be^{-L\sqrt{\gamma}} = 0.$$

There will exist a nontrivial solution (A, B) only if the determinant of this system is zero for some value of γ. But this determinant is $-2\sinh(L\sqrt{\gamma})$, which is never zero for $\gamma \neq 0$. Thus there are no solutions of the eigenvalue problem (4.6) for $\lambda < 0$. The trivial solution $\lambda = 0$ yields only the solution $\psi(x) \equiv 0$ which cannot be an eigenfunction.

Consequently our building-block solutions of the problem (4.1) without the initial conditions are given by

$$\varphi_n(x)\psi_n(t) = \sin(\frac{n\pi x}{L})\exp(-k(\frac{n\pi}{L})^2 t)$$

$n = 1, 2, 3, \cdots$. Because the equation is linear, any linear combination of these is again a solution:

$$u(x, t) = \sum_{n=1}^{N} A_n\varphi_n(x)\psi_n(t)$$

which has initial value

$$u(x, 0) = \sum_{n=1}^{N} A_n\varphi_n(x).$$

Now if $f(x)$ is a finite linear combination of the φ_n, then we have found a solution of the IBVP (4.1). It was Fourier who ventured that we might get a solution of (4.1) for general f by looking at infinite sums:

$$u(x, t) = \sum_{n=1}^{\infty} A_n\varphi_n(x)\psi_n(t) \tag{4.8}$$

which would have initial value

$$u(x, 0) = \sum_{n=1}^{\infty} A_n\varphi_n(x) = f(x). \tag{4.9}$$

This form of the solution is called an *eigenfunction expansion.*

Several questions must be considered:

(1) Is there a broad class of functions f such that it is possible to find coefficients A_n (depending on f) such that (4.9) is true?

(2) In what sense do we understand that the infinite series converges to f?

(3) With this choice of A_n, does the series (4.8) converge to a strict solution of the problem (4.1)?

To address these questions, we first make some observations about the eigenfunctions $\varphi_n(x) = \sin(n\pi x/L)$:

$$\int_0^L \varphi_n(x)\varphi_m(x)dx = 0 \text{ for } m \neq n \tag{4.10}$$

and

$$\int_0^L \varphi_n^2(x)dx = \frac{L}{2} \text{ for all } n.$$

Equation (4.10) follows from a trig formula

$$\sin(ax)\sin(bx) = \frac{1}{2}[\cos((a-b)x) - \cos((a+b)x)].$$

Thus, for $m \neq n$,

$$\int_0^L \varphi_n(x)\varphi_m(x)dx = \frac{1}{2}\int_0^L \left[\cos(\frac{(m-n)\pi x}{L}) - \cos(\frac{(m+n)\pi x}{L})\right]dx$$

$$= \frac{1}{2}\left[\frac{\sin((m-n)(\pi x/L))}{(m-n)(\pi/L)} - \frac{\sin((m+n)(\pi x/L))}{(m+n)(\pi/L)}\right]\Big|_0^L = 0$$

and for $m = n$,

$$\int_0^L \varphi_n^2(x)dx = \int_0^L \sin^2(\frac{n\pi x}{L})dx = \frac{1}{2}\int_0^L \left[1 - \cos(\frac{2n\pi x}{L})\right]dx = \frac{L}{2}.$$

We want to exploit these special properties of the eigenfunctions. To do this we must add some structure to the vector space framework we have developed so far. We need to be able to talk about angles and distances in the vector space. In the finite-dimensional context of R^n, this is done by introducing the *scalar product* (inner product) of two vectors. We shall do the same here.

Definition   Let X be the vector space C[0, L]. We put a scalar product on C[0, L] as follows:
For $f, g \in X$

$$\langle f, g \rangle = \int_0^L f(x)g(x)dx. \tag{4.11}$$

This scalar product on X has all the properties of the scalar product on R^n. It is easy to verify that

(1) $\langle f, f \rangle \geq 0$ for all $f \in X$ and $\langle f, f \rangle = 0$ only when $f = 0$;

(2) $\langle f, g \rangle = \langle g, f \rangle$ for all $f, g \in X$;

(3) $\langle f + g, h \rangle = \langle f, h \rangle + \langle g, h \rangle$ for all $f, g, h \in X$; and

(4) $\langle af, g \rangle = a \langle f, g \rangle$ for all $f, g \in X, a \in R$.

We will say that $f, g \in X$ are *orthogonal* if $\langle f, g \rangle = 0$. The eigenfunctions φ_n form an *orthogonal set* in X because they are pairwise orthogonal,

$$\langle \varphi_n, \varphi_m \rangle = 0 \quad \text{for } m \neq n. \tag{4.12}$$

In addition we saw that

$$\langle \varphi_n, \varphi_n \rangle = \frac{L}{2} \quad \text{for all } n. \tag{4.13}$$

We can use the orthogonality of the eigenfunctions to determine the coefficients in the series (4.9). Suppose for the moment that the series (4.9) converges sufficiently well to make the following calculation valid. Then, using the orthogonality, we have

$$\langle f, \varphi_m \rangle = \langle \sum_{n=1}^{\infty} A_n \varphi_n, \varphi_m \rangle = \sum_{n=1}^{\infty} A_n \langle \varphi_n, \varphi_m \rangle = A_m \langle \varphi_m, \varphi_m \rangle.$$

Then property (4.13) implies that

$$A_m = \frac{\langle f, \varphi_m \rangle}{\langle \varphi_m, \varphi_m \rangle} = \frac{2}{L} \langle f, \varphi_m \rangle = \frac{2}{L} \int_0^L f(x) \sin(\frac{m\pi x}{L}) dx. \tag{4.14}$$

4.2   Convergence of the eigenfunction expansions

We must determine in what sense the series (4.9) converges to f, and in what sense the series (4.8) yields a solution of the PDE.

There are several measures of convergence of a series of functions. We will use one that comes naturally from the scalar product (4.11).

Definition   *The* norm *of a function $f \in X = C[0, L]$ is defined as*

$$\|f\| = [\int_0^L f^2(x) dx]^{1/2} = \langle f, f \rangle^{1/2}. \tag{4.15}$$

Properties of the norm

(1) $\|f\| \geq 0$ and $\|f\| = 0$ only when $f = 0$;

(2) $\|af\| = |a|\|f\|$ *for all* $f \in X, a \in R$; *and*
(3) $\|f + g\| \leq \|f\| + \|g\|$ *for all* $f, g \in X$.

The first two properties are easy to verify, and the third follows from the next inequality.

Cauchy-Schwarz inequality   For all $f, g \in X$,

$$|\langle f, g \rangle| \leq \|f\|\|g\|. \tag{4.16}$$

You may be familiar with this inequality for vectors in R^n. The proof is the same. For $f, g \in X$, $f \neq 0$, and $t \in R$, the quadratic polynomial

$$p(t) = t^2\|f\|^2 + 2t\langle f, g \rangle + \|g\|^2 = \langle tf + g, tf + g \rangle \quad \geq 0 \quad \text{for all } t.$$

It follows that its discriminant in the quadratic formula

$$4\langle f, g \rangle^2 - 4\|f\|^2\|g\|^2 \leq 0.$$

Take the square roots to deduce (4.16).

Now property (3) of the norm follows from (4.16) because

$$\|f + g\|^2 = \langle f + g, f + g \rangle = \|f\|^2 + 2\langle f, g \rangle + \|g\|^2$$

$$\leq \|f\|^2 + 2\|f\|\|g\| + \|g\|^2 = (\|f\| + \|g\|)^2.$$

Taking the square roots yields (3).

The norm (4.15) is a measure of the magnitude of the function f, and $\|f - g\|$ is a measure of the "distance" between f and g. There are many other norms satisfying (1), (2), and (3) which can be put on $C[0, L]$ which do not come from the scalar product (4.11). The norm (4.15) is special in that it satisfies the following Pythagorean property:

If $f, g \in X$ with $\langle f, g \rangle = 0$, then

$$\|f + g\|^2 = \|f\|^2 + \|g\|^2. \tag{4.17}$$

This can be seen by expanding the left side and using the orthogonality of f and g :

$$\|f + g\|^2 = \langle f + g, f + g \rangle = \langle f, f \rangle + 2\langle f, g \rangle + \langle g, g \rangle = \|f\|^2 + \|g\|^2.$$

Similarly if $f_1, \cdots, f_n$ are an orthogonal set in X, then

$$\| f_1 + \cdots + f_n \|^2 = \| f_1 \|^2 + \cdots + \| f_n \|^2. \tag{4.18}$$

Now let $f \in X = C[0, L]$, and let Λ_n be determined by (4.14). Fix an integer $N > 0$. Then

$$f = f - \sum_{n=1}^{N} A_n \varphi_n + \sum_{n=1}^{N} A_n \varphi_n.$$

Using the orthogonality of the φ_n, it is not hard to verify that $f - \sum_1^N A_n \varphi_n$ and $\sum_1^N A_n \varphi_n$ are orthogonal. Hence by (4.17),

$$\| f \|^2 = \| f - \sum_{n=1}^{N} A_n \varphi_n \|^2 + \| \sum_{n=1}^{N} A_n \varphi_n \|^2 \tag{4.19}$$

$$\geq \sum_{n=1}^{N} | A_n |^2 \| \varphi_n \|^2 = \frac{L}{2} \sum_{1}^{N} | A_n |^2.$$

This is called Bessel's inequality.   Since the bound for the partial sums $\sum_{n=1}^{N} |A_n|^2 \| \varphi_n \|^2$ is independent of N, we conclude that the series

$$\sum_{n=1}^{\infty} | A_n |^2 \| \varphi_n \|^2 < \infty.$$

However it is not immediately clear that the series (4.9) converges to f. This depends on there being "enough" of the orthogonal functions φ_n. In this case with $\varphi_n(x) = \sin(n\pi x / L)$, it is true that for each $f \in X = C[0, L]$, the series (4.9) converges to f in the sense that

$$\| f - \sum_{n=1}^{N} A_n \varphi_n \| \to 0, \quad \text{as } N \to \infty. \tag{4.20}$$

In terms of the integrals,

$$\int_0^L | f(x) - \sum_{n=1}^{N} A_n \varphi_n(x) |^2 dx \to 0, \quad \text{as } N \to \infty.$$

The series is said to converge in the *mean square*. Furthermore, it can be shown
that

$$\| f \|^2 = \frac{L}{2} \sum_{n=1}^{\infty} |A_n|^2 \tag{4.21}$$

because $\| \varphi_n \|^2 = \frac{L}{2}$ for all n.

In fact, the scalar product $\langle f, g \rangle$ and the norm can be defined on the larger class
of piecewise continuous functions on $[0, L]$. This class is defined as follows:

Definition   *A function f on $[0, L]$ is piecewise continuous if f is continuous at
all but a finite number of points $c_j \in [0, L]$, and at each c_j, f has a simple jump
discontinuity. This means that $\lim_{x \uparrow c} f(x) = f(c-0)$ and $\lim_{x \downarrow c} f(x) = f(c+0)$
both exist at each point of discontinuity $c = c_j$.*

The set of piecewise continuous functions on $[0, L]$ forms a vector space which
includes the vector space $X = C[0, L]$.

Whenever there are enough orthogonal eigenfunctions to expand every f in the
manner above, we say that the set of eigenfunctions is *complete*. This is analogous
to the situation in R^n when we have a set of orthogonal vectors which form a
basis for R^n. In the case of $X = C[0, L]$, we require an infinite number of
orthogonal vectors to span the space. In fact these eigenfunctions span the larger
space of piecewise continuous functions on $[0, L]$. In general, it is not easy to prove
the completeness of the eigenfunctions. We usually appeal to abstract operator
theory (see [RR]), although in some specific situations, such as the boundary value
problem (4.3), (4.4), one can use a geometric argument which depends on the fact
that the eigenvalues $\lambda_n \to \infty$, as $n \to \infty$ (see [Str1]).

A second measure of convergence of a series of functions is *uniform* convergence.
Uniform convergence and the Weierstrass test for uniform convergence of a series
of functions is discussed in Section 1.1.3. It is not hard to show that uniform
convergence implies mean square convergence.

From (4.21) it is clear that the coefficients $A_n \to 0$, as $n \to \infty$, for any $f \in X$.
The rate at which the $A_n \to 0$ depends on two properties of f:

(1) the regularity of f; and
(2) the degree to which f satisfies the boundary conditions.

By regularity we mean how many continuous derivatives f possesses. We can think
of $A_n \varphi_n(x)$ as the component of f with frequency $n\pi/L$. A more regular function
has "smaller" high-frequency components. A function with rapid oscillations or
discontinuities in its derivatives needs "larger" high-frequency components in its
synthesis.

We shall illustrate in examples how the second factor influences the rate at which
$A_n \to 0$.

Examples

(1) First take $f(x)$ to be a constant $C \neq 0$. f is C^∞, but does not satisfy the boundary conditions.

$$A_n = \frac{2C}{L} \int_0^L \sin(\frac{n\pi x}{L})dx = \frac{2C}{n\pi}[1 - \cos(n\pi)] = O(\frac{1}{n}), \quad \text{as } n \to \infty.$$

(2) Next let

$$f(x) = \begin{cases} 0, & 0 \le x \le L/4 \\ C, & L/4 < x < 3L/4 \\ 0, & 3L/4 \le L \end{cases}.$$

Here f and all its derivatives vanish at $x = 0$ and $x = L$, but f is discontinuous at $x = L/4$ and at $x = 3L/4$. f is piecewise continuous.

$$A_n = \frac{2C}{L} \int_{L/4}^{3L/4} \sin(\frac{n\pi x}{L})dx = \frac{2C}{n\pi}[\cos(n\pi/4) - \cos(3n\pi/4)]$$

$$= O(\frac{1}{n}), \quad \text{as } n \to \infty.$$

Note that in these two examples that $\sum A_n^2 < \infty$, as predicted by (4.21). (Recall from Section 1.1.3 that $\sum n^{-p}$ converges for $p > 1$ and diverges for $p \le 1$.)

(3)
$$f(x) = \begin{cases} x & 0 \le x \le L/2 \\ L - x & L/2 \le x \le L \end{cases}.$$

This f is continuous, but not C^1, and $f(0) = f(L) = 0$. Here

$$A_n = \frac{2}{L} \int_0^{L/2} x \sin(\frac{n\pi x}{L})dx + \frac{2}{L} \int_{L/2}^L (L - x) \sin(\frac{n\pi x}{L})dx$$

$$= \frac{4L}{(n\pi)^2} \sin(\frac{n\pi}{2}) = O(\frac{1}{n^2}), \quad \text{as } n \to \infty.$$

(4) Finally let $f(x) = x^2(x - L)^2$. This $f \in C^\infty$, with $f(0) = f'(0) = f(L) = f'(L) = 0$.

$$A_n = \frac{2}{L} \int_0^L f(x) \sin(\frac{n\pi x}{L})dx = -\frac{2L}{(n\pi)^2} \int_0^L f''(x) \sin(\frac{n\pi x}{L})dx$$

because there are no boundary terms when we integrate by parts twice. In the last integral, $f'' = 12x^2 - 12x + 2L^2$ and it can be calculated to be $O(\frac{1}{n})$. Thus $A_n = O(n^{-3})$.

In examples (3) and (4), the convergence to f is stronger than mean square convergence because the coefficients $|A_n| \le C(n^{-2})$, and hence

$$\sum_{n=1}^{\infty} |A_n| < \infty. \tag{4.22}$$

By the Weierstrass test (Theorem 1.5), this summability condition implies that the series (4.9) converges uniformly to f.

Next we turn to the question of convergence of the series (4.8), and in what sense its sum gives a solution of the PDE. Certainly if we are allowed to differentiate the series term by term, then (4.8) yields a solution:

$$u_t - k u_{xx} = (\partial_t - k \partial_x^2) \sum_{n=1}^{\infty} u_n(x, t) = \sum_{n=1}^{\infty} A_n (\partial_t - k \partial_x^2) \varphi_n \psi_n = 0.$$

The mathematical result we need to use here is Theorem 1.6 c). If the differentiated series converges uniformly, then the term by term differentiation is valid. Let us examine the terms of the series for $t > 0$:

$$|u_n(x, t)| = |A_n \varphi_n(x) \psi_n(t)| \le |A_n| \exp[-k(\frac{n\pi}{L})^2 t] = |A_n| \gamma^{n^2},$$

where $\gamma = \exp(-kt(\frac{\pi}{L})^2) < 1$ for $t > 0$. Even without using the fact that $\sum |A_n|^2 < \infty$, we see that there is a constant M such that

$$|u_n(x, t)| \le M_n \equiv M \gamma^{n^2}.$$

Now γ^{n^2} goes to zero faster than n^{-p} for any $p > 0$ because $\gamma < 1$. Hence $\sum M_n < \infty$. Thus for $t > 0$, the terms of the series satisfy the summability condition of the Weierstrass test, and hence, for $t > 0$, the series for u converges uniformly. The same argument shows that the differentiated series also converge uniformly for $t > 0$. We see that for $t > 0$, there are constants C and $\tilde{M}$, such that

$$|\partial_t^m u_n(x, t)| = |k^m (\frac{n\pi}{L})^{2m} u_n(x, t)| \le C n^{2m} \gamma^{n^2} \le \tilde{M} \tilde{\gamma}^{n^2}$$

for some $\tilde{\gamma}$, $\gamma < \tilde{\gamma} < 1$. Similarly,

$$|\partial_x^m u_n(x, t)| \le C n^m \gamma^{n^2}.$$

Thus the differentiated series

$$\partial_t^m u(x, t) = \sum_{n=1}^{\infty} \partial_t^m u_n(x, t)$$

and

$$\partial_x^m u(x, t) = \sum_{n=1}^{\infty} \partial_x^m u_n(x, t)$$

converge uniformly for $0 \le x \le L$ and $t \ge \delta > 0$ so that term by term differentiation is valid for $t > 0$. The precise result can be stated as follows:

Theorem 4.1
Let $f \in C[0, L]$ with $f(0) = f(L) = 0$. Then there is a unique solution $u(x, t)$ of the IBVP

$$u_t - ku_{xx} = 0 \qquad 0 < x < L, \ t > 0, \tag{4.23}$$

$$u(x, 0) = f(x) \qquad 0 \le x \le L, \tag{4.24}$$

$$u(0, t) = u(L, t) = 0 \qquad t \ge 0, \tag{4.25}$$

with u continuous on $[0, L] \times [0, \infty)$. In fact $u \in C^\infty((0, L) \times (0, \infty))$. u is given by the eigenfunction expansion (4.8), and the coefficients A_n are given in (4.14).

The argument we have given so far proves that u given by (4.8) is a solution for $t > 0$ and is in class C^∞. However, we have not shown that u is in fact continuous right up to the initial line $t = 0$. This is not hard to see if we make a stronger hypothesis on the initial data f. Assume that $f \in C^1[0, L]$. Then the series of coefficients $\sum |A_n|$ converges. By the Weierstrass test (Theorem 1.5), this means that the series (4.8) converges uniformly for $t \ge 0$. Since the partial sums of (4.8) are continuous functions in x, t, this implies that $u(x, t)$ is continuous for $t \ge 0$. A proof (not given here) that u is continuous at $t = 0$, under the hypothesis of the theorem that f is only continuous, requires more information about the convergence of Fourier series and an application of the maximum principle.

However, we can prove the uniqueness of the solution in the class of continous functions. Suppose that u and v are two solutions as described in the theorem. Then $w = u - v$ solves the same equation with zero initial conditions. Now by the maximum principle, for any $T > 0$,

$$\max_{0 \le x \le L, \ 0 \le t \le T} w(x, t) = \max_{\Gamma} w(x, t) = 0,$$

where

$$\Gamma = \{x = 0, 0 \le t \le T\} \cup \{x = L, 0 \le t \le T\} \cup \{0 \le x \le L, t = 0\}.$$

Thus $u(x, t) - v(x, t) = w(x, t) \le 0$. When we reverse the roles of u and v, we deduce that $v(x, t) - u(x, t) \le 0$. Consequently $u = v$. $\square$

We can also use the maximum principle to deduce a result about the dependence of the solution on the initial data:

Let $f, g \in C[0, L]$, $f(0) = f(L) = g(0) = g(L) = 0$, and let u and v be the corresponding solutions of (4.23) - (4.25) with $u(x, 0) = f(x)$, $v(x, 0) = g(x)$. Then

$$\max_{0 \le x \le L, \, t \ge 0} |u(x, t) - v(x, t)| = \max_{0 \le x \le L} |f(x) - g(x)|.$$

Thus we have shown that the IBVP (4.23) - (4.25) is well posed, as described in Chapter 2.

We make two additional comments about the solutions of the IBVP (4.23) - (4.25). First we note that each term in the series solution (4.8) has a factor which decays exponentially, as $t \to \infty$. This implies that the solution $u(x, t) \to 0$, as $t \to \infty$. Physically this makes sense because the zero boundary condition allows all of the heat to flow from the end points of the bar.

Now suppose that instead of the zero boundary condition (4.25) we prescribe the temperature at either end of the interval. For instance, we might require that

$$u(0, t) = \alpha \qquad u(L, t) = \beta, \qquad \text{for } t \ge 0.$$

We shall see in Section 4.4 that the solution of (4.23), (4.24) with this inhomogeneous boundary condition tends toward a steady-state solution $U(x)$, as $t \to \infty$. What do we mean by a steady-state solution? It should not depend on time, and hence satisfy

$$U_{xx} = 0, \quad 0 < x < L,$$

$$U(0) = \alpha \qquad U(L) = \beta.$$

The solution of this time-independent equation is the line

$$U(x) = \alpha + \frac{\beta - \alpha}{L} x.$$

You will see the convergence of the solution to this steady-state solution in the computer exercises.

Exercises 4.2

1. Let $u(x, t)$ be the solution of (4.23), (4.24). Let

$$Q_T = \{(x, t) : 0 \le x \le L, \ 0 \le t \le T\}.$$

   (a) Because u solves (4.23),

   $$0 = \int \int_{Q_T} (u_t - k u_{xx}) dx dt.$$

   Integrate the derivatives, and use the fact that $u(x, t) \to 0$, as $t \to \infty$, to show that

   $$\int_0^L f(x) dx = k \int_0^\infty u_x(0, t) dt - k \int_0^\infty u_x(L, t) dt.$$

   (b) What physical interpretation does this equation have in the context of heat flow?

2. Let u be a solution of (4.23)-(4.25) with initial data $f(x) \ge 0$ and $f(0) = f(L) = 0$. Use the maximum (or minimum) principle to show that $u(x, t) \ge 0$ for $0 \le x \le L, t \ge 0$.

3. Show that the eigenvalue problem (4.3), (4.4) has no complex solutions λ with $Im(\lambda) \ne 0$.

4. Consider the solution of the IBVP (4.23)-(4.25) with initial data $f(x) = 5\sin(2\pi x/L) - 2\sin(3\pi x/L) + 3\sin(5\pi x/L)$.

   (a) Find the solution. It should consist of three terms.

   (b) Set $L = \pi$ and $k = 1$. Make an mfile u.m for the solution $u(x, t)$,

   ```
   function w = u(x,t)
   w = . . .
   ```

   Plot snapshots of the solution for $t = 0, \ .5, 1, 2, 5$ with the instructions

   ```
   x = 0: pi/100: pi;
   u0 = u(x,0); u1 = u(x,.5); u2 = u(x,1);
   u3 = u(x,2); u4 = u(x,5);
   plot(x,u0,x,u1,x,u2,x,u3,x,u4)
   ```

   Which term in the sum is a good approximation to the solution when $t = 5$? Why?

5. Compute the coefficients A_n of the eigenfunction expansion for the function $f(x) = x$ on $[0, L]$. How fast do the coefficients tend to zero, as $n \to \infty$? Keeping in mind that the partial sums of the series all vanish at the points $x = 0$ and $x = L$, and the examples, do you expect the series to converge uniformly?

6. Compute the coefficients A_n of the eigenfunction expansion for the function $f(x) = 1$ for $0 \leq x < L/2$, $f(x) = 0$ for $L/2 < x \leq L$. This function has a jump at $x = L/2$. Will the series converge uniformly?

7. Write a short MATLAB program to sum the first N terms of the eigenfunction expansions given in examples (1), (2), and (3) of this section. Take $L = \pi$. Plot the partial sum s_N against the function f to see the convergence. Try $N = 2, 5, 10, 20$. How do the plots of examples (1) and (2) differ from those of (3) ? Remember example (3) converges uniformly, while (1) and (2) do not. In examples (1) and (2), pay special attention to the behavior at $x = 0$ and $x = L/2$. This is called the Gibb's phenomenon. Here is a model program for example (1) with $C = 1$.

```
N = input('enter the number of terms to be summed  ')
x = 0: pi/100:pi;
sum = 0;
for n = 1:N
    A = (2/(n*pi))*(1-cos(n*pi));
    sum = sum + A*sin(n*x);
end
y = ones(size(x));
plot(x,y,x,sum)
```

For example (2), replace the line y = . . .   with
```
        y = (x <= 3*pi/4) - (x< pi/4);
```
and for example (3) use
```
        y = (x<pi/2).*x + (x>= pi/2).*(pi-x); .
```
Of course you must change the line defining A for each example.

8. Suppose that the initial data $f(x)$ is C^1, with $f(0) = f(L) = 0$, so that the coefficients A_n satisfy the summability condition

$$\sum_1^\infty |A_n| < \infty.$$

Show that the solution $u(x, t)$ of (4.23) - (4.25) tends to zero exponentially as, $t \to \infty$:

$$|u(x, t)| \le C e^{-\lambda_1 kt},$$

where C is a constant. Can you show this type of result even when f is only piecewise continuous? Hint: Let $t_0 > 0$, and show that the estimate can be made for $t \ge t_0$.

9. Let u be a solution of (4.23) - (4.25) given by the eigenfunction series

$$u(x, t) = \sum_1^\infty A_n \sin(n\pi x/L) e^{-(\frac{n\pi}{L})^2 kt}.$$

The average in the spatial variable x for each t is

$$\bar{u}(t) = \frac{1}{L} \int_0^L u(x, t) dx.$$

Show that

$$\bar{u}(t) = \sum_1^\infty \frac{2A_n}{n\pi} e^{-(\frac{n\pi}{L})^2 kt},$$

where the sum is taken over n odd.

10. Take initial data

$$f(x) = \begin{cases} x, & 0 \le x \le L/2 \\ L - x, & L/2 \le x \le L \end{cases}.$$

The eigenfunction series for the solution with this initial data is given in example (3) of this section.

(a) We shall take the first term

$$u_1(x, t) = \frac{4L}{\pi^2} \sin(\frac{\pi x}{L}) e^{-(\frac{\pi}{L})^2 kt}$$

as an approximation to the solution. Show that

$$|u(x, t) - u_1(x, t)| \le \frac{L}{2\pi^2} (\pi^2 - 8) e^{-9(\frac{\pi}{L})^2 kt}.$$

You may use the fact that

$$\sum_{1}^{\infty} \frac{1}{n^2} = \frac{\pi^2}{8},$$

where the sum is taken over n odd.

(b) Find a bound for the error relative to $u_1(\frac{L}{2}, t)$.

The program heat3 integrates the heat equation on $[0, 10]$ with Dirichlet boundary conditions. $\Delta x = \Delta t = .05$ are set in the program. Invoke MATLAB and type help heat3 to see information about this program. You must write an mfile, f.m, for the initial data which is array-smart. You must also write mfiles left.m and right.m for the boundary data at $x = 0$ and at $x = 10$.

11. This exercise uses program heat3 with initial data

$$f(x) = \begin{cases} x, & 0 \le x < 5 \\ 10 - x, & 5 < x \le 10 \end{cases}.$$

The boundary conditions are $u = 0$ at $x = 0$ and at $x = 10$. This is the same data as in exercise 10 with $L = 10$. Write the file f.m as follows

```
function y = f(x)
y = (x< 5).*x + (x>=5).*(10-x);
```

The boundary data files left.m and right.m should be

```
function y = left(t)
   y = 0;
```

```
function y = right(t)
   y = 0;
```

(a) Run heat3 with this initial data, $k = 1$, and small values of t, say $t = .05, .1, .2, .5$, which correspond to $n1 = n2 = 1$, $n3 = 2$, and $n4 = 6$. How does the regularity (smoothness) of the initial data compare with that of the solution at these times? What do you think causes the kink in the graph of the solution for small t, when analytically, it should be perfectly smooth?

(b) We want to compare the solution $u(x, t)$ with the first term $u_1(x, t)$ in the eigenfunction expansion of exercise 10. Determine the value of λ_1

and use $k = 1$. Make an array-smart mfile for $u_1 = (40/\pi^2) \sin(\pi x/10)$ $\exp(-\lambda_1 kt)$ which begins

```
function w = u1(x,t)
     w = . . .
```

Run program heat3 for times $t_1 = .5, t_2 = 1, t_3 = 2,$ and $t_4 = 5$. Plot snap1 together with $x \to u_1(x, .5)$, etc. Do these plots confirm the estimates you made in exercise 10?

12. Run program heat3 with $f(x) = -\sin(\pi x/5)\exp(x/2)$, $u = 0$ at $x = 0$ and $x = 10$ and $k = 1$. Modify the mfile f.m that you wrote for exercise 11 for this data. Notice here that the data is not symmetric with respect to $x = 5$ but that the solution tends toward a symmetric temperature distribution.

   (a) Why does this happen? Check the flux at either end of the bar in plots at different times.

   (b) Run heat3 with the same data and $k = 1$ up to $t = 10$. Then run again with $k = 2$ and $t = 5$. Why are the plots with $k = 1, t = 10$ and $k = 2, t = 5$ identical?

   (c) Compute A_1 using the numerical integration tool quad of MATLAB. It computes

   $$\int_a^b f(x)dx.$$

   The command for quad is

   ```
   >> quad('ff',a,b)
   ```

   where $f(x)$ is given in a function mfile ff.m. You will have to rename your mfile ff.m to use quad. Then graph the first term $u_1(x, t) = A_1 \sin(\pi x/10)\exp(-\lambda_1 kt)$ against the solution at time t. Observe how the approximation improves as t increases.

13. Run heat3 with $k = 1$, with initial data $f(x) = x/2 - 2 + \sin(4\pi x/10)$, and boundary conditions $u(0, t) = -2$, $u(10, t) = 3$. In addition to rewriting the mfile f.m, you will also have to modify the mfiles left.m and right.m.

   (a) By running up to a large value of t, determine the steady-state solution $U(x)$ toward which the time-dependent solution is tending.

   (b) How large must t be so that max $|u(x, t) - U(x)| \le 1/10$?

14. Run heat3 with initial condition $f(x) = 0$ and boundary conditions

$$u(0, t) = \frac{t}{1+t} + te^{-t/5}, \qquad u(10, t) = 0.$$

The array-smart way to write the mfile for f is

```
function y = f(x)
    y = zeros(size(x));
```

You will also have to modify the mfiles left.m and right.m.

(a) Does the solution $u(x, t)$ tend toward a steady-state $U(x)$, as $t \to \infty$? What is $U(x)$? How large must t be so that $|u(x, t) - U(x)| \le .1$?

(b) Do the experiment with $k = 1, t = 10$, and with $k = 2, t = 5$. Are the plots identical? Why or why not?

(c) Make the left boundary function left.m array-smart. Then plot the left boundary function on $[0, 20]$. What is the max $u(0, t)$? Where does it occur? Verify that the maximum principle holds by plotting several snapshots of the solution.

15. Another way to prove the uniqueness of the solutions of (4.23) - (4.25).

(a) Multiply the equation by u, and integrate in x from 0 to L:

$$0 = \int_0^L uu_t dx - k \int_0^L u_{xx} u dx.$$

Integrate by parts in x, and use the boundary conditions $u(0, t) = u(L, t) = 0$ to deduce that

$$\frac{1}{2} \frac{d}{dt} \int_0^L u^2 dx + k \int_0^L u_x^2 dx = 0.$$

Finally integrate in t to arrive at

$$\int_0^L u^2(x, t) dx + 2k \int_0^T \int_0^L u_x^2(x, t) dx = \int_0^L u^2(x, 0) dx.$$

(b) Use this equation to prove the uniqueness of the solutions.

16.   Use separation of variables to solve the IBVP

$$u_t - ku_{xx} = \gamma u, \qquad u(0,t) = u(L,t) = 0$$

$$u(x,0) = f(x).$$

When $\gamma > 0$, the right side of the equation represents a heat source which is proportional to the temperature. How large must γ be so that the solution grows with t instead of decaying to zero?

17.   Show that, if a sequence of functions $f_n(x)$ converges uniformly on $[0, L]$ to $f(x)$, then f_n converges to f in the mean square sense.

4.3   Symmetric boundary conditions

We wish to find a class of boundary conditions for which it is possible to construct solutions of the heat equation by eigenfunction expansions, as we did in Section 4.2, and to develop a general approach to these problems. The IBVP for the heat equation is

$$u_t - ku_{xx} = 0, \qquad u(x,0) = f(x) \tag{4.26}$$

$$+ \text{ homogeneous boundary conditions.}$$

The boundary conditions will determine a subspace W of $C^2[0, L]$. In Sections 4.1 and 4.2 we chose

$$W = \{w \in C^2[0, L] : w(0) = w(L) = 0\},$$

but we also want to consider other situations. For example, if the ends of the bar are insulated (no heat flux through the ends), we need

$$W = \{w \in C^2[0, L] : w'(0) = w'(L) = 0\}.$$

If we repeat the procedure of Section 4.1 and separate variables in (4.26), we shall arrive at an ODE in the time variable

$$\psi'(t) + k\lambda\psi(t) = 0$$

and an eigenvalue problem

$$M\varphi = \lambda\varphi.$$

Here M is the differential operator $-d^2/dx^2$, *together* with a subspace $W \subset C^2[0, L]$ determined by the boundary conditions. The subspace W is the *domain* of the operator M.

The following class of boundary conditions is important for this eigenvalue problem.

Definition   *A subspace $W \subset C^2[0, L]$ determined by boundary conditions is* symmetric *for the operator $-d^2/dx^2$ if*

$$\langle Mw, z \rangle = \langle w, Mz \rangle \qquad \text{for all } w, z \in W. \tag{4.27}$$

W is nonnegative *for $-d^2/dx^2$ if*

$$\langle Mw, w \rangle \geq 0 \qquad \text{for all } w \in W.$$

To find which boundary conditions are symmetric, we use Green's second identity (integrate by parts twice):

$$\int_0^L -w''z\,dx = (-w'z + wz')\Big|_0^L + \int_0^L wz''dx.$$

- If the boundary terms

$$(-w'z + wz')\Big|_0^L = 0 \qquad \text{for all } w, z \in W \tag{4.28}$$

then the subspace is symmetric for $-d^2/dx^2$.

If we integrate by parts only once, we have Green's first identity:

$$\int_0^L -w''z\,dx = (-w'z)\Big|_0^L + \int_0^L w'z'dx.$$

Set $w = z$ in this identity. We see that

$$\int_0^L -w''w\,dx = (-w'w)\Big|_0^L + \int_0^L (w')^2dx.$$

- If the boundary conditions of W imply that

$$(-w'w)\Big|_0^L \geq 0 \qquad \text{for all } w \in W,$$

then the subspace W is nonnegative for $-d^2/dx^2$.

The reader should verify that each pair of boundary conditions in the following list determines a subspace $W \subset C^2[0, L]$ which is symmetric and nonnegative for $-d^2/dx^2$.

$$w'(0) = w'(L) = 0 \tag{4.29}$$

$$w(0) = w(L) = 0 \tag{4.30}$$

$$w'(0) = w(L) = 0 \tag{4.31}$$

$$w(0) = w'(L) = 0 \tag{4.32}$$

$$w'(0) - hw(0) = 0, \quad w(L) = 0, \quad h > 0 \tag{4.33}$$

$$w'(0) = 0, \quad w'(L) + hw(L) = 0 \quad h > 0 \tag{4.34}$$

From the general theory of linear operators in complex vector spaces, it follows that if M satisfies (4.27), then all the eigenvalues of M are real. Furthermore, if W is also nonnegative for $-d^2/dx^2$, all the eigenvalues of M are nonnegative. This second point is easy to see. If λ is an eigenvalue of M with eigenfunction φ, then $M\varphi = \lambda\varphi$ and

$$\lambda\|\varphi\|^2 = \lambda\langle\varphi, \varphi\rangle = \langle M\varphi, \varphi\rangle \geq 0,$$

which implies that $\lambda \geq 0$. In fact, for each of the boundary conditions (4.30) - (4.34), the eigenvalues are strictly positive. Suppose that φ is an eigenfunction of $-d^2/dx^2$ for any of the boundary conditions (4.29) - (4.34). Then using Green's first identity,

$$\lambda\langle\varphi, \varphi\rangle = \lambda\int_0^L \varphi^2 dx = \int_0^L -\varphi''\varphi dx = (-\varphi'\varphi)\Big|_0^L + \int_0^L (\varphi')^2 dx \geq \int_0^L (\varphi')^2 dx.$$

Therefore

$$\lambda \geq \frac{\int_0^L (\varphi')^2 dx}{\int_0^L \varphi^2 dx}. \tag{4.35}$$

If $\lambda = 0$ is an eigenvalue, then the corresponding eigenfunction φ must satisfy $\int_0^L (\varphi')^2 dx = 0$, which implies $\varphi \equiv$ constant. If $\varphi(x)$ is a *nonzero* constant, it will be a legitimate eigenfunction for the boundary condition (4.29). However, if the constant function φ is to satisfy any of the other boundary conditions (4.30) - (4.34), then $\varphi = 0$, so that it is not an eigenfunction. Only the boundary condition (4.29) admits $\lambda = 0$ as an eigenvalue.

Next we deduce the orthogonality of the eigenfunctions from (4.27). Suppose that W is symmetric and that φ and $\psi \in W$ are eigenfunctions of M with different eigenvalues λ and μ. φ and ψ satisfy

$$M\varphi = \lambda\varphi \quad \text{and} \quad M\psi = \mu\psi, \qquad \lambda \neq \mu.$$

Then,

$$\langle \varphi, \psi \rangle = 0.$$

This is so because (4.27) implies that

$$\lambda\langle \varphi, \psi \rangle = \langle M\varphi, \psi \rangle = \langle \varphi, M\psi \rangle = \mu\langle \varphi, \psi \rangle$$

so that

$$(\lambda - \mu)\langle \varphi, \psi \rangle = 0.$$

Hence $\lambda \neq \mu$ implies $\langle \varphi, \psi \rangle = 0$.

Now two more ingredients from the general theory are needed. When M is defined by any of the boundary conditions (4.29) - (4.34), the eigenspace for each eigenvalue is one-dimensional. This means that, if φ and $\tilde{\varphi}$ are eigenfunctions corresponding to the same eigenvalue λ, then $\varphi = \alpha\tilde{\varphi}$ for some constant α. Furthermore, the eigenvalues form an increasing sequence. In the case of (4.29),

$$0 = \lambda_0 < \lambda_1 < \lambda_2 < \cdots < \lambda_n < \cdots,$$

and for (4.30) - (4.34),

$$0 < \lambda_1 < \lambda_2 < \cdots \lambda_n < \cdots.$$

In both cases, $\lambda_n \to +\infty$ as $n \to \infty$. The set of eigenfunctions is *complete* in the sense that, if f is piecewise continuous on $[0, L]$, then

$$f = \sum_{n=1}^{\infty} A_n \varphi_n,$$

where the convergence is in the mean square and

$$A_n = \frac{\langle f, \varphi_n \rangle}{\langle \varphi_n, \varphi_n \rangle} = \frac{\int_0^L f(x)\varphi_n(x)dx}{\int_0^L \varphi_n^2(x)dx}.$$

When the boundary condition is (4.29), the sum begins at $n = 0$. The completeness of the eigenfunctions for each of the boundary conditions (4.30) - (4.34) follows from the theory of symmetric, integral operators (see [RR]). The theory is applied to the inverse of the operator M, which in each of these cases is an integral operator with the same eigenfunctions φ_n and eigenvalues λ_n^{-1}. The boundary condition (4.29) is treated in a similar fashion by considering the operator $M + I$ which is invertible.

Consequently, for any of the boundary conditions (4.29) - (4.34), we can solve the IBVP

$$u_t - ku_{xx} = 0, \quad 0 < x < L, t > 0, \qquad u(x, 0) = f(x), \quad 0 < x < L$$

$$+B.C.$$

in the form

$$u(x, t) = \sum_{n=1}^{\infty} A_n \varphi_n(x)\psi_n(t)$$

where

$$\psi_n(t) = \exp(-\lambda_n kt).$$

Example A

Consider first the boundary condition (4.29). The eigenfunctions must satisfy

$$-\varphi'' = \lambda\varphi, \qquad \varphi'(0) = \varphi'(L) = 0.$$

The general solution of the differential equation is given by

$$\varphi(x) = A \cos(\sqrt{\lambda}x) + B \sin(\sqrt{\lambda}x).$$

Now $\varphi'(0) = 0$ implies that $B = 0$. Setting $A = 1$, we require that

$$\sin(L\sqrt{\lambda}) = 0,$$

which is true for $L\sqrt{\lambda} = n\pi$. The eigenvalues are

$$\lambda_n = (\frac{n\pi}{L})^2, \tag{4.36}$$

and the eigenfunctions are

$$\varphi_n(x) = \cos(\frac{n\pi x}{L}), \qquad n = 0, 1, 2, \cdots. \tag{4.37}$$

It is not hard to calculate that

$$\langle \varphi_n, \varphi_n \rangle = \int_0^L \varphi_n^2(x)dx = \frac{L}{2}$$

for $n = 1, 2, 3, \cdots$, and that taking $\varphi_0 \equiv 1$,

$$\langle \varphi_0, \varphi_0 \rangle = \int_0^L dx = L.$$

The expansion of f in these eigenfunctions is usually written

$$f(x) = \frac{A_0}{2} + \sum_{n=1}^{\infty} A_n \varphi_n(x), \tag{4.38}$$

where

$$A_n = \frac{2}{L} \int_0^L f(x) \cos(\frac{n\pi x}{L})dx \qquad n = 0, 1, 2, \cdots.$$

Note that the leading term

$$\frac{A_0}{2} = \frac{1}{L} \int_0^L f(x)dx = \frac{\langle f, \varphi_0 \rangle}{\langle \varphi_0, \varphi_0 \rangle},$$

is the average value of f. The solution of the IBVP is given by the infinite sum

$$u(x, t) = \frac{A_0}{2} + \sum_{n=1}^{\infty} A_n e^{-\lambda_n kt} \varphi_n(x), \tag{4.39}$$

where λ_n and φ_n are given in (4.36) and (4.37).

Example B

For a second example consider the boundary conditions (4.34). We must solve

$$-\varphi''(x) = \lambda\varphi(x)$$

$$\varphi'(0) = 0, \qquad \varphi'(L) + h\varphi(L) = 0, \quad h > 0.$$

The general solution $\varphi(x) = A\cos(\sqrt{\lambda}x) + B\sin(\sqrt{\lambda}x)$ and the condition $\varphi'(0) = 0$ implies that $B = 0$. Setting $A = 1$, the boundary condition at $x = L$ becomes

$$-\sqrt{\lambda}\sin(L\sqrt{\lambda}) + h\cos(L\sqrt{\lambda}) = 0$$

or

$$-\frac{s}{L}\sin(s) + h\cos(s) = 0, \qquad s = L\sqrt{\lambda}.$$

This becomes

$$\tan s = \frac{Lh}{s}. \tag{4.40}$$

The location of the solutions of (4.40) can be determined from the graphs of $\tan s$ and Lh/s (see Figure 4.1). It can be seen that there is an infinite sequence of roots $s_n \to \infty$, $n = 1, 2, 3, \cdots$, with $s_n \approx (n-1)\pi$ as $n \to \infty$. The eigenvalues

$$\lambda_n = (\frac{s_n}{L})^2 \approx (\frac{(n-1)\pi}{L})^2 \text{ as } n \to \infty.$$

Finally, the eigenfunctions are $\varphi_n(x) = \cos(\sqrt{\lambda_n}x)$.

Remarks

In the examples we have treated so far, the eigenvalues of

$$M\varphi = \lambda\varphi$$

have always been nonnegative. However, this need not be the case. When $h < 0$ in Example B, there may be negative eigenvalues. This will be explored in exercise 10 to follow.

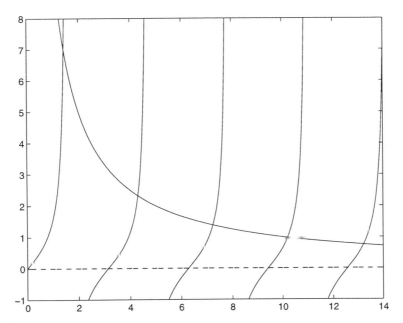

FIGURE 4.1
Graphs of left and right sides of (4.40) with $Lh = 10$.

Finally we note that the structure sketched in this section is also adequate to handle the situation when the material properties of the bar are not constant. For instance if the heat conductivity $\kappa = \kappa(x)$, ρ, $c = 1$, the heat equation has variable coefficients,

$$u_t(x, t) = \partial_x(\kappa(x)u_x(x, t)).$$

The condition that the boundary conditions must satisfy to be symmetric for the differential operator $-\partial_x(\kappa \partial_x)$ becomes

$$[-\kappa(x)u_x(x)v(x) + \kappa(x)u(x)v_x(x)]\Big|_0^L = 0.$$

Thus (4.29) - (4.34) are all symmetric boundary conditions for the operator $-\partial_x(\kappa \partial_x)$. These operators are called Sturm-Liouville operators and there is an extensive theory for their eigenvalues and eigenfunctions, see [CL].

Exercises 4.3

1. Let u be a solution of $u_t - ku_{xx} = 0$ with the Neumann boundary conditions (4.29) (Example A).

    (a) Show that the average

    $$\bar{u}(t) = \frac{1}{L} \int_0^L u(x, t)dx = \frac{1}{L} \int_0^L f dx = \bar{f}$$

    for all $t \geq 0$.

    (b) Show that $u(x, t) - \bar{f}$ tends to zero exponentially as $t \to \infty$.

2. Let $u(x, t)$ be the solution of $u_t - ku_{xx} = 0$, with the Neumann boundary conditions (4.29), and the initial data

    $$f(x) = \begin{cases} 1, & 0 \leq x < L/2 \\ 0, & L/2 < x \leq L \end{cases}.$$

    (a) Find the coefficients A_n in the appropriate eigenfunction expansion.

    (b) Verify the conclusions of exercise 1.

3. Let u solve $u_t - ku_{xx} = 0$ with the boundary conditions $u_x(0, t) = u(L, t) = 0$. Use the technique of exercise 1 of Section 4.2 to show that

    $$-k \int_0^\infty u_x(L, t)dt = \int_0^L f(x)dx.$$

    Give a physical interpretation of this equation in the context of heat flow.

4. (a) Find the eigenvalues λ_n and the eigenfunctions φ_n for the boundary conditions (4.31). Verify directly the orthogonality conditions $\langle \varphi_n, \varphi_m \rangle = 0$ for $m \neq n$, and calculate the normalizing constants $\langle \varphi_n, \varphi_n \rangle$.

    (b) Solve the IBVP with these boundary conditions and the initial condition

    $$f(x) = x.$$

    (c) Let $\tilde{\lambda}_n$ be the eigenvalues of the boundary condition (4.30). Verify that $\tilde{\lambda}_n \geq \lambda_n$ for all n. What physical reason can you give for this inequality ?

5. Same as exercise 4 but with the boundary conditions (4.32). Why are the eigenvalues the same? How are the eigenfunctions related? Are the coefficients A_n the same? Why or why not?

6. Let M be $-d/dx^2$ with any of the boundary conditions (4.29) - (4.34). We have seen that M is symmetric and $\langle Mu, u \rangle \geq 0$ for all u in the domain W of M. Let $u(x, t)$ be the solution of

$$u_t - ku_{xx} = 0, \qquad u(x, 0) = f(x).$$

(a) Show that

$$\int_0^L u^2(x, t)dx \leq \int_0^L f^2(x)dx$$

for all $t \geq 0$. Hint: Look at exercise 15 of Section 4.2.

(b) Use this result to show the uniqueness of the solutions of the IBVP for any of the boundary conditions (4.29) - (4.34).

The program heat4 computes the solution of $u_t - ku_{xx} = q$ on the interval $0 \leq x \leq 10$ with the homogeneous Neumann boundary conditions (4.29). $\Delta x = \Delta t = .05$ is set in the program. You will need to write array-smart mfiles f.m for the initial data $f(x)$ and q.m for the source term $q(x)$. More information on the program can be found by typing help heat4.

7. Run heat4 with initial data $f(x) = \exp[-(x-5)^2]$ and $q(x) = 0, k = 1$ and $k = 5$. Write the mfile q.m as

```
function  z = q(x)
   z = zeros(size(x));
```

(a) Compute the integral $(1/10) \int_0^{10} f(x)dx$ (using the error function). Does this value agree with the asymptotic value, as $t \to \infty$, of the solution as seen in the graphs of the solution?

(b) For which value of k is the convergence more rapid? Does this show up in the estimate of exercise 1?

8. Run heat4 with initial data

$$f(x) = .01(x^4/4 - 17x^3/3 + 35x^2) - 1,$$

$q(x) = 0$ and $k = 1, 5$. Same questions (a) and (b) as in exercise 7.

(c)   We can use the first *two* terms of the eigenfunction series to make a better approximation to the solution u:

$$u(x, t) \approx \frac{A_0}{2} + A_1 \cos(\pi x/10)e^{-(\frac{\pi}{10})^2 kt}.$$

Can you see the presence of these two terms in the graphs of u as t gets larger? You should compute $A_0/2$, and you must at least determine the sign of A_1.

9.   In example B, the eigenvalues of boundary condition (4.34) were associated with the roots of equation (4.40). Let $L = 10$ and $h = 1$. Then the eigenvalues are given by $\lambda_n = s_n^2/100$, where s_n is the nth root of $\tan s = 10/s$. The MATLAB function `fzero` uses the secant method to find the roots of such an equation. First write an mfile, `sigma.m`, for the function $y = \sigma(s) = \tan s - 10/s$.

```
function y = sigma(s)
      y = tan(s) - 10./s;
```

The call for fzero to find the zeros of the function $y = \sigma(s)$ is

```
>> fzero('sigma', s0)
```

where s0 is a first guess for the desired root. Use Figure 4.1 to guide your first guesses. In this situation you must make a very good first guess, or `fzero` will give a spurious answer because of the singularity nearby. For more information on fzero, type `help fzero`. Find the first four eigenvalues, and graph the first four eigenfunctions on $[0, 10]$.

10.   Consider the IBVP

$$u_t - ku_{xx} = 0, \qquad u(x, 0) = f(x),$$

$$u_x(0, t) = 0, \qquad u_x(L, t) + hu(L, t) = 0.$$

The boundary condition at $x = L$ is the radiating boundary condition.

(a)   Use Green's first identity to show that all of the eigenvalues of this IBVP are strictly positive when $h > 0$. This means that for any initial condition f, the solution $u \to 0$ as $t \to \infty$.

(b)   Now consider what happens when $h < 0$. We can no longer conclude that the corresponding operator M has only positive eigenvalues. When $h = 0$, this boundary condition reduces to the Neumann condition and has $\lambda_0 = 0$ as an eigenvalue. This raises the possibility that when $h < 0$ there might be some negative eigenvalues in addition to an infinite sequence of positive eigenvalues. First find the positive eigenvalues by following the procedure of Example B. Locate the roots on the appropriate graphs. Find the first three positive eigenvalues using MATLAB with $L = 10$ and $h = -1$.

(c) Now look for the negative eigenvalue(s). If $\lambda = -\gamma < 0$ the eigenvalue problem for M is

$$\varphi'' = \gamma\varphi, \qquad \varphi'(0) = 0, \quad \varphi'(L) + h\varphi(L) = 0.$$

Instead we shall seek a solution in the form $\varphi = \cosh(x\sqrt{\gamma})$. Chosen this way, $\varphi'(0) = 0$. It remains to find γ so that the boundary condition at $x = L$ is satisfied. How many roots γ are there? You may use `fzero` to find these root(s), for example, with $L = 10$, $h = -1$.

(d) Write the solution in terms of these eigenfunctions:

$$u(x, t) = A_0 \cosh(x\sqrt{\gamma})e^{\gamma kt} + \sum_1^\infty A_n \cos(x\sqrt{\lambda_n})e^{-\lambda_n kt}.$$

(e) What condition can you impose on the initial data f so that the solution u remains bounded as $t \to \infty$?

(f) What is the significance of the negative eigenvalue? Do you think this kind of behavior is reasonable on physical grounds in the context of heat flow?

4.4 Inhomogeneous problems and asymptotic behavior

We wish to consider problems with sources of heat in the interior of the bar, or with inhomogeneous boundary conditions. First we consider the heat equation with a heat source.

$$u_t - ku_{xx} = q(x, t), \qquad u(x, 0) = f(x), \qquad (4.41)$$

$$+\text{homogeneous boundary conditions.}$$

The boundary conditions are assumed to be symmetric and nonnegative, such as those in the list (4.30) - (4.34). We shall assume that the desired solution $u(x, t)$

and the given source $q(x, t)$ can be expanded in terms of the eigenfunctions $\varphi_n(x)$ for each t, that is,

$$u(x, t) = \sum_1^\infty u_n(t)\varphi_n(x), \qquad (4.42)$$

and

$$q(x, t) = \sum_1^\infty q_n(t)\varphi_n(x), \qquad (4.43)$$

where

$$q_n(t) = \frac{\langle q_n(t, x), \varphi_n \rangle}{\langle \varphi_n, \varphi_n \rangle}.$$

We shall try to compute the coefficient functions $u_n(t)$ in terms of the coefficients $q_n(t)$. Substitute the expansion for u in the left side of (4.41). Since $-\varphi_n''(x) = \lambda_n \varphi_n(x)$, it follows that

$$u_t - ku_{xx} = \sum_1^\infty u_n'(t)\varphi_n(x) - ku_n(t)\varphi_n''(x) = \sum_1^\infty [u_n'(t) + \lambda_n ku_n(t)]\varphi_n(x).$$

If we equate this expansion on the left of (4.41) with the expansion for q in the right side of (4.41), we deduce that

$$\sum_1^\infty [u_n'(t) + \lambda_n ku_n(t) - q_n(t)]\varphi_n(x) = 0.$$

Because the eigenfunctions φ_n are orthogonal, this can be the case only if for each $n = 1, 2, \ldots$

$$u_n' + \lambda_n ku_n(t) = q_n(t).$$

Thus we have reduced the PDE (4.41) to a family of ODE's. We can find the initial values for each u_n by expanding the initial data f in terms of the eigenfunctions as well:

$$f(x) = \sum_1^\infty A_n \varphi_n(x),$$

where as usual

$$A_n = \frac{\langle f, \varphi_n \rangle}{\langle \varphi_n, \varphi_n \rangle}.$$

From our expansion of u, we see that

$$u(x, 0) = \sum_1^\infty u_n(0)\varphi_n(x) = f(x) = \sum_1^\infty A_n\varphi_n(x).$$

Thus the initial values for the u_n are

$$u_n(0) = A_n.$$

Now the solution of this first-order linear ODE is well known. For each $n = 1, 2, \ldots$,

$$u_n(t) = A_n e^{-\lambda_n k t} + \int_0^t e^{-\lambda_n k(t-s)} q_n(s)ds.$$

Substituting these expressions for $u_n(t)$ we see that the solution u is given by

$$u(x, t) = \sum_{n=1}^\infty [A_n e^{-\lambda_n k t} + \int_0^t q_n(s)e^{-\lambda_n k(t-s)}ds]\varphi_n(x). \tag{4.44}$$

Observe that the solution can be written conveniently as $u = v + w$ where

$$v(x, t) = \sum_1^\infty A_n e^{-\lambda_n k t}\varphi_n(x)$$

solves

$$v_t - kv_{xx} = 0, \qquad v(x, 0) = f(x),$$

and

$$w(x, t) = \sum_1^\infty \int_0^t e^{-\lambda_n k(t-s)}q_n(s)dx\, \varphi_n(x)$$

solves

$$w_t - kw_{xx} = q, \qquad w(x, 0) = 0.$$

Example A

Take the boundary conditions to be $u(0) = u(L) = 0$. Set $f(x) = 0$ and suppose that $q(x, t) = \exp(-t)r(x)$ where

$$r(x) = \begin{cases} x, & 0 \leq x \leq \frac{L}{2} \\ L - x, & \frac{L}{2} \leq x \leq L \end{cases}.$$

Then from (4.44),

$$u(x, t) = \sum_{n=1}^{\infty} \int_0^t q_n(s) \exp[-\lambda_n k(t - s)] ds \, \varphi_n(x)$$

with $\varphi_n(x) = \sin(n\pi x/L)$. From example (3) of Section 4.2,

$$q_n(s) = e^{-s} \frac{\langle r(x), \varphi_n(x) \rangle}{\langle \varphi_n, \varphi_n \rangle} = e^{-s} \frac{4L}{(n\pi)^2} \sin(\frac{n\pi}{2}),$$

and

$$\int_0^t q_n(s) e^{-\lambda_n k(t-s)} ds = \frac{4L}{(n\pi)^2} \sin(\frac{n\pi}{2}) e^{-\lambda_n k t} \int_0^t e^{(\lambda_n k - 1)s} ds$$

$$= \frac{4L}{(n\pi)^2} \sin(\frac{n\pi}{2}) \left(\frac{1}{\lambda_n k - 1} \right) (e^{-t} - e^{-\lambda_n k t}).$$

Therefore

$$u(x, t) = \sum_{n=1}^{\infty} \left\{ \frac{4L}{(n\pi)^2} \sin(\frac{n\pi}{2}) \left(\frac{1}{\lambda_n k - 1} \right) (e^{-t} - e^{-\lambda_n k t}) \right\} \sin(\frac{n\pi x}{L}),$$

assuming, of course, that $\lambda_n k \neq 1$ for all n.

Example B

Next we consider a situation where the temperatures are prescribed at the ends of the bar. We consider

$$u_t = k u_{xx}, \qquad u(x, 0) = f(x), \tag{4.45}$$

$$u(0, t) = l(t), \qquad u(L, t) = r(t).$$

To obtain a continuous solution on the rectangle $[0, L] \times [0, \infty)$, we impose the compatibility conditions

$$f(0) = l(0), \qquad f(L) = r(0).$$

We solve this problem by converting it to a problem of the form (4.41). Let $g(x, t)$ be a C^2 function such that $g(0, t) = l(t)$, and $g(L, t) = r(t)$. We can take for instance

$$g(x, t) = l(t) + \frac{x}{L}(r(t) - l(t)).$$

Then let $v(x, t) = u(x, t) - g(x, t)$ and observe that since $g_{xx} = 0$, v solves

$$v_t = k v_{xx} - g_t, \qquad v(x, 0) = f(x) - g(x, 0), \qquad (4.46)$$

$$v(0, t) = v(L, t) = 0.$$

We expand $g(x, t)$ in the eigenfunctions φ_n

$$g(x, t) = \sum_{1}^{\infty} g_n(t)\varphi_n(x),$$

and, under some natural assumptions on r and l, it follows that g_t can be expanded

$$g_t(x, t) = \sum_{1}^{\infty} g_n'(t)\varphi_n(x).$$

We can solve (4.46) as we did (4.41) and then write the solution u as

$$u(x, t) = v(x, t) + g(x, t),$$

where

$$v(x, t) = \sum_{1}^{\infty} [A_n e^{-\lambda_n kt} - \int_0^t e^{-\lambda_n k(t-s)} g_n'(s) ds] \varphi_n(x) \qquad (4.47)$$

and

$$A_n = \frac{\langle f(x) - g(x, 0), \varphi_n(x) \rangle}{\langle \varphi_n, \varphi_n \rangle}.$$

When $l(t) \equiv \alpha$ and $r(t) \equiv \beta$, we may take $g(x, t) = U(x) = \alpha + (\beta - \alpha)x/L$ so that $g_n' = 0$ for all n. In this case the integral terms in (4.47) are zero, so that the solution becomes simply

$$u(x, t) = U(x) + \sum_1^\infty A_n e^{-\lambda_n k t} \varphi_n(x).$$

The exponential factors all tend to zero as $t \to \infty$ so that $u(x, t)$ tends towards the steady state $U(x)$ as $t \to \infty$. The steady state U satisfies

$$U_{xx} = 0, \qquad U(0) = \alpha, \quad U(L) = \beta.$$

More generally, if for some $T > 0$, $l(t) = \alpha$, and $r(t) = \beta$ for $t \geq T$, it follows that $g(x, t) = U(x)$ for $t \geq T$. Then $g'_n(s) = 0$ for $s \geq T$ and and we can estimate the integral term of (4.47). Assume that there is a constant M such that $|g'_n(s)| \leq M$ for $0 \leq s \leq T$ and all $n = 1, 2, \ldots$. Then

$$\left| \int_0^t e^{-\lambda_n k(t-s)} g'_n(s) ds \right| \leq M \int_0^T e^{-\lambda_n k(t-s)} ds$$

$$\leq \frac{M}{\lambda_n k} e^{-\lambda_n k t} [e^{\lambda_n k T} - 1]$$

which tends to zero as $t \to \infty$. Thus we have $v(x, t) \to 0$ as $t \to \infty$ and $g(x, t) = U(x)$ for $t \geq T$, so that $u(x, t)$ converges to the steady state

$$u(x, t) = g(x, t) + v(x, t) \to U(x).$$

A slight extension of this argument shows that if $l(t)$ and $r(t)$ have limiting values α and β as $t \to \infty$, then $u(x, t)$ again converges to the steady state $U(x)$.

Example C

As a last example of an inhomogeneous equation with inhomogeneous boundary conditions, consider

$$u_t = k u_{xx} + q, \qquad u(x, 0) = f(x), \qquad (4.48)$$

$$u_x(0, t) = \alpha, \qquad u_x(L, t) = \beta.$$

We assume that $q = q(x)$ does not depend on time.

How do we solve this problem and what is the asymptotic behavior as $t \to \infty$? Since α, β, and q do not depend on time, we should consider the possibility that u

converges to a steady state U as $t \to \infty$. However for this to be true, the source q must be compatible with the boundary data. Indeed, if $U(x)$ is a steady-state solution, then it must satisfy

$$-kU_{xx} = q, \qquad U_x(0) = \alpha, \quad U_x(L) = \beta. \qquad (4.49)$$

Integrate the differential equation on $[0, L]$ to get

$$0 = \int_0^L (q + kU_{xx})dx = \int_0^L qdx + k[\beta - \alpha]. \qquad (4.50)$$

This equation says that for a steady state to exist, the internal source q must be balanced by the flow of heat through the ends of the bar.

Return to the time-dependent problem (4.48). Convert it to a problem with homogeneous boundary conditions by finding a function $p(x)$ such that

$$p_x(0) = \alpha, \qquad p_x(L) = \beta.$$

It is easy to check that

$$p(x) = (\frac{\beta - \alpha}{2L})x^2 + \alpha x$$

does the trick. Then write $u(x, t) = v(x, t) + p(x)$. v must solve

$$v_t - kv_{xx} = q(x) + kp''(x) = q(x) + \frac{k(\beta - \alpha)}{L}, \qquad (4.51)$$

$$v(x, 0) = f(x) - p(x),$$

$$v_x(0, t) = v_x(L, t) = 0.$$

The eigenfunctions in this case are $\varphi_n(x) = \cos(n\pi x/L)$, $\lambda_n = (n\pi/L)^2$. We expand v in terms of these eigenfunctions.

$$v(x, t) = v_0(t) + \sum_{n=1}^{\infty} v_n(t)\varphi_n(x).$$

Here

$$v_0(t) = \frac{1}{L} \int_0^L v(x, t)dx.$$

Now substitute in the left side of (4.51). We obtain

$$v_t - k v_{xx} = v_0'(t) + \sum_{n=1}^{\infty} [v_n'(t) + k \lambda_n v_n(t)] \varphi_n(x).$$

Then expand the right side of (4.51) as

$$q(x) + k p'' = q_0 + k \frac{(\beta - \alpha)}{L} + \sum_{n=1}^{\infty} q_n \varphi_n(x),$$

where

$$q_0 = \frac{1}{L} \int_0^L q(x) dx \quad \text{and} \quad q_n = \frac{2}{L} \int_0^L q(x) \varphi_n(x) dx \text{ for } n \geq 1.$$

In addition we also expand $v(x, 0) = f(x) - p(x)$ as

$$f(x) - p(x) = \sum_{n=0}^{\infty} (f_n - p_n) \varphi_n(x),$$

where

$$f_0 - p_0 = \frac{1}{L} \int_0^L [f(x) - p(x)] dx, \qquad f_n - p_n = \frac{2}{L} \int_0^L [f(x) - p(x)] \varphi_n(x) dx.$$

Next we compare the expansions of the left and right sides of (4.51). This yields the following ODE initial value-problems for the functions $v_n(t)$:

$$v_0'(t) = q_0 + k \frac{(\beta - \alpha)}{L}, \qquad v_0(0) = f_0 - p_0,$$

because $\lambda_0 = 0$ and

$$v_n'(t) + \lambda_n k v_n(t) = q_n, \qquad v_n(0) = f_n - p_n \text{ for } n \geq 1.$$

Integrate these equations to obtain

$$v_0(t) = t \left(q_0 + k \frac{(\beta - \alpha)}{L} \right) + f_0 - p_0, \tag{4.52}$$

$$v_n(t) = (f_n - p_n) \exp(-\lambda_n kt) + \frac{q_n}{\lambda_n k}[1 - \exp(-\lambda_n kt)] \quad \text{for } n \geq 1.$$

Therefore v will have a limit as $t \to \infty$ if and only if the coefficient of t in $v_0(t)$ is zero. But this is exactly the source-flux balance condition (4.50). If (4.50) is satisfied, then $v_0(t) \equiv f_0 - p_0$, and

$$v(x, t) \to f_0 - p_0 + \sum_{n=1}^{\infty} \frac{q_n}{\lambda_n k} \varphi_n(x).$$

Hence

$$u(x, t) = v(x, t) + p(x) \to U(x)$$

where

$$U(x) = f_0 + p(x) - p_0 + \sum_{n=1}^{\infty} \frac{q_n}{\lambda_n k} \varphi_n(x). \tag{4.53}$$

We want to verify that U does indeed solve (4.49). In fact,

$$-kU''(x) = -\frac{k(\beta - \alpha)}{L} + k \sum_{n=1}^{\infty} \frac{q_n}{\lambda_n k} \varphi_n''(x)$$

$$= q_0 + \sum_{n=1}^{\infty} q_n \varphi_n(x) = q(x),$$

where we have used (4.50) and the fact that $-\varphi_n'' = \lambda_n \varphi_n$. In addition

$$U'(x) = \frac{(\beta - \alpha)x}{L} + \alpha + \sum_{n=1}^{\infty} \frac{q_n}{\lambda_n k} \varphi_n'(x)$$

so $U'(0) = \alpha$ and $U'(L) = \beta$ because $\varphi_n'(0) = \varphi_n'(L) = 0$.

Finally, we should note that the formula (4.53) is really independent of the choice of p. Indeed, if $\tilde{p} = p + \theta$, where $\theta'(0) = \theta'(L) = 0$, the contribution of θ in (4.53) will drop out.

Exercises 4.4

1. Consider the IBVP (4.41) with $f(x) = g(x) = 0$, boundary conditions $u(0, t) = u(L, t) = 0$, and

    $$q(x, t) = e^{-t} \sin(\pi x / L) - \sin(3\pi x / L).$$

    (a) Solve this problem. Note that the solution will contain only a few terms.

    (b) Change the equation to
    $$u_t - ku_{xx} + 2u = q,$$

    and solve with the same boundary conditions, initial conditions, and the same q as in part (a).

2. Solve the problem
    $$u_t - ku_{xx} = q, \qquad u(x, 0) = 0$$

    with $q(x, t) = (\sin t)r(x)$ and with boundary condition $u(0, t) = u(L, t) = 0$. How does the solution behave as $t \to \infty$? Does it oscillate, or does it tend to zero?

3. Consider the heat problem
    $$u_t - ku_{xx} = x, \qquad 0 < x < L, \ t > 0,$$

    $$u(0, t) = 1, \qquad u(L, t) = 2, \qquad t \geq 0,$$

    $$u(x, 0) = 1 + (x/L)^2, \qquad 0 \leq x \leq L.$$

    (a) Find the steady-state solution $U(x)$ to this problem. Look for U in the form $U(x) = x^2/2 + Ax + B$; find the coefficients A and B.

    (b) Write the desired solution $u(x, t) = v(x, t) + U(x)$, and determine the IBVP satisfied by v. What is the initial condition of v?

    (c) Find v in terms of an eigenfunction expansion. You do not need to calculate the coefficients A_n. Show that $v(x, t) \to 0$ as $t \to \infty$. Finally, write the complete expression for u.

4. Consider the problem with inhomogeneous boundary conditions
    $$u_t - ku_{xx} = 0, \qquad u(x, 0) = 0,$$

    $$u(0, t) = l(t), \qquad u_x(L, t) = r(t).$$

(a) Find a function $p(x, t)$, such that $p(0, t) = l(t)$ and $p_x(L, t) = r(t)$. Then write the solution in the form

$$u(x, t) = v(x, t) + p(x, t).$$

(b) Find the equation, initial condition and boundary conditions satisfied by v and solve for v.

(c) Suppose that $l(t) \to \alpha$ as $t \to \infty$ and $r(t) \to \beta$ as $t \to \infty$. Will the solution u converge toward a steady-state solution $U(x)$? Give physical reasons. If $l(t) = \alpha$ and $r(t) = \beta$ for $t \geq T$, use the formula of part (b) to show that $u(x, t)$ does converge to a steady state $U(x)$ as $t \to \infty$. Find $U(x)$.

5. Run heat4 with initial data $f(x) = 0$, and source

$$q(x) = .01(x^4 - 20.5x^3 + 127.5x^2 - 225x).$$

Set $k = 1$. Compute the average value of q. Is condition (4.50) satisfied? Does the solution appear to tend to a steady state $U(x)$ as $t \to \infty$?

6. Run heat4, this time with $f(x) = 0$, and $q(x) = \exp[-(x-2)^2]$. Set $k = 1$.

(a) Does the solution appear to tend to a steady-state solution as $t \to \infty$?

(b) If the solution grows as t increases, how fast does $\max_x u(x, t)$ grow as $t \to \infty$? Let n1 = 500, n2 = n3 = n4 = 20. Then compute the rate of growth of $\max_x u(x, t)$ with the command

>> max(snap2) - max(snap1)

Does this rate of growth agree with that predicted by the first term in (4.52)?

7. Return to example C. Why is the steady-state solution $U(x)$ independent of the choice of p? Hint: Look at equation (4.53).

The program heat5 solves the IBVP on $[0, 10]$ with Dirichlet boundary conditions at $x = 0$ and at $x = 10$, but with a discontinuous diffusion coefficient k. Imagine two bars of different materials joined together at $x = 5$. Both materials have the same values of $c = 1$ and $\rho = 1$. Thus the conductivity $\kappa = k$ is discontinuous with $k = k_l$ for $x < 5$ and $k = k_r$ for $x > 5$. The jump condition in Chapter 3, equation (3.6) becomes

$$k_l u_x(5^-, t) = k_r u_x(5^+, t).$$

In program heat5, $k_r = 1$, and k_l is an input parameter. The program uses a time step $\Delta t = .025$. You will need to write an array-smart mfile f.m for the initial data $f(x)$, and mfiles left.m, and right.m for the boundary data. For more information type help heat5 .

8. Run heat5 with initial data $f(x) = \sin(\pi x/10)$, $u(0, t) = u(10, t) = 0$.
    Set $k_l = 2$. Start with $t_1 = 2, t_2 = 4, t_3 = 6, t_4 = 8$.
    (a) The initial data is symmetric about $x = 0$. Why does the solution lose this symmetry? Run again with $k_l = 5$. Same questions.
    (b) Plot the graph of $x \to u(x, t)$ for $k_l = 5$ at $t = 2$ (80 time steps). Use the zoom feature on the graph around $x = 5$. From the graph, determine the value of u_x from either side. Is the jump condition satisfied? Repeat at $t = 4$ (160 time steps).

9. (a) Run heat5 with initial data

    $$f(x) = x/2 - 2 + 3e^{-(x-5)^2},$$

    and boundary data $u(0, t) = -2$, $u(10, t) = 3$. Set $k_l = 5$. First run the program with values $t = .5, 1, 1.5, 2$. Observe how the smooth initial data develops a "corner" at $x = 5$. Then choose values of t large enough so that you can guess what the asymptotic behavior is. What is the limiting behavior of the solution as $t \to \infty$?

    (b) The steady-state problem (with $u_t = 0$) is

    $$U_{xx} = 0, \quad 0 < x < 5, \qquad U(0) = -2,$$

    $$U_{xx} = 0, \quad 5 < x < 10, \qquad U(10) = 3,$$

    with the jump conditions

    $$U(5^-) = U(5^+), \qquad k_l U_x(5^-) = k_r U_x(5^+).$$

    Look for the solution in the form

    $$U_l(x) = ax+b, \quad 0 < x < 5, \qquad U_r(x) = cx+d, \quad 5 < x < 10.$$

    Find equations to determine the coefficients a, b, c, and d and solve for a, b, c, and d. Do the plots of $u(x, t)$ for large t appear to tend toward this steady-state solution?

10. Two layers of heat conducting material, each 5 cm thick, are glued together to form the wall of a building. The interior layer on the left has $k = 5$

while the exterior layer on the right has $k = 1$ (insulation). Initially the temperature is zero, inside and outside. Then a 100° heat source is placed on the inside surface.

(a) How long does it take for the temperature half-way through the wall to reach 50° ?

(b) What is the steady-state temperature profile toward which the temperature tends as $t \to \infty$?

(c) What is the rate at which heat flows through the exterior surface of the wall in the steady state?

Run heat5 with initial data $f(x) = 0$, boundary data $u(0, t) = 100$, $u(10, t) = 0$, and set $k_l = 5$ to answer (a). Find the answers to (b) and (c) analytically.

11. Find the eigenfunctions for the IBVP

$$u_t - k_l u_{xx} = 0, \quad -L \leq x < 0, \qquad u_t - k_r u_{xx} = 0, \quad 0 < x \leq L,$$

$$u(-L, t) = 0, \qquad u(L, t) = 0,$$

$$u(0^-, t) = u(0^+, t), \qquad k_l u_x(0^-, t) = k_r u_x(0^+, t).$$

Look for a building-block solution in the form $u(x, t) = \exp(-\lambda t)\varphi(x)$ where,

$$\varphi(x) = \begin{cases} \sin[(x + L)\sqrt{\lambda}], & x < 0 \\ \gamma \sin[(x - L)\sqrt{\lambda}], & x > 0 \end{cases}.$$

You will get two nonlinear equations for γ and λ. There should be an infinite number of solutions.

4.5 Projects

1. Consider the boundary value problem for steady-state heat flow on the interval $[0, 10]$ with a source $q(x)$:

$$-ku''(x) = q(x), \qquad u(0) = u(10) = 0.$$

The goal is to write a MATLAB program to solve this problem numerically for given data q. Make mesh points $x_j = j\Delta x$, $j = 1, \ldots, J$, where $\Delta x = 10/J$. Then write a finite difference scheme by replacing the second derivative by a centered difference.

$$-k\frac{u(x_j + \Delta x) - 2u(x_j) + u(x_j - \Delta x)}{(\Delta x)^2} \approx q(x_j),$$

$j = 1, \ldots, J - 1$. Taking into account the boundary conditions, we arrive at the $(J - 1) \times (J - 1)$ system

$$kT\mathbf{u} = \mathbf{q}, \tag{4.54}$$

where $\mathbf{u} = [u_1, \ldots, u_{J-1}]$ is the vector of unknowns, $\mathbf{q} = [q(x_1), \ldots, q(x_{J-1})]$, and $T = (\Delta x)^{-2}S$ where S is the sparse matrix

$$S = \begin{bmatrix} 2 & -1 & 0 & 0 & 0 \\ -1 & 2 & -1 & 0 & 0 \\ 0 & -1 & 2 & -1 & 0 \\ . & . & . & . & . \\ . & . & . & . & . \\ 0 & 0 & 0 & -1 & 2 \end{bmatrix}.$$

You will need to write a function mfile q.m for q. Then use the sparse matrix operations of MATLAB to enter the matrix S. For information enter help sparse.

To see if your program works correctly, try $q(x) \equiv 1$. The exact solution of the BVP is $u(x) = .5x(10 - x)/k$. The centered difference formula for u'' is exact on quadratic polynomials. Thus the vector of values $\mathbf{u}$ should agree exactly with the values of u at the grid points. Try $J = 5$ to begin. When you have gotten your program working correctly, try several different choices of q with $J = 100$. Plot q and the solution together on the interval $[0, 10]$. The vector of values $\mathbf{u}$, which you get from solving (4.54), does not include the zero values at the end points $x = 0$ and $x = 10$. You must add these values to make a vector of $(J + 1)$ components for plotting. Next determine the rate of convergence of the computed solution values to the values of an exact solution. Take $q(x) = \sin(\pi x/10)$. q is the first eigenfunction $\varphi_1(x)$ of this BVP. Hence the exact solution of $-ku'' = q = \varphi_1$ is given by

$$u(x) = \frac{1}{k\lambda_1}\varphi_1(x) = \frac{1}{k}(\frac{10}{\pi})^2 \sin(\frac{\pi x}{10}).$$

Now take the number of intervals to be $J = 5, 10, 20, 40, 80$. For each value of J, find the maximum error between the the calculated values and the values of the exact solution at the grid points. What happens to this error when J is doubled (that is, Δx is halved)?

The matrix operator T is an approximation to the differential operator $Mu = -u''$ with the boundary condtions $u(0) = u(10) = 0$. You can find the eigenvalues of the matrix T with the single MATLAB command `eig(T)`. Put them in order of increasing size and compare with the eigenvalues of M. What do you see happening as J gets larger ?

2. Consider the same boundary value problem as above, but with a piecewise constant diffusion function $k = k(x)$. For example, take $k(x) = k_l$ for $0 \le x < 5$ and $k(x) = 1$ for $5 < x \le 10$, and make k_l an input parameter. The linear system to solve will use the jump conditions in Section 3.1.

Chapter 5

Waves Again

In this chapter we shall study the linear wave equation in one space variable. Solutions of this equation have many features in common with the first-order equation $u_t + cu_x = 0$ that we studied in Chapter 2. The linear wave equation arises in many contexts which include gas dynamics (acoustics), vibrating solids, and electromagnetism.

5.1 Acoustics

5.1.1 The equations of gas dynamics

We begin by sketching the derivation of the equations of gas dynamics in much the same way as we derived the nonlinear conservation laws in Chapter 2. Imagine a gas occupying a straight, narrow tube, so that the velocity of the gas particles, u, the linear density, ρ, and the pressure, P, depend only on the axial coordinate x, and the time t, that is, u, ρ, and P are constant on cross sections of the tube. Now consider a portion of the gas contained in the *moving* interval $a(t) < x < b(t)$. We assume that the ends of the interval move with the gas particles, so that $a'(t) = u(a(t), t)$ and $b'(t) = u(b(t), t)$. Then conservation of mass says that the mass of gas contained in this moving interval of the tube is constant:

$$\frac{d}{dt} \int_{a(t)}^{b(t)} \rho(x, t) dx = 0.$$

Assuming that ρ and u are C^1 functions, we can use the differentiation formula (1.2) for integrals to deduce that

$$0 = \int_{a(t)}^{b(t)} \rho_t(x, t) dx + b'(t)\rho(b(t), t) - a'(t)\rho(a(t), t)$$

$$= \int_{a(t)}^{b(t)} \rho_t(x,t)dx + (\rho u)(b(t),t) - (\rho u)(a(t),t)$$

$$= \int_{a(t)}^{b(t)} [\rho_t(x,t) + (\rho u)_x(x,t)]dx.$$

We have used the fact that $a'(t) = u(a(t),t)$ and $b'(t) = u(b(t),t)$. Since this holds for all moving portions of gas, we conclude that

$$\rho_t + (\rho u)_x = 0.$$

The momentum of this moving portion of gas is given by

$$\int_{a(t)}^{b(t)} \rho(x,t)u(x,t)dx.$$

By Newton's second law, the rate of change of momentum, in the absence of external forces, is equal to the difference in the forces due to pressure at the ends of this moving portion of gas:

$$\frac{d}{dt} \int_{a(t)}^{b(t)} \rho(x,t)u(x,t)dx = AP(a(t),t) - AP(b(t),t) = - \int_{a(t)}^{b(t)} AP_x(x,t)dx.$$

This balance law expresses conservation of momentum. Here A is the (constant) area of a cross section of the tube. To avoid excess notation, we shall assume units chosen so that $A = 1$. Another application of (1.2) to differentiate the integral on the left yields the equation

$$(\rho u)_t + (\rho u^2)_x + P_x = 0.$$

The conservation of mass equation can be combined with this second equation to yield the pair of equations

$$\rho_t + \rho_x u + \rho u_x = 0,$$

$$\rho[u_t + uu_x] + P_x = 0.$$

This is a system of two equations in three functions ρ, u and P. To get a closed system in u and ρ assume that the flow is *isentropic*. That is, $P = P(\rho)$ so that $P_x = P'(\rho)\rho_x$. (In an ideal gas $P(\rho) = G\rho^\gamma$, $\gamma > 1$. For air, $\gamma = 5/3$.) With this assumption, we arrive at a system of two equations in the two unknown functions u and ρ.

$$\rho_t + \rho u_x + \rho_x u = 0, \tag{5.1}$$

$$\rho[u_t + u u_x] + P'(\rho)\rho_x = 0.$$

5.1.2 The linearized equations

Note that for any $\rho_0 > 0$, the constant functions $\rho(x, t) = \rho_0$, $u(x, t) = 0$ form a solution pair of (5.1). We wish to consider the equations satisfied by a small perturbation of this constant solution. The term 'small' must be in a relative sense for the density. Thus we look for a perturbed density of the form

$$\rho(x, t) = (1 + \delta(x, t))\rho_0.$$

$\delta(x, t) = (\rho(x, t) - \rho_0)/\rho_0$ is called the *condensation*, and we shall assume that

$$\Gamma = \max |\delta(x, t)| << 1.$$

We also wish to think of the particle velocity u as 'small' in some relative sense. Experience with acoustic waves, which are small perturbations of the constant solutions indicated above, shows that particles of gas do not move much and that the particle velocity u is small relative to the speed at which sound waves propagate in the gas. Let $c_0 = \sqrt{P'(\rho_0)}$. c_0 has the dimensions of a velocity, called the sound speed of the gas. In the case of an ideal gas, $c_0 = A\gamma\rho_0^{\gamma-1}$. Let $U = \max |u(x, t)|$ be the maximum particle velocity of the gas. We shall assume that

$$M \equiv \frac{U}{c_0} << 1.$$

M is the Mach number. Thus we shall write our perturbed velocity in the form $u(x, t) = c_0 v(x, t)$ where v is a small dimensionless quantity on the order of the Mach number. Substitute $\rho = (1 + \delta)\rho_0$ and $u = c_0 v$ into (5.1) and divide out factors of ρ_0. The resulting system is

$$\delta_t + c_0 v_x + c_0 \delta v_x + c_0 \delta_x v = 0 \tag{5.2}$$

$$(1 + \delta)[c_0 v_t + c_0^2 v v_x] + P'((1 + \delta)\rho_0)\delta_x = 0. \tag{5.3}$$

Since $1 + \delta \approx 1$ and $P'((1 + \delta)\rho_0) \approx P'(\rho_0) = c_0^2$ because δ is small, (5.3) becomes (after cancelling out a factor of c_0)

$$v_t + c_0 \delta_x + c_0 v v_x \approx 0. \tag{5.4}$$

Look at the leading linear terms in (5.2) and (5.4). For consistency, we must assume that δ_t, $c_0 v_x$, v_t, $c_0 \delta_x$ are all of the same order of magnitude. Thus in (5.2) the terms $c_0 \delta v_x$ and $c_0 \delta_x v$ are small compared to the linear terms δ_t and $c_0 v_x$. Hence the nonlinear terms can be dropped. Similarly the term $c_0 v_x v$ is small compared to the linear terms v_t and $c_0 \delta_x$. We drop the nonlinear terms in (5.4). This leaves the linear system

$$\delta_t + c_0 v_x = 0 \tag{5.5}$$

$$v_t + c_0 \delta_x = 0.$$

Differentiate the first equation of (5.5) with respect to t and the second with respect to x. Then multiply the second equation by c_0, and subtract from the first to arrive at

$$\delta_{tt} - c_0^2 \delta_{xx} = 0 \tag{5.6}$$

Proceeding in almost the same way, we also find that v satisfies

$$v_{tt} - c_0^2 v_{xx} = 0. \tag{5.7}$$

Equations (5.6) and (5.7) are the linear wave equations. In the context of gas dynamics, solutions of the linear wave equation are called acoustic waves. Of course the original δ and v do not satisfy (5.5) exactly, and we should use different letters $\bar{\delta}$, $\bar{v}$ to denote solutions of (5.5). However it is well accepted abuse of notation to continue to use δ and v.

In acoustics it is often preferable to deal with a *velocity potential*

$$\phi(x, t) = \int u(x, t) dx.$$

ϕ will satisfy the same linearized equation (5.7).

For a thorough classical treatment of one-dimensional gas dynamics see [CF].

5.2   The vibrating string

5.2.1   The nonlinear model

The linear wave equation is also used to describe the vibrations of strings and membranes. For this reason the linear wave equation in one space dimension is often called the equation of the vibrating string.

We assume a uniform, perfectly flexible string which offers no resistance to bending. Let ρ_0 be the density (mass /unit length) when the string is in its stretched equilibrium state along the x axis. The string is fixed at the points $x = 0$ and $x = L$. We use the traditional notation for the unit vectors in R^3, $\mathbf{i} = (1, 0, 0)$, $\mathbf{j} = (0, 1, 0)$, and $\mathbf{k} = (0, 0, 1)$. Now let

$$\mathbf{r}(x, t) = (x + u(x, t))\mathbf{i} + v(x, t)\mathbf{j} + w(x, t)\mathbf{k}$$

be the displacement of the point $(x, 0, 0)$ of the string at time t. u is the longitudinal displacement, while v and w are the transverse displacements. $\mathbf{r}(0, t) = \mathbf{0}$ and $\mathbf{r}(L, t) = \mathbf{i}L$ for all t. Let $\Gamma(a, b)$ be the portion of the string which occupies the interval $[a, b]$ on the axis in the equilibrium position (see Figure 5.1).

The element of arc length of the string is

$$ds = |\mathbf{r}_x(x, t)|dx = \sqrt{(1 + u_x(x, t))^2 + v_x^2(x, t) + w_x^2(x, t)}dx.$$

while

$$s(b, t) = \int_0^b |\mathbf{r}_x(x, t)|dx$$

is the arc length from the left end point. The arc length of $\Gamma(a, b)$ is $s(b, t) - s(a, t)$. Let $\rho(x, t)$ be the density (mass/actual unit length) of the string when displaced to the point $\mathbf{r}(x, t)$ (see Figure 5.1). The mass of $\Gamma(a, b)$ is constant, always equal to its mass $\rho_0(b - a)$ in the equilibrium position. Thus

$$\rho_0(b - a) = \int_{s(a,t)}^{s(b,t)} \rho\, ds = \int_a^b \rho(x, t)|\mathbf{r}_x(x, t)|dx. \tag{5.8}$$

Since this is true for all intervals $[a, b]$, we deduce that

$$\rho(x, t)|\mathbf{r}_x(x, t)| = \rho_0. \tag{5.9}$$

Hence if we know $|\mathbf{r}_x|$, then we can determine $\rho(x, t)$ from ρ_0. Now using (5.9), we see that the momentum of $\Gamma(a, b)$ is given by

$$\int_a^b \mathbf{r}_t(x, t)\rho(x, t)|\mathbf{r}_x(x, t)|dx = \int_a^b \rho_0\mathbf{r}_t(x, t)dt$$

and the rate of change of the momentum of $\Gamma(a, b)$ is

$$\frac{d}{dt}\int_a^b \rho_0\mathbf{r}_t(x, t)dx = \rho_0\int_a^b \mathbf{r}_{tt}(x, t)dx. \tag{5.10}$$

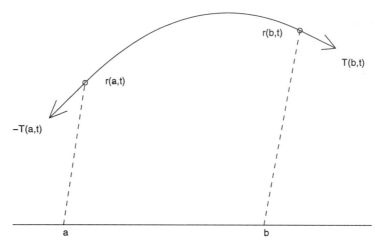

FIGURE 5.1
Portion of the stretched string with tension.

Note that $\mathbf{r}_{tt} = u_{tt}\mathbf{i} + v_{tt}\mathbf{j} + w_{tt}\mathbf{k}$, and we have assumed that the motion is smooth enough so that $\mathbf{r}$ is C^2 in each of its components.

Next we describe the forces acting on the string. There are internal forces, and external forces. First the internal forces. Let $\mathbf{T}(x, t)$ be the force exerted by the portion of the string to the right of $u(x, t)$ on the portion to the left of $u(x, t)$. Because the string offers no resistance to bending, this force is tangential to the string:

$$\mathbf{T}(x, t) = \hat{T}(x, t)\frac{\mathbf{r}_x(x, t)}{|\mathbf{r}_x(x, t)|}. \tag{5.11}$$

The *tension* $\hat{T}(x, t)$ is the magnitude of $\mathbf{T}(x, t)$ and $\mathbf{r}_x(x, t)/|\mathbf{r}_x(x, t)|$ is the unit tangent vector to the deformed string at the point $\mathbf{r}(x, t)$. We will assume that the force exerted by the portion of the string to the left of $u(x, t)$ on the portion to the right is $-\mathbf{T}(x, t)$, although this can be easily deduced.

The external forces, such as gravity, acting on $\Gamma(a, b)$ have the form

$$\rho_0 \int_a^b \mathbf{q}(x, t, \mathbf{r}, \mathbf{r}_t)dx.$$

Here $\mathbf{q} = q_1\mathbf{i} + q_2\mathbf{j} + q_3\mathbf{k}$.

Now we state Newton's law for $\Gamma(a, b)$. The rate of change of momentum (5.10) is equal to the forces acting on $\Gamma(a, b)$:

$$\rho_0 \int_a^b \mathbf{r}_{tt}(x, t)dx = -\mathbf{T}(a, t) + \mathbf{T}(b, t) + \int_a^b \rho_0\mathbf{q}(x, t, \mathbf{r}, \mathbf{r}_t)dx.$$

We assume that $(x, t) \to \mathbf{T}(x, t)$ is C^1 in all components. Then we can rewrite this equation as

$$\rho_0 \int_a^b \mathbf{r}_{tt} dx = \int_a^b [\frac{\partial}{\partial x} \mathbf{T} + \rho_0 \mathbf{q}] dx.$$

Since this is true for all intervals $[a, b]$, we deduce the vector equation

$$\rho_0 \mathbf{r}_{tt} = \frac{\partial}{\partial x} \mathbf{T} + \rho_0 \mathbf{q}. \tag{5.12}$$

To get a closed system for the three components of the displacement, we must state a *constitutive relation* which expresses the tension in terms of $\mathbf{r}$. We assume that the magnitude $\hat{T}$ of the tension depends smoothly on the amount of stretching of the string:

$$\hat{T}(x, t) = \hat{T}(|\mathbf{r}_x(x, t)|),$$

and $s \to \hat{T}(s)$ is C^1. Hence,

$$\mathbf{T}(x, t) = \hat{T}(|\mathbf{r}_x(x, t)|) \frac{\mathbf{r}_x(x, t)}{|\mathbf{r}_x(x, t)|}. \tag{5.13}$$

The amount of stretch is indicated by $|\mathbf{r}_x|$. In the equilibrium state $|\mathbf{r}_x| = 1$. The tension in the equilibrium state is denoted $T_0 = \hat{T}(1)$. When the string is stretched out of the equilibrium position, $|\mathbf{r}_x| > 1$.

With the addition of this constitutive relation, equations (5.12) become a non-linear system of three PDE's in the three components u, v, w of the displacement. These equations are extremely hard to analyze. We must make some simplifying assumptions to arrive at a decoupled system of linear equations.

5.2.2 The linearized equation

Many derivations of the linear equation of the vibrating string begin by limiting the discussion to purely transverse vibrations ($u = 0$). This is too strong an assumption and frequently leads to a derivation which is logically flawed. To see this, assume that the motion is purely transverse. Then $|\mathbf{r}_x| = \sqrt{1 + v_x^2 + w_x^2}$, and the first equation of (5.12) and (5.13) with $q_1 = 0$ reduces to

$$0 = \frac{\partial}{\partial x} [\hat{T}(|\mathbf{r}_x|) \frac{1}{|\mathbf{r}_x|}].$$

This equation implies that $\hat{T}(s)$ is linear, $\hat{T}(s) = T_0 s$. The second two equations then become, after dividing by ρ_0,

$$v_{tt} = (\frac{T_0}{\rho_0}) v_{xx} + q_2$$

$$w_{tt} = (\frac{T_0}{\rho_0})w_{xx} + q_3.$$

The assumption of purely transverse motion is so strong that it forces the tension function to be linear, a far too restrictive result. Furthermore, we arrived at linear equations for v and w (assuming q_2 and q_3 are linear in v and w) without assuming that the motions are small perturbations of the equilibrium state.

We give a more systematic derivation of the linearized equations by making our first assumption that the motions we consider have the form

$$\mathbf{r}(x, t) = (x + \varepsilon \bar{u}(x, t))\mathbf{i} + \varepsilon \bar{v}(x, t)\mathbf{j} + \varepsilon \bar{w}(x, t)\mathbf{k},$$

where $\varepsilon > 0$ is a small parameter and $\bar{u}, \bar{v}, \bar{w}$ and their derivatives are $O(1)$. We shall approximate the expressions $|\mathbf{r}_x|$, $\hat{T}(|\mathbf{r}_x|)$, and $(\partial/\partial x)[\hat{T}(|\mathbf{r}_x|)\mathbf{r}_x/|\mathbf{r}_x|]$ which arise when (5.13) is substituted in (5.12). To begin, note that Taylor expansions around $\varepsilon = 0$ imply that

$$\sqrt{(1 + \varepsilon a)^2 + (\varepsilon b)^2 + (\varepsilon c)^2} = 1 + \varepsilon a + O(\varepsilon^2)$$

and

$$\frac{1}{\sqrt{(1 + \varepsilon a)^2 + (\varepsilon b)^2 + (\varepsilon c)^2}} = 1 - \varepsilon a + O(\varepsilon^2).$$

Consequently, to first order in ε,

$$|\mathbf{r}_x| = 1 + \varepsilon \bar{u}_x + O(\varepsilon^2) \approx 1 + \varepsilon \bar{u}_x$$

and

$$\frac{\mathbf{r}_x}{|\mathbf{r}_x|} \approx [(1 + \varepsilon \bar{u}_x)\mathbf{i} + \varepsilon \bar{v}_x\mathbf{j} + \varepsilon \bar{w}_x\mathbf{k}][1 - \varepsilon \bar{u}_x]$$

$$\approx \mathbf{i} + \varepsilon \bar{v}_x\mathbf{j} + \varepsilon \bar{w}_x\mathbf{k}.$$

Furthermore, using another Taylor expansion for $\hat{T}$ around $s = 1$, we see that

$$\hat{T}(|\mathbf{r}_x|) = \hat{T}(1 + \varepsilon \hat{u}_x + O(\varepsilon^2)) \approx \hat{T}(1) + \hat{T}'(1)\varepsilon \bar{u}_x.$$

Combining these approximations, we see that the tension (5.13) can be expressed, to within order ε, by

$$\hat{T}(|\mathbf{r}_x|)\frac{\mathbf{r}_x}{|\mathbf{r}_x|} \approx (\hat{T}(1) + \hat{T}'(1)\varepsilon \bar{u}_x)(\mathbf{i} + \varepsilon \bar{v}_x\mathbf{j} + \varepsilon \bar{w}_x\mathbf{k}) \qquad (5.14)$$

$$\approx \varepsilon T_0'\bar{u}_x\mathbf{i} + \varepsilon T_0(\bar{v}_x\mathbf{j} + \bar{w}_x\mathbf{k}).$$

$T_0 = \hat{T}(1)$ is the tension in the equilibrium state while $T_0' = \hat{T}'(1)$ is the tensile stiffness at $s = 1$. Now we assume that the exterior forces $\mathbf{q} = \varepsilon\bar{\mathbf{q}}$. Then substitute $\mathbf{r}_{tt} = \varepsilon(\bar{u}_{tt}\mathbf{i} + \bar{v}_{tt}\mathbf{j} + \bar{w}_{tt}\mathbf{k})$ in the left side of (5.12) and use (5.14) in the right side of (5.12). We arrive at

$$\rho_0\varepsilon(\bar{u}_{tt}\mathbf{i} + \bar{v}_{tt}\mathbf{j} + \bar{w}_{tt}\mathbf{k}) = \varepsilon T_0'\bar{u}_{xx}\mathbf{i} + \varepsilon T_0(\bar{v}_{xx}\mathbf{j} + \bar{w}_{xx}\mathbf{k}) + \varepsilon\rho_0(\bar{q}_1\mathbf{i} + \bar{q}_2\mathbf{j} + \bar{q}_3\mathbf{k}).$$

This vector equation is equivalent to the system

$$\bar{u}_{tt} = c_1^2\bar{u}_{xx} + \bar{q}_1 \tag{5.15}$$

$$\bar{v}_{tt} = c^2\bar{v}_{xx} + \bar{q}_2 \tag{5.16}$$

$$\bar{w}_{tt} = c^2\bar{w}_{xx} + \bar{q}_3, \tag{5.17}$$

where $c = \sqrt{T_0/\rho_0}$ and $c_1 = \sqrt{T_0'}$. The system is decoupled if we assume that

$$q_1 = q_1(x, t, u, u_t), \quad q_2 = q_2(x, t, v, v_t), \quad q_3 = q_3(x, t, w, w_t).$$

Equation (5.15) describes the longitudinal motion of the string, while (5.16) and (5.17) describe the transverse part of the motion. c is the propagation speed of the transverse waves, and c_1 is the propagation speed of the longitudinal waves. Typical forms for the q_j might be

$$q_2(x, t, v, v_t) = -dv_t - kv$$

where $d, k \geq 0$. The $-dv_t$ term represents friction which dissipates the mechanical energy of the string, while the $-kv$ term represents an additional elastic restoring force.

For more information on this subject, the reader is referred to the excellent article of S. Antman [A].

5.3 The wave equation without boundaries

5.3.1 The initial-value problem and d'Alembert's formula

First we consider the wave equation without boundary conditions to understand how waves are propagated before they interact with boundaries.

What are the appropriate initial conditions for the wave equation

$$u_{tt} = c^2 u_{xx} \qquad x, t \in R \tag{5.18}$$

that will yield a well-posed problem? Recall that for a second-order ODE

$$y'' = f(t, y, y'),$$

we usually specify $y(0)$ and $y'(0)$. Thus we are led by analogy to pose the following IVP for (5.18):

$$u(x, 0) = f(x), \qquad u_t(x, 0) = g(x) \qquad \text{for } x \in R. \tag{5.19}$$

In the context of the vibrating string, we are specifying the initial displacement of the string $f(x)$ and the initial velocity $g(x)$.

Since (5.18) is supposed to govern the propagation of waves, our work in Chapter 2 should be of some relevance here. In fact we can factor (5.18), writing it as

$$(\partial_t + c\partial_x)(\partial_t - c\partial_x)u = 0.$$

It is easy to verify that (5.18) has solutions of the form

$$u = \phi(x - ct) \qquad \text{and} \qquad u = \psi(x + ct).$$

Thus (5.18) will have waves which move in both directions (left and right) and thus two sets of characteristics:

$$x - ct = \text{constant} \qquad \text{and} \qquad x + ct = \text{constant}.$$

To obtain the general solution of (5.18), we make a change of variable which takes the characteristics into coordinate axes. Set

$$\xi = x - ct, \qquad \eta = x + ct.$$

Let $w(\xi, \eta)$ be u expressed in ξ, η coordinates. Thus

$$u(x, t) = w(x - ct, x + ct).$$

To determine the equation satisfied by w, we use the chain rule to calculate

$$u_t = -cw_\xi + cw_\eta,$$

$$u_{tt} = c^2 w_{\xi\xi} - 2c^2 w_{\xi\eta} + c^2 w_{\eta\eta},$$

and

$$c^2 u_{xx} = c^2 w_{\xi\xi} + 2c^2 w_{\xi\eta} + c^2 w_{\eta\eta},$$

so that

$$u_{tt} - c^2 u_{xx} = -4c^2 w_{\xi\eta}.$$

Thus u solves (5.18) if and only if

$$w_{\xi\eta} = 0. \tag{5.20}$$

We can solve (5.20) by first integrating in η,

$$w_{\xi}(\xi, \eta) = f(\xi),$$

and then in ξ to yield

$$w(\xi, \eta) = \int f(\xi)d\xi + G(\eta) = F(\xi) + G(\eta),$$

where F and G are arbitrary C^2 functions. Transforming back, we see that the general solution of (5.18) is given by

$$u(x, t) = F(x - ct) + G(x + ct). \tag{5.21}$$

Thus solutions of (5.18) are a sum of two waves, both moving with speed c, one to the right ($F(x - ct)$) and one to the left ($G(x + ct)$).

We want to choose the functions F and G so as to satisfy the initial conditions (5.19). The first condition of (5.19) becomes

$$F(x) + G(x) = f(x) \tag{5.22}$$

while the second becomes

$$-F'(x) + G'(x) = \frac{1}{c}g(x). \tag{5.23}$$

If we differentiate (5.22) and first subtract from (5.23) and then add, we get

$$F'(x) = \frac{1}{2}(f'(x) - \frac{1}{c}g(x))$$

and

$$G'(x) = \frac{1}{2}(f'(x) + \frac{1}{c}g(x)).$$

It follows that

$$F(x) = \frac{1}{2}f(x) - \frac{1}{2c}\int_0^x g(y)dy + C_1,$$

and

$$G(x) = \frac{1}{2}f(x) + \frac{1}{2c}\int_0^x g(y)dy + C_2,$$

where C_1 and C_2 are arbitrary constants. We set $C_1 + C_2 = 0$ so that (5.22) is satisfied. Then

$$u(x,t) = F(x - ct) + G(x + ct)$$

$$= \frac{1}{2}[f(x + ct) + f(x - ct)] + \frac{1}{2c}\int_{x-ct}^{x+ct} g(y)dy. \qquad (5.24)$$

This is *d'Alembert's formula* for the solution of (5.18), (5.19).

Theorem 5.1
Let $f \in C^2(R)$, $g \in C^1(R)$. Then there is a unique C^2 solution of (5.18), (5.19) given by d'Alembert's formula. We can estimate the magnitude of the solution in terms of the initial data. For each fixed t,

$$\max_{x \in R} |u(x,t)| \leq \max_{x \in R} |f(x)| + t \max_{x \in R} |g(x)|,$$

provided both $\max |f|$ and $\max |g|$ are finite. If $\int_R |g(x)|dx$ is finite, a better estimate is

$$|u(x,t)| \leq \max_{x \in R} |f(x)| + \frac{1}{2c}\int_R |g(x)|dx.$$

The existence and uniqueness are a direct consequence of the derivation. The estimates follow easily from d'Alembert's formula. We remark again that these estimates imply that the solution depends continuously on the initial data. If u is a solution with initial data f_1, g_1 and v is a solution with initial data f_2, g_2, then $w = u - v$ is a solution with initial data $f_1 - f_2, g_1 - g_2$, and the estimates apply to w. Thus we have shown that the IVP (5.18) (5.19) is a well-posed problem.

Examples

1) If $f(x) = 0$ for $|x| \geq a > 0$ and $g \equiv 0$, the solution is simply

$$u(x, t) = \frac{1}{2} f(x + ct) + \frac{1}{2} f(x - ct).$$

We can visualize this as one bump $f(x)$ splitting into two bumps of one-half the amplitude, one moving to the right with speed c, $(1/2) f(x - ct)$, and one moving the left with speed c, $(1/2) f(x + ct)$.

2) $f(x) \equiv 0$ and $g(x) = 0$ for $|x| \geq a$. In this case

$$u(x, t) = \frac{1}{2c} \int_{x-ct}^{x+ct} g(y) dy.$$

For each x, as t increases, the interval $(x - ct, x + ct)$ of length $2ct$ will eventually include the interval $[-a, a]$ where g is nonzero. Thus for each x,

$$u(x, t) \to \frac{1}{2c} \int_{-a}^{a} g(y) dy \qquad \text{as} \quad t \to \infty.$$

It is interesting to note that d'Alembert's formula (5.24) still makes sense even when f and g are not differentiable. For instance if $g \equiv 0$ and

$$f(x) = \begin{cases} 1 & \text{for } |x| \leq 1 \\ 0 & \text{for } |x| > 1 \end{cases},$$

(5.24) yields two square waves of height $1/2$ moving in opposite directions. This function is obviously not a strict solution of the PDE. It is another example of a *weak* solution. Using the same techniques as in Section 2.5.2, it can be shown that

$$u(x, t) = \frac{1}{2} [f(x + ct) + f(x - ct)]$$

with f defined above solves the integral equation

$$\int \int u(x, t)[\varphi_{tt}(x, t) - c^2 \varphi_{xx}(x, t)] dx dt = 0$$

for all functions $\varphi \in C^2(R^2)$ such that $\varphi = 0$ outside some rectangle.

If either f, f' or f'', or g or g' is discontinous at x_0, we say that the initial data has a *singularity* at x_0. (Recall the discussion in Chapter 2 following the definition of weak solutions of $u_t + cu_x = 0$.) D'Alembert's formula (5.24) shows that a singularity in the initial data will propagate along the characteristic line $x = x_0 + ct$, or the line $x = x_0 - ct$, or perhaps along both. For example, in the case just above, the initial data has singularities at $x = \pm 1$ because it is discontinuous there. The weak solution has discontinuities along both characteristics $x = 1 \pm ct$ emanating from $x = 1$ and along both characteristics $x = -1 \pm ct$ emanating from $x = -1$. In the exercises we will see how this property of weak solutions is helpful in understanding their structure.

5.3.2   Domains of influence and dependence

Two important concepts which embody the idea of a finite speed of propagation emerge from an inspection of d'Alembert's formula. They are the *domain of dependence* of a point (x_0, t_0), and the *domain of influence* of an interval $[a, b]$ of the x axis. Pick a point (x_0, t_0), $t_0 > 0$. Which parts of the initial data f, g can influence the solution u at the point (x_0, t_0)? From (5.24) we see that $u(x_0, t_0)$ involves the values of f at $x_0 + ct_0$ and $x_0 - ct_0$ and the values of g in the interval $[x_0 - ct_0, x_0 + ct_0]$. This is the domain of dependence of (x_0, t_0) (see Figure 5.2). If we change f away from these two points, and g outside this interval, it will not change the value of u at (x_0, t_0).

Furthermore if $f = g = 0$ for x outside some interval $[a, b]$, we see that $u = 0$ for $x \geq b + ct$ and for $x \leq a - ct$. This expresses clearly the fact that information travels no faster than speed c to the left or right. The domain of influence of the interval $[a, b]$ at time $t_0 > 0$ is the interval $[a - ct_0, b + ct_0]$ (see Figure 5.3).

5.3.3   Conservation of energy on the line

Our next topic in this section is conservation of energy. To motivate this idea, let us first look at the ODE which describes a harmonic oscillator,

$$m\varphi'' + k\varphi = 0.$$

The constant $k > 0$ is the spring constant and m is the mass. If we multiply the equation by φ', we see that

$$\frac{1}{2}\frac{d}{dt}[m(\varphi')^2 + k\varphi^2] = m\varphi''\varphi' + k\varphi\varphi' = 0.$$

The quantity

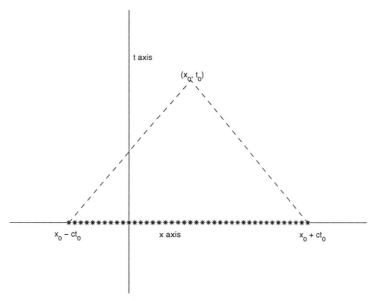

FIGURE 5.2
Domain of dependence of the point (x_0, t_0) shown on the x axis.

$$\frac{m}{2}(\varphi')^2 + \frac{k}{2}\varphi^2$$

is conserved during the motion of the oscillator. It is the total energy of the oscillator. The first term is easily recognized as the kinetic energy, and the second as the potential energy of the spring.

In the context of the vibrating string, (5.18) has a similar conserved quantity. If we think of each particle of the string as the infinitesimal mass of an oscillator vibrating up and down, we can expect the conserved energy to take the form of an integral.

Definition   *For a function $u(x, t)$ we define the* energy *at time t as the integral*

$$e(t) \equiv \frac{1}{2}\int_R [u_t^2(x, t) + c^2 u_x^2(x, t)]dx,$$

provided this quantity is finite.

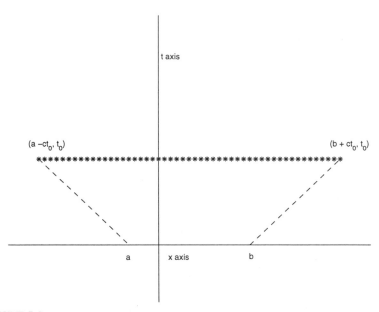

FIGURE 5.3
Domain of influence of the interval $[a, b]$ at time $t = t_0$.

Theorem 5.2
Conservation of energy for the wave equation. Suppose that $f \in C^2(R)$ and
$g \in C^1(R)$ *with*

$$\int_R (f')^2(x)dx < \infty \qquad and \qquad \int_R g^2(x)dx < \infty.$$

*Let $u(x, t)$ be the unique solution of (5.18), (5.19). Then for each $t \in R$, $e(t)$ is
finite and is constant:*

$$e(t) = e(0) = \frac{1}{2} \int_R [g^2(x) + c^2(f')^2(x)]dx.$$

The result can be found by a direct computation using d'Alembert's formula,
but we shall use another method which can be generalized to higher dimensions.
In the following we shall assume that we are dealing with a solution of the wave
equation u such that all the integrals involved converge uniformly with respect to
t and that u_x and $u_t \to 0$ as $x \to \pm\infty$. This would be the case, for example, if
the initial data f and g are both zero outside some finite interval $|x| \leq a$.

Now differentiate the expression for $e(t)$, passing the derivative under the integral sign:

$$\frac{d}{dt}e(t) = \int_R u_{tt}(x, t)u_t(x, t) + c^2 u_x(x, t)u_{xt}(x, t)dx \qquad (5.25)$$

Next integrate by parts in the second term on the right of (5.25).

$$\int_R u_x u_{xt}dx = u_x u_t\Big|_{-\infty}^{\infty} - \int_R u_{xx}u_t dx = -\int_R u_{xx}u_t dx$$

because of our assumptions about the behavior of u_x and u_t as $x \to \pm\infty$. Thus

$$\frac{d}{dt}e(t) = \int_R (u_{tt} - c^2 u_{xx})u_t dx = 0$$

because u solves (5.18). This shows that $e(t)$ is constant. $\square$

Remarks

(1) The quantity $(1/2)[u_t^2(x, t) + c^2 u_x^2(x, t)]$ is called the energy density. This quantity is not conserved. Rather it is the *integral* of this quantity that is conserved. When this integral exists (is finite), we say that such a solution is a *finite-energy* solution.

In the context of the vibrating string, when $c^2 = T_0/\rho_0$, we can rewrite the conserved energy quantity as

$$\frac{1}{2}\int_R (\rho_0 u_t^2 + T_0 u_x^2)dx.$$

Then

$$\frac{1}{2}\int_R \rho_0 u_t^2 dx$$

is the kinetic energy of the string and

$$\frac{1}{2}\int_R T_0 u_x^2 dx$$

is the potential energy of the string. Both terms have the dimension of length $\times$ force which is the dimension of work, or energy.

(2) We can redo the computation for the telegraph equation

$$u_{tt} - c^2 u_{xx} + ku = 0, \qquad k > 0.$$

and deduce that the quantity

$$e_k(t) \equiv \frac{1}{2} \int_R [u_t^2 + c^2 u_x^2 + ku^2] dx$$

is conserved, provided the integrals of the initial data f, g

$$\int_R f^2(x) dx, \quad \int_R (f'(x))^2 dx, \quad \text{and} \int_R g^2(x) dx$$

are all finite.

5.3.4   An inhomogeneous problem

The third topic of this section is the inhomogeneous equation

$$u_{tt} - c^2 u_{xx} = q(x, t) \qquad x, t \in R, \tag{5.26}$$

$$u(x, 0) = 0, \qquad u_t(x, 0) = 0 \tag{5.27}$$

Here the term q on the right hand-side can represent either an acoustical source or an external force in the case of the vibrating string. Again we turn to the principle of Duhamel to represent the solution as

$$u(x, t) = \int_0^t z(x, t, s) ds, \tag{5.28}$$

where z satisfies the IVP

$$z_{tt} - c^2 z_{xx} = 0 \qquad \text{for } t \geq s,$$

$$z(x, s, s) = 0, \qquad z_t(x, s, s) = q(x, s).$$

To verify that (5.28) is indeed the solution of (5.26), (5.27), we first calculate

$$u_t(x, t) = z(x, t, t) + \int_0^t z_t(x, t, s) ds = \int_0^t z_t(x, t, s) ds$$

and then

$$u_{tt} = z_t(x, t, t) + \int_0^t z_{tt}(x, t, s)ds = q(x, t) + \int_0^t c^2 z_{xx}(x, t, s)ds$$

$$= q(x, t) + c^2 \partial_x^2 \int_0^t z(x, t, s)ds = q(x, t) + c^2 u_{xx}(x, t).$$

Now to put (5.28) in a more explicit form, use d'Alembert's formula to express z. We use (5.24) with $f = 0$, $g(y) = q(y, s)$ and since we are solving for $t \geq s$, we replace t by $t - s$. We obtain

$$z(x, t, s) = \frac{1}{2c} \int_{x-c(t-s)}^{x+c(t-s)} q(y, s)dy,$$

so that

$$u(x, t) = \frac{1}{2c} \int_0^t \int_{x-c(t-s)}^{x+c(t-s)} q(y, s)dyds = \frac{1}{2c} \int \int_{D(x,t)} q(y, s)dyds, \quad (5.29)$$

where $D(x, t)$ is the triangular region with vertex at (x, t):

$$D(x, t) = \{(y, s) : |y - x| \leq c(t - s), 0 \leq s \leq t\}.$$

The representation (5.29) allows us to draw some conclusions about the region of influence of a source $q(x, t)$. For instance, if $q(x, t) = 0$ for $|x| \geq a > 0$ for all $t \geq 0$, then the solution $u = 0$ for $|x| \geq a + ct$. The waves produced by the source spread no faster than the propagation speed c.

Exercises 5.3

1. Consider IVP (5.18) - (5.19) with $c = 1$ and with data $f(x) = \exp(-x^2)$, $g = 0$. Make an mfile for f and call it f.m, making sure it is array-smart. Plot the solution $u(x, t)$ as given by d'Alembert's formula on $[-10, 10]$ for $t = 1, 2, 3, 4$. Superimpose the five plots on one graph, with different colors. You can do this by making vectors y0 = f(x), y1 = .5*(f(x+1) + f(x-1)), etc. Then use the command plot(x,y0, x,y1, x,y2, x,y3, x,y4).
   In the context of the vibrating string, what does the solution of this IVP represent? Describe what you see in the plots.

2. Consider the IVP (5.18) - (5.19) with $f(x) = 0$ and

$$g(x) = xe^{-x^2}.$$

Find the solution using d'Alembert's formula and sketch snapshots for several values of $t > 0$.

3. Consider the IVP (5.18) - (5.19) with $f(x) = 0$ and

$$g(x) = \begin{cases} e^{-x}, & x > 0 \\ -e^x, & x < 0 \end{cases}.$$

(a) Sketch the graph of g. How is it similar to the data of exercise 2)? Do you expect the solutions to be similar?

(b) The data g has a singularity at $x = 0$. In the x, t plane, sketch the two characteristics emanating from $x = 0$. They divide the x, t plane (for $t \geq 0$) into three regions. The singularity in the initial data will propagate along these lines. You can see that the solution has a jump in u_{xx} there.

(c) Find the solution using d'Alembert's formula. You will need to evaluate the integral in three different ways, yielding three different formulas, one for each of the three regions.

(d) Show that the solution of this IVP can also be represented

$$u(x, t) = F(x - ct) + G(x + ct),$$

where

$$F(x) = \begin{cases} \frac{1}{2c}(e^{-x} - 1), & x \geq 0 \\ \frac{1}{2c}(e^x - 1), & x < 0 \end{cases}$$

and

$$G(x) = -F(x).$$

(e) Assume that $c = 1$. Write an mfile bigf.m for F as follows:

```
function y = bigf(x)
y = .5*(x>=0).*(exp(-x) -1)
   + .5*(x< 0).*(exp(x) -1);
```

Then use the command `plot(x,bigf(x-t)-bigf(x+t))` to plot snapshots of the solution on $[-5, 5]$ for several values of $t > 0$.

4. Consider the IVP (5.18) - (5.19) with $c = 1$, $f = 0$, and

$$g(x) = \begin{cases} 1, & |x| < 2 \\ 0, & \text{otherwise} \end{cases} .$$

(a) The data g has singularities at $x = \pm 2$. Make a sketch showing the characteristics emanating from $x = \pm 2$. These lines divide up the x, t plane , $t \geq 0$, into six regions, $R_1, R_2, \ldots, R_6$.

(b) For (x, t) in each of the regions, show that the integral formula reduces to the following collection of formulas for the solution (when you label the regions appropriately):

R_1, R_6 where $u(x, t) = 0$. R_2 where $u(x, t) = t$. R_3 where $u(x, t) = 2$.

R_4 where $u(x, t) = (x+t+2)/2$. R_5 where $u(x, t) = (t+2-x)/2$.

(c) Show that the solution of this IVP can also be represented

$$u(x, t) = F(x - t) + G(x + t),$$

where

$$F(x) = \begin{cases} 1, & x < -2 \\ -x/2, & -2 \leq x \leq 2 \\ -1, & x > 2 \end{cases}$$

and

$$G(x) = -F(x).$$

(d) Change the mfile `bigf.m` to read as follows:

```
function y = bigf(x)
y1 = (x < -2);
y2 = -.5*x.*((x<=2)-(x < -2));
y3 = -(x > 2);
y = y1 + y2 + y3;
```

Use the command `plot(x,bigf(x-t)-bigf(x+t))` to plot snapshots of the solution at $t = 1, 2, 3, 4$ on $[-8, 8]$. Relate these snapshots with your sketch in the x, t plane. In particular, compare the points in the snapshots where the solution has a jump in the derivative with the position of the singularities at that time.

5. Let $F(x) = \exp[-(x+5)^2]$ and $G(x) = -2\exp[-(x-5)^2]$. Assuming that $c = 1$, let $u(x, t) = F(x - t) + G(x + t)$.

   (a) What is the data pair $f(x) = u(x, 0)$, and $g(x) = u_t(x, 0)$?

   (b) Write the mfile bigf.m for this $F(x)$. Plot the solution on $[-10, 10]$ for values of t ranging from 1 to 7. In particular observe snapshots at $t = 4.5, 5$, and 5.5. What happens as the two waves interact? Are the waves intact after this interaction?

6. Let $F(x) = \exp(-x/10)\sin x$, $G = 0$. Assuming $c = 1$, set $u(x, t) = F(x - t)$.

   (a) What is the initial data f, g of this solution?

   (b) Write the mfile bigf.m for this F. Then plot snaphots of $u(x, t)$ on $[-10, 10]$ for several values of $t > 0$. Put them on the same graph in different colors. Describe what you see.

7. (a) If a solution of the wave equation is in the form $u(x, t) = F(x - ct) + G(x + ct)$, show that the energy

$$e(t) = e(0) = c^2 \int [(F'(x))^2 + (G'(x))^2]dx,$$

   provided this integral converges.

   (b) Which of the solutions in exercises 1 - 6 is a finite- energy solution?

8. Let D be the parallelogram in Figure 5.4 with vertices $P_j = (x_j, t_j)$, $j = 1, 2, 3, 4$. The sides of the parallelogram are characteristics with speed c. Thus the side C_1 is parameterized by $x(t) = x_1 + c(t - t_1)$ for $t_1 \leq t \leq t_2$, while the side C_2 is parameterized by $x(t) = x_2 - c(t - t_2)$ for $t_2 \leq t \leq t_3$, etc. Let u satisfy $u_{tt} = c^2 u_{xx}$, and let $M = c^2 u_x$ and $N = u_t$. Then Green's theorem says that

$$0 = \int\int_D (N_x - M_t)dxdt = \int_C Mdx + Ndt$$

   where $C = \partial D = C_1 \cup C_2 \cup C_3 \cup C_4$ oriented in the counterclockwise direction. Evaluate the line integral to show that

$$u(P_1) + u(P_3) = u(P_2) + u(P_4).$$

9. Consider the equation

$$u_{tt} + 2du_t - u_{xx} + d^2u = 0 \quad \text{for } x, t \in R$$

with initial conditions

$$u(x, 0) = f(x), \quad u_t(x, 0) = g(x) \quad \text{for } x \in R.$$

(a) Let $v(x, t) = \exp(dt)u(x, t)$. Show that v satisfies $v_{tt} - v_{xx} = 0$.
     What is the initial data of v?

(b) Solve for v using d'Alembert's formula. Then find $u(x, t) = \exp(-dt)v(x, t)$.

(c) If $\int_R |f(x)|dx$, $\max |f(x)|$, and $\int_R |g(x)|dx$ are all finite, show that
     $|u(x, t)| \leq C \exp(-dt)$, where C is a constant determined by the
     initial data f, g.

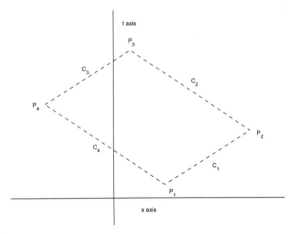

FIGURE 5.4
Parallelogram with sides consisting of characteristics.

10. Let
$$q(x, t) = \begin{cases} (\sin t)(1 - x^2) & \text{for } |x| \leq 1 \\ 0 & \text{for } |x| > 1. \end{cases}$$

(a) Use formula (5.29) to determine $u(0, t)$ for $t \geq 0$, where u is the
     solution of (5.26), (5.27).

(b) Show that $u(x, t) = 0$ for $|x| \geq ct + 1$, $\quad t \geq 0$.

11. Repeat the calculation of Theorem 5.2 for $e(t)$, but this time assume that
     u solves the damped equation

$$u_{tt} + du_t - c^2 u_{xx} = 0,$$

where $d > 0$. Show that $de(t)/dt \le 0$, so that energy may be decreasing.

12. **An energy inequality over cones.** Let K_T be the truncated cone of height T:

$$K_T = \{(x, t) : |x| \le a + c(T - t), 0 \le t \le T\}.$$

(a) Verify that for any C^2 function $u(x, t)$,

$$(u_{tt} - c^2 u_{xx})u_t = div\ \mathbf{f},$$

where $\mathbf{f} = (-c^2 u_x u_t, \frac{1}{2}(u_t^2 + c^2 u_x^2))$.
Now assume that u solves (5.18). By the divergence theorem

$$0 = \int\int_{K_T} div\ \mathbf{f} = \int_{\partial K_T} \mathbf{n} \cdot \mathbf{f} ds,$$

where $\mathbf{n}$ is the exterior unit normal to the boundary ∂K_T. The boundary consists of four parts: the top where $t = T$, the bottom where $t = 0$, the left side sloping side Σ_l, and the right sloping side Σ_r.

(b) Show that the integrals

$$\int_{\Sigma_l} \mathbf{n} \cdot \mathbf{f} ds \ge 0, \qquad \text{and} \qquad \int_{\Sigma_r} \mathbf{n} \cdot \mathbf{f} ds \ge 0.$$

Deduce that

$$\frac{1}{2} \int_{top} (u_t^2 + c^2 u_x^2) dx \le \frac{1}{2} \int_{bottom} (u_t^2 + c^2 u_x^2) dx.$$

This inequality expresses the domain of dependence in terms of the energy.

13. Consider the IVP (5.26) - (5.27) for the inhomogeneous equation. Assume that q has the form

$$q(x, t) = q_0(x)e^{i\omega t} = q_0(x)[\cos(\omega t) + i \sin(\omega t)].$$

(a) Take $q_0(x) = 1$ for $|x| < a$ and $q_0(x) = 0$ for $|x| > a$. Put this q into (5.29) and evaluate. Does the solution have growing oscillations or bounded oscillations as $t \to \infty$?

(b) Now use $q_0(x) = x$ for $|x| < a$ and $q_0(x) = 0$ for $|x| > a$. Evaluate (5.29). Show that for each x,

$$u(x, t) \to v(x)e^{i\omega t} \quad \text{as} \quad t \to \infty,$$

where $v(x)$ satisfies the Helmholtz equation (with $k = \omega/c$)

$$-v''(x) - k^2 v(x) = q_0(x)/c^2.$$

14. (a) Consider the equation

$$u_{tt} + 2u_{xt} - u_{xx} = 0.$$

Find the speeds of propagation c by seeking solutions of the form $u(x, t) = F(x - ct)$. Sketch the two families of characteristics in the x, t plane.

(b) Now let the equation be

$$u_{tt} + 3u_{xt} + 2u_{xx} = 0.$$

Same questions as in part (a).

5.4 Boundary value problems on the half-line

5.4.1 d'Alembert's formula extended

We consider several boundary value problems for the wave equation on the half-line. The first boundary value problem is

$$u_{tt} - c^2 u_{xx} = 0 \qquad \text{in } x > 0, t \in R, \tag{5.30}$$

$$u(x, 0) = f(x), \qquad u_t(x, 0) = g(x) \qquad \text{for } x \geq 0,$$

$$u(0, t) = 0 \qquad \text{for } t \in R. \tag{5.31}$$

The solution to this problem may be expressed in closed form using d'Alembert's formula (5.24). First we make some observations about the solutions of (5.18), (5.19). Recall that $f(x)$ defined on R is an *odd* function if $f(-x) = -f(x)$ for all x. In particular this implies that $f(0) = 0$. We say that f is an *even* function if $f(-x) = f(x)$ for all x. If $f \in C^1$ and f is even, then f' is odd, and hence $f'(0) = 0$.

Now suppose that the initial data f, g of the IVP (5.18), (5.19) are both odd functions. Then it follows from d'Alembert's formula that for each t, $x \to u(x, t)$ is odd, and hence that $u(0, t) = 0$ for all t. If both f and g are even functions, then the solution $u(x, t)$ of (5.18), (5.19) is even in x for each t so that $u_x(0, t) = 0$. The solutions of the IVP (5.18), (5.19) preserve the symmetry of the initial data.

This suggests the following method of solving (5.30), (5.31). We think of the data f and g, which are originally only defined for $x \geq 0$, as defined to be zero for $x < 0$. Then as we did in Chapter 3, we define the *odd extensions* of f and g, as

$$\tilde{f}(x) = f(x) - f(-x) \qquad \tilde{g}(x) = g(x) - g(-x).$$

$\tilde{f}$ and $\tilde{g}$ are both odd functions on R and

$$\tilde{f}(x) = \begin{cases} f(x) & \text{for } x > 0 \\ -f(-x) & \text{for } x < 0 \end{cases}, \qquad \tilde{g} = \begin{cases} g(x) & \text{for } x > 0 \\ -g(-x) & \text{for } x < 0 \end{cases}.$$

Let $\tilde{u}(x, t)$ be the solution on the whole line of (5.18) with initial data $\tilde{f}, \tilde{g}$. Because $x \to \tilde{u}(x, t)$ is odd, $\tilde{u}(0, t) = 0$ for all t. Then the desired solution u of (5.30)-(5.31) is the restriction of $\tilde{u}$ to $x \geq 0$. We can get a closed form representation of u in terms of the data f and g from d'Alembert's formula. We restrict our attention to the quarter-plane $\{(x, t) : x, t \geq 0\}$. For $x, t \geq 0$,

$$\tilde{f}(x + ct) = f(x + ct) - f(-(x + ct)) = f(x + ct)$$

because $f(x) = 0$ for $x < 0$ and

$$\tilde{f}(x - ct) = f(x - ct) - f(ct - x).$$

When we substitute $\tilde{f}$ and $\tilde{g}$ into d'Alembert's formula (5.24) we obtain

$$u(x, t) = \tilde{u}(x, t) = \frac{1}{2}[\tilde{f}(x + ct) + \tilde{f}(x - ct)] + \frac{1}{2c} \int_{x-ct}^{x+ct} \tilde{g}(y) dy$$

or

$$u(x,t) = \frac{1}{2}[f(x+ct) - f(ct-x) + f(x-ct)] + \frac{1}{2c}\int_{x-ct}^{x+ct}[g(y) - g(-y)]dy.$$
(5.32)

This formula simplifies somewhat if we divide up the quarter-plane $\{(x,t):x,t \geq 0\}$ into two regions (see Figure 5.5),

$$\mathcal{R} = \{(x,t) : 0 \leq x \leq ct, t \geq 0\} \qquad \mathcal{Q} = \{(x,t) : x > ct, t \geq 0\}.$$

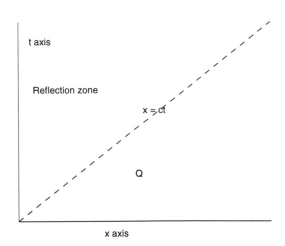

FIGURE 5.5
The reflection zone in the quarter-plane $x,t \geq 0$.

The region $\mathcal{R}$ is called the reflection zone. It may be thought of as the domain of influence of the part of the boundary $x = 0$ where $t \geq 0$. $\mathcal{Q}$ is the region in space-time where the boundary has no influence. For $(x,t) \in \mathcal{Q}$, $x + ct \geq 0$, and $x - ct \geq 0$, so that for $(x,t) \in \mathcal{Q}$, $f(ct-x) = 0$, and $g(y) - g(-y) = g(y)$ for $x - ct \leq y \leq x + ct$. Hence for $(x,t) \in \mathcal{Q}$, (5.32) reduces to d'Alembert's formula,

$$u(x,t) = \frac{1}{2}[f(x+ct) + f(x-ct)] + \frac{1}{2c}\int_{x-ct}^{x+ct} g(y)dy.$$

For $(x, t) \in \mathcal{R}$, $x - ct \leq 0$, but $x + ct \geq 0$. Hence

$$\frac{1}{2}[f(x + ct) + f(x - ct) - f(xt - x)] = \frac{1}{2}[f(x + ct) - f(ct - x)],$$

and

$$\frac{1}{2c}\int_{x-ct}^{x+ct}[g(y) - g(-y)]dy = \frac{1}{2c}\int_0^{x+ct} g(y)dy - \frac{1}{2c}\int_{x-ct}^0 g(-y)dy$$

$$= \frac{1}{2c}\int_0^{x+ct} g(y)dy - \frac{1}{2c}\int_0^{ct-x} g(y)dy.$$

Thus for $(x, t) \in \mathcal{R}$,

$$u(x, t) = \frac{1}{2}[f(x+ct) - f(ct-x)] + \frac{1}{2c}[\int_0^{x+ct} g(y)dy - \int_0^{ct-x} g(y)dy] \quad (5.33)$$

Theorem 5.3
Let f be C^2 on $x \geq 0$ with $f(0) = f''(0) = 0$. Let g be C^1 on $x \geq 0$ with $g(0) = 0$. Then the unique solution of (5.30) - (5.31) is given by d'Alembert's formula for $x \geq ct$ and by (5.33) for (x, t) in the reflection zone $\mathcal{R}$.

For the IBVP (5.30) with the Neumann boundary condition $u_x(0, t) = 0$, we substitute the even extensions of the given data f and g into (5.24). As in the case of the Dirichlet boundary condition, the solution in the region Q is given by d'Alembert's formula. In the reflection zone $\mathcal{R}$,

$$u(x, t) = \frac{1}{2}[f(x+ct) + f(ct-x)] + \frac{1}{2c}[\int_0^{x+ct} g(y)dy + \int_0^{ct-x} g(y)dy]. \quad (5.34)$$

Incident and reflected wave profiles

An alternate way of finding solutions to (5.30), with either Dirichlet or Neumann boundary conditions, is to represent them as an incident wave $G(x + ct)$ moving to the left, and a reflected wave $F(x - ct)$ moving to the right. We assume that $G(x) = 0$ for $x \leq 0$. Then the incident wave $G(x + ct)$ has not reached the boundary $x = 0$ for $t \leq 0$. The reflected wave $F(x - ct)$ will appear on the half-line $x \geq 0$ for $t > 0$. The reflected wave profile F will be determined by G and the boundary condition. If we write

$$u(x, t) = F(x - ct) + G(x + ct)$$

and impose the boundary condition $u(0, t) = 0$, it follows that

$$F(-ct) + G(ct) = u(0, t) = 0 \qquad \text{for all } t.$$

Thus if $G(z)$ is the given wave profile moving to the left, the reflected wave form is $F(z) = -G(-z)$. The desired solution is then the superposition of these two wave forms which cancel exactly at $x = 0$:

$$u(x, t) = F(x - ct) + G(x + ct) = G(x + ct) - G(ct - x).$$

Note that $x \rightarrow u(x, t)$ is an odd function for each t.

In the case of the Neumann boundary condition, we choose F so that $x \rightarrow u(x, t)$ is an even function for each t, that is, we choose

$$F(x) = G(-x).$$

Then

$$u(x, t) = G(x + ct) + F(x - ct) = G(x + ct) + G(ct - x)$$

satisfies $u_x(0, t) = 0$, and $u(x, t) = G(x + ct)$ for $x \geq 0$ and $t \leq 0$.

The Dirichlet and Neumann boundary conditions have different physical interpretations depending on the context. When u represents the transverse displacement of a string, the homogeneous Dirichlet condition $u(0, t) = 0$ corresponds to the end of the string being fixed. The Neumann condition $u_x(0, t) = 0$ corresponds to a free end in which the end of the string is attached to a massless rider that moves up and down freely in a vertical channel at $x = 0$.

In one-dimensional acoustics, the solutions of the linear wave equation represent the longitudinal motion of a tube of gas, and the preferred dependent variable is the velocity potential $\phi(x)$. In this case, the Neumann condition $\phi_x(0, t) = 0$ says that the velocity of the gas particles $v(0, t) = \phi_x(0, t) = 0$, which happens when the end of the tube is capped. In acoustics, the Neumann condition is called the "hard boundary condition." The Dirichlet boundary condition is applied to the small perturbation from the background state of the pressure or density. At the end of an open tube, the pressure inside the tube is the same as outside, so that the small perturbation is zero there. In acoustics, the Dirichlet condition is called the "soft boundary condition." A standard reference is [IM].

5.4.2 A transmission problem

Proceeding in the same vein, we consider a transmission problem next. Here we have two strings of different materials joined at the origin. There is a constant tension T_0 but differing densities ρ_l on the left and ρ_r on the right. The speeds of propagation are $c_l = \sqrt{T_0/\rho_l}$ and $c_r = \sqrt{T_0/\rho_r}$. We assume given an incident wave $G(x+c_r t)$ moving to the left, with $G(x) = 0$ for $x \leq 0$. In addition to a wave $F(x - c_r t)$ reflected back from the joint, we also expect there to be a transmitted wave $H(x + c_l t)$ moving to the left. Thus we seek a solution in the form

$$u(x, t) = \begin{cases} F(x - c_r t) + G(x + c_r t) & \text{for } x > 0 \\ H(x + c_l t) & \text{for } x < 0 \end{cases}.$$

The condition linking the two parts of the solution on either side of the joint is

$$u \text{ continuous at } x = 0, \quad u_x \text{ continuous at } x = 0.$$

This yields the two equations

$$H(c_l t) = F(-c_r t) + G(c_r t) \text{ for all } t \tag{5.35}$$

and

$$H'(c_l t) = F'(-c_r t) + G'(c_r t) \text{ for all } t. \tag{5.36}$$

We differentiate (5.35) to obtain

$$c_l H'(c_l t) = -c_r F'(-c_r t) + c_r G'(c_r t). \tag{5.37}$$

Then multiply (5.36) by c_r and add to (5.37) :

$$(c_r + c_l) H'(c_l t) = 2c_r G'(c_r t).$$

If we let $z = c_l t$, then $c_r t = (c_r/c_l)z$ and

$$H'(z) = (\frac{2c_r}{c_r + c_l}) G'(\frac{c_r z}{c_l}).$$

This implies that

$$H(z) = (\frac{2c_l}{c_r + c_l}) G(\frac{c_r z}{c_l}). \tag{5.38}$$

Now that H is determined, we return to (5.35) and use (5.38).

$$F(-c_r t) = H(c_l t) - G(c_r t) = (\frac{c_l - c_r}{c_l + c_r}) G(c_r t).$$

Thus

$$F(z) = (\frac{c_l - c_r}{c_l + c_r}) G(-z). \qquad (5.39)$$

With both H and F determined in terms of G, we have solved the problem. Set

$$R = \frac{c_l - c_r}{c_l + c_r} \quad \text{and} \quad T = \frac{2c_l}{c_r + c_l}. \qquad (5.40)$$

R is called the reflection coefficient and T is called the transmission coefficient. The solution can be expressed entirely in terms of the incident wave form G:

$$u(x, t) = \begin{cases} G(x + c_r t) + RG(c_r t - x) & \text{for } x > 0 \\ TG(\frac{c_r}{c_l}(x + c_l t)) & \text{for } x < 0 \end{cases}. \qquad (5.41)$$

5.4.3 Inhomogeneous problems

Finally we solve the problem where a source is placed on the boundary. We consider

$$u_{tt} - c^2 u_{xx} = 0 \qquad \text{for } x > 0, \qquad (5.42)$$

$$u(0, t) = h(t) \qquad \text{for } t \geq 0,$$

$$u(x, 0) = u_t(x, 0) = 0.$$

$h(t)$ is a given function which prescribes the motion of the left-hand end point. We assume that $h(t) = 0$ for $t \leq 0$. Since the initial displacement and velocity are both zero, we expect the solution to consist of a single wave form $F(x - ct)$ which moves down the string to the right. To satisfy the boundary condition, we set $F(-ct) = h(t)$ which implies that

$$F(z) = \begin{cases} h(\frac{-z}{c}), & z < 0 \\ 0, & z \geq 0 \end{cases}.$$

Hence our solution is

$$u(x, t) = F(x - ct) = h(t - \frac{x}{c}).$$ (5.43)

We note that $u(x, t) = 0$ for $x \geq ct$, and that u is nonzero only in the reflection zone $\mathcal{R}$, that is, for $\{0 \leq x \leq ct, t \geq 0\}$. Furthermore u is constant on the lines $x - ct = $ constant.

A similar method of solution works for the boundary condition $u_x(0, t) = h(t)$.

One can also use the principle of Duhamel to obtain an integral formula for the solution of the inhomogeneous half-line problem

$$u_{tt} - c^2 u_{xx} = q(x, t) \qquad \text{for } x, t > 0,$$ (5.44)

$$u(0, t) = 0, \qquad u(x, 0) = u_t(x, 0) = 0.$$

It takes the form

$$u(x, t) = \int_0^t z(x, t, s) ds$$ (5.45)

where now z solves the homogeneous boundary value problem (5.30), (5.31) with initial condition $z(x, t, t) = 0$, $z_t(x, t, t) = q(x, t)$.

Exercises 5.4

1.   Consider the IBVP (5.30), (5.31), and $g(x) \equiv 0$. In this case the formula (5.32) becomes

$$u(x, t) = \frac{1}{2}[f(x + ct) + f(x - ct) - f(ct - x)].$$

Let the function f be

$$f(x) = \begin{cases} 0 & 0 \leq x < 2 \\ 1 & 2 \leq x < 4 . \\ 0 & x > 4 \end{cases}$$

The initial data has singularities at $x = 2$ and $x = 4$.

(a)   In the quarter-plane $\{(x, t) : x > 0, t > 0\}$, sketch the characteristics emanating from $x = 2$ and from $x = 4$. At what times do the leftward leaning characteristics reach the boundary $x = 0$? Draw rightward leaning characteristics from these points on the t axis. These are the reflected characteristics.

(b) The collection of characteristics emanating from $x = 2$, $x = 4$, and the reflected characteristics emanating from the t axis divide the quarter-plane into nine regions. Use the formula above to find the values of the (weak) solution u in each of these nine regions. When using using this formula, keep in mind that you are assuming that $f(x) = 0$ when $x < 0$.

(c) If you are an observer standing at $x = 6$, what motion of the string do you see as t increases?

2. (a) Derive the formulas for the solution of the half-line problem with the Neumann boundary condition. You can imitate the argument used to derive the formulas (5.32) and (5.33) for the Dirichlet condition.

(b) Take $g = 0$ and f as given in exercise 1. Sketch the characteristics. Use the formula derived in part (a) to determine the solution in each of the nine regions. As an observer at $x = 6$, what do you see?

3. Let the incident wave form be

$$G(x) = (x - 5)(x - 6)e^{-(x-5)^2}.$$

G is not symmetric, so that the leading edge is different from the trailing edge. Write an mfile bigg.m for $G(x)$.

(a) Assume $c = 1$. Plot the solution of (5.30) with $u(0, t) = 0$,

$$u(x, t) = G(x + t) - G(t - x),$$

for $t = 2, 4, 6, 8$. Put all the plots on one graph in different colors. Describe how the shape of the incident wave is changed by the reflection. How long does it take for the reflected wave to leave the boundary completely ?

(b) Same questions for the solution of (5.30), with the Neumann condition,
$$u(x, t) = G(x + t) + G(t - x).$$

(c) The program film_rfl makes a 40-frame movie of the reflection with Dirichlet or Neumann boundary conditions of this same incident wave form. After running the program, you can see the movie with the matlab command movie(Reflect, 1, 6). The frames of the movie are loaded one at a time (this appears as a slow motion version of the movie). Then the movie will run once at the rate of 6 frames per second. If you want to view the film twice, use movie(Reflect, 2,6).

4. (a) Make a calculation similar to that made in Section 5.3 to show that the energy

$$e(t) = \frac{1}{2} \int_0^\infty [u_t^2 + c^2 u_x^2] dx$$

is conserved for solutions of (5.30), (5.31), and of (5.30), with the Neumann boundary condition.

(b) What energy quantity is conserved by solutions of (5.30) with the Robin condition?

5. Let $u(x, t) = G(x + t)$ be a wave incident from the right on the boundary $x = 0$ where we impose the Robin condition $u_x(0, t) = hu(0, t)$. To solve this problem use the same idea we saw in Chapter 4. Let $v = u_x - hu$. Then we want v to solve (5.30), (5.31).

(a) What is the incident wave corresponding to v?

(b) Solve the problem for v.

(c) Recover u from v by the formula

$$u(x, t) = - \int_x^\infty e^{h(x-y)} v(y, t) dy.$$

(d) Show that for $h = 0$ the solution u solves (5.30) with the Neumann condition.

(e) Show that as $h \to \infty$, the solution u converges to the solution of (5.30) with the Dirichlet condition (5.31).

6. Program robin computes the solution of (5.30) with $c = 1$ for the Robin boundary condition as found in exercise 5). The incident wave form G is that of exercise 3 which you have written in the mfile bigg.m.

(a) Run with $h = .1, .5, 1, 2, 5$, and times $t = 3, 5, 7, 9$. How does the reflected wave change as h increases? What is the limiting behavior as $h \to 0$ and as $h \to \infty$?

(b) Run with $h = -.1$. What happens to the reflected wave now?

(c) Now look at your calculated solution of exercise 5 with $h < 0$. What happens to $u(0, t)$ as $t \to \infty$?

7. We consider the transmission problem of the two strings joined together at $x = 0$. Suppose that you are standing at a point $x > 0$, and know the speed

c_r. Furthermore suppose that, by sending certain waveforms G toward the origin, you can measure the reflection coefficient R. Can you determine the speed c_l, and hence the density ρ_l ? What would the formula be?

The program trans computes solutions to two transmission problems, as described in Section 5.4. In this program the speed c_r is set equal to 1 and the speed c_l is an input parameter. In the first problem there is a wave $G(x + t)$ incident from the right, where G is the wave form of exercises 3 and 6. It should be written in the mfile bigg.m. Program trans takes the function $G(x)$ and normalizes it to have maximum height 1. Snapshots of the solution at times t_1, t_2 are written into the vectors snap1, snap2. As usual snap0 is the snapshot at $t = 0$.

In the second problem the incident wave is sinusoidal. This problem is discussed, along with program trans, in exercise 9.

8. Run program trans with the first choice of data and with values $c_l = .1, .5, 1, 1.5, 2.0$, and times $t = 5, 9$.

    (a) Measure the amplitude (height) of the reflected and transmitted wave from the plots. Since the incident wave has height 1, the amplitudes of the reflected and transmitted wave are the reflection and transmission coefficients. Compare your measurements with the values predicted by the formula (5.40).

    (b) What happens when $c_l = 0$? What kind of reflection is occurring? Explain what is happening in physical terms ($c_l = 0$ means $\rho_l = \infty$).

    (c) What happens when c_l is very large? Same questions as in part (b).

9. Let the incident wave from the right be sinuoidal, $u(x, t) = \sin(\omega t) \sin(k_r x)$. The wave number k_r and the angular frequency ω are given. They are related by $k_r = \omega / c_r$. The solution consists of the incident wave for $x < 0$, and a transmitted wave, with the same angular frequency ω, for $x > 0$. Thus u has the form

$$u(x, t) = \begin{cases} \sin(\omega t) \sin(k_r x) & x > 0 \\ A \sin(\omega t) \sin(k_l x) & x < 0 \end{cases}.$$

k_l is determined by the relation $k_l = \omega / c_l$.

    (a) By imposing the condition that u_x must be continuous at $x = 0$, determine the constant A. It will be independent of ω.

    (b) The spatial factor of the solution is given by

$$v(x) = \begin{cases} \sin(k_r x), & x > 0 \\ A \sin(k_l x), & x < 0 \end{cases}.$$

Program trans with the second choice of data calculates the values of A and k_l and plots $v(x)$ on $[-10, 10]$. At run time enter c_l as before and in addition frequency ω. Now run program trans with $\omega = 2$, and $c_l = .5$ and $c_l = 2$. Plot the spatial factor of the solution. Describe what happens as c_l changes. What happens as c_l tends to zero? What happens as c_l becomes much larger than 1 ?

10. Use (5.43) to find the solution u of the inhomogeneous IBVP

$$u_{tt} - c^2 u_{xx} = 0 \qquad x > 0, \qquad u(x, 0) = u_t(x, 0) = 0,$$

$$u(0, t) = te^{-t} \qquad \text{for } t \geq 0.$$

(a) Find max $h(t)$.

(b) At what time t does the top of the hump reach $x = 4$?

11. Consider the IBVP

$$u_{tt} - c^2 u_{xx} = 0 \qquad x > 0, \qquad u(x, 0) = u_t(x, 0) = 0,$$

$$u_x(0, t) = h(t).$$

(a) Again look for the solution in the form of a wave moving to the right down the string, $u(x, t) = F(x - ct)$. Impose the boundary condition, deriving a simple ODE that F must satisfy. Solve the ODE for F as an integral of h.

(b) Suppose that $h(t) = t/(1 + t^2)$ for $t > 0$ and $h(t) = 0$ for $t < 0$. Find $u(x, t)$. What is $\lim_{t \to \infty} u(0, t)$?

5.5 Boundary value problems on a finite interval

5.5.1 A geometric construction

The first and simplest boundary value problem we shall consider is that of transverse motion of the vibrating string with both ends fixed.

$$u_{tt} - c^2 u_{xx} = 0 \qquad \text{in } 0 < x < L, t \in R, \qquad (5.46)$$

$$u(0, t) = u(L, t) = 0 \qquad \text{for } t \in R, \qquad (5.47)$$

$$u(x, 0) = f(x), \qquad u_t(x, 0) = g(x) \qquad \text{for } 0 < x < L. \qquad (5.48)$$

To develope an intuitive feeling for properties of the solutions of this IBVP, we consider a special solution u with initial data

$$u(x, 0) = G(x), \qquad u_t(x, 0) = cG'(x),$$

where G is a hump contained in the subinterval $[a, b] \subset [0, L]$. The initial velocity $g(x) = cG'(x)$ is chosen so that for $t \le a/c, u(x, t) = G(x+ct)$ is a hump moving to the left. The characteristics and profiles for this solution are displayed in Figures 5.6a and 5.6b. We know from Section 5.4.1 that when the hump reaches the left boundary at $x = 0$, it will be reflected, and for $a/c \le t \le b/c$, it will have the form

$$u(x, t) = G(x + ct) - G(ct - x).$$

When $t = b/c, ct + x \ge b$, so the first term vanishes, and $u(x, t) = -G(ct - x)$. This is the same hump, upside down, reversed in left-right symmetry, and moving to the right. When this hump reaches the right boundary $x = L$, it will be reflected again, this time back to its original shape, moving to the left. At time $t = 2L/c$, the solution looks exactly like it did at time $t = 0$. Thus the motion is periodic with period $T = 2L/c$.

This graphical construction can be extended to more general initial data, but becomes very complicated. However, it will always be the case that the solution has period $T = 2L/c$ in t. This suggests that the solution may have an expansion in terms of vibrating modes with periods that divide T.

5.5.2 Modes of vibration

We return to the technique of separation of variables. We look for simple building-block solutions in the form of products:

$$u(x, t) = \varphi(x)\psi(t).$$

Substitution in (5.46) yields

$$\frac{\psi''}{c^2\psi} = \frac{\varphi''}{\varphi} = -\lambda$$

and the ODE problems

$$\psi'' + c^2\lambda\psi = 0, \tag{5.49}$$

$$-\varphi'' = \lambda\varphi, \qquad \varphi(0) = \varphi(L) = 0. \tag{5.50}$$

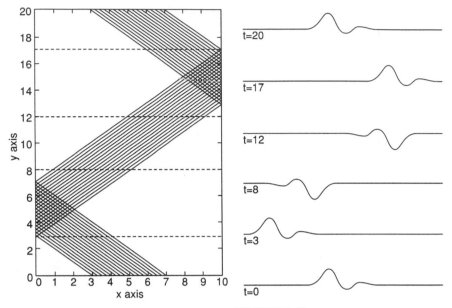

FIGURE 5.6a

Characteristics emanating from the subin-
terval [3, 7]. *Here* $c = 1$. *Solution u is*
constant along characteristics where they
are not crosshatched.

FIGURE 5.6b

Snapshots of the solution at times $t =$
0, 3, 8, 12, 17, 20. *Times correspond*
to dotted lines on Figure 5.6a.

Equation (5.50) defines the eigenvalue problem that we studied in Chapter 4 for
the heat equation. The eigenvalues are

$$\lambda_n = (\frac{n\pi}{L})^2, \qquad n = 1, 2, \ldots$$

and the eigenfunctions are

$$\varphi_n(x) = \sin(\frac{n\pi x}{L}).$$

Consequently, the solutions of (5.49) are

$$\psi_n(t) = A_n \cos(\omega_n t) + B_n \sin(\omega_n t).$$

$\omega_n = c\sqrt{\lambda_n} = c(n\pi/L)$ is the angular frequency. Consequently we look for a
solution of (5.46) - (5.48) in the form of a series (again an eigenfunction expansion)

$$u(x,t) = \sum_1^\infty \psi_n(t)\varphi_n(x) = \sum_1^\infty [A_n \cos(\omega_n t) + B_n \sin(\omega_n t)] \sin(\frac{n\pi x}{L}). \quad (5.51)$$

Now we have two sequences of coefficients A_n and B_n to be determined by the two initial conditions (5.48):

$$f(x) = u(x,0) = \sum_1^\infty A_n \varphi_n(x),$$

and

$$g(x) = u_t(x,0) = \sum_1^\infty \omega_n B_n \varphi_n(x).$$

Hence as in Chapter 4,

$$A_n = \frac{2}{L} \int_0^L f(x) \sin(\frac{n\pi x}{L}) dx \quad (5.52)$$

and

$$\omega_n B_n = \frac{2}{L} \int_0^L g(x) \sin(\frac{n\pi x}{L}) dx. \quad (5.53)$$

The frequency in cycles per second $v_n = \omega_n/2\pi = c(n/2L)$. In acoustics $v_1 = c/2L$ is called the fundamental frequency of the string, and $\sin(\pi x/L)$ is called the fundamental mode. The fundamental period of the motion $T = 2L/c$. $v_2 = 2v_1 = c/L$ is the frequency of the first overtone. Since $v_2 = 2v_1$, this tone sounds one octave higher than the fundamental tone. The corresponding mode is $\sin(2\pi x/L)$ which now has a node (point where the string does not move) in the interior of the interval at $L/2$. The next frequency $v_3 = 3c/2L$ sounds a fifth above v_2 and the corresponding mode $\sin(3\pi x/L)$ has two interior nodes at $x = L/3$ and $2L/3$. $v_4 = 4c/2L = 2v_2 = 4v_1$ sounds two octaves above the fundamental frequency v_1, and the mode has three interior nodes (see Figure 5.7).

How well does the series representation of the solution converge? We notice that in (5.51) we do not have the decaying exponentials $\exp(-\lambda_n kt)$ that made the series for solutions of the heat equation converge so rapidly. The solutions of the wave equation do not become smoother as time increases. If the initial data has a "corner," the solution will continue to have a "corner" at later times. A standard example of this kind of behavior is that of the string plucked at its midpoint. Take the initial data (5.48) to be $g \equiv 0$ and

$$f(x) = \begin{cases} x & \text{for } 0 < x < L/2 \\ L - x & \text{for } L/2 < x < L \end{cases}. \quad (5.54)$$

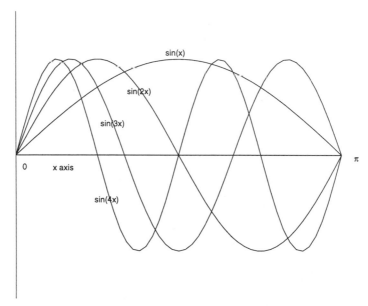

FIGURE 5.7
Fundamental modes of the vibrating string on $[0, \pi]$.

All the coefficients $B_n = 0$, and

$$A_n = \frac{2}{L} \int_0^{L/2} x \sin(\frac{n\pi x}{L})dx + \frac{2}{L} \int_{L/2}^{L} (L - x) \sin(\frac{n\pi x}{L})dx$$

$$= \frac{4L}{n^2\pi^2} \sin(\frac{n\pi}{2}), \qquad n = 1, 2, \ldots.$$

Note that $A_n = 0$ for n even. Thus the sound of the plucked string has none of the overtones $\nu_2, \nu_4, \ldots$.

5.5.3 Conservation of energy for the finite interval

We can get a better idea how well the series (5.51) converges by considering conservation of energy. The energy of a solution is defined as it as on the whole line as

$$e(t) = \frac{1}{2} \int_0^L [u_t^2(x, t) + c^2 u_x^2(x, t)]dx \qquad (5.55)$$

When we wish to show the dependence of the energy on the particular solution, we may write $e = e(t, u)$.

We assume that u is smooth enough so that we can differentiate under the integral sign. Then

$$\frac{de(t)}{dt} = \int_0^L [u_{tt}u_t + c^2 u_{xt}u_x]dx. \tag{5.56}$$

We integrate by parts in the second term to find

$$\int_0^L u_{xt}u_x dx = u_t u_x \Big|_0^L - \int_0^L u_t u_{xx} dx.$$

From the boundary condition $u(0, t) = u(L, t) = 0$, we deduce that $u_t(0, t) = u_t(L, t) = 0$ so that the boundary terms vanish. Thus (5.56) becomes

$$\frac{de(t)}{dt} = \int_0^L [u_{tt} - c^2 u_{xx}]u_t dx = 0.$$

so that $e(t)$ is conserved. Note that the boundary terms also vanish when u satisfies the boundary condtion $u_x(0, t) = u_x(L, t) = 0$, so that energy is also conserved in this case. This suggests that a good class of initial data for this problem consists of functions f, g with the following properties:

$$f \text{ is continuous}, \ f(0) = f(L) = 0; \tag{5.57}$$

f' exists except at a finite number of points and is piecewise continuous;

and

$$g \text{ is piecewise continuous.}$$

For such f and g, the energy integral (5.55) is well defined and the sums $\sum A_n^2 < \infty$ and $\sum B_n^2 < \infty$. Therefore the series converges at least in the mean square to $u(x, t)$ for each t. However the condition (5.57) ensures that the differentiated series for u_t also converges in the mean square, as does the series for u_x. In general the conditions (5.57) are not sufficient to make the series converge to a strict solution. In fact what we are getting is a weak solution, which has all the structure of a strict solution, but not the regularity. The solution for the plucked string (5.54) is an example of a weak solution.

We can summarize our results for the boundary value problem (5.46) - (5.48) in

Theorem 5.4
Suppose that the initial f, g satisfy the conditions (5.57). Then there exists a unique (weak) solution $u(x, t)$ of (5.46) - (5.48). The energy is conserved. u is given by the eigenfunction expansion (5.51).

This theorem says that (5.46) - (5.48) is a well-posed problem in the sense of the energy for initial data which satisfy conditions (5.57). In fact, if u and v are two such solutions, then $w = u - v$ is another solution whose energy is conserved. Thus

$$e(t, u - v) = e(t, w) = e(0, w) = e(0, u - v).$$

This equation says that if initially the energy of the difference $u - v$, $e(0, u-v) \leq \delta$, then $e(t, u - v) \leq \delta$ for all t.

5.5.4 Other boundary conditions

The boundary conditions

$$u_x(0, t) = u_x(L, t) = 0 \tag{5.58}$$

are best interpreted in the context of linear acoustics. Here we think of u as the velocity potential of a gas in a tube. Thus $u_x(x, t)$ is the particle velocity and (5.58) says that the tube is stopped at both ends because the particle velocity is zero at both ends. In the context of the vibrating string, these boundary conditions are said to represent the string with free ends. The solutions of (5.46), (5.48) together with (5.58) are found using the eigenfunctions for the eigenvalue problem

$$-\varphi'' = \lambda\varphi, \qquad \varphi'(0) = \varphi'(L) = 0.$$

We have seen before that the eigenfunctions are $\varphi_n(x) = \cos(n\pi x/L)$, with eigenvalues $\lambda_n = (n\pi/L)^2$, $n = 0, 1, \ldots$. The differential equations for the temporal factors are as before

$$\psi'' + c^2\lambda_n\psi = 0.$$

For $n = 1, 2, \ldots$, the general solution is

$$\psi_n(t) = A_n \cos(\omega_n t) + B_n \sin(\omega_n t).$$

Since $\lambda_0 = 0$, the solution for $n = 0$ is different,

$$\psi_0(t) = \frac{1}{2}(A_0 + B_0 t).$$

It follows that the solution of the wave equation with the Neumann boundary conditions (5.58) can be written

$$u(x, t) = \frac{1}{2}(A_0 + B_0 t) + \sum_{n=1}^{\infty}[A_n \cos(\omega_n t) + B_n \sin(\omega_n t)]\cos(\frac{n\pi x}{L})$$

with

$$A_n = \frac{2}{L}\int_0^L f(x)\cos(\frac{n\pi x}{L})dx, \qquad n = 0, 1, \ldots \qquad (5.59)$$

and

$$\omega_n B_n = \frac{2}{L}\int_0^L g(x)\cos(\frac{n\pi x}{L})dx \qquad n = 1, 2, \ldots, \qquad (5.60)$$

$$B_0 = \frac{2}{L}\int_0^L g(x)dx.$$

As we observed earlier, conservation of energy also holds for this boundary condition. Note that, even though the zeroth-order term in the solution grows with t, its energy is constant. We shall study other boundary conditions and their interpretations in the exercises.

5.5.5 Inhomogeneous equations

Consider now the inhomogeneous equation

$$u_{tt} - c^2 u_{xx} = q(x, t), \qquad (5.61)$$

$$u(x, 0) = f(x), \qquad u_t(x, 0) = g(x), \qquad (5.62)$$

+ homogeneous boundary conditions.

We impose any of the homogeneous symmetric boundary conditions studied in Chapter 4. Thus we can assume that there is a sequence of eigenfunctions $\varphi_n(x)$ which is complete. We shall assume that for these boundary conditions, all the eigenvalues $\lambda_n > 0$. A slight modification of the formulas is needed to treat the case of Neumann boundary conditions, where $\lambda_0 = 0$ is an eigenvalue.

As in Chapter 4, we assume that the solution of the problem (5.61), (5.62) can be expanded in terms of the eigenfunctions:

$$u(x, t) = \sum_1^{\infty} u_n(t)\varphi_n(x).$$

Substituting this expansion in the left side of (5.61), we find that

$$u_{tt} - c^2 u_{xx} = \sum_1^\infty [u_n''(t) + \omega_n^2 u_n(t)]\varphi_n(x).$$

We have used the fact that the eigenfunctions satisfy $-\varphi_n(x) = \lambda_n \varphi_n(x)$ and that the frequencies $\omega_n = c\sqrt{\lambda_n}$. We also assume that the right-hand side q can be expanded

$$q(x, t) = \sum_1^\infty q_n(t)\varphi_n(x),$$

where for each t

$$q_n(t) = \frac{\langle q(x, t), \varphi_n(x)\rangle}{\langle \varphi_n, \varphi_n\rangle}.$$

If we equate the coefficients in the expansions of the left and right sides of (5.61) we find the family of second-order ODE's

$$u_n''(t) + \omega_n^2 u_n(t) = q_n(t). \tag{5.63}$$

The initial conditions for these ODE's are found by expanding the initial data, f and g, in terms of the eigenfunctions. We require

$$\sum_1^\infty u_n(0)\varphi_n(x) = u(x, 0) = f(x) = \sum_1^\infty A_n \varphi_n(x)$$

and

$$\sum_1^\infty u_n'(0)\varphi_n(x) = u_t(x, 0) = g(x) = \sum_1^\infty \omega_n B_n \varphi_n(x),$$

where

$$A_n = \frac{\langle f, \varphi_n\rangle}{\langle \varphi_n, \varphi_n\rangle}, \qquad \omega_n B_n = \frac{\langle g, \varphi_n\rangle}{\langle \varphi_n, \varphi_n\rangle}.$$

Therefore the initial conditions are

$$u_n(0) = A_n, \qquad u_n'(0) = \omega_n B_n. \tag{5.64}$$

Now the solution of the ODE initial value problem

$$\psi''(t) + \omega^2 \psi(t) = g(t), \qquad \psi(0) = A, \qquad \psi'(0) = \omega B$$

is given by

$$\psi(t) = A \cos(\omega t) + B \sin(\omega t) + \frac{1}{\omega} \int_0^t \sin(\omega(t-s))g(s)ds.$$

Hence the solution of (5.61), (5.62) is

$$u(x, t) = \sum_1^\infty u_n(t)\varphi_n(x) \tag{5.65}$$

$$= \sum_1^\infty [A_n \cos(\omega_n t) + B_n \sin(\omega_n t) + \frac{1}{\omega_n} \int_0^t \sin(\omega_n(t-s))q_n(s)ds]\varphi_n(x).$$

As in Section 4.4 we see that the solution of (5.61) can be written $u = v + w$, where

$$v(x, t) = \sum_1^\infty [A_n \cos(\omega_n t) + B_n \sin(\omega_n t)]\varphi_n(x)$$

solves $v_{tt} - c^2 v_{xx} = 0$, with $v(x, 0) = f(x)$, $v_t(x, 0) = g(x)$, and

$$w(x, t) = \sum_1^\infty \frac{1}{\omega_n} \int_0^t \sin(\omega_n(t-s))q_n(s)ds\,\varphi_n(x)$$

solves $w_{tt} - c^2 w_{xx} = q$ with $w(x, 0) = w_t(x, 0) = 0$.

5.5.6 Boundary forcing and resonance

Finally, in this section we consider the case of forced vibrations of the string where the inhomogeneous source is placed on the boundary. We consider the IBVP

$$u_{tt} - c^2 u_{xx} = 0, \quad 0 < x < L, \tag{5.66}$$

$$u(0, t) = 0, \qquad u(L, t) = h(t),$$

$$u(x, 0) = f(x), \qquad u_t(x, 0) = g(x), \qquad 0 < x < L.$$

The function $h(t)$ represents the forcing on the boundary. First we convert this problem to one of the form (5.61). We let $v(x, t) = u(x, t) - h(t)(x/L)$. Then v satisfies the inhomogeneous problem

$$v_{tt} - c^2 v_{xx} = -h''(t)(x/L),$$

$$v(0, t) = v(L, t) = 0,$$

$$v(x, 0) = f(x) - h(0)(x/L), \qquad v_t(x, 0) = g(x) - h'(0)(x/L).$$

By subtracting off $h(t)(x/L)$, we have made the boundary conditions homogeneous at the expense of adding a forcing term to the right side of the equation. However the solution to this problem is known from the previous section. The eigenfunctions are $\varphi_n(x) = \sin(n\pi x/L)$ and the frequencies are $\omega_n = cn\pi/L$. With $q(x, t) = -h''(t)(x/L)$, use (5.65). Then

$$v(x, t) = \sum_1^\infty [A_n \cos(\omega_n t) + B_n \sin(\omega_n t)]\varphi_n(x) + \sum_1^\infty R_n(t)\varphi_n(x), \qquad (5.67)$$

where

$$A_n = \frac{2}{L} \int_0^L [f(x) - h(0)(x/L)]\varphi_n(x)dx,$$

$$\omega_n B_n = \frac{2}{L} \int_0^L [g(x) - h'(0)(x/L)]\varphi_n(x)dx,$$

$$R_n(t) = -\frac{r_n}{\omega_n} \int_0^t \sin(\omega_n(t - s))h''(s)ds, \qquad (5.68)$$

and

$$r_n = \frac{2}{L} \int_0^L (x/L)\varphi_n(x)dx = \frac{2}{n\pi L}[1 - (-1)^n].$$

Now we specifically choose $h(t) = \sin(\omega t)$. Are there certain values of the forcing frequencing ω which produce a stronger response from the string than others? These values are "resonant" values. $h''(t) = -\omega^2 \sin(\omega t)$, so that the answer to this question can be found by substituting this expression for h'' into (5.68) and evaluating

$$R_n(t) = \frac{\omega^2 r_n}{\omega_n} \int_0^t \sin(\omega_n(t-s))\sin(\omega s)ds.$$

There are two cases: If $\omega \neq \omega_n$, then two integrations by parts shows that

$$R_n(t) = (\frac{r_n\omega^2}{\omega_n}) \frac{\omega_n \sin(\omega t) - \omega \sin(\omega_n t)}{\omega_n^2 - \omega^2}. \tag{5.69}$$

In this case it is clear that $R_n(t)$ is a bounded oscillation. However, if for some index m, $\omega = \omega_m$, then

$$R_m(t) = \frac{r_m}{2}[\sin(\omega_m t) - (\omega_m t)\cos(\omega_m t)], \tag{5.70}$$

an oscillation which grows linearly with time. Note however that $r_n = 0$ for n even, so that resonance occurs when $\omega = \omega_m$ for some m odd.

Exercises 5.5

1. Solve the IBVP (5.46) - (5.48) with initial data

    $$f(x) = 7\sin(2\pi x/L) - 2\sin(5\pi x/L), \qquad g(x) = -4\sin(3\pi x/L).$$

    Think!! You will not need the full expansion of the solution. What frequencies are present in the solution? What is the period of the motion?

2. Solve the IBVP (5.46), (5.48) with the homogeneous Neumann boundary conditions $u_x(0, t) = u_x(L, t) = 0$. Take $f(x) = 0$ and $g(x) = x$. Use formulas (5.59), (5.60). What frequencies ω_n are present in the solution?

Program wave1 uses the finite difference scheme of the next section to integrate the telegraph equation (with damping)

$$u_{tt} + 2du_t - u_{xx} + ku = 0$$

on the interval $[0, 10]$ with $u = 0$ at $x = 0$ and $x = 10$. The spatial step $\Delta x = .025$ and time step $\Delta t = .025$ are set in the program. Input parameters are the damping constant d and the spring constant k. Two choices of the initial conditions are built into the program. They are the data for the "plucked string," and for the "hammer blow. "

(1) $f(x) = 0$,

$$g(x) = \begin{cases} 0, & 0 \le x < 2.5 \\ -1, & -2.5 \le x \le 7.5 \\ 0, & 7.5 < x \le 10 \end{cases}.$$

(2) $f(x) = \begin{cases} x/5 & 0 \le x \le 5 \\ 1 - (x-5)/5, & 5 \le x \le 10, \end{cases}$ $g = 0.$

When you run program wave1 and select data choice 3, it looks for initial data f and g in function mfiles f.m and g.m. You must provide these mfiles. More information about this program can be found by first invoking MATLAB and then typing help wave1 .

3.   Consider the IBVP (5.46) - (5.48) with initial data $f(x) = \sin(2\pi x/10)$ and $g(x) = 0$.

   (a)   What is the period of the motion?

   (b)   Write mfiles f.m and g.m for f and g which are array-smart. For g use

```
function y = g(x)
y = zeros(size(x));
```

   Now run program wave1 with data choice 3, $d = k = 0$. The time step for wave1 is $\Delta t = .025$. Choose the number of time steps n1, n2, n3, n4 so that snap1 is 1/8 period, snap2 is 1/4 period, etc. Make a number of plots so that you get a complete picture of the motion. Where are the node(s)?

   (c)   Do the same for initial data of exercise 1.

4.   Solve the equation of the vibrating string (with $c = 1$) on the interval $[0, L]$ with fixed ends at $x = 0$ and L, and with initial data

$$u(x, 0) = 0, \qquad u_t(x, 0) = \begin{cases} 0, & 0 \le x < L/4 \\ -U, & L/4 \le x \le 3L/4 \\ 0, & 3L/4 < x \le L \end{cases}.$$

This solution can be interpreted as the motion of the string after it is hit with a hammer in the center of the string.

5.   To see what the profiles of the solution of exercise 4 look like, run program wave1 with data choice (1). The solution is computed on $[0, 10]$ so that

the $L = 10$ in exercise 1. Notice that the initial data has singularities at $x = 2.5$ and at $x = 7.5$. We expect to see singularities propagated along the characteristics emanating from $x = 2.5$ and from $x = 7.5$. The singularities will show up in the profiles of the solution as "corners," that is, places where the first derivative has a jump.

(a) Run with $d = k = 0$, and n1 = n2 = n3 = n4 = 40. How fast do the "corners" move? What happens at time $t = 2.5$ (100 time steps)? How many corners are there for $2.5 < t < 5$? Draw a sketch in the x, t plane which shows the characteristics emanating from $x = 2.5$ and $x = 7.5$, and their reflections from the left and right boundaries. Use this sketch to explain the motion of the "corners" in the profiles.

(b) What is the period of the motion? This is the same as asking when the string will return to the initial conditions $f = 0$ and g as given.

6. Consider the IBVP for the telegraph equation (displayed above) with boundary conditions $u(0, t) = u(L, t) = 0$, and initial conditions $u(x, 0) = f(x), u_t(x, 0) = g(x)$.

(a) Separate variables. Find the building-block solutions in the form $u(x, t) = \varphi(x)\psi(t)$, where $-\varphi''(x) = \lambda\varphi(x)$. Show that ψ must satisfy the ODE

$$\psi'' + 2d\psi' + (k + \lambda)\psi = 0.$$

(b) Find the general solution of this ODE in the following three cases:

case I	$d^2 < k + \lambda$	underdamped
case II	$d^2 = k + \lambda$	critically damped
case III	$d^2 > k + \lambda$	over damped.

(c) Taking $\lambda = \lambda_n = (n\pi/L)^2$ and case I for all n, show that the solution of the PDE is

$$u(x, t) = \sum_1^\infty [A_n \cos(\omega_n t) + B_n \sin(\omega_n t)]e^{-dt} \sin(\frac{n\pi x}{L}).$$

What are the frequencies ω_n? What is the time behavior of each mode? Will the motion of the string be periodic when there is no damping, but $k > 0$?

(d) Now assume that $k + \lambda_1 < d^2 < k + \lambda_2 < \dots$ How does the time behavior of the first mode differ from the time behavior of the modes with $n \geq 2$?

7. Now run program wave1 with data choice (2) (plucked string). Again the
    initial data is singular, this time at $x = 5$.

    (a) Run with $d = k = 0$, n1 $=$ n2 $=$ n3 $=$ n4 $= 40$. The solution has
        corners (i.e., points where the derivative has a jump). How fast do
        the singularities (corners) move horizontally? Make a sketch in the
        rectangle $\{(x, t) : 0 \le x \le 10, t \ge 0\}$ showing the path in the
        (x, t) plane followed by the corners (singularities). When do the
        singularities reflect from the boundaries? What is the period of the
        motion?

    (b) Run with $d = .2, k = 0$, n1 $=$ n2 $=$ n3 $=$ n4 $= 40$. Compare the profiles
        of this run with those of part a) at $t = 1, 2, 3, 4$. What is the effect
        of the damping? How fast do the singularities move horizontally ?
        Now run with n1 $=$ n2 $=$ n3 $=$ n4$= 100$. What happens to the motion
        as $t \to \infty$?

    (c) Run with $d = 0, k = .2$ and n1 $=$ n2 $=$ n3 $=$ n4 $= 40$. Compare the
        profiles of this run with those of part a) at times $t = 1, 2, 3, 4$. How
        fast do the singularities move horizontally ? How do the profiles
        differ? What is the effect of the added term ku in the equation? Is
        the motion periodic? You can check by looking at the frequencies
        ω_n that you computed in exercise 3.

    (d) Run with $d = .2, k = .2$. Use n1 $=$ n2 $=$ n3 $=$ n4 $= 40$. Compare
        with the graphs of part (a). How do you explain the difference in the
        graphs? Notice how fast the singularities move.

8. Write the function mfile f.m for $f(x) = x(x - 5)(x - 10)/50$, and the
    mfile g.m for $g(x) = 0$. Run program wave1 with data choice (3).

    (a) Set $d = k = 0$. Here there are no corners, the initial data has no
        singularities. What is the period of the motion for this data? Why?

    (b) Now run with $d = .2, k = 0$. Describe what you see, comparing
        with the profiles of part (a). Is the motion periodic?

    (c) Run with $d = 0, k = .5$. What is the difference in profiles? Is the
        motion periodic?

9. Now change the file f.m to be $f(x) = \exp(-3(x - 5)^2)$, and keep the
    mfile g.m to be $g(x) = 0$. Run program wave1 with data choice (3)

    (a) First with $d = k = 0$, n1 $=$ n2 $=$ n3 $=$ n4 $= 40$. The initial data is zero
        outside the interval $[4, 6]$.

    (b) Now run with $d = 0$ and $k = .1, .2, .5, 1$, same choices of n1, n2, n3
        and n4. How is the spreading bump deformed with the presence of
        the $k > 0$? We shall see in Chapter 7 that this is caused by dispersion.
        How fast does the leading edge of the bumps travel? What about the
        peaks?

(c)   Run with $n1 = n2 = n3 = n4 = 100$ and with $d = 0, k = .5$. Label the reflected waves on the plots. How do they differ from the reflected waves when $k = 0$? To get a detailed picture of the reflection, you can use $n1 = 180, n2 = n3 = n4 = 10$.

10.   Consider the vibrating string equation $u_{tt} - c^2 u_{xx} = 0$ with the boundary conditions $u(0, t) = 0$, and $u_x(L, t) = 0$. Show that energy is conserved for this boundary condition.

11.   Continue with the IBVP of exercise 10.

(a)   Find the eigenfunctions and eigenvalues for this boundary condition.

(b)   Solve the IBVP with these boundary conditions and the initial data $f(x) = x$ for $0 \leq x \leq L/2$, $f(x) = L/2$ for $L/2 \leq x \leq L$, and $g(x) = 0$.

12.   Solve the inhomogeneous equation (5.61), (5.62) with boundary conditions $u(0, t) = u(L, t) = 0$.

(a)   Set $f(x) = g(x) = 0$ and

$$q(x, t) = e^t \sin(2\pi x/L).$$

You will not need a full eigenfunction expansion.

(b)   Again set $f(x) = g(x) = 0$ and now set

$$q(x, t) = \sin(2\pi x/L) \text{ for } 0 \leq t \leq T, \quad q(x, t) = 0 \text{ for } t > T.$$

Solve the equation. What happens to the solution as $t \to \infty$? Does it damp out, or does it continue to oscillate?

13.   Suppose that we have a string which is fixed at $-L$ and L, and which is composed of two different materials joined at $x = 0$. We suppose a constant tension T_0 but differing densities ρ_l and ρ_r. The speeds of propagation on each half are $c_l = \sqrt{T_0/\rho_l}$ and $c_r = \sqrt{T_0/\rho_r}$. A solution u and its derivative u_x are both continuous at the joint.

(a)   Formulate this situation as an IBVP.

(b)   Look for the building-block solutions in the form $u_n(x, t) = \psi_n(t)\varphi_n(x)$. What is the eigenvalue problem?

(c)   Find the eigenfunctions and eigenvalues.

14.   Calculate the conserved energy quantity $e(t)$ for the solution of

$$u_{tt} - c^2 u_{xx} = 0$$

with boundary conditions

$$u_x(0, t) = hu(0, t), \qquad u_x(L, t) = 0, \quad t \in R.$$

Follow the derivation of (5.56), but pay close attention to the boundary terms when you integrate by parts in x.

5.6   Numerical methods

Just as for the heat equation, we introduce a mesh or grid of points in the x, t plane:

$$x_j = j\Delta x \qquad \text{and} \qquad t_n = n\Delta t.$$

We shall compute values $u_{j,n}$ which will approximate the true value of the solution $u(x_j, t_n)$. We replace u_{tt} by the centered difference in the t direction, centered at (x_j, t_n):

$$u_{tt}(x_j, t_n) = \frac{u_{j,n+1} - 2u_{j,n} + u_{j,n-1}}{(\Delta t)^2} + O((\Delta t)^2).$$

Likewise

$$u_{xx}(x_j, t_n) = \frac{u_{j+1,n} - 2u_{j,n} + u_{j-1,n}}{(\Delta x)^2} + O((\Delta x)^2).$$

Substituting these expressions for the derivatives into the wave equation, we arrive at

$$\frac{u_{j,n+1} - 2u_{j,n} + u_{j,n-1}}{(\Delta t)^2} = c^2 \frac{u_{j+1,n} - 2u_{j,n} + u_{j,n-1}}{(\Delta x)^2}.$$

The truncation error is $O((\Delta t)^2) + O((\Delta x)^2)$. Solving for $u_{j,n+1}$, we get

$$u_{j,n+1} = 2(1 - s)u_{j,n} + \rho[u_{j+1,n} + u_{j-1,n}] - u_{j,n-1} \qquad (5.71)$$

where $s = c^2(\Delta t/\Delta x)^2$. The computational diagram is shown in Figure 5.8. Note that when $s = 1$, the scheme becomes

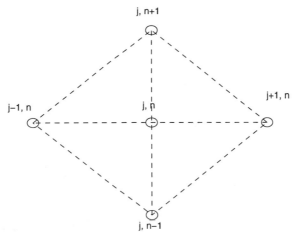

FIGURE 5.8
Computational diagram for (5.71).

$$u_{j,n+1} = u_{j+1,n} + u_{j-1,n} - u_{j,n-1},$$

which is exactly the formula of exercise 8, Section 5.3. Thus this numerical scheme is exact when $c = \Delta x / \Delta t$.

To compute $u_{j,n+1}$, we see that we need to have already computed all the values $u_{j,n}$ and $u_{j,n-1}$. Thus to get started, we need the values computed on the first two lines, $t = 0$ and $t = t_1 = \Delta t$. We know that

$$u_{j,0} = f(x_j) = f(j\Delta x).$$

What about the values on the second line? By Taylor's Theorem we know that

$$u(x, \Delta t) = u(x, 0) + u_t(x, 0)\Delta t + O((\Delta t)^2). \tag{5.72}$$

Thus we could use the formula

$$u_{j,1} = f(x_j) + g(x_j)\Delta t = f(j\Delta x) + g(j\Delta x)\Delta t$$

to get the data on the second line. However a more accurate formula is possible if we keep an additional term in (5.72):

$$u(x, \Delta t) = u(x, 0) + u_t(x, 0)\Delta t + (1/2)u_{tt}(x, 0)\Delta t^2 + O((\Delta t)^3). \tag{5.73}$$

Now using the fact that $u_{tt} = c^2 u_{xx}$, we find that

$$u(x, \Delta t) = u(x, 0) + u_t(x, 0)\Delta t + (c^2/2)u_{xx}(x, 0)(\Delta t)^2 + O((\Delta t)^3)$$

$$= f(x) + g(x)\Delta t + (c^2/2)f''(x)(\Delta t)^2 + O((\Delta t)^3).$$

We can approximate the second derivative of f by a centered difference

$$f''(x_j) = \frac{f(x_{j+1}) - 2f(x_j) + f(x_{j-1})}{(\Delta x)^2} + O((\Delta x)^2).$$

Thus our formula for the second line is

$$u_{j,1} = f(x_j) + g(x_j)\Delta t + (s/2)[f(x_{j+1}) - 2f(x_j) + f(x_{j-1})]. \qquad (5.74)$$

Recall that in Chapter 2 when we considered numerical methods for the equation

$$u_t + cu_x = 0, \qquad u(x, 0) = f(x),$$

we had to restrict the grid to satisfy the CFL condition:

$$\frac{c}{(\Delta x/\Delta t)} = \frac{c\Delta t}{\Delta x} \le 1.$$

This inequality expresses the idea that the domain of dependence of the numerical scheme must include the domain of dependence of the exact solution. Specifically, suppose that $u_{j,n+1}$ is determined by the values $u_{i,n}$ at points x_i on the nth line. Then the interval spanned by the x_i must include those points (x, t_n) on the nth line which determine the exact solution at the point (x_j, t_{n+1}) on the $(n+1)$st line.

According to formula (5.71), to compute $u_{j,n+1}$, we need the values of $u_{i,n}$ on the nth line at $x_i = x_{j-1} = x_j - \Delta x$ and at $x_i = x_j + \Delta x$. Hence to satisfy the CFL condition, the interval $[x_j - \Delta x, x_j + \Delta x]$ must contain the domain of dependence of the exact solution at (x_j, t_{n+1}). This is the interval $[x_j - c\Delta t, x_j + c\Delta t]$. For this to be true we again require that the CFL condition be satisfied,

$$c\Delta t \le \Delta x.$$

Note that then $s = (c\Delta t/\Delta x)^2 \le 1$, which makes the coefficient of $u_{j,n}$ nonnegative. If $s > 1$, large unstable oscillations can develop.

The method (5.71) is an explicit method. An implicit method can be developed by centering the difference approximation for u_{xx} at (x_j, t_{n+1}), that is,

$$u_{xx} \approx \frac{u_{j+1,n+1} - 2u_{j,n+1} + u_{j-1,n+1}}{(\Delta x)^2}.$$

The resulting set of difference equations must be solved at each time step. This method is quite stable. One can also take an average of the explicit and implicit methods as in Crank-Nicholson.

For a detailed discussion of these methods see [RM] or [St].

5.7 A nonlinear wave equation

In this section we add a nonlinear term to the wave equation and find travelling waves. The equation we study (we set $c = 1$) is called the nonlinear Klein-Gordon equation.

$$u_{tt} - u_{xx} + u^3 = 0 \qquad \text{for } x, t \in R. \tag{5.75}$$

It comes from quantum field theory and is one of the simplest models for nonlinear wave interaction.

Note that if the amplitude of a solution is small, $|u(x, t)| \leq \varepsilon < 1$, then the nonlinear term $|u^3(x, t)| \leq \varepsilon^3 \ll \varepsilon$. Thus for small amplitude solutions, the nonlinear term does not play a large role.

There is a conserved energy quantity associated with this equation. Multiply (5.75) by u_t and integrate in x, assuming that all of the integrals converge. The result is that the energy quantity

$$e(t) = \frac{1}{2} \int_R [u_t^2(x, t) + u_x^2(x, t)]dx + \frac{1}{4} \int_R u^4(x, t)dx \tag{5.76}$$

is conserved.

Now we look for special solutions which are travelling waves in the form

$$u(x, t) = \varphi(x - \theta t), \tag{5.77}$$

where φ is the wave profile, and θ is the speed of propagation. We substitute this form into (5.75) and find that φ must satisfy a nonlinear ODE

$$\varphi'' + \frac{1}{\theta^2 - 1}\varphi^3 = 0. \tag{5.78}$$

For the moment, let us assume that $\theta > 1$, and write (5.78) as a system in the phase plane:

$$\varphi' = \psi$$
$$\psi' = -(\tfrac{1}{\theta^2-1})\varphi^3 \cdot \qquad (5.79)$$

The orbits are the level curves

$$\frac{1}{2}\psi^2 + \frac{\varphi^4}{4(\theta^2 - 1)} = E, \qquad (5.80)$$

where E is a constant. To see this, multiply the first equation of (5.79) by ψ' and the second equation by φ':

$$\varphi'\psi' = \psi\psi' = (1/2)\frac{d}{dt}\psi^2$$

$$\psi'\varphi' = -(\frac{1}{\theta^2 - 1})\varphi^3\varphi' = -\frac{1}{4(\theta^2 - 1)}\frac{d}{dt}\varphi^4.$$

Subtract the second equation from the first to get

$$\frac{d}{dt}[\frac{1}{2}\psi^2 + \frac{1}{4(\theta^2 - 1)}\varphi^4] = 0,$$

which implies (5.80). Several of the level curves are plotted for $\theta = 1.5$ in Figure 5.9.

Although these travelling waves do not have finite energy in the sense of (5.76), we can think of the conserved quantity (5.80) as the energy of the travelling wave. The orbits of the travelling wave are the level curves of this energy funtion.

For any value of $\theta > 1$, the solution φ of (5.78) is periodic, and the travelling wave is thus a periodic wave train moving to the right at speed $\theta > 1$, which is faster than the speed of propagation of the linear equation (with $c = 1$). The same wave form can also propagate to the left if φ solves (5.78) because then

$$v(x, t) = \varphi(x + \theta t)$$

is also a solution of (5.75). We leave the case $\theta < 1$ to the exercises.

An important question we must raise about these special travelling wave solutions is how likely we are to find them in a real physical setting. By way of analogy, recall that the equation for the pendulum is

$$\theta'' + \sqrt{\frac{g}{l}}\sin\theta = 0.$$

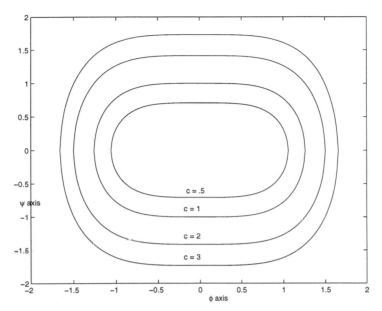

FIGURE 5.9
Level curves $(1/2)\psi^2 + (1/5)\varphi^4 = c/2$. Here $\theta = 1.5$.

We know that there is a solution of this equation in which the bob of the pendulum approaches the inverted position, never quite reaching it. But the smallest perturbation of this solution can have wild oscillations. We would say that this special solution is *unstable*.

Now consider a small perturbation of our travelling wave. Look for a solution of (5.75) in the form

$$u(x, t) = \varphi(x - \theta t) + p(x, t).$$

Since φ solves (5.75) exactly, we are left with

$$0 = p_{tt} - p_{xx} + (\varphi + p)^3 - \varphi^3$$

$$= p_{tt} - p_{xx} + 3\varphi^2 p + 3\varphi p^2 + p^3.$$

Now assuming that p is small, we drop the terms involving p^2 and p^3 to obtain the linearized equation for p:

$$p_{tt} - p_{xx} + 3\varphi^2(x - \theta t)p = 0.$$

Solutions of this linearized equation are studied to determine the stability of the travelling wave.

Questions of stability for travelling waves and other solutions of nonlinear wave equations are active areas of research. See [Str2] for a summary of problems and results in this area.

Exercises 5.7

1. Let K_T be the trapezoid

$$K_T = \{(x,t) : |x - x_0| \leq T + a - t, 0 \leq t \leq T\}.$$

(a) Verify that for any C^2 function u, the quantity

$$(u_{tt} - u_{xx} + u^3)u_t = div\mathbf{v},$$

where
$$\mathbf{v} = (-u_x u_t, \frac{1}{2}(u_t^2 + u_x^2) + \frac{1}{4}u^4).$$

(b) Let u be a solution of (5.75). As in exercise 12, Section 5.3,

$$0 = \int\int_{K_T} div\mathbf{v} dx dt = \int_{\partial K_T} \mathbf{n} \cdot \mathbf{v} ds.$$

Show that the integrals on the lateral boundaries are both nonnegative and hence that

$$\frac{1}{2}\int_{|x-x_0|\leq a}[u_t^2+u_x^2+\frac{u^4}{2}](x,T)dx \leq \frac{1}{2}\int_{|x-x_0|\leq a+T}[u_t^2+u_x^2+\frac{u^4}{2}](x,0)dx.$$

(c) Deduce that if $u(x,0) = u_t(x,0) = 0$ for $|x| \geq r$, then $u(x,t) = 0$ for $|x| \geq r + t$, all $t \geq 0$. This says that the nonlinear equation (5.75) has the same finite speed of propagation as the linear equation (without the u^3 term).

2. Now add a damping term to (5.75):

$$u_{tt} + 2du_t - u_{xx} + u^3 = 0.$$

Repeat the analysis for travelling waves with $\theta > 1$. What is the behavior of the trajectories in the (φ, ψ) plane as $s \to \pm\infty$? How does it depend on the sign of d? How does this translate into behavior of the solution $u(x, t) = \varphi(x - \theta t)$ at a fixed point x?

3. Numerical solutions of $u_{tt} - u_{xx} + u^3 = 0$.

    (a) Write down an explicit finite difference scheme for this equation.

    (b) How would you use the initial data $u(x, 0) = f(x)$ and $u_t(x, 0) = g(x)$ to calculate the numerical solution on the line $t = \Delta t$?

Program wave2 uses a finite difference scheme to numerically integrate solutions of the IBVP

$$u_{tt} - u_{xx} + \gamma u^3 = 0, \quad 0 < x < 10, \ t > 0, \qquad (5.81)$$

$$u(0, t) = u(10, t) = 0,$$

$$u(x, 0) = \delta f(x), \qquad u_t(x, 0) = \delta g(x).$$

The first choice for the initial data is that of the plucked string. For the second choice, the user must provide mfiles f.m and g.m. The parameters γ and δ are entered at run time. Again $\Delta x = \Delta t = .025$.

4. In this exercise we shall see how scaling affects the solutions of the IBVP (5.81).

    (a) Let $u(x, t)$ be the solution of (5.81) with $\gamma = 0$ (the linear equation) and $\delta = 1$. Let $v(x, t) = 2u(x, t)$. Show that v solves the same linear equation and initial data with $\delta = 2$.

    (b) Now let $u(x, t)$ be the solution of (5.81) with $\gamma = 1$ and $\delta = 1$, and let $v(x, t) = 2u(x, t)$. v has the initial data with $\delta = 2$, but does it satisfy the same nonlinear equation? What nonlinear equation does v satisfy?

    (c) Run program wave2 with data choice (1). First set $\gamma = 0$, $\delta = 1.0$, and run with n1 = n2 = n3 = n4 = 40. Then run again with $\gamma = 1.0$, $\delta = 1.0$. What is the effect of the nonlinear term?

    (d) Run again, same values for n1, n2, etc., but with $\gamma = 0$ and $\delta = 2.0$. How does the solution compare with the case $\gamma = 0$ and $\delta = 1.0$?

    (e) Now run again with $\gamma = 1.0$ and $\delta = 2.0$. How does it compare with the case $\gamma = 1.0$, $\delta = 1.0$?

5. Make mfiles f.m and g.m for the initial data

$$f(x) = e^{-5(x-3)^2} + .5e^{-5(x-7)^2}, \qquad g(x) = 0.$$

The data consists of two humps centered at $x = 3$ and $x = 7$.

(a) Run with $\gamma = 0, \delta = 1.0$, n1 = n2 = n3 = n4 = 40. Observe the parts of the bumps that move toward the origin. When do they meet? What happens after they meet? Why?

(b) Now do the same with $\gamma = 1.0, \quad \delta = 1.0$. What happens when these waves meet? Compare with the results of part (a).

6. Make mfiles f.m and g.m for initial data

$$f(x) = \sin(\pi x/10), \qquad g(x) = 0.$$

(a) Compare the linear and nonlinear solutions with $\delta = 1$, n1 = n2 = n3 = n4 = 100. Why does the solution of the nonlinear equation develop extra ripples?

(b) Finally, run with $\delta = 1.0$ and $\gamma = -1.0$. What is the direction of the nonlinear force term when $\gamma = -1$? Set n1 = 64, n2 = n3 = 4, and n4 = 2. Run again with n1, n2, and n3 the same, but with n4 = 4. What has happened? How can your explain this?

7. The equation $u_{tt} - u_{xx} = 0$ is the linearization of the nonlinear equation $u_{tt} - u_{xx} + u^3 = 0$ about the trivial solution $u \equiv 0$. For small amplitude initial data, the solution of the linear equation ($\gamma = 0$) should not differ too much from that of the nonlinear equation ($\gamma = 1$), at least for a while. To test this idea out, run program wave2 with the initial conditions $f(x) = \sin(2\pi x/10)$, and $g(x) = 0$ and with scale factor $\delta = .5$.

(a) Do this first over a half period of the linear problem ($t = 5$) by taking n1 = n1 = n1 = n4 = 50. Move the output vectors snap1, snap2, snap3, and snap4 into new vectors lsnap1, lsnap2, lsnap3, lsnap4 with commands lsnap1 = snap1; ,etc. Then run the nonlinear problem and take snapshots at the same times. Plot the snapshots in pairs. For example use the command plot(x,lsnap1, x,snap1) . Use the same vertical scale, $-.5 \le u \le .5$, for these four graphs. Does the solution of the nonlinear problem get ahead or lag behind the solution of the linear problem ?

(b) Compare the results of part (a) with what happens when $\delta = .1$. Use the vertical scale $-.1 \le u \le .1$ for these graphs. When δ drops from $\delta = .5$ to $\delta = .1$, how does the maximum size of the nonlinear term change? Does this explain what you see?

5.8 Projects

1. Modify the program wave1 to include a boundary forcing function $h(t)$, as discussed in section 5.5.5. Use $h(t) = \sin(\omega t)$, making ω an input parameter. Vary the choices of ω and d to see what happens when resonance occurs.

2. Modify the program wave1 to model two strings with different speeds of propagation on the interval $[0, 10]$, joined together at $x = 5$. Take $c = 1$ for $5 < x \leq 10$ and $c = c_l$ for $0 \leq x < 5$, where c_l is an input parameter. The finite difference scheme (5.71) must be modified to take into account the jump conditions of Section 5.3, which assert that u and u_x must be continous across the joint.

3. Use the implicit finite difference scheme for the wave equation

$$\frac{u(x, t + \Delta t) - 2u(x, t) + u(x, t - \Delta t)}{(\Delta t)^2}$$

$$\approx c^2 \frac{u(x + \Delta x, t + \Delta t) - 2u(x, t + \Delta t) + u(x - \Delta x, t + \Delta t)}{(\Delta x)^2}.$$

Use the boundary conditions $u(0, t) = u(10, t) = 0$. This scheme will yield a system of linear equations to be solved at each time step. Write a MATLAB program to implement this scheme, and plot snapshots of the solution.

4. Make a movie using any of the programs for the vibrating string. See Appendix B for information or enter help getframe and help movie.

5. The equation

$$u_{tt} - u_{xx} + \sin(u) = 0$$

is called the "sine Gordon" equation (instead of Klein-Gordon).

(a) Find the ODE satisfied by φ if $u(x, t) = \varphi(x - \theta t)$ is to be a travelling wave solution of this equation. You should be reminded of the ODE for the motion of a pendulum.

(b) Show that the solutions φ satisfy

$$\frac{1}{2}(1 - \theta^2)\psi^2 + \cos(\varphi) = E, \quad \text{constant},$$

where $\psi = \varphi'$. Use MATLAB to plot the level curves of this function in the (φ, ψ) plane. You must distinguish between the cases when $\theta > 1$ and $\theta < 1$.

(c) For the case $\theta > 1$, describe the several different kinds of travelling waves. For which values of E are the waves periodic? For which values of E do they have a limit as $x \to \pm\infty$? What are the limits? Is there a third kind of wave?

(d) Repeat the discussion of part (c), now assuming that $0 < \theta < 1$.

6. Consider the damped sine Gordon equation

$$u_{tt} + 2du_t - u_{xx} + \sin(u) = 0.$$

The travelling waves $\varphi(x - \theta t)$ are given by functions $\varphi(s)$ and $\psi(s)$, which satisfy the system of ODE's

$$\varphi' = \psi$$

$$\psi' = \frac{1}{\theta^2 - 1}[-\sin(\varphi) + 2d\theta\psi].$$

(a) Use MATLAB to integrate this system with initial values $\varphi(0) = \varphi_0$, $\psi(0) = 0$ for a range of values of φ_0 sufficient to give you a picture of the phase-plane portrait of the system. Consider the different cases $0 < \theta < 1$ and $\theta > 1$.

(b) Plot $s \to \varphi(s)$ for several different trajectories of different kinds. This will be a snapshot of the travelling wave at time $t = 0$. In each case, describe the behavior of the travelling wave at a fixed point x as t increases.

Chapter 6

Fourier Series and Fourier Transform

In this chapter we look at some of the eigenfunction expansions in terms of Fourier series. We develop the Fourier transform and use it to solve the heat equation again. We also give a brief treatment of the discrete Fourier transform (DFT) and the fast Fourier transform (FFT).

6.1 Fourier series

We already have seen in Chapter 4 that we can expand functions in terms of sines and cosines. Let $f(x)$ be a piecewise continuous function given on $[0, L]$. Then the expansions are

$$f(x) = \frac{A_0}{2} + \sum_{n=1}^{\infty} A_n \cos(\frac{n\pi x}{L}), \tag{6.1}$$

$$\text{with } A_n = \frac{2}{L} \int_0^L f(x) \cos(\frac{n\pi x}{L}) dx \tag{6.2}$$

or

$$f(x) = \sum_{n=1}^{\infty} B_n \sin(\frac{n\pi x}{L}), \tag{6.3}$$

$$\text{with } B_n = \frac{2}{L} \int_0^L f(x) \sin(\frac{n\pi x}{L}) dx \tag{6.4}$$

Both of these expansions can be thought of as special cases of the full Fourier series which uses both sines and cosines. We define the full Fourier series as follows. Let $f(x)$ be a piecewise continous, real valued function on $[-L, L]$. The real Fourier series for f is

$$f(x) = \frac{A_0}{2} + \sum_{n=1}^{\infty} A_n \cos(\frac{n\pi x}{L}) + B_n \sin(\frac{n\pi x}{L}), \qquad (6.5)$$

where

$$A_n = \frac{1}{L} \int_{-L}^{L} f(x) \cos(\frac{n\pi x}{L}) dx \qquad (6.6)$$

and

$$B_n = \frac{1}{L} \int_{-L}^{L} f(x) \sin(\frac{n\pi x}{L}) dx. \qquad (6.7)$$

We claim that the expansions (6.1) and (6.3) are just special cases of the Fourier series (6.5). In fact let $f(x)$ be a function given on the interval $[0, L]$. Then extend f as an odd function to $[-L, L]$. Call this odd extension $\tilde{f}(x)$. Also extend f to $[-L, L]$ as an even function $\check{f}$.

Now calculate the Fourier series (6.5) of $\tilde{f}$. The product $x \to \tilde{f}(x) \cos(\frac{n\pi x}{L})$ is an odd function and hence the coefficients A_n given by (6.6) all vanish. Furthermore, the product $x \to \tilde{f}(x) \sin(\frac{n\pi x}{L})$ is even so that the formula (6.7) reduces to (6.4). Thus the full Fourier series for $\tilde{f}$ is the same as the sine series (6.3), (6.4). Similarly, the full Fourier series for the even extension $\check{f}$ reduces to the cosine series (6.1), (6.2).

We can also think of (6.5) as an eigenfunction expansion. Let the operator M be

$$Mu = -u''$$

with domain $W \subset C^2(R)$ being the subspace of functions which have period $2L$. It is easy to check that the eigenfunctions and eigenvalues of M are

$$\cos(\frac{n\pi x}{L}), \quad n = 0, 1, 2 \ldots \qquad \text{and} \qquad \sin(\frac{n\pi x}{L}), \quad n = 1, 2, \ldots$$

with eigenvalues $\lambda_n = (n\pi/L)^2$ for $n \geq 0$. Note that for $n \geq 1$, the eigenspace associated with λ_n is two-dimensional. For each $\lambda_n, n \geq 1$, there are two independent real eigenfunctions.

Complex Fourier series

The full Fourier series (6.5) can itself be thought of as a special case of the complex form of Fourier series. Recall that if $z = x + iy$ is a complex number, then $Re(z) = x$ and $Im(z) = y$. Furthermore, the complex conjugate of z is $\bar{z} = x - iy$, and the modulus $|z|$ is defined by $|z|^2 = x^2 + y^2 = z\bar{z}$. We shall be dealing with functions $f(x)$ which are complex-valued functions of the real variable x. We write

$$f(x) = u(x) + iv(x),$$

where for each x, $u(x) = Ref(x)$, and $v(x) = Imf(x)$. For example, the Euler formula for the complex exponential function is

$$e^{i\lambda x} = \cos(\lambda x) + i \sin(\lambda x). \tag{6.8}$$

Using this formula, it is easy to verify that for real λ, μ,

$$e^{i\lambda x} \cdot e^{i\mu x} = e^{i(\lambda+\mu)x}$$

and

$$\overline{e^{i\lambda x}} = e^{-i\lambda x}.$$

A complex-valued function is piecewise continuous, continuous, or of class C^k if both real and imaginary parts have this property. The scalar product on $C[-L, L]$ for complex-valued functions f, g is

$$\langle f, g \rangle = \int_{-L}^{L} f(x)\bar{g}(x)dx. \tag{6.9}$$

The norm associated with this scalar product is

$$\|f\| = \langle f, f \rangle^{1/2} = [\int_{-L}^{L} |f(x)|^2 dx]^{1/2},$$

where $|f(x)|^2 = f(x)\overline{f(x)} = [Ref(x)]^2 + [Imf(x)]^2$. The Schwarz inequality also holds for this complex scalar product:

$$|\langle f, g \rangle| \le \|f\| \|g\|.$$

This scalar product and norm reduce to the scalar product (4.11) and the norm (4.15) when f and g are real-valued. The complex eigenfunctions of M are

$$\varphi_n(x) = \exp(i\frac{n\pi x}{L}), \, n = 0, \pm 1, \pm 2, \ldots.$$

They are pairwise orthogonal in the scalar product (6.9). Indeed for $m \neq n$,

$$\langle \varphi_n, \varphi_m \rangle = \int_{-L}^{L} \varphi_n(x)\bar{\varphi}_m(x)dx = \int_{-L}^{L} e^{i(n-m)\pi x/L}dx$$

$$= \int_{-L}^{L} \cos((n-m)\pi x/L)dx + i \int_{-L}^{L} \sin((n-m)\pi x/L)dx = 0.$$

Furthermore

$$\langle \varphi_n, \varphi_n \rangle = \int_{-L}^{L} \varphi_n(x)\bar{\varphi}_n(x)dx = \int_{-L}^{L} e^{in\pi x/L} e^{-in\pi x/L}dx = 2L.$$

Consequently, the complex Fourier series expansion for a complex-valued function is given by

$$f(x) = \sum_{-\infty}^{\infty} c_n e^{in\pi x/L}, \tag{6.10}$$

where

$$c_n = \frac{\langle f, \varphi_n \rangle}{\langle \varphi_n, \varphi_n \rangle} = \frac{1}{2L} \int_{-L}^{L} f(x)e^{-in\pi x/L}dx.$$

When $f(x)$ is a real-valued function of period $2L$, $c_{-n} = \bar{c}_n$, and the complex exponential form of the series (6.10) reduces to the real form (6.5). In this case, the coefficients c_n are related to the real coefficients,

$$c_{\pm n} = \frac{1}{2}[A_n \mp i B_n], \quad n = 1, 2, \ldots \tag{6.11}$$

$$c_0 = \frac{A_0}{2}.$$

The terms in the complex series (when f is real) combine to yield the terms of the real series,

$$c_n e^{in\pi x/L} + c_{-n}e^{-in\pi x/L} = A_n \cos(\frac{n\pi x}{L}) + B_n \sin(\frac{n\pi x}{L}).$$

Exercises 6.1

1. Make the even and odd extensions to $[-L, L]$ of the following functions and then extend them as functions of period $2L$ on all of R. Which of the periodic extensions are continuous, and which are piecewise C^1?

   (a) $f(x) = (x/L)^2$.

   (b) $f(x) = 1 - x/L$.

   (c) $f(x) = x(x - L)$.

   (d) $f(x) = x^3/3 - Lx^2/2 + L^3/6$.

2. Verify equation (6.11).

3. Suppose that $f(x)$ is a C^1 function with period $2L$. Let d_n be the complex Fourier coefficient of f' and c_n the complex Fourier coefficent of f. Show that

$$d_n = (\frac{in\pi}{L})c_n.$$

6.2 Convergence of Fourier series

In this section we state (mostly) without proof some of the standard convergence results for Fourier series in one variable. We shall state them for the complex series (6.10).

A sequence of complex-valued functions $s_N(x)$ defined on $[a, b]$ converges *pointwise* on $[a, b]$ if both the real and imaginary parts of s_N converge pointwise, as defined in Section 1.1.3. Similary, s_N converges *uniformly* on $[a, b]$ means that both real and imaginary parts converge uniformly on $[a, b]$. Finally

$$s_N(x) \to s(x) \text{ in the mean square on } [a, b] \tag{6.12}$$

if

$$\int_a^b |s_N(x) - s(x)|^2 dx \to 0 \text{ as } N \to \infty.$$

We have already seen the notions of uniform convergence and mean square convergence in Chapter 4. It is easy to show that uniform convergence implies both pointwise and mean square convergence. The converse is not true, as seen in the following examples.

Example A

The sequence of real functions $s_N(x) = x^N$ on $[0, 1]$ converges pointwise to the discontinuous function $s(x) = 0$ for $0 \le x < 1$, $s(1) = 1$, and

$$\int_0^1 |s_N(x) - s(x)|^2 dx = \int_0^1 x^{2N} dx = \frac{1}{2N + 1} \to 0,$$

as $N \to 0$, so that s_N also converges to s in the mean square. However, s_N does not converge uniformly to s because for all N,

$$\max_{(0,1)} |s_N(x) - s(x)| = 1.$$

Example B

The sequence

$$s_N(x) = \begin{cases} 0 & \text{for } x = 0 \\ \sqrt{N} & \text{for } 0 < x < \frac{1}{N} \\ 0 & \text{for } \frac{1}{N} < x \le 1 \end{cases}$$

converges to zero pointwise on $(0, 1)$, but does not converge to zero in the mean square sense because

$$\int_0^1 |s_N(x)|^2 dx = \int_0^{1/N} N dx = 1$$

for all N.

We have used this term before, but we restate it here for convenience. A function f on $[a, b]$ is *piecewise continuous* if f is continuous at all but a finite number of points p_j, and at each p_j, f has a simple jump discontinuity. That is, $\lim_{x \uparrow p} f(x) = f(p - 0)$ and $\lim_{x \downarrow p} f(x) = f(p + 0)$ both exist at each point of discontinuity $p = p_j$.

We shall use these definitions of convergence on the sequences s_N, where the s_N are the partial sums of the complex Fourier series

$$s_N(x) = \sum_{-N}^{N} c_n e^{in\pi x/L}.$$

Theorem 6.1
Let f be piecewise continuous on $[-L, L]$. Then the Fourier series for f converges to f in the mean square sense:

$$\int_{-L}^{L} |s_N(x) - f(x)|^2 dx \to 0 \quad \text{as } N \to \infty;$$

and

$$\int_{-L}^{L} |f(x)|^2 dx = 2L \sum_{n=-\infty}^{\infty} |c_n|^2. \tag{6.13}$$

Equation (6.13) is known as Parseval's equality.

The *mean square error* is defined as

$$\sigma_N^2 = \frac{1}{2L} \int_{-L}^{L} |f(x) - s_N(x)|^2 dx = \frac{1}{2L} \|f - s_N\|^2.$$

It follows from (6.13) that

$$\sigma_N^2 = \sum_{|n| \geq N+1} |c_n|^2.$$

We can estimate how fast $\sigma_N^2 \to 0$ by comparing the sum with an improper integral. For instance, if $|c_n| = O(1/n^2)$, then

$$\sigma_N^2 \leq C \sum_{N+1}^{\infty} \frac{1}{n^4} \approx C \int_{N}^{\infty} \frac{1}{x^4} dx = O(N^{-3}).$$

Theorem 6.2
Suppose that f is continuous on $[-L, L]$, $f(-L) = f(L)$, and that f' exists at all but a finite number of points and is piecewise continuous. Then the Fourier series for f converges uniformly on $[-L, L]$:

$$\max_{-L \leq x \leq L} |s_N(x) - f(x)| \to 0 \quad \text{as } N \to \infty.$$

Remark   This theorem is often stated for the $2L$ periodic extension, $\tilde{f}$ of f. In this case the hypothesis is that $\tilde{f}$ is continuous, and that the condition on the derivative holds on each finite interval.

The proof of Theorem 6.2 is quite short, so we include it here. We assume that $L = \pi$ to simplify writing. To show uniform convergence, we shall appeal to the

Weierstrass test (Section 1.1.3). It will suffice to show that $\sum_{-\infty}^{\infty} |c_n| < \infty$. Since f is continuous and f' is piecewise continuous, we can calculate the coefficients of f':

$$\frac{1}{2\pi} \int_{-\pi}^{\pi} f'(x)e^{-inx}dx = \frac{1}{2\pi}[fe^{-inx}\Big|_{-\pi}^{\pi} + in \int_{-\pi}^{\pi} f(x)e^{-inx}dx] = inc_n.$$

The boundary terms cancel because $f(\pi) = f(-\pi)$. By Theorem 6.1, applied to f',

$$2L \sum_{-\infty}^{\infty} n^2|c_n|^2 = 2L \sum_{-\infty}^{\infty} |inc_n|^2 = \int_{-\pi}^{\pi} |f'(x)|^2dx < \infty.$$

Now the Schwarz inequality holds for complex sequences $\{a_n\}, \{b_n\}_{n=-\infty}^{n=\infty}$,

$$\left|\sum_{-\infty}^{\infty} a_n b_n\right| \le \left[\sum_{-\infty}^{\infty} |a_n|^2\right]^{1/2} \left[\sum_{-\infty}^{\infty} |b_n|^2\right]^{1/2}.$$

Hence

$$\sum_{n\neq 0} |c_n| = \sum_{n\neq 0} n|c_n|\frac{1}{n} \le \left[\sum_{n\neq 0} n^2|c_n|^2\right]^{1/2} \left[\sum_{n\neq 0} \frac{1}{n^2}\right]^{1/2} < \infty,$$

which completes the proof of Theorem 6.2. $\square$

We recall a basic theorem of analysis from Chapter 1.

Theorem 6.3
Let $f_n(x)$ be a sequence of continuous functions which converges uniformly on $[a, b]$ to the function $f(x)$. Then the limiting function $f(x)$ must be continuous on $[a, b]$.

We shall often apply this theorem when the f_n are the partial sums of a Fourier series. In this case the f_n, being the sum of sines and cosines, are clearly continuous. Thus if the Fourier series converges uniformly to its sum f, f must be continuous. On the other hand, if f is not continuous, then the Fourier series of f cannot converge uniformly.

Since uniform convergence implies pointwise convergence, we see that the hypotheses of Theorem 6.2 also imply pointwise convergence of the Fourier series to f. We can relax the hypothesis of continuity of f to get the following result, which is intermediate to the results stated in Theorems 6.1 and 6.2.

Theorem 6.4

Suppose that f is piecewise continuous on $[-L, L]$, *that f is differentiable at all but a finite number of points, and that f' is piecewise continuous. Then the Fourier series of f converges to* $f(x)$ *at each point* $x \in (-L, L)$ *where f is continuous, and if f is discontinuous at a point* $c \in (-L, L)$, *then*

$$s_N(c) \to \frac{1}{2}[f(c-0) + f(c+0)] \quad as \ N \to \infty.$$

At $x = \pm L$, *the series converges to* $(1/2)[f(L-0) + f(-L+0)]$.

To get convergence information about the cosine series (6.1), where f is given on $[0, L]$, extend f as an even function on $[-L, L]$, and then apply Theorems 6.1, 6.2, or 6.3. For the sine series (6.3), extend f as an odd function on $[-L, L]$, and apply Theorems 6.1, 6.2, or 6.3.

Example C

We consider the simple step function

$$f(x) = \begin{cases} -1/2 & \text{for } -\pi \leq x < 0 \\ 1/2 & \text{for } 0 < x \leq \pi \end{cases}.$$

Since f is real and odd, we can see that the Fourier series for f will contain only sine terms:

$$f(x) = \sum_{n=1}^{\infty} B_n \sin(nx),$$

with

$$B_n = \frac{1}{\pi} \int_{-\pi}^{\pi} f(x) \sin(nx) dx = \frac{1}{\pi} \int_{0}^{\pi} \sin(nx) dx$$

$$= \begin{cases} 0 & n \text{ even} \\ \frac{2}{n\pi} & n \text{ odd} \end{cases}.$$

Thus the Fourier series for f is

$$f(x) = \frac{2}{\pi}[\sin x + \frac{\sin 3x}{3} + \frac{\sin 5x}{5} + \cdots].$$

All of the terms in the series are zero when $x = 0$, so that the series converges to zero at $x = 0$ in agreement with Theorem 6.3.

However the partial sums S_N, which are continuous, try to make the discontinuous jump at $x = 0$ and overshoot by about 9% on either side. This overshoot at a discontinuity is called the *Gibbs phenomenon*. You will see this phenomenon very clearly when you run the program fseries in the exercises.

For an excellent treatment of Fourier series and related topics see [Be].

Exercises 6.2

1. Recall exercise 1 of section 6.1.

   (a) For which functions does the sine series converge uniformly on $[0, L]$?

   (b) For which does the cosine series converge uniformly $[0, L]$?

2. Let f be real-valued, piecewise continuous, of period $2L$ with real Fourier series

$$f(x) = \frac{A_0}{2} + \sum_{1}^{\infty} A_n \cos(\frac{n\pi x}{L}) + B_n \sin(\frac{n\pi x}{L}).$$

Deduce the real form of Parseval's equality from (6.11) and (6.13),

$$\frac{1}{2L} \int_{-L}^{L} |f(x)|^2 dx = \frac{A_0^2}{4} + \frac{1}{2} \sum_{1}^{\infty} (A_n^2 + B_n^2).$$

3. Let $\tilde{\varphi}_n(x) = \frac{1}{\sqrt{2L}} \exp[(in\pi x/L)]$ be the normalized eigenfunctions (note that $\|\tilde{\varphi}_n\|^2 = 1$ for all n). Let

$$\tilde{c}_n = < f, \tilde{\varphi}_n > = \int_{-L}^{L} f(x)\overline{\tilde{\varphi}_n(x)}dx.$$

Use Parseval's equality (Theorem 6.1) to show that

$$\|f\|^2 = \sum_{-\infty}^{\infty} |\tilde{c}_n|^2.$$

4. Use the real form of Parseval's equality (exercise 2) to show that when f is real-valued, the mean square error

$$\sigma_N^2 = \frac{1}{2} \sum_{N+1}^{\infty} (A_n^2 + B_n^2).$$

Exercises 5 - 9 deal with Fourier series of the following functions. The first two are given on $[0, L]$ and the last two are given on $[-L, L]$.

(1)
$$f(x) = \begin{cases} x, & 0 \le x \le L/2 \\ L - x, & 1 \le x \le L \end{cases}.$$

(2) $f(x) = 1 - x/L$ on $[0, L]$.

(3) $f(x)$ given on $[-L, L]$ is $f(x) = 1$ for $0 \le |x| \le L/2$ and $f(x) = 0$ for $1 < |x| \le L$.

(4) $f(x)$ given on $[-L, L]$ is $f(x) = x(x + L)(L - x)$.

The program fseries sums the real Fourier series of these functions with $L = 2$. You must choose which function, and for choices (1) and (2), whether you want the even or the odd extension to $[-2, 2]$. After you enter the number N of terms to be summed, the program computes the partial sum $s_N(x)$ and graphs it together with $f(x)$. In addition, the mean square error σ_N is computed and displayed on the screen.

5. Use data choice (1). Recall that in example 3 of Section 4.2 we expanded f in terms of a sine series:

$$f(x) = \sum_{1}^{\infty} B_n \sin(\frac{n\pi x}{L}), \qquad B_n = \frac{4L}{(n\pi)^2} \sin(\frac{n\pi}{2}),$$

and that the series converged uniformly.

(a) Instead expand f in a cosine series

$$f(x) = \frac{A_0}{2} + \sum_{1}^{\infty} A_n \cos(\frac{n\pi x}{L}).$$

Calculate the coefficients A_n.

(b) Does the cosine series converge uniformly?

(c)   Now run program `fseries` with data choice (1). This will sum the
      Fourier series (with $L = 2$) of the odd or even extension of f to
      $[-2, 2]$. Sum the series with the number of terms $N = 4, 8, 16, 32$
      and the odd extension to get the sine series. How rapidly does σ_N^2
      decrease? Make an analytic estimate of p, such that $\sigma_N^2 \approx N^{-p}$.
      Then check your power p with the numbers computed by the program
      `fseries`.

(d)   Now run program `fseries` with data choice (1) and sum the cosine
      series by choosing the even extension. Let the number of terms
      $N = 4, 8, 16, 32$. Does this series fit the data differently? How
      rapidly does σ_N^2 decrease?

6.   Calculate the Fourier sine coefficients of $f(x) = 1 - x/L$ given on $[0, L]$
     (data choice (2)).

(a)   Is the convergence uniform on $[0, L]$? To what value does the series
      converge at $x = 0$?

(b)   Put in the value $x = L/2$, and deduce the result

$$\frac{\pi}{4} = 1 - \frac{1}{3} + \frac{1}{5} - \dots .$$

(c)   Run the program `fseries` with data choice (2), the odd extension.
      Use $N = 4, 8, 16, 32$ You can see the Gibbs phenomenon at $x = 0$.
      What is the percentage of the overshoot?

(d)   How rapidly does σ_N^2 decrease?

7.   Calculate the Fourier cosine coefficients of $f(x) = 1 - x/L$ on $[0, L]$
     (again data choice (2)).

(a)   Is the convergence uniform on $[0, L]$?

(b)   Run program `fseries` with data choice (2), the even extension. Does
      the series appear to converge better than the sine series of exercise 6?

(c)   How rapidly does σ_N^2 decrease?

8.   Calculate the real Fourier coefficients of data choice (3).

(a)   Is the convergence uniform on $[-L, L]$?

(b)   Run program `fseries` with data choice (3). To what value does the
      series converge at $x = \pm 1$?

(c)   What is the percentage of overshoot at $x = \pm 1$?

(d)   How rapidly does σ_N^2 decrease?

9.   (a)   Verify that $f(-L) = f(L)$ and that $f'(-L) = f'(L)$ for data choice
            (4).

(b) Run program `fseries` with data choice (4), $N = 4, 8, 16, 32$. Compare the fit of the partial sums of this series with that of exercise 1. Which converges more rapidly?

(c) How rapidly is σ_N^2 decreasing?

6.3 The Fourier transform

We have used Fourier series and eigenfunction expansions extensively to solve problems for the wave equation and heat equation on a finite interval. One could (and should) wonder if there is a similar treatment possible for problems on the half-line or whole line. In fact there is if we change from a discrete index n to a continuous variable. The result is the Fourier transform.

To show the transition from Fourier series and motivate the form of the Fourier integral, we begin with the complex form of the Fourier series. Let $f(x)$ be given on R with $f(x) \to 0$ rapidly as $x \to \pm\infty$. We expand the restriction of $f(x)$ to $[-L, L]$ in the complex Fourier series

$$f(x) = \sum_{n=-\infty}^{\infty} c_n e^{in\pi x/L}, \tag{6.14}$$

where

$$c_n = \frac{1}{2L} \int_{-L}^{L} f(x) e^{-in\pi x/L} dx. \tag{6.15}$$

Substitute (6.15) into (6.14) and let $\xi_n = n\pi/L$. Then for $|x| \le L$,

$$f(x) = \sum_{-\infty}^{\infty} [\frac{1}{2L} \int_{-L}^{L} f(x) e^{-i\xi_n x} dx] e^{i\xi_n x}. \tag{6.16}$$

Let us set $\hat{f}_L(\xi) = \int_{-L}^{L} f(x) e^{-i\xi x} dx$. The distance between the points ξ_n is $\Delta\xi = \pi/L$. Hence (6.16) may be written

$$f(x) = \frac{1}{2\pi} \sum_{-\infty}^{\infty} \hat{f}_L(\xi_n) e^{i\xi_n x} \Delta\xi. \tag{6.17}$$

Equation (6.17) looks like a Riemann sum for the integral of the function $\hat{f}_L(\xi) e^{i\xi x}$ over the whole line. As $L \to \infty$, we expect the function $\hat{f}_L(\xi)$ to converge to the function

$$\hat{f}(\xi) = \int_{-\infty}^{\infty} f(x) e^{-ix\xi} dx. \tag{6.18}$$

At the same time, $\Delta \xi \to 0$ as $L \to \infty$. Thus we expect (6.17) to converge to

$$f(x) = \frac{1}{2\pi} \int_{-\infty}^{\infty} \hat{f}(\xi) e^{ix\xi} d\xi. \tag{6.19}$$

This is a delicate limiting process, but these two formulas can be proved rigorously when f satisfies certain fairly general hypotheses. Formula (6.18) defines $\hat{f}(\xi)$, called the Fourier transform of f, and (6.19) shows how to recover f from $\hat{f}$. It is called the Fourier inversion formula, or the inverse Fourier transform. ξ is called the transform variable.

There are tables of Fourier transforms available, for example, in [AS]. Two important functions whose Fourier transforms we shall need are the exponential,

$$f(x) = e^{-|x|} \quad \text{with transform} \quad \hat{f}(\xi) = \frac{2}{1+\xi^2}, \tag{6.20}$$

and the Gaussian,

$$g(x) = e^{-x^2/2} \quad \text{with transform} \quad \hat{g}(\xi) = \sqrt{2\pi} e^{-\xi^2/2}. \tag{6.21}$$

We shall prove an extended version of (6.21) in Section 4.

The Fourier transform is a decomposition of a function f into frequency components $\hat{f}(\xi)$. Formula (6.19) expresses the synthesis of f from its frequency components. Rapid oscillations in f, or sharp peaks, require larger high-frequency components. For a sharply peaked or irregular function f, $\hat{f}(\xi)$ does not decay rapidly as $\xi \to \pm\infty$. The more regular f is, the more rapidly $\hat{f}(\xi)$ decays as $\xi \to \pm\infty$. For example, the Fourier transform of the discontinuous function

$$f(x) = \begin{cases} 0, & |x| > 1 \\ 1, & |x| < 1 \end{cases}$$

is $\hat{f}(\xi) = 2\sin(\xi)/\xi = O(1/\xi)$ as $\xi \to \infty$. However, the transform of the more regular continuous function

$$g(x) = \begin{cases} 0, & |x| > 1 \\ 1 - x^2, & |x| \leq 1 \end{cases}$$

is $\hat{g}(\xi) = 4[\sin(\xi)/\xi - \cos(\xi)]/\xi^2 = O(1/\xi^2)$ as $\xi \to \infty$.

This relationship is also seen in the important scaling rule. Let $a > 0$ and set $f_a(x) = f(ax)$. Then making the change of variable $y = ax$ (whence $dx = (1/a)dy$) we conclude that

$$\hat{f}_a(\xi) = \int_{-\infty}^{\infty} f(ax) e^{-ix\xi} dx = \frac{1}{a} \int_{-\infty}^{\infty} f(y) e^{-iy(\xi/a)} dy.$$

Hence

$$\hat{f}_a(\xi) = \frac{1}{a}\hat{f}(\xi/a). \tag{6.22}$$

Let us apply this rule to the Gaussian. We see that

$$e^{-(ax)^2/2} \quad \text{transforms into} \quad \frac{\sqrt{2\pi}}{a}e^{-(\xi/a)^2/2}.$$

When a is large, $g_a(x) = \exp[-(ax)^2/2]$ is sharply peaked, centered at $x = 0$. The transform of g_a has a small maximum value, $\sqrt{2\pi}/a$, and decays less rapidly. This can be stated intuitively as saying that the more sharply peaked Gaussian has larger high-frequency components.

Another property related to the scaling rule is stated in the following theorem.

Theorem 6.5 (Plancherel Theorem)

$$\int_R |f(x)|^2 dx = \frac{1}{2\pi}\int_R |\hat{f}(\xi)|^2 d\xi, \tag{6.23}$$

and

$$\int_R f(x)\overline{g(x)}dx = \frac{1}{2\pi}\int_R \hat{f}(\xi)\overline{\hat{g}(\xi)}d\xi. \tag{6.24}$$

Note here that we are allowing f and $\hat{f}$ to be complex-valued functions of a real variable, and $|f(x)|$ and $|\hat{f}(\xi)|$ are the moduli.

The Plancherel theorem (6.23) is the continuous analogue of (6.13) which expresses the completeness of the Fourier series components $\exp(in\pi x/L)$.

A third property of the Fourier transform is the way in which derivatives are transformed.

$$\widehat{f'}(\xi) = i\xi\hat{f}(\xi). \tag{6.25}$$

We can see (6.25) as follows:

$$\widehat{f'}(\xi) = \int_R f'(x)e^{-ix\xi}dx = f(x)e^{-ix\xi}\Big|_{-\infty}^{\infty} + i\xi\int_R f(x)e^{-ix\xi}dx$$

$$= i\xi\hat{f}(\xi).$$

We have assumed that $\hat{f}(\xi)$ decays rapidly enough so that $\xi \to \xi\hat{f}(\xi)$ is integrable.

Convolution

Finally we consider the Fourier transform and convolution. The *convolution* of two functions f and g, denoted $f * g$, is defined as

$$(f * g)(x) = \int_R f(x - y)g(y)dy. \tag{6.26}$$

We have seen the convolution before in the formula (3.16) for the solution of the heat equation. The change of variable $z = x - y$ in this integral shows that it is equal to

$$\int_R f(z)g(x - z)dz.$$

Thus the convolution is commutative

$$(f * g)(x) = (g * f)(x). \tag{6.27}$$

Now apply the Fourier transform to (6.26):

$$(\widehat{f * g})(\xi) = \int_R (f * g)(x)e^{-ix\xi}dx = \int_R \int_R f(x - y)g(y)e^{-ix\xi}dydx. \tag{6.28}$$

Furthermore $\exp(-ix\xi) = \exp(-i(x - y)\xi)\exp(-iy\xi)$ so that if we interchange the order of integration in (6.28), we obtain

$$(\widehat{f * g})(\xi) = \int_R \int_R f(x - y)e^{-i(x-y)\xi}dxg(y)e^{-iy\xi}dy \tag{6.29}$$

$$= \int_R f(z)e^{-iz\xi}dz \int_R g(y)e^{-iy\xi}dy = \hat{f}(\xi)\hat{g}(\xi).$$

Thus convolution is transformed into the pointwise product of the transforms.

Fourier transformation rules

We summarize some of these important properties of the Fourier transform:

(1) $\hat{f}_a(\xi) = (1/a)\hat{f}(\xi/a).$

(2) $\widehat{f'}(\xi) = i\xi\hat{f}(\xi).$

(3) $(\widehat{f * g})(\xi) = \hat{f}(\xi)\hat{g}(\xi).$

(4) $\int_R |f(x)|^2dx = (1/2\pi)\int_R |\hat{f}(\xi)|^2d\xi.$

(5) $\widehat{\tau_a f}(\xi) = \exp(-ia\xi)\hat{f}(\xi).$

Here $\tau_a f(x) = f(x - a)$ is the translation of f.

In modern analysis, the Fourier transform is often used in the context of distribution theory. A recent introductory treatment can be found in [Strichartz].

Exercises 6.3

1. Show that f is even and real-valued if and only if $\hat{f}$ is even and real valued.

2. Verify rule (5) above.

3. Let

$$f(x) = \begin{cases} 1/2, & |x - a| \le b \\ 0, & |x - a| > b \end{cases}.$$

   Find $\hat{f}$. What is the rate of decay of $\hat{f}(\xi)$ as $\xi \to \infty$?

4. Let

$$f(x) = \begin{cases} 0, & |x - a| > b \\ x + b - a, & -b + a \le x \le a \\ b + a - x, & a \le x \le a + b \end{cases}.$$

   Find $\hat{f}$. Compare the rate of decay as $\xi \to \infty$ of this transform with that found in exercise 3.

5. Let

$$f_n(x) = \begin{cases} 0, & |x| > 1/n \\ n/2, & |x| < 1/n \end{cases}.$$

   (a) Find $\hat{f}_n(\xi)$.

   (b) Show that $\lim_{n \to \infty} \hat{f}_n(\xi) = 1$ for all ξ.

   Since formally $f_n(x) \to \delta(x)$ as $n \to \infty$, this calculation indicates that we may define the Fourier transform of the generalized function $\delta(x)$ by

   (i) $\hat{\delta}(\xi) = \lim_{n \to \infty} \hat{f}_n(\xi) \equiv 1$.

   This is consistent with another formal definition

   (ii) $\hat{\delta}(\xi) = \int \delta(x) \exp(-ix\xi) dx \equiv 1$.

   (c) Find the Fourier transform of the generalized function $\delta(x - a)$ using both (i) and (ii) above and show they are again consistent, and also consistent with the general rule (5) above.

6.4   The heat equation again

We shall use the Fourier transform to give a second derivation of the fundamental solution of the heat equation. Formula (6.21) is adequate for this purpose but we want to prove an extended version to use with the Schrödinger equation as well.

Let μ be complex with $Re\mu > 0$. Then the Fourier transform of $\exp(-\mu x^2)$ is

$$\int_R e^{-\mu x^2} e^{-ix\xi} dx = \sqrt{\frac{\pi}{\mu}} e^{-\xi^2/4\mu} \tag{6.30}$$

where we have taken the square root $\sqrt{\mu}$ with positive real part. Note that $Re\mu > 0$ means that

$$|e^{-\mu x^2}| = e^{-(Re\mu)x^2}$$

is rapidly decreasing as $x \to \pm\infty$.

To prove (6.30), we let $y(\xi)$ denote the unknown Fourier transform and observe that

$$y(\xi) \equiv \int_R e^{-\mu x^2} e^{-ix\xi} dx = \int_R e^{-\mu x^2} (\cos(x\xi) - i\sin(x\xi)) dx$$

$$= \int_R e^{-\mu x^2} \cos(x\xi) dx$$

because $x \to \exp(-\mu x^2)\sin(x\xi)$ is odd. We integrate by parts once:

$$y(\xi) = (2\mu/\xi) \int_R x e^{-\mu x^2} \sin(x\xi) dx$$

$$= -(2\mu/\xi)\frac{\partial}{\partial\xi} \int_R e^{-\mu x^2} \cos(x\xi) dx = -(2\mu/\xi) y'(\xi).$$

Thus y satisfies the differential equation

$$y'(\xi) + (\xi/2\mu) y(\xi) = 0,$$

which is easily integrated:

$$y(\xi) = y(0) e^{-\xi^2/4\mu}. \tag{6.31}$$

Then it remains to calculate the constant

$$y(0) = \int_R e^{-\mu x^2} dx = 2 \int_0^\infty e^{-\mu x^2} dx.$$

We use the standard trick employed to evaluate this integral:

$$\left[\int_0^\infty e^{-\mu x^2} dx \right]^2 = \int_0^\infty e^{-\mu x^2} dx \int_0^\infty e^{-\mu y^2} dy$$

$$= \int_0^\infty \int_0^\infty e^{-\mu(x^2+y^2)} dx dy = \int_0^{\pi/2} \int_0^\infty e^{-\mu r^2} r dr d\theta$$

$$= (\pi/2) \int_0^\infty r e^{-\mu r^2} dr = \frac{\pi}{4\mu}.$$

Thus $y(0) = \sqrt{\pi/\mu}$, where we are careful to take $\sqrt{\mu}$ so that it has positive real part. Substitution for $y(0)$ in (6.31) yields (6.30).

Now we turn to the solution of the IVP for the heat equation on R.

$$u_t = k u_{xx} \qquad x \in R, \ t > 0, \tag{6.32}$$

$$u(x, 0) = f(x) \qquad \text{for } x \in R.$$

We assume that $\int_R |f(x)| dx < \infty$ and that $x \to u(x, t)$ has a Fourier transform for each t, which we denote by

$$\hat{u}(\xi, t) = \int_R u(x, t) e^{-ix\xi} dx.$$

Now

$$\widehat{u_t}(\xi, t) = \partial_t \hat{u}(\xi, t)$$

and

$$\widehat{k u_{xx}}(\xi, t) = -k\xi^2 \hat{u}(\xi, t),$$

so that (6.32) is transformed into an ordinary differential equation in t with ξ playing the role of a parameter:

$$\partial_t \hat{u}(\xi, t) = -k\xi^2 \hat{u}(\xi, t), \tag{6.33}$$

with initial condition

$$\hat{u}(\xi, 0) = \hat{f}(\xi).$$

The solution of (6.33) is given by

$$\hat{u}(\xi, t) = \hat{u}(\xi, 0)e^{-k\xi^2 t} = \hat{f}(\xi)e^{-k\xi^2 t}. \tag{6.34}$$

We want to apply the inverse Fourier transform to the rightmost term in (6.34). We see that it is a product of $\hat{f}(\xi)$ and the Fourier transform in x of some function $S(x, t)$, such that

$$\hat{S}(\xi, t) = e^{-k\xi^2 t}.$$

To find $S(x, t)$, we go to (6.30) with $\mu = 1/(4kt)$. Thus

$$S(x, t) = \frac{1}{\sqrt{4\pi kt}} e^{-x^2/4kt},$$

and

$$u(x, t) = (S * f)(x, t) = \frac{1}{\sqrt{4\pi kt}} \int_R e^{-\frac{(x-y)^2}{4kt}} f(y) dy. \tag{6.35}$$

$S(x, t)$ is the fundamental solution of the heat equation we found in Chapter 3, and (6.35) is the solution of the IVP (6.32), which we recognize now as a convolution.

6.5   The discrete Fourier transform

In this section we introduce the discrete Fourier transform (DFT). We will use the DFT to calculate approximations to the Fourier coefficients given by (6.10) and the Fourier transform (6.18)

6.5.1   The DFT and Fourier series

Let $f(x)$ be a periodic function of period 2π. The complex Fourier coefficients of f are given by the integrals (6.10) with $L = \pi$. They can also be written

$$c_k = \frac{1}{2\pi} \int_0^{2\pi} f(x) e^{-ikx} dx \tag{6.36}$$

$k = 0, \pm 1, \pm 2, \ldots$, because the integrand has period 2π, and the integral will have the same value if the integration is taken over any interval $[c, c + 2\pi]$. In all but a few examples the integrals cannot be done exactly, or the function f might be given in tabular form. Thus we must find ways to approximate the values of the c_k. One way would be to use some numerical integration routine to estimate the integrals. However, for large k, the integrand is rapidly oscillating which makes the integral difficult to estimate numerically. Accurate computations can be very time consuming and we may need to compute thousands of coefficients.

We need to have an efficient way to compute the Fourier coefficients, or at least, good approximations to them. Recall that the Fourier expansion for f is

$$f(x) = \sum_{-\infty}^{\infty} c_k e^{ikx}. \tag{6.37}$$

In this way we represent f as an infinite sum of complex exponentials. However, in practice, even if we could compute the Fourier coefficients exactly, we would only sum a finite number of terms, thereby getting an approximation to f. Instead of the Fourier coefficients we could look for coefficients $\tilde{c}_k$, such that the finite sum

$$\sum_{k=-N/2}^{k=N/2} \tilde{c}_k e^{ikx}$$

agrees with f at $(N+1)$ equally spaced points in the interval $[0, 2\pi]$. For notational and computational reasons, we choose to use the complex exponentials $\exp(ikx)$, $k = 0, 1, \ldots, N - 1$, and seek coefficients d_k, such that

$$\sum_{k=0}^{N-1} d_k e^{ikx}$$

agrees with f at N equally spaced points in the interval $[0, 2\pi]$. We shall see that finding the $\tilde{c}_k$ or the d_k are equivalent problems. The d_k will approximate the Fourier coefficients when the index is adjusted.

Let $\Delta x = 2\pi/N$ and $x_j = j\Delta x$, $j = 0, 1, \ldots, N - 1$. Δx is the *sampling interval*. The requirement that the sum of exponentials match f at the points x_j yields the N equations for the d_k:

$$\sum_{k=0}^{N-1} d_k e^{ikj\Delta x} = f(x_j) \qquad j = 0, 1, \ldots, N - 1.$$

To get a more concrete feeling for these equations we write them out for the case $N = 4$. In this case $\Delta x = \pi/2$, and

$$e^{ikj\Delta x} = e^{ijk\pi/2} = W^{jk}.$$

$W = e^{i\pi/2} = i$ is a complex fourth root of unity, $W^4 = 1$. The equations become

$$
\begin{aligned}
d_0 + d_1 + d_2 + d_3 &= f(x_0) \\
d_0 + d_1 W + d_2 W^2 + d_3 W^3 &= f(x_1) \\
d_0 + d_1 W^2 + d_2 W^4 + d_3 W^6 &= f(x_2) \\
d_0 + d_1 W^3 + d_2 W^6 + d_3 W^9 &= f(x_3)
\end{aligned}
$$

In more compact matrix notation

$$F\mathbf{d} = \mathbf{f},$$

where

$$
F = \begin{bmatrix}
1 & 1 & 1 & 1 \\
1 & W & W^2 & W^3 \\
1 & W^2 & W^4 & W^6 \\
1 & W^3 & W^6 & W^9
\end{bmatrix}
$$

is the Fourier matrix, while

$$
\mathbf{d} = \begin{bmatrix} d_0 \\ d_1 \\ d_2 \\ d_3 \end{bmatrix}
\quad \text{and} \quad
\mathbf{f} = \begin{bmatrix} f(x_0) \\ f(x_1) \\ f(x_2) \\ f(x_3) \end{bmatrix}.
$$

How can we solve these equations for the d_k and how do we relate them to the Fourier coefficients c_k ? We shall exploit the fact that $W^4 - 1 = 0$ which can be factored

$$0 = (W - 1)(1 + W + W^2 + W^3).$$

Since $W \neq 1$, this implies that

$$1 + W + W^2 + W^3 = 0. \tag{6.38}$$

But W^2 and W^3 are also fourth roots of unity, so that we can deduce that

$$1 + W^2 + W^4 + W^6 = 0 \tag{6.39}$$

and

$$1 + W^3 + W^6 + W^9 = 0. \tag{6.40}$$

Using these facts about W and some row manipulations, we find that the inverse of F is given by

$$F^{-1} = \frac{1}{4}\bar{F} = \frac{1}{4}\begin{bmatrix} 1 & 1 & 1 & 1 \\ 1 & \bar{W} & \bar{W}^2 & \bar{W}^3 \\ 1 & \bar{W}^2 & \bar{W}^4 & \bar{W}^4 \\ 1 & \bar{W}^3 & \bar{W}^6 & \bar{W}^9 \end{bmatrix}.$$

To verify this, multiply out $\bar{F}F/4$. The first row of this product is

$$\frac{1}{4}[1+1+1+1, \quad 1+W+W^2+W^3, \quad 1+W^2+W^4+W^6, \quad 1+W^3+W^6+W^9]$$

$$= [1, 0, 0, 0]$$

using (6.38), (6.39), and (6.40). The second row is

$$\frac{1}{4}[1+\bar{W}+\bar{W}^2+\bar{W}^3, \quad 1+1+1+1, \quad 1+\bar{W}+\bar{W}^2+\bar{W}^3, \quad 1+\bar{W}^2+\bar{W}^4+\bar{W}^6]$$

$$= [0, 1, 0, 0]$$

where we have used the fact that $W\bar{W} = 1$ and that (6.38), (6.39), and (6.40) hold for $\bar{W}$ as well.

For general N, the Fourier matrix is

$$F = \begin{bmatrix} 1 & 1 & \cdots & 1 \\ 1 & W & \cdots & W^{(N-1)} \\ 1 & W^2 & \cdots & W^{2(N-1)} \\ . & . & \cdots & . \\ . & . & \cdots & . \\ . & . & \cdots & . \\ 1 & W^{N-1} & \cdots & W^{(N-1)(N-1)} \end{bmatrix}$$

with $W = \exp(2\pi i/N)$ a complex root of unity. F is written in abbreviated notation as

$$(F)_{j,k} = W^{jk}, \qquad j, k = 0, 1, \ldots, N-1, \tag{6.41}$$

and the inverse of F is

$$F^{-1} = \frac{1}{N}\bar{F}. \tag{6.42}$$

The mapping of complex N vectors into complex N vectors

$$\mathbf{f} \to \mathbf{d} = F^{-1}\mathbf{f}$$

is called the *discrete Fourier transform*. The DFT coefficients are given by

$$d_k = \frac{1}{N} \sum_{j=0}^{N-1} f(x_j) \bar{W}^{kj}, \qquad k = 0, 1, \ldots, N-1. \tag{6.43}$$

Next we ask how the DFT coefficients d_k are related to the Fourier coefficients c_k defined by (6.36). Suppose f is expanded as in (6.37), and that $\sum |c_k| < \infty$. Substitute this expansion into (6.43). We obtain

$$d_k = \frac{1}{N} \sum_{j=0}^{N-1} (\sum_{n=-\infty}^{\infty} c_n e^{inx_j}) W^{-kj}$$

$$= \frac{1}{N} \sum_{n=-\infty}^{\infty} c_n \sum_{j=0}^{N-1} W^{(n-k)j}.$$

Remember that $W^N = 1$, so that

$$1 + W + W^2 + \ldots + W^{(N-1)} = 0.$$

Since W^p is also an Nth root of unity, we have as well

$$1 + W^p + W^{2p} + \ldots + W^{(N-1)p} = 0,$$

provided $p \neq mN$ for some integer m. Thus the inner sum is

$$\sum_{j=0}^{N-1} W^{(n-k)j} = \begin{cases} 0, & n - k \neq mN \\ N, & n - k = mN \end{cases}.$$

Consequently,

$$d_k = \sum_{m=-\infty}^{\infty} c_{k+mN}, \qquad k = 0, 1, \ldots, N-1. \tag{6.44}$$

For $0 \leq k < N/2$,

$$d_k = c_k + \sum_{m=-\infty}^{\infty} c_{k+mN} \tag{6.45}$$

and for $1 \le k < N/2$,

$$d_{N-k} = c_{-k} + \sum_{m=-\infty}^{\infty} c_{-k+mN}. \tag{6.46}$$

In (6.45) and (6.46) the sum is taken over $m \ne 0$. Finally,

$$d_{N/2} = c_{N/2} + c_{-N/2} + \sum_{m=-\infty}^{\infty} c_{N/2+mN}, \tag{6.47}$$

where the sum in (6.47) is taken over $m \ne 0, -1$.

In general d_k is not equal to c_k. Equations (6.45) and (6.46) show that in addition to c_k, there may be contributions from higher frequencies. This phenomenon is known as *aliasing*. However for fixed k, as N gets larger, eventually $N/2 > k$, and therefore $|k + mN| > (|m| - 1/2)N \to \infty$ as $N \to \infty$. Assuming that the series of Fourier coefficients $\sum c_n$ is absolutely convergent, we see that the contribution of the higher frequencies in (6.45) and (6.46) tends to zero as $N \to \infty$. Hence, for fixed $k \ge 0$,

$$d_k \to c_k \qquad \text{as } N \to \infty$$

and

$$d_{N-k} \to c_{-k} \qquad \text{as } N \to \infty.$$

This means that as the sampling interval $\Delta x = 2\pi/N \to 0$, the DFT coefficients d_k become better and better approximations to the Fourier coefficients c_k, depending on how rapidly the sum $\sum c_n$ converges. However (6.47) shows that the coefficient $d_{N/2}$ does not provide useful information. Put another way, for a given sampling frequency $2\pi/\Delta x = N$, we can only expect the DFT coefficients to approximate $c_{\pm k}$ for $k < N/2$.

To further illustrate this idea, suppose that f has a finite Fourier series

$$f(x) = \sum_{k=-M}^{M} c_k e^{ikx}.$$

It follows from (6.45) and (6.46) that if $M < N/2$, then

$$d_k = c_k \quad \text{and} \quad d_{N-k} = c_{-k} \qquad \text{for } 0 \le k \le M. \tag{6.48}$$

For this to be the case we must choose the sampling interval $\Delta x = 2\pi/N < P/2$, where $P = 2\pi/M$ is the period of the highest frequency present in f.

We further remark that if the given function f has general period $2a$ instead of 2π and the sample values are taken at $x_j = j(2a/N)$, $j = 0, \ldots, N-1$, the Fourier matrix and its inverse are the same, and the relations (6.45), (6.46), and (6.47) are also unchanged. The frequencies in the Fourier expansion are $n\pi/a$.

6.5.2   The DFT and the Fourier transform

The DFT can also be used to approximate the Fourier transform

$$\hat{f}(\xi) = \int_R f(x)e^{-ix\xi}\,dx.$$

To simplify the discussion, we shall suppose that there is an $a > 0$ such that $f(x) \approx 0$ for $x < 0$ and $x > 2a$. Then

$$\hat{f}(\xi) \approx \int_0^{2a} f(x)e^{-ix\xi}\,dx. \tag{6.49}$$

Now let $\xi_k = k\pi/a$ where k is an integer. We see that

$$\hat{f}(k\pi/a) \approx \int_0^{2a} f(x)e^{-ik\pi x/a}\,dx = 2ac_k,$$

where c_k is the Fourier coefficient. We could use our previous results about the DFT approximation to c_k, but we prefer to take a direct route to the DFT by using a Riemann sum approximation to the integral (6.49). Let $\Delta x = 2a/N$ and $x_j = j\Delta x$, $j = 0, \ldots N-1$ be N mesh points in the interval $[0, 2a]$. Then we may approximate the integral (6.49) by Riemann sums

$$\hat{f}(\xi) \approx h_N(\xi) \equiv \sum_{j=0}^{N-1} f(x_j)e^{-ix_j\xi}\,\Delta x \tag{6.50}$$

$$= \Delta x \sum_{j=0}^{N-1} f(j\Delta x)e^{-ij\Delta x\xi}.$$

Now in general $\xi \to \hat{f}(\xi)$ is not periodic, but we are trying to approximate $\hat{f}(\xi)$ by $h_N(\xi)$, which has period $P = 2\pi/\Delta x = \pi N/a$. As $N \to \infty$ the period of h_N becomes longer and longer. On any fixed interval $|\xi| \leq \xi_0$, $h_N(\xi)$ converges to the integral (6.49). However, we may only need to know a good approximation to $\hat{f}(\xi)$ at a sequence of points $\xi_k = k\pi/a$, k, integer. From (6.50), we see that

$$\hat{f}(k\pi/a) \approx h_N(k\pi/a)$$

$$= \Delta x \sum_{j=0}^{N-1} f(j\Delta x)e^{-ij\Delta xk\pi/a}$$

$$= 2a(\frac{1}{N} \sum_{j=0}^{N-1} f(j\Delta x)\bar{W}^{jk})$$

because $\Delta x = 2a/N$, whence $j\Delta x k\pi/a = jk(2\pi/N)$. From (6.43) we see that the last sum is the DFT of the vector of sampled values

$$\mathbf{f} = (f(0), f(\Delta x), \ldots, f(2a - \Delta x)).$$

Consequently,

$$\hat{f}(k\pi/a) \approx h_N(k\pi/a) = 2ad_k, \tag{6.51}$$

where d_k are the DFT coefficients of f restricted to the interval $[0, 2a]$. Because $\xi \to h_N(\xi)$ has period $\pi N/a$,

$$h_N(-k\pi/a) = h_N((N-k)\pi/a).$$

Thus

$$\hat{f}(-k\pi/a) \approx h_N(-k\pi/a) = h_N((N-k)\pi/a) = 2ad_{N-k} \tag{6.52}$$

for $1 \le k \le N$.

We shall take $-N\pi/2a \le \xi \le N\pi/2a$ as a representative period interval for $h_N(\xi)$. The points $\xi_k = k\pi/a$, $|k| \le N/2$, span this interval. Therefore we take the collection of values

$$h_N(\frac{k\pi}{a}), \quad k = -\frac{N}{2}, \ldots, 0, \ldots, \frac{N}{2}$$

to approximate $\hat{f}(\xi)$ over the interval $[-N\pi/(2a), N\pi/(2a)]$. By (6.51) and (6.52), this is the same as the arrangement of DFT coefficients

$$2ad_{N/2}, \ldots, 2ad_{N-1}, 2ad_0, 2ad_1, \ldots, 2ad_{N/2}. \tag{6.53}$$

In practice the last coefficient on the right is not needed because it duplicates the one on the far left. This rearrangement of the DFT coefficients is a simple swap of the left and right halves of the vector $\mathbf{d}$ of DFT coefficients. For example, if $\mathbf{d} = [d_0, d_1, \ldots, d_7]$, then the swapped vector is $\mathbf{dd} = [d_4, d_5, d_6, d_7, d_0, d_1, d_2, d_3]$.

For $|\xi| > N\pi/2a = \pi \Delta x$, $h_N(\xi)$ repeats itself, and so we do not expect $h_N(\xi)$ to be a good approximation to $\hat{f}$ there. This interference of the periodic repetition of h_N with the approximation to $\hat{f}$ is again called aliasing.

We get better results when we assume that f is *band-limited*. This means that f contains no frequencies higher than a certain cutoff frequency, denoted ξ_c:

$$\hat{f}(\xi) = 0 \qquad \text{for } |\xi| \geq \xi_c.$$

f being band-limited is the analogue for nonperiodic functions of the condition that a periodic function has a finite Fourier series. The *Nyquist frequency* for a band-limited function is $\xi_* = 2\xi_c$. For a sampling interval Δx, the sampling frequency is $2\pi/\Delta x$. By choosing N sufficiently large, we can make the sampling interval Δx small enough so that the sampling frequency

$$\frac{\pi}{\Delta x} > \xi_* = 2\xi_c.$$

This is the same as choosing N so large that

$$\frac{2a}{N} = \Delta x < \frac{\pi}{\xi_c}. \tag{6.54}$$

In this case, the period interval of $h_N(\xi)$, $|\xi| \leq N\pi/2a$, contains the interval $|\xi| \leq \xi_c$ where $\hat{f}(\xi) \neq 0$. Thus if the sampling frequency is greater than the Nyquist frequency, there will be no aliasing in the approximation of $h_N(\xi)$ to $\hat{f}(\xi)$.

This result is related to the well known sampling theorem for band-limited functions which says that if f is band limited, then it can be recreated from its sampled values $f(x_j)$ when $x_j = j\Delta x$, $j = 0, \pm 1, \pm 2, \ldots$ and $\Delta x < \pi/\xi_c$ (which is just (6.54)). Specifically,

$$f(x) = \sum_{-\infty}^{\infty} f(x_j) \frac{\sin(\xi_c x - j\pi)}{\xi_c x - j\pi}.$$

Symmetric sampling range In many cases $f(x)$ is concentrated in a symmetric interval $[-a, a]$. We can reduce this situation to that treated in our previous discussion by introducing the translate $f_a(x) \equiv f(x - a)$. Then

$$\hat{f}(\xi) \approx \int_{-a}^{a} f(x)e^{-ix\xi}\,dx = e^{ia\xi}\int_{0}^{2a} f_a(x)e^{-ix\xi}\,dx,$$

so that

$$\hat{f}(\xi) \approx e^{ia\xi}h_N(\xi).$$

h_N is constructed from the values of f_a in the interval $[0, 2a]$. As before, $\xi \to h_N(\xi)$ has period $N\pi/a$, so that $h_N(-k\pi/a) = h_N((N-k)\pi/a) = 2ad_{N-k}$ for $1 \le k \le N/2$. We see that

$$\hat{f}(k\pi/a) \approx 2ae^{ia\xi}\Big|_{\xi=k\pi/a} d_k, \quad \text{for } 0 \le k \le N/2 - 1,$$

$$\hat{f}(-k\pi/a) \approx 2ae^{ik\xi}\Big|_{\xi=-k\pi/a} d_{N-k}, \quad \text{for } 1 \le k \le N/2.$$

Thus the DFT approximation for $\hat{f}$ on $[-N\pi/(2a), N\pi/(2a)]$ is generated by the following sequence of operations
  (i) sampling f on $[-a, a]$;
  (ii) computing the vector $\mathbf{d}$ of DFT coefficients for the sampled vector

$$\mathbf{f}_a = (f_a(0), \dots, f_a(2a - \Delta x)) = (f(-a), \dots, f(a - \Delta x)) = \mathbf{f};$$

  (iii) Swapping the halves of the vector $\mathbf{d}$ as in (6.53); and
  (iv) Multiplying by the phase shift $\exp(ia\xi)$.

This sequence of operations will be implemented in the program `ftrans.m` in Section 6.6. Program `ftrans` was used to produce Figure 6.2.

Since there are several parameters involved, we suggest an order in which to choose them. First choose an $a > 0$ such that $f(x) \approx 0$ for $|x| > a$. Then specify a plotting range $[-\xi_0, \xi_0]$ on which you wish to view $\hat{f}(\xi)$. If f is band-limited, choose $\xi_0 = \xi_c$. Finally choose $N = 2^p$ so large that $N\pi/(2a) > \xi_0$. To get a smoother plot in the viewing range, you must make $\Delta\xi = \pi/a$ smaller by choosing a larger. If you double a, then you must also double N, so that the ratio $N\pi/(2a)$ is still larger than ξ_0. On the other hand, if you fix a, and increase N, then the values of $h_N(\xi)$ become better approximations to $\hat{f}(\xi)$ at the points $\xi = k\Delta\xi$, without making the plot smoother. Do not confuse the accuracy of the approximation with smoothness of the plot. In Figure 6.1 we see displayed the graph of $f(x) = \sin(x)\exp(-|x - 2|/4)$, and in Figure 6.2, the graphs of the real and imaginary parts of the Fourier transform, $\hat{f}$.

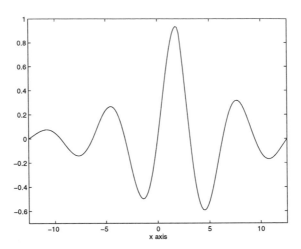

FIGURE 6.1
The function $f(x) = \sin(x)\exp(-|x - 2|/4)$.

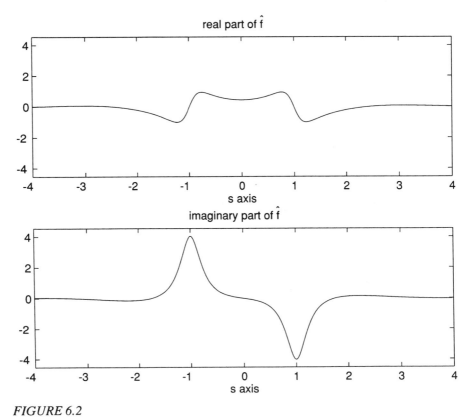

FIGURE 6.2

Real and imaginary parts of $\hat{f}$. For this figure, $a = 160$ and $N = 8192$.

Exercises 6.5

1. Show that if $f(x)$ has period 2π and is real-valued, then the Fourier coefficients satisfy $c_{-k} = \bar{c}_k$, and the DFT coefficients satisfy $d_{N-k} = \bar{d}_k$ for $0 \le k < N/2$.

2. Let
    $$f(x) = 2e^{-ix} + 1 + 3e^{ix}.$$

    Here the Fourier coefficients are $c_1 = 3$, $c_0 = 1$, and $c_{-1} = 2$. Use (6.45) to find the DFT coefficients for $N = 2$, then for $N = 4$.

    For the next three exercises, you will need to write an mfile, h.m, for the function h_N given in (6.50). Write it as a function h(s,N). You will need a for loop to execute the sum. It is also helpful to have h.m call another mfile, f.m, for the function values.

3. Let $f(x) = 1$ for $0 < x < 2$, and $f(x) = 0$ elsewhere.

    (a) Compute the Fourier transform of f, and plot it on the interval $-4\pi \le \xi \le 4\pi$. Let the transform variable be s.

    (b) The function mfile for this kind of discontinuous function is

    ```
    function y = f(x)
    y = (x< 2) - (x <= 0)
    ```

    Using the mfiles h.m and f.m, plot the graphs of h_N for various values of N against the graph of $\hat{f}$. Describe how the period of h_N lengthens, and how the fit gets better as N increases. Note that, unless told otherwise, MATLAB plots the real part of a complex-valued function.

4. Let $f(x) = \exp(-|x|)$. The Fourier transform $\hat{f}$ is given in formula (6.20). It decays like $|\xi|^{-2}$ as $|\xi| \to \infty$ while the transform of exercise 3 decays like $|\xi|^{-1}$. Since $\exp(-8) \approx 3 \times 10^{-4}$, we shall agree that $f \approx 0$ for $|x| > 8$.

    (a) Plot the Fourier transform $\hat{f}$ on $[-4\pi, 4\pi]$.

    (b) Use the function mfile h.m written for exercise 3 to construct h_N based on the translate $f_8(x) = f(x - 8)$. What is the period of $h_n(\xi)$? How large does N have to be for the period interval of h_N to include $[-4\pi, 4\pi]$?

    (c) $\hat{f}(\xi) \approx \exp(i8\xi)h_N(\xi)$. Plot both functions on $[-4\pi, 4\pi]$ for $N = 4, 8, 16, 32, 64$. Compare how fast the fit improves as N increases with the results of exercise 3.

5. Let $f(x) = \exp(-x^2/2)$. The Fourier transform is given in formula (6.21). It is also a Gaussian, and we can consider it band-limited with a cut off frequency $\xi_c = 4$. In constructing the approximation h_N to $\hat{f}$, what should you take as a? What is the Nyquist frequency? Plot h_N against $\hat{f}$ for various values of N.

6.6   The fast Fourier transform (FFT)

We have seen that the coefficients of the DFT can be computed by a single matrix multiplication on the vector of sampled values of the function f. However the Fourier matrix F_N is a full $N \times N$ matrix, having no zero elements. The direct evaluation of $\bar{F}_N \mathbf{f}$ requires N^2 multiplications. In applications, one deals with situations where N is quite large, on the order of 1000, or even 10^6. Fortunately there is an algorithm, the FFT, which dramatically reduces the number of operations needed to evaluate $\bar{F}_N \mathbf{f}$. The FFT has made possible many of the advances in signal processing.

The FFT takes advantage of the special form of the Fourier matrix F_N, in particular, the relationship between F_N and $F_{N/2}$. In what follows we shall assume that N is even and we let $M = N/2$. Recall that

$$(F_N)_{j,k} = W_N^{jk}, \qquad j, k = 0, 1, \dots, N-1,$$

where $W = \exp(2\pi i/N)$. Note that

$$W_N^2 = e^{4\pi i/N} = e^{2\pi i/M} = W_M.$$

Let $\mathbf{u} = (u_0, u_1, \dots, u_{N-1})$ be a complex N vector and let $\mathbf{v} = F_N \mathbf{u}$. We shall show that the calculation of $\mathbf{v}$ can be reduced to multiplications by the matrix F_M ($M = N/2$). Now the jth component of $\mathbf{v} = F_N \mathbf{u}$ is

$$(F_N \mathbf{u})_j = v_j = \sum_{k=0}^{N-1} W_N^{jk} u_k$$

$$= \sum_{k=0,even}^{N-2} W_N^{jk} u_k + \sum_{k=1,odd}^{N-1} W_N^{jk} u_k.$$

In the first sum, we write the index as $k = 2l$, and in the second as $k = 2l + 1$. In both cases, $l = 0, \dots, M-1$. Thus

$$v_j = \sum_{l=0}^{M-1} W_N^{j(2l)} u_{2l} + \sum_{l=0}^{M-1} W_N^{j(2l+1)} u_{2l+1}$$

$$= \sum_{l=0}^{M-1} W_M^{jl} u_{2l} + W_N^{j} \sum_{l=0}^{M-1} W_M^{jl} u_{2l+1},$$

using the fact that $W_N^{j(2l)} = (W_N^2)^{jl} = W_M^{jl}$. We can rewrite v_j as

$$v_j = (F_M \mathbf{u}_{even})_j + W_N^j (F_M \mathbf{u}_{odd})_j \qquad \text{for} \quad j = 0, \dots, M-1, \qquad (6.55)$$

where $\mathbf{u}_{even} = (u_0, u_2, \dots, u_{N-2})$ and $\mathbf{u}_{odd} = (u_1, u_3, \dots, u_{N-1})$ are M vectors. For $j = M, \dots, N-1$, we write $j = j' + M$ where $j' = 0, \dots, M-1$ and observe that

$$W_M^{jl} = W_M^{(j'+M)l} = W_M^{j'l},$$

while

$$W_N^j = W_N^{j'+M} = W_N^{N/2} W_N^{j'} = -W_N^{j'}.$$

Thus,

$$v_{j'+M} = (F_M \mathbf{u}_{even})_{j'} - W_N^{j'} (F_M \mathbf{u}_{odd})_{j'}. \qquad \text{for} \quad j' = 0, \dots, M-1. \quad (6.56)$$

Now we count the number of multiplications needed to compute $F_N \mathbf{u}$, letting this number be $r(N)$. From (6.55) and (6.56) we see that we must compute $F_M \mathbf{u}_{even}$ and $F_M \mathbf{u}_{odd}$, each requiring $r(M)$ multiplications. In addition we require M multiplications to compute $W_N^j (F_M \mathbf{u}_{odd})_j$, $j = 0, \dots, M-1$. Thus

$$r(N) = 2r(M) + M. \qquad (6.57)$$

But there is no reason to stop here. If M is even, we can use the same decomposition to reduce the number of multiplications required to evaluate $F_M \mathbf{u}_{even}$ and $F_M \mathbf{u}_{odd}$. Assuming $N = 2^p$ for some integer $p > 0$, we can repeat this decomposition over and over again until we reduce the evaluations to multiplications by F_2. Nominally taking $r(2) = 1$, and using (6.57), we have

p	N	$r(N)$
1	2	1
2	4	4
3	8	12
4	16	32
.	.	.
.	.	.

We surmise that if $N = 2^p$, then

$$r(N) = \frac{Np}{2} = \frac{1}{2}N \log_2 N.$$

This formula is easily proved true by induction for all p.

Now see what a tremendous savings this is. If $N = 1024 = 2^{10}$, then a direct calculation of $F_N \mathbf{u}$ requires $N^2 = 2^{20} \approx 10^6$ multiplications while the algorithm outlined above requires only $Np/2 = 5120$ multiplications. We have reduced the number of multiplications by a factor of 200.

The decomposition (6.55), (6.56) can also be viewed as a matrix factorization of F_N. We illustrate for $N = 4$.

$$F_4 = \begin{bmatrix} 1 & 1 & 1 & 1 \\ 1 & i & -1 & -i \\ 1 & -1 & 1 & -1 \\ 1 & -i & -1 & i \end{bmatrix}$$

$$= \begin{bmatrix} 1 & 0 & 1 & 0 \\ 0 & 1 & 0 & i \\ 1 & 0 & -1 & 0 \\ 0 & 1 & 0 & -i \end{bmatrix} \begin{bmatrix} 1 & 1 & 0 & 0 \\ 1 & -1 & 0 & 0 \\ 0 & 0 & 1 & 1 \\ 0 & 0 & 1 & -1 \end{bmatrix} \begin{bmatrix} 1 & 0 & 0 & 0 \\ 0 & 0 & 1 & 0 \\ 0 & 1 & 0 & 0 \\ 0 & 0 & 0 & 1 \end{bmatrix}.$$

The first matrix on the right shuffles the components of $\mathbf{u}$

$$\mathbf{u} = (u_0, u_1, u_2, u_3) \rightarrow (u_0, u_2, u_1, u_3).$$

We recognize the middle factor as

$$\begin{bmatrix} F_2 & 0 \\ 0 & F_2 \end{bmatrix}.$$

Finally the matrix on the left recombines the components of $F_2\mathbf{u}_{even}$ and $F_2\mathbf{u}_{odd}$ according to (6.55) and (6.56).

Of course, all the previous results apply to the DFT, which is multiplication by $\frac{1}{N}\bar{F}_N$. There is an extra division by N.

A recent, very readable, treatment of the DFT and the FFT is [BH].

Exercises 6.6

MATLAB has an easily used fast Fourier transform. It can be used with any integer value for N but works fastest when $N = 2^p$. For a given complex N vector u, the command

>> fft(u)

computes the N vector $\mathbf{v} = \bar{F}\mathbf{u}$. This is not exactly the discrete Fourier transform (6.43) which is $\frac{1}{N}\bar{F}\mathbf{u}$ in our notation. To agree with our usage, we must use the command

>> fft(u)/N

The inverse of the discrete Fourier transform, which is given by $\mathbf{u} = F\mathbf{v}$ in our discussion, is rendered by the command

>> N*ifft(v).

One must also take note of the fact that in MATLAB the index i in a vector $\mathbf{v} = (v_i)$ always starts with $i = 1$. The standard presentation of the DFT and FFT that we have followed starts the index with $i = 0$. Thus if v = fft(u)/N, u(1) is the value of the sampled function at $x_0 = 0$, and v(1) is the DC component, corresponding to d_0 in our notation.

The program fast computes the DFT coefficients of a function $f(x)$ given on $[0, 2\pi]$ and plots the real part of the partial sum using the DFT coefficients. It consists of the following sequence of instructions. You must provide an mfile f.m for the function $f(x)$. The number of sample points should be a power of 2.

```
N = input('Enter the number of sample points    ')
delx = 2*pi/N;
xsample = 0:delx:2*pi-delx ;
u = f(xsample)
v = fft(u)/N;
d0 = v(1); d = v(2:J);
x = 0:pi/100:2*pi;
sum = d0*ones(size(x));
for k = 1:N/2-1
```

```
sum = sum + d(k)*exp(i*k*x) + d(N-k)*exp(-i*k*x);
end
M = max(real(sum)) + .5;   m = min(real(sum)) - .5;
plot(x,f(x),x,sum)
axis([0 2*pi m M])
d0
```

In this manner of indexing, $d0$ is d_0 while $d(k)$ is d_k. Keep in mind that the DFT coefficients approximate the complex Fourier coefficients with

$$d_k \approx c_k, \quad 0 \leq k < N/2,$$

$$d_{N-k} \approx c_{-k}, \quad 1 \leq k < N/2.$$

1. Let $f(x) = x$ on $[0, 2\pi]$, and then assume f is extended periodically to R.

   (a) Calculate the complex Fourier coefficients c_k of f according to (6.36).

   (b) Make the mfile f.m for this function. Run program fast with $N = 4, 8, 16, 32, 64, 256$. For each N compare the DFT coefficients with the Fourier coefficients. The program automatically prints out d_0. To see the four DFT coefficients d_1, d_2, d_3, d_4, enter the command d(1:4). Note the rate of convergence of d_k to c_k for a fixed k as $N \to \infty$. In particular note how slowly $Re(d_k) \to 0$.

   (c) Because the periodic extension of f is discontinuous, the Fourier series for f will converge (at $x = 0$) to the average value $(f(0+) + f(2\pi-))/2 = \pi$. After the command u = f(xsample) in the script mfile above, insert the command u(1) = pi. Does this make the DFT coefficients converge more rapidly to the Fourier coefficients?

2. Now let $f(x) = x(2\pi - x)$.

   (a) Again calculate the complex Fourier coefficients c_k. How fast do they tend to zero as $k \to \pm\infty$? Faster or slower than the coefficients of exercise 1? Why ?

   (b) Write the mfile f.m and run program fast as in part (b) of exercise 1. Compare the rate of convergence of the DFT coefficients in this exercise with the rate of convergence in exercise 1. What is different, and why?

3. Finally let $f(x) = \exp(-(x - \pi)^2)$.

   (a) Compute the value of c_0 using the error function.

(b) Write the mfile f.m for this function, and run program fast. Compare the values of d_0 for various values of N. How large does N have to be for the values of d_0 and c_0 to agree to 5 decimal places?

The mfile ftrans.m implements the FFT approximation to the Fourier transform when the function f is concentrated in the range $-a \leq x \leq a$. You must write an mfile f.m for the function $f(x)$. We use s instead of ξ for the transform variable. Thus the viewing range in the transform variable is $[-s_0 \leq s \leq s_0]$. The N sampling points in the range $[-a, a]$ are given by

```
xsample = -a: delx :a -delx ;
```

The key commands are

```
u = f(xsample);
d = fft(u)/N;
dd = fftshift(d);
fhat = 2a*exp(i*a*s).*dd;
```

The MATLAB command ffshift swaps the left and right halves of the vector **d** of DFT coefficients. Multiplication by exp(i*a*s) is the same as multiplication by $(-1)^k$ because s runs through the discrete values $s = k\pi/a$. The program plots both real and imaginary parts of the approximation to $\hat{f}$ on the viewing range $[-s_0, s_0]$.

4. Let $f(x) = 1$ for $0 < x < 2$, $f(x) = 0$ elsewhere. This is the same function used in exercise 3 of Section 6.5. Use the mfile written for this function in that exercise.

(a) Run program ftrans with this function. Choose $a = 20$, $s0 = 4\pi$, and $N = 32, 128, 256, 1024$. You will see plotted the real and imaginary parts of the approximation to $\hat{f}(s)$. How large does N have to be to fill up the viewing range $[-s0, s0]$?

(b) Calculate the exact Fourier transform $\hat{f}(s)$ of this function by hand. Make an mfile g.m for the exact transform. Plot the exact transform and the approximate transform on the same graph as follows. Run program ftrans, then enter the commands

```
>> clf
>> plot(s,g(s),s,fhat)
>> axis([-4*pi 4*pi -M M])
```

Do this for $N = 32, 128, 256, 1024$. Describe how the fit gets better as N increases.

5. Let $f(x) = \cos(x)\exp(-|x|/5)$. We shall assume that $f(x) \approx 0$ for $|x| > a = 20$.

(a) Make an mfile f.m for this f. Then make an mfile h.m for the Riemann sum approximation $h_N(s)$ with s and N as variables. Use the form for the symmetric sample range $[-a, a]$. Let $N = 1024$, so that $\Delta x = 40/1024$. Plot on the viewing range $-s0 \le s \le s0$, with $s_0 = 4\pi$ and $\Delta s = \pi/20$. We shall count the number of floating point operations necessary to evaluate h_N at these 160 points. Use the sequence of commands

```
>> dels = pi/20;
>> s = -4*pi:dels:4*pi- dels;
>> flops(0)
>> plot(s,h(s,1024))
>> flops
```

Now use program ftrans to make the same plot of 160 points. Use $a = 20$, $s0 = 4\pi$, and $N = 1024$ for data inputs for ftrans. Then count the operations needed for ftrans.

```
>> flops(0)
>> ftrans
>> flops
```

In both cases we sampled the data at 1024 points in the range $[-20, 20]$, and calculated the result at 160 points in the viewing interval $[-4\pi, 4\pi]$. What is the difference in the operation count?

6. Now let

$$f(x) = (\frac{2}{\pi})(\frac{\sin x}{x^3} - \frac{\cos x}{x^2}) \quad \text{for } x \ne 0.$$

(a) How should f be defined at $x = 0$ to make it continuous there? Use the Taylor expansions for $\sin x$ and $\cos x$.

(b) The Fourier transform of this function is given by

$$\hat{f}(\xi) = \begin{cases} 1 - \xi^2, & |\xi| \le 1 \\ 0, & |\xi| > 0 \end{cases}.$$

f is band-limited with $\xi_c = 1$. Make an array-smart function mfile for this function. Then use program ftrans to compute an approximation to $\hat{f}(\xi)$. Choose a so that $\Delta\xi = \pi/a \approx .02$. For this choice of a, what is the value N_* such that (6.54) is satisfied ? Run the program with this value of a and values of N (in the form $N = 2^p$) larger and smaller than N_*. Describe what happens when $< N_*$. What phenomenon are you seeing? Run for $N > N_*$, and plot the results of ftrans on the same graph with the exact Fourier transform. How good is the fit for $N > N_*$?

6.7 Projects

1. We have seen that the Fourier series partial sums S_N can develop oscillations near points x where f has a discontinuity. In the following procedure, due to Lanczos, we find a smoother approximation g_δ to f, and compute the Fourier series coefficients of g_δ.

   Let $f(x)$ have period $2L$, and let δ be a number which is small compared to L. Define

   $$g_\delta(x) = \frac{1}{2\delta} \int_{x-\delta}^{x+\delta} f(y)dy.$$

   If f is continous, then g_δ is C^1, and $g_\delta \to f$ uniformly as $\delta \to 0$.

   (a) Show that the complex Fourier coefficients d_n of g_δ are related to the Fourier coefficients c_n of f by

   $$d_n = c_n \frac{\sin(n\delta)}{n\delta} \quad \text{for } n \neq 0$$

   and

   $$d_0 = c_0.$$

   (b) Use the FFT to compute approximate coefficients $\tilde{c}_n$ for f and then define coefficients

   $$\tilde{d}_0 = \tilde{c}_0, \quad \text{and} \quad \tilde{d}_n = \tilde{c}_n \frac{\sin(n\delta)}{n\delta} \quad \text{for } n \neq 0.$$

   Write a short program using the fft of MATLAB which computes $\tilde{c}_n$ and $\tilde{d}_n$, and graphs the partial sums

   $$\sum_{M}^{M} \tilde{c}_n e^{inx} \quad \text{and} \quad \sum_{M}^{M} \tilde{d}_n e^{inx}.$$

   Compare the convergence.

2. Consider the boundary value problem in one dimension

$$-u''(x) = q(x), \qquad 0 < x < 2\pi,$$

$$u(0) = u(2\pi) = 0.$$

(a) If $q(x) = \sum q_n \sin(nx)$ and $u(x) = \sum u_n \sin(nx)$, show that $u_n = q_n/n^2$.

(b) Write a program that takes input data a function $q(x)$ and uses the FFT to find approximate coefficients $\tilde{q}_n$. Then define $\tilde{u}_n = \tilde{q}_n/n^2$. Graph partial sums of $\tilde{u} = \sum \tilde{u}_n \sin(nx)$. These will be approximate solutions of the boundary value problem.

3. Now assume that $q(x)$ has period 2π on R. Consider the problem of finding a solution u of $-u'' = q$, with the condition that u also has period 2π.

(a) Using complex Fourier series expansions of u and q, show that solutions may not always exist. What condition must be placed on q to ensure the existence of solutions? When solutions exist, are they unique?

(b) Write a short program using FFT which takes as input data a function $q(x)$, computes an approximate solution, and then graphs it.

4. Write a program which solves the IVP (6.32) for the heat equation. Use the FFT to calculate $\hat{f}$. Then multiply $\hat{f}$ as in formula (6.34), and transform back using the inverse FFT. Write the program so that it takes the time t as an input variable, and then graphs the solution.

Chapter 7

Dispersive Waves and the Schrödinger Equation

After a short section in which we discuss the method of stationary phase, we treat dispersive waves. Then we turn to quantum mechanics, using the Fourier transform to establish the Heisenberg uncertainty principle. Finally we derive the Schrödinger equation. We solve the free Schrödinger equation and the Schrödinger equation with a square well potential.

7.1 Oscillatory integrals and the method of stationary phase

By an oscillatory integral, we mean an integral of the form

$$F(t) = \int_a^b f(x)e^{-it\phi(x)}dx. \tag{7.1}$$

The Fourier transform is of this type when $\phi(x) = x$, and $a = -\infty, b = \infty$. The coefficients of the complex Fourier series are also of this form with t taking on discrete values $1, 2, 3, \ldots$.

We shall investigate how $F(t)$ behaves as $t \to \infty$. First we consider the case of the linear phase $\phi(x) = x$. Assume $f \in C^1$. Then we integrate by parts in the definition of $F(t)$ to obtain

$$F(t) = \frac{1}{-it}[f(b)e^{-ibt} - f(a)e^{-iat}] + \frac{1}{it}\int_a^b f'(x)e^{-itx}dx = O(\frac{1}{t}).$$

Now suppose that $f \in C^2$ and that $f(a) = f(b) = 0$. Then we may integrate by parts a second time to obtain

$$F(t) = \frac{1}{(it)^2}[f'(a)e^{-iat} - f'(b)e^{-ibt} + \int_a^b f''(y)e^{-ixt}dx] = O(\frac{1}{t^2}).$$

More generally we can show by induction that if $f \in C^{n+2}[a, b]$ with $f^{(k)}(a) = f^{(k)}(b) = 0$ for $0 \le k \le n$, then

$$F(t) = O(\frac{1}{t^{n+2}}). \tag{7.2}$$

The same asymptotic behavior holds when $a = -\infty, b = \infty$ if we make the assumptions that $f \in C^{n+2}(R)$, $f^{(k)}$ is integrable on R for $0 \le k \le n+2$, and $f^{(k)}(y) \to 0$ as $y \to \pm\infty$ for $0 \le k \le n$.

Exercises 7.1

1. Prove the estimate (7.2) using induction.

2. Let $g(x) = (x - 2)^2(x + 2)^2$ for $|x| \le 2$. Let

    $$F(t) = \int_{-2}^2 g(x)e^{-itx}dx.$$

    (a) For which value of n does the asymptotic estimate (7.2) hold?

    (b) Compute $F(t)$ explicitly to verify.

3. Suppose that $f \in C^\infty[a, b]$ and $f^{(k)}(a) = f^{(k)}(b) = 0$ for all k. Show that $F(t) = O(t^{-n})$ as $t \to \infty$ for all n. This means that $|t^n F(t)| \le C$ for all n. The value of the constant C may depend on n.

Now we shall compare the integrands of (7.1) for two different choices of phase function. We take $f(x) = \exp(-3x^2)$ with phase functions $\phi_1(x) = x$, the linear phase, and $\phi_2(x) = (x - x_0)^2$, a nonlinear phase. The function mfiles intgnd1.m and intgnd2.m are, respectively,

$$f(x)e^{-it\phi_1(x)} = e^{-3x^2}e^{-itx}$$

and

$$f(x)e^{-it\phi_2(x)} = e^{-3x^2}e^{-it(x-x_0)^2}.$$

Use the following sequence of commands to plot the integrand with the linear phase at, say, $t = 4$. MATLAB plots the real part of a complex-valued function unless otherwise specified. The last command computes the integral.

```
>> x = -2:.02: 2;
>> t = 4;
>> plot(x,intgnd1(x,t))
>> quad8('intgnd1', -2, 2, [], [], t))
```

Note that the integrator `quad8` acts only on functions of one variable, but that one can pass on other other variables to the integrand in the form of parameters. To plot and integrate for another value of t, you only need to repeat the last three commands.

4.   Use the sequence of commands above for ϕ_1, and $t = 0, 5, 10, 15, 20$. What happens to the graph of the integrand as t gets larger? How does the change in the form of the graph correspond to the change in the value of the integral as t increases? Why?

The script mfile `osc.m` combines several of the commands listed above to calculate the value of $F(t)$ for both the linear phase function $\phi_1(x) = x$ and the nonlinear phase function $\phi_2(x) = (x - x_0)^2$. At run time you must choose between the linear and nonlinear phase. Then you must enter a vector of times $t_1, t_2, \ldots, t_N$. For example, if we wish to calculate $F(t)$ at times $t = 0, 1, 2, \ldots, 20$, we enter [1:20]. For more information, enter `help osc`. The values of $F(t)$ are placed in a vector F. Then the real part of F is plotted against t.

5.   Run progam `osc.m` for the linear phase, and enter the times $t = 0, 1, \ldots, 20$. Because the function $f(x) = \exp(-3x^2)$ decays so rapidly, along with all its derivatives, we see that $f^{(k)}(\pm 2) \approx 0$ for many values of k. How does this show up in the graph of $ReF(t)$?

To plot the integrand for the nonlinear phase, with $t = 4$, and say, $x_0 = .2$, and compute the integral, use the sequence:

```
>> x = -2:.01:2;
>> t = 4;
>> global x0
>> x0 = .2;
>> plot(x,intgnd2(x,t))
>> quad8('intgnd2',-2, 2,[], [], t)
```

The variable x_0 is treated as a parameter in the function mfile `intgnd2.m`, and so must be declared a global variable that is shared in the general workspace, and in the workspace of the mfile `intgnd2.m`.

6.   Run this sequence of commands with $x_0 = .5$ Now observe how the graph changes as t increases. Will there be more or less cancellation in the integral now compared with what happened in exercise 4 ?

7. Run the program osc.m to calculate $F(t)$ with the nonlinear phase ϕ_2. Again enter times $t = 0, 1, \ldots, 20$. Choose $x_0 = .2$.

(a) How does this graph of $F(t)$ compare with that seen in exercise 5? You can put both curves on the same graph by first running osc with the linear phase, and saving the output vector with the command Flin = F. Then run osc with the same values of t and the nonlinear phase. Plot both results together with the command plot(t,Flin,t,F).

(b) Run osc with values of $t = 21, 22, \ldots, 40$, and the nonlinear phase with $x_0 = .2$. Superimpose the graph of Ct^{-p} for various values of p and C as in exercise 5. What choice gives the best fit on the interval $[21, 40]$?

(c) You can find C and p analytically by solving the two equations $ReF(t_1) = Ct_1^{-p}$ and $ReF(t_2) = Ct_2^{-p}$ for two different values t_1 and t_2.

Finally we give an analytical account of what has been observed in exercise 7. The method used here is called the *Method of Stationary Phase*. We consider the oscillatory integral

$$F(t) = \int_R e^{-it\phi(x)} f(x)dx. \qquad (7.3)$$

We assume that $\phi'(x) = 0$ for exactly one point $x = x_0$, and that $\phi''(x_0) \neq 0$. Now we approximate ϕ for x near x_0 by the first two terms of the Taylor expansion:

$$\phi(x) \approx \phi(x_0) + \frac{\phi''(x_0)}{2}(x - x_0)^2.$$

As we saw in the plots of the integrand in exercise 6, the principal contribution to the integral comes from the values of the integrand near the stationary point x_0. Thus we can approximate $F(t)$ by substituting the two term approximation for ϕ and we can approximate f by its value at $x = x_0$. We deduce that

$$F(t) \approx e^{-it\phi(x_0)} f(x_0) \int_R e^{-\frac{it}{2}\phi''(x_0)(x-x_0)^2} dx.$$

This last integral can be evaluated using contour integration, and it can be shown that it is a (complex) multiple of the error integral

$$\int_R e^{-\frac{it}{2}\phi''(x)(x-x_0)^2} dx = e^{-\frac{i\pi}{4}(\pm 1)} \int_R e^{-\beta t x^2} dx = e^{-\frac{i\pi}{4}(\pm 1)} \sqrt{\frac{\pi}{\beta t}},$$

where

$$\beta = \sqrt{\frac{|\phi''(x_0)|}{2}}$$

and we take the plus sign when $\phi''(x_0) > 0$, and the minus sign otherwise. Thus we have the asymptotic approximation

$$F(t) \approx f(x_0)\sqrt{\frac{2\pi}{t|\phi''(x_0)|}}e^{-i[\phi(x_0)t\pm\frac{\pi}{4}]} \tag{7.4}$$

as $t \to \infty$. If f is real-valued, the expression becomes

$$Re\,F(t) \approx f(x_0)\cos(\phi(x_0)t \pm \frac{\pi}{4})\sqrt{\frac{2\pi}{t|\phi''(x_0)|}}. \tag{7.5}$$

8. Now back to the program osc. Remember that $f(x) = \exp(-3x^2)$ and the nonlinear phase $\phi_2(x) = (x - x_0)^2$. Write a function mfile for the function given in formula (7.4), making x_0 a global variable. Then plot this function on $[10, 40]$ together with the output vector F of program osc for $x_0 = 0$ and again for $x_0 = .5$. In each case how well does the function given by the formula (7.4) compare with the values computed by program osc? Do the two curves fit uniformly well for all $t \in [10, 40]$? Say why or why not.

7.2 Dispersive equations

7.2.1 The wave equation

In Chapter 5 we dealt with the linear wave equation $u_{tt} - c^2 u_{xx} = 0$ and found that all signals propagated with speed c and maintained their shape. In this chapter we shall study equations where waves of different wave lengths propagate with different speeds and waves change shape as they are propagated. The Fourier transform will allow us to see these phenomena clearly. First we use the Fourier transform to construct solutions of

$$u_{tt} - c^2 u_{xx} = 0.$$

We look for solutions in the form of a complex exponential

$$u(x, t) = e^{i(kx - \omega t)}, \tag{7.6}$$

where ω is the angular frequency of the wave and $\nu = \omega/2\pi$ is the frequency in cycles per second. k is called the wave number in this context, while $\lambda = (2\pi)/k$ is the wavelength. We see that $k = (2\pi)/\lambda$ counts the number of wave cycles occurring in the spatial length 2π. If we substitute (7.6) in the wave equation, we see that a function of the form (7.6) will be a solution if and only if

$$-\omega^2 + c^2 k^2 = 0$$

or

$$\omega = \pm ck.$$

Thus for any wave number k,

$$u(x, t) = e^{ik(x \pm ct)} \tag{7.7}$$

is a solution of the wave equation. Each of these waves travels with speed c, regardless of the wave number k. A new solution can be built up from the solutions (7.7) by integrating over k:

$$u(x, t) = \frac{1}{2\pi} \int_R [U(k)e^{ik(x - ct)} + V(k)e^{ik(x + ct)}]dk. \tag{7.8}$$

This superposition of waves is a Fourier integral where we see the solution u synthesized from its Fourier components U and V. In the exercises we shall see how to choose U and V to solve the IVP.

7.2.2 Dispersion relations

Now we consider *dispersive* equations in which the speed of propagation of a wave (7.6) depends on the wave number k. In this case we shall see that a wave changes shape as it propagates because some components travel faster than others.

Example

The linear Klein-Gordon equation is important in quantum mechanics. It is derived from the relativistic energy $E = c^2 p^2 + c^4 m^2$ for a particle with rest mass m. Using the correspondence principle (see Section 7.4.2), we replace E by the operator $i\hbar \partial_t$ and p by the operator $-i\hbar \partial_x$. Then after choosing units in which $c = \hbar = 1$ we obtain the linear Klein-Gordon equation

$$u_{tt} - u_{xx} + m^2 u = 0. \tag{7.9}$$

Substitute a function of the form (7.6) into this equation. We find that for u to be a solution of (7.9), ω and k must satisfy.

$$-\omega^2 + k^2 + m^2 = 0,$$

so that

$$\omega = \pm\sqrt{k^2 + m^2}.$$

Taking $\omega(k)$ to denote the plus root, we see that waves of the form

$$u(x, t) = e^{i(kx \pm \omega(k)t)} \tag{7.10}$$

solve (7.9). The relation

$$\omega(k) = \sqrt{k^2 + m^2} \tag{7.11}$$

is called the *dispersion relation* for equation (7.9). The speed of propagation of this plane wave is

$$\frac{\omega(k)}{k} = \frac{\sqrt{k^2 + m^2}}{k}.$$

Later we shall refer to this velocity as the *phase velocity*.

Back to the general case. Suppose some dispersion relation $\omega(k) > 0$ like (7.11) is given. We assume that $\omega(k)$ is *nontrivial*, by which we mean that

$$\omega''(k) \neq 0.$$

We consider an integral of waves of the form (7.10):

$$u(x, t) = \frac{1}{2\pi} \int U(k) e^{i(kx - \omega(k)t)} dk. \tag{7.12}$$

Note that

$$u(x, 0) = \frac{1}{2\pi} \int U(k) e^{ikx} dk,$$

so that $U(k)$ is the Fourier transform of $u(x, 0)$. Because $\omega(k) > 0$, u is a wave which moves to the right. We want to describe the behavior of this function as $t \to \infty$. Because of the dispersion relation, this wave will change shape as t increases and the frequency components will tend to separate, or disperse. To see this, let us restrict u to a space-time line $x = \sigma t, \sigma > 0$:

$$u(t\sigma, t) = \int U(k) e^{it\phi(k)} dk$$

where $\phi(k) = k\sigma - \omega(k)$. This is an oscillatory integral of the form (7.1), and we may apply the method of stationary phase to get an asymptotic description as $t \to \infty$. We see that

$$\phi'(k) = \sigma - \omega'(k) \quad .$$

The phase $\phi(k)$ has a stationary point at k_0 if $\omega'(k_0) = \sigma$, and there can be at most one such root because

$$\phi''(k) = -\omega''(k)$$

and we have assumed that $\omega(k)$ is a nontrivial dispersion relation. Thus applying the result (7.4), we see that

$$u(\sigma t, t) \approx U(k_0) \sqrt{\frac{2\pi}{|\omega''(k_0)|t}} e^{i(k_0 \sigma t - \omega(k_0)t - \frac{\pi}{4})}.$$

Along the space-time line $x = \sigma t$, u looks like the wave $U(k_0) \exp(i(k_0 x - \omega(k_0)t - \pi/4))$, multiplied by a time-dependent factor which decays like $t^{-1/2}$. We see that along the different space-time lines $x = \sigma t$, waves of different wave number will separate out of u. These are called the *group lines*. For reasons given shortly, the speed

$$\sigma = \omega'(k)$$

is called the group velocity. Back to the

Example (continued)

The dispersion relation for the Klein-Gordon equation (7.9) is

$$\omega(k) = \sqrt{k^2 + m^2},$$

so that the group velocity is

$$\omega'(k) = \frac{k}{\sqrt{k^2 + m^2}}.$$

Note that $\omega'(k)$ is a strictly increasing function with

$$\lim_{k \to \infty} \omega'(k) = 1.$$

This means that the wave components corresponding to large k (small λ) travel faster than those corresponding to small k (large λ). The short wavelength components, travelling with speed close to 1 will move to the front of the wave train and the longer wavelength components will lag further and further behind.

7.2.3 Group velocity and phase velocity

Back to the general case of a dispersive wave defined by (7.12). If $U(k)$ is concentrated about a single value k_*, $u(x, t)$ is called a *wave packet*, and will tend to move with speed $\sigma_* = \omega'(k_*)$, which is called the *group velocity* of the packet. Of course dispersion will occur and the packet will broaden in space because some components will move at a speed $\sigma > \sigma_*$ and some will move at a speed $\sigma < \sigma_*$. The wavelength $\lambda = 2\pi/k$ is not uniform across the packet, so that we can speak only of a *local wavelength* and a *local wave number*.

How do the crests and troughs of a dispersive wave move? The group line from the origin $(0, 0)$ through the point (x, t) has speed $\sigma = x/t$. For large x, t the wave number of the wave component that travels at speed σ is determined by the group velocity equation

$$\omega'(k) = \sigma = x/t, \tag{7.13}$$

which now expresses the relevant wave number as a function $k = k(x/t)$. This in turn defines the frequency via the dispersion relation $\omega = \omega(k(x/t))$. We set

$$\theta(x, t) = k(x/t)x - \omega(k(x/t))t.$$

Thus along a group line $x = \sigma t$, we can write

$$u(x, t) \approx \sqrt{\frac{2\pi}{t|\omega''(k(x/t))|}} \cdot e^{i(\theta(x,t) - \frac{\pi}{4})}.$$

$\theta(x, t)$ is the local phase function of the wave. The level curves $\theta(x, t) = \theta_0$ are the space-time curves of constant phase. They are the curves in space-time which track a particular crest or trough. Let $x(t)$ be a curve of constant phase. Thus

$$\theta_0 = \theta(x(t), t).$$

The speed of this curve is found by implicit differentiation. We see that

$$0 = \frac{d}{dt}\theta(x(t), t) = \theta_x \frac{dx}{dt} + \theta_t,$$

whence

$$\frac{dx}{dt} = -\frac{\theta_t}{\theta_x}.$$

However

$$\theta_x = k'(x/t)\frac{x}{t} + k(x/t) - \omega(k(x/t))k'(x/t)$$

$$= k(x/t) + k'(x/t)[\frac{x}{t} - \omega'(k(x/t))] = k(x/t)$$

because of the defining relation for $k(x/t)$. Likewise,

$$\theta_t = -\omega(x/t).$$

Hence

$$\frac{dx}{dt} = -\frac{\theta_t}{\theta_x} = \frac{\omega(x,t)}{k(x,t)}.$$

The quotient ω/k is called the *phase velocity*. The wave number k to be used in this expression is the local wave number. Thus the phase velocity is not constant across a dispersing wave packet. If we think of k as a function of x/t defined by (7.11), then both the group velocity and the phase velocity are constant on the group lines. When the equation is not dispersive, that is, $\omega(k) = ck$, the phase velocity

$$\frac{\omega}{k} = c = \frac{d\omega}{dk}$$

is the same as the group velocity.

Example (continued further)

We compute the curves of constant phase for the Klein-Gordon equation (7.9). The expression for the group velocity is

$$\omega'(k) = \frac{k}{\sqrt{m^2 + k^2}} = \sigma = x/t.$$

Solving for k in terms of x/t, we find that

$$k(x,t) = \frac{m(x/t)}{\sqrt{1-(x/t)^2}},$$

whence

$$k^2(x,t) = \frac{m^2(x/t)^2}{1-(x/t)^2}$$

and $\omega(k) = \sqrt{k^2 + m^2}$ becomes

$$\omega(x,t) = \frac{m}{\sqrt{1-(x/t)^2}}.$$

Hence

$$\theta(x,t) = k(x,t)x - \omega(x,t)t$$

$$= -mt\sqrt{1-(x/t)^2}.$$

The equations for the curves of constant phase therefore are

$$\theta_0^2 = m^2(t^2 - (x/t)^2)$$

or

$$t = \sqrt{\frac{\theta_0^2}{m^2} + x^2}.$$

Several of the group lines and curves of constant phase are sketched in Figure 7.1 for $m = 1$.. If we follow any crest in the evolving wave, that is the same as following a curve of constant phase $\theta(x,t) = \theta_0$. Along this curve $dx/dt \nearrow 1$ as $t \to \infty$. We see that if we follow a curve of constant phase in the direction of increasing t, it crosses group lines of larger and larger σ. Because $k(\sigma) = m\sigma/\sqrt{1-\alpha^2}$ is increasing in σ, this means that the curve of constant phase crosses group lines of larger and larger wave number k. Thus a surfer riding the same crest will see the wave lengths becoming shorter and shorter. On other hand, an observer following waves of the same wavelength, will see crests overtake and

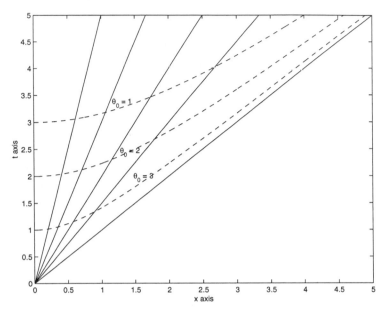

FIGURE 7.1
Group lines and curves of constant phase of the linear Klein-Gordon equation with
$m = 1, \theta_0 = 1, 2, 3.$

pass him. Remember, however, that this picture is only an approximation whose
validity improves as x and t get large.

For a more complete discussion of dispersion, see the book [Wh].

Exercises 7.2

1. Find U and V in equation (7.8) so that u solves the initial conditions

$$u(x, 0) = f(x), \qquad u_t(x, 0) = g(x).$$

Hint: The initial conditions imply that

$$\hat{u}(k, 0) = \hat{f}(k), \qquad \hat{u}_t(k, 0) = \hat{g}(k).$$

Take the Fourier transform of (7.8).

2. The linearized Korteweg-deVries (LKdV) equation is

$$u_t + \alpha u_x + \beta u_{xxx} = 0,$$

where α and β are given positive constants.

(a) Find the dispersion relation, group velocity, and phase velocity.

(b) Sketch the group lines and the curves of constant phase.

(c) State in words how the wave number will change if we follow a particular crest.

3. The equation which describes flexural vibrations in a beam is

$$u_{tt} + \gamma^2 u_{xxxx} = 0.$$

Same questions (a), (b), and (c) as in exercise 2.

4. In the integral (7.12) suppose that $U(k)$ is nonzero for $a \le k \le b$, where $1 \le a < b$. What is the range of values for the group velocity $\omega'(k)$

(a) for the linearized Klein-Gordon equation;

(b) for the linearized Korteweg-deVries equation; and

(c) for the beam equation ?

(d) for which of these equations would you expect a wave packet to spread the most (least) as it evolves?

Program `disper` will be used to illustrate dispersion. It uses the MATLAB fast Fourier transform to evaluate the integral

$$u(x,t) = Re\{\frac{1}{2\pi} \int U(k)e^{i(kx-\omega(k)t)}dk\} = \frac{1}{2\pi} \int U(k)\cos(kx - \omega(k)t)dk,$$

where $U(k) = \sqrt{24\pi} \exp(-6(k-2)^2)$. The range of k is approximately $1 \le k \le 3$. That is to say, U is effectively zero for k outside this interval. U is the Fourier transform of

$$u(x,0) = e^{-x^2/24} \cos(2x).$$

We have chosen $\omega(k) > 0$ so that u is a wave packet moving to the right as t increases. There are three different choices of equation and corresponding dispersion relation.

(i) linearized Klein-Gordon (with $m = 2$); $\omega(k) = \sqrt{4+k^2}$.

(ii) linearized Korteweg-deVries (with $\alpha = 10$ and $\beta = 1/3$); $\omega(k) = 10k - k^3/3$.

(iii) vibrating beam equation (with $\gamma = 1$) ; $\omega(k) = k^2$.

The part of the wave packet with wave number $k = 2$ (near the "center") travels with group velocity $\omega'(2)$.

To run program `disper` you must enter the choice of equation, then enter the time t when you want to view the evolved packet. The solution at time t is put into the vector u, and the original profile at time $t = 0$ is put into the vector u0. The program plots both u and u0 on the interval $[-200, 200]$. On this scale of x values, the graphs of $u(x, t)$ and $u_0(x)$ are both quite compressed. If you want to see some detail of the original packet u_0, use the command `axis([-10 10 -1 1])`. For a "half-wave" profile, use the command `axis([-10 10 0 1])`. To get back to a graph on the original interval $[-200, 200]$, use the command `axis('auto')`. To get some detail on the packet at time t, make a rough estimate of the range of values where u (in green) is nonzero. If this appears to be in the interval $[10, 40]$, then use the command `axis([10 40 -1 1])` . You can further blow up a particular area of the graph with the `zoom` feature.

Program `disper` also calculates a vector v which is the original profile u0 translated by the quantity $t\omega'(2)$. To compare the original profile with that of the evolved wave packet at time t, enter `plot(x,u,x,v)` followed by the appropriate `axis` command.

5. The local wavelength λ of a dispersing wave is twice the distance between two sucessive zeros. Remember that the local wave length and the local wave number k are related by $k = 2\pi/\lambda$.

    Run `disper` for Klein-Gordon with $t = 0$. Only the graph of u_0 will appear. To determine the distance between two sucessive zeros, look at a half-wave profile on the interval $[-10, 10]$. Next use the `ginput` feature of matlab to read off the coordinates of two successive zeros with the following command

    ```
    >> [x,y] = ginput(2)
    ```

    Hit return, and then move the mouse so that the white cross which appears in the figure window is at the intersection of the curve with the x axis. Click once with the left button. Then move to the next zero, and click again. The two x coordinates and the two y coordinates (which should be very close to zero) will appear on the screen.

    Is the wavelength constant across the packet? According to the given data $U(k)$ and the rules for Fourier transform, what should the wavelength be?

6. In this exercise we want to determine the group velocity from the graphs, and see if it agrees with the value predicted by the dispersion relation for the linearized Klein-Gordon equation.

    (a) What is the expression for the group velocity for the linearized Klein-Gordon equation? According to this expression, do the components of the packet with wave number $k > 2$ travel faster or slower than the components with $k < 2$? Restate this result in terms of the wavelengths $\lambda = 2\pi/k$.

    (b)   Run program `disper` with the Klein-Gordon equation and $t = 100$. Look at the wavelengths as they vary across the packet. Are the shorter wavelengths in front or in the rear of the packet? Does this agree with your calculation in part (a)?

    (c)   Calculate $x_2 = 100\omega'(2)$. Use the technique of exercise 5 to estimate the wavelength λ of the packet close to x_2. Then calculate the local wave number k by the formula $k = 2\pi/\lambda$. What should k be? Why?

    (d)   Calculate $x_3 = 100\omega'(3)$, and answer the same questions.

7.   In this exercise we wish to compare the phase velocity, as measured from the graphs, with that predicted by the analytic expression for the phase velocity.

    (a)   Calculate the phase velocity for the linearized Klein-Gordon equation in terms of the local wave number for $m = 2$.

    (b)   Again run program `disper` with $t = 100$. There is a zero at approximately $x = 70.2$. Use the `ginput` feature to determine the local wavelength λ and hence the local wave number $k = 2\pi/\lambda$. Use the command `hold on`. Then run program `disper` again with $t = 100.1$. The profile at time $t = 100.1$ will be superimposed on the profile at time $t = 100$. Blow up the part of graph near 70.2 using the `zoom` feature, and then enter `zoom off`. Next use the `ginput` feature to measure how far the zero at $x = 70.2$ has moved to the right over the time interval $\Delta t = .1$. Determine the approximate velocity of the zero. How does this compare with the phase velocity predicted by the expression in part (a) when evaluated at the local wave number k near $x = 70$?

8. (a)   Repeat exercise 6 for the LKdV equation using $t = 10$. The wave packet now broadens much more rapidly than it did for the linear Klein-Gordon equation. Why?

    (b)   Repeat exercise 7 for LKdV with $t = 10$. Determine the phase velocity at time $t = 10$ for the zero at approximately 60.4.

9. (a)   Repeat exercise 6 for the beam equation. When you run program `disper`, set $t = 20$.

    (b)   Repeat exercise 7 for the beam equation with $t = 20$. Calculate the phase velocity of the zero at approximately 80.4.

7.3   Quantum mechanics and the uncertainty principle

We begin by discussing a property of the Fourier transform that we have noticed before. Then we restate this property in the language of quantum mechanics as the Heisenberg uncertainty principle.

In Section 6.3, we have seen that the scaling rule for the Fourier transform applied to a Gaussian implied that a sharply peaked Gaussian $f(x)$ transforms into a broadly spread Gaussian $\hat{f}(k)$. We shall make a precise statement of this property that applies to a wide range of functions. For the moment let us assume that we have a complex-valued function $f(x)$ such that

$$\int_R |f(x)|^2 dx = 1.$$

By Plancherel's theorem it follows that

$$\int_R |\hat{f}(k)|^2 \frac{dk}{2\pi} = 1.$$

We can think of $x \to |f(x)|^2$ and $k \to (1/2\pi)|\hat{f}(k)|^2$ as probability densities. We introduce

$$m_x = \int_R x|f(x)|^2 dx \qquad \text{and} \qquad m_k = (1/2\pi)\int_R k|\hat{f}(k)|^2 dk. \qquad (7.14)$$

m_x is the *expected value* of the x coordinate or the "center of mass" of the probability density $|f(x)|^2$. m_k is the expected value of the k coordinate or the center of mass of the probability density $|\hat{f}(k)|^2/2\pi$. A measure of how much the function f is concentrated around m_x is given by

$$\sigma_x = \left[\int_R (x - m_x)^2 |f(x)|^2 dx\right]^{1/2}. \qquad (7.15)$$

On the other hand,

$$\sigma_k = \left[\int_R (k - m_k)^2 |\hat{f}(k)|^2 \frac{dk}{2\pi}\right]^{1/2} \qquad (7.16)$$

measures the spread of $\hat{f}$ around m_k. σ_x and σ_p are the *standard deviations* of the probabililty densities $|f(x)|^2$ and $|\hat{f}(k)|^2/2\pi$. $m_x, m_k, \sigma_x,$ and σ_k are all real quantities.

Theorem 7.1

$$\sigma_x \sigma_k \geq \frac{1}{2}.$$ (7.17)

We give the proof. Recall that the Fourier transform takes differentiation in x into multiplication by ik. Thus

$$-i\frac{df}{dx} \leftrightarrow k\hat{f}(k).$$

Hence

$$\sigma_k = \left[\int_R |-if'(x) - m_k f(x)|^2 dx\right]^{1/2}.$$ (7.18)

The Schwarz inequality is also valid for complex-valued functions on R:

$$\left|\int_R f(x)\overline{g(x)}dx\right| \leq \left[\int_R |f(x)|^2 dx\right]^{1/2}\left[\int_R |g(x)|^2 dx\right]^{1/2}.$$ (7.19)

Applying this inequality to

$$J \equiv \int_R (x - m_x) f(x)\overline{(-if'(x) - m_k f(x))}dx,$$

we obtain

$$|J| \leq \left[\int_R (x - m_x)^2 |f(x)|^2 dx\right]^{1/2}\left[\int_R |-if'(x) - m_k f(x)|^2 dx\right]^{1/2} = \sigma_x \sigma_k$$
(7.20)

by (7.19). Now to show that a complex number $z = x + iy$ satisfies $|z| \geq 1/2$ it suffices to show that $|Imz| = |y| \geq 1/2$. If we multiply out the integrand of J, we get four terms:

$$J = \int_R ixf(x)\overline{f'(x)}dx - m_x \int_R f(x)\overline{(-if'(x))}dx$$ (7.21)

$$-m_k \int_R xf(x)\overline{f(x)}dx + m_x m_k \int_R f(x)\overline{f(x)}dx.$$

Now by (6.26)

$$-m_x \int_R f(x)\overline{(-if'(x))}dx = -m_x \int_R \hat{f}(k)k\overline{\hat{f}(k)}\frac{dk}{2\pi}$$

$$= -m_x m_k,$$

and

$$-m_k \int_R x f(x) \overline{f(x)} dx = -m_k \int_R x |f(x)|^2 dx = -m_x m_k.$$

Hence the four terms of (7.21) sum to

$$J = i \int_R x f(x) \overline{f'(x)} dx - m_x m_k \qquad (7.22)$$

because in the last term $\int |f(x)|^2 dx = 1$. We use the fact that

$$2Re(f(x)\overline{f'(x)}) = f(x)\overline{f'(x)} + \overline{f(x)}f'(x)$$

$$= \frac{d}{dx}(f(x)\overline{f(x)}) = \frac{d}{dx}|f(x)|^2.$$

Because m_k and m_x are real,

$$Im(J) = Im(i \int_R x f(x) \overline{f'(x)} dx - m_x m_k) = Re \int_R x f(x) \overline{f'(x)} dx$$

$$= (1/2) \int_R x \frac{d}{dx} |f(x)|^2 dx = (1/2) x |f(x)|^2 \Big|_{-\infty}^{\infty} - (1/2) \int_R |f(x)|^2 dx = -\frac{1}{2}.$$

Thus $|J| \geq |Im(J)| = 1/2$, which is the assertion of Theorem 7.1.$\square$

Now we give a quantum mechanical interpretation of this property. In quantum mechanics, subatomic particles, such as the electron, are described by a *wave function* $x \to \psi(x)$ which is a complex-valued function, such that $|\psi(x)|^2$ is a probability density. The probability of finding a particle with position x in the interval $[a, b]$ is

$$\int_a^b |\psi(x)|^2 dx$$

and, of course, we assume

$$\int_R |\psi(x)|^2 dx = 1.$$

One of the fundamental principles of quantum mechanics is the de Broglie relation for matter waves

$$p = \hbar k$$

where k is the wave number, p is the momentum, and $\hbar = h/2\pi$, with h being Planck's constant. We replace k by $p/\hbar$ in the definition of the Fourier transform of ψ, introduce a normalizing constant, and define φ by

$$\varphi(p) = \frac{1}{\sqrt{2\pi\hbar}}\hat{\psi}(\frac{p}{\hbar}) = \frac{1}{\sqrt{2\pi\hbar}}\int_R \psi(x)e^{-ixp/\hbar}dx. \qquad (7.23)$$

φ is the representation of the particle in momentum space. With this choice of normalizing constants,

$$\psi(x) = \frac{1}{\sqrt{2\pi\hbar}}\int_R \varphi(p)e^{ixp/\hbar}dp, \qquad (7.24)$$

and Plancherel's theorem becomes

$$1 = \int_R |\psi(x)|^2 dx = \int_R |\varphi(p)|^2 dp. \qquad (7.25)$$

Thus $p \to |\varphi(p)|^2$ is also a probability density and

$$\int_{p_1}^{p_2} |\varphi(p)|^2 dp$$

is the probability that the particle has momentum lying in the interval $p_1 \leq p \leq p_2$. As before, the expected value of the position of the particle is

$$m_x = \int_R x|\psi(x)|^2 dx,$$

and the expected value of momentum is

$$m_p = \int_R p|\varphi(p)|^2 dp.$$

The standard deviations of the position and momentum are

$$\sigma_x = \left[\int_R (x - m_x)^2 |\psi(x)|^2 dx\right]^{1/2}$$

and

$$\sigma_p = \left[\int_R (p - m_p)^2 |\varphi(p)|^2 dp \right]^{1/2}.$$

Restated in terms of these variables, Theorem 7.1 becomes the

Heisenberg uncertainty principle $\sigma_x \sigma_p \geq \frac{\hbar}{2}.$

This inequality expresses the idea that it is impossible to determine both the position and the momentum of a particle with arbitrarily good accuracy. This limitation is important for subatomic particles but is negligible for larger objects whose product of position times momentum is of much greater magnitude than $\bar{h} = 1.054 \times 10^{-27}$ erg-sec.

A more thorough treatment of these questions is given in [Be].

Exercises 7.3

1. Here is an intuitive example to illustrate Theorem 7.1. Let

$$f(x) = \begin{cases} 1, & |x| < a \\ 0, & |x| > a \end{cases}$$

be the square pulse of width $2a$.

(a) Find $\hat{f}(k)$, and graph it using MATLAB for several values of a. What happens as a gets larger?

(b) Let $\Delta x = 2a$ be the width of the square pulse. Assign a width Δk to $\hat{f}(k)$ by taking Δk as the distance between the two zeros of $\hat{f}$ that lie closest to $k = 0$.

(c) Verify that $\Delta x \Delta k = 4\pi$, independent of a.

7.4   The Schrödinger equation

7.4.1   The dispersion relation of the Schrödinger equation

Now we turn to the question of the time evolution of the wave function. We consider ψ as a function $\psi(x, t)$ where for each $t, x \rightarrow \psi(x, t)$ is the wave function of the particle. We seek a partial differential equation with the wave functions as solutions. This PDE must satisfy at least two conditions:

(i) It must be first order in t so that specification of the wave function at $t = 0$ determines it uniquely for all times.

(ii) The wave function must satisfy the normalizing condition

$$\int_R |\psi(x,t)|^2 dx = 1$$

for all t.

Recall that we wrote the wave function as an integral over momentum space

$$\psi(x) = \frac{1}{\sqrt{2\pi\hbar}} \int_R \varphi(p) e^{ixp/\hbar} dp.$$

In an attempt to model the wave-particle nature of matter, we shall write $\psi(x,t)$ as an integral of plane waves (called a wave packet):

$$\psi(x,t) = \frac{1}{\sqrt{2\pi\hbar}} \int_R \varphi(p) e^{i(xp/\hbar - \omega t)} dp,$$

where ω should depend on p. This dependence will be a dispersion relation which determines the PDE satisfied by ψ. We want to choose $\omega(p)$ so that the group velocity of this wave packet agrees with the expected velocity of a classical particle. Assume that the momentum probability density φ is concentrated around the value $p = p_0$. We write the exponent as

$$i(xp - \omega\hbar t)/\hbar.$$

Then the group velocity of the packet is

$$v_g = \frac{d}{dp}(\omega\hbar)\bigg|_{p=p_0}.$$

We want this expression to agree with the velocity of a particle with mass m and momentum p_0, that is, we require that

$$\frac{d}{dp}(\omega(p)\hbar)\bigg|_{p=p_0} = \frac{p_0}{m}.$$

Since p_0 is arbitrary, we deduce the more general equation

$$\frac{d\omega}{dp}(p) = \frac{1}{\hbar}\frac{p}{m}.$$

for all p. This requirement will be satisfied if we take

$$\omega(p) = \frac{1}{\hbar} \frac{p^2}{2m},$$

and yields the desired dispersion relation.

With this dispersion relation, what PDE do the plane-wave components

$$w(x, p) \equiv \exp(i(xp/\hbar - \omega(p)t)$$

satisfy? We see that

$$\partial_t w = -i\omega(p)w = -\frac{i}{\hbar} \frac{p^2}{2m} w$$

and

$$\partial_x^2 w = -\frac{p^2}{\hbar^2} w,$$

so that

$$i\hbar w_t = -\frac{\hbar^2}{2m} w_{xx}.$$

Thus the same is true for the wave functions ψ which are integrals (with respect to p) of these plane waves:

$$i\hbar\psi_t = -\frac{\hbar^2}{2m} \psi_{xx}. \qquad (7.26)$$

This is the Schrödinger equation for a wave function which describes the time evolution of a free particle of mass m (no potential).

We note that the dispersion relation we have just found for matter waves is the same (modulo constants) as that found for the vibrating beam $u_{tt} + u_{xxxx} = 0$. Thus the PDE that the plane-wave components satisfy is not uniquely determined by the dispersion relation. However our requirement that the wave function $\psi(x, t)$ be determined by its initial state $\psi(x, 0)$ forces us to choose an equation which is first order in t, that is, the Schrödinger equation (7.26).

7.4.2 The correspondence principle

The plane-wave representation of the wave function is often written in terms of the energy. The Einstein hypothesis assigns an energy E to any matter wave that has a time harmonic behavior $\exp(-i\omega t)$ by the relation

$$E = \hbar\omega.$$

Each plane wave component of the wave function has frequency ω given above. Hence for these plane-wave components,

$$E = E(p) = \frac{p^2}{2m}. \tag{7.27}$$

This agrees with the classical energy of a particle with mass m and momentum p.

Using (7.27) and the dispersion relation we can express the wave function as the integral

$$\psi(x,t) = \frac{1}{\sqrt{2\pi\hbar}} \int_R \varphi(p) e^{i(xp-Et)/\hbar} dp. \tag{7.28}$$

We can think of equation (7.26) as arising (at least formally) by substitution of the operators $i\hbar\partial_t$ for E and $-i\hbar\partial_x$ for p in the relation (7.27). This is an illustration of the *correspondence principle* of quantum mechanics.

The equation for a particle in the presence of a potential is derived via the correspondence principle as follows. The equation for a classical particle of mass m in the presence of a potential is

$$mx'' = -g(x),$$

where $-g(x)$ is the force and $V(x) = \int^x g(y)dy$ is the potential. Now the conserved energy is

$$E = \frac{m}{2}(x')^2 + V(x) = \frac{p^2}{2m} + V(x). \tag{7.29}$$

Substituting $i\hbar\partial_t$ for E and $-i\hbar\partial_x$ for p in (7.29) (which extends (7.27)), we arrive at the Schrödinger equation with a potential

$$i\hbar\psi_t = -\frac{\hbar^2}{2m}\psi_{xx} + V(x)\psi. \tag{7.30}$$

We have replaced the observables E and p by differential operators, and the potential $V(x)$ has been associated with the multiplication operator $\psi \to V(x)\psi$. The conserved energy quantity (7.29) is called the Hamiltonian of the classical system, written

$$H(x, p) = \frac{p^2}{2m} + V(x).$$

Equation (7.30) is often written symbolically

$$i\hbar\psi_t = H(x, -i\hbar\partial_x)\psi.$$

The Schrödinger equation (7.30) is first order in time which ensures that condition (i) is satisfied. Now we verify that condition (ii) is satisfied by solutions of (7.30). Observe that for a complex-valued function

$$\frac{d}{dt}|\psi|^2 = \frac{d}{dt}(\psi\bar\psi) = \psi_t\bar\psi + \psi\bar\psi_t = 2Re(\psi_t\bar\psi).$$

Now multiply (7.30) by $\bar\psi$ and use the previous observation to deduce

$$i(\hbar/2)\frac{d}{dt}|\psi|^2 = i\hbar Re(\psi_t\bar\psi) = Im(i\hbar\psi_t\bar\psi)$$

$$= -\frac{\hbar^2}{2m}Im(\psi_{xx}\bar\psi)$$

because $Im(V(x)|\psi|^2) = 0$. Finally integrate in x to find

$$i(\hbar/2)\frac{d}{dt}\int_R |\psi|^2 dx = i(\hbar/2)\int_R \frac{d}{dt}|\psi|^2 dx$$

$$= -\frac{\hbar^2}{2m}Im\int_R \psi_{xx}\bar\psi dx = 0$$

because

$$\int_R \psi_{xx}\bar\psi dx = -\int_R \psi_x\bar\psi_x dx = -\int_R |\psi_x|^2 dx$$

is real. Hence $\int |\psi|^2 dx$ is constant which implies that solutions of (7.30) satisfy condition (ii).

7.4.3 The initial-value problem for the free Schrödinger equation

We turn our attention to solving the IVP for the free Schrödinger equation (7.26):

$$i\hbar\psi_t = -\frac{\hbar^2}{2m}\psi_{xx}, \qquad \psi(x,0) = \psi_0(x),$$

where $\psi_0(x)$ is the wave function specified at $t = 0$. According to our heuristic arguments, the solution should be given by (7.28) with $\varphi(p) = \varphi_0(p)$ being the momentum space representaton of the initial wave function ψ_0. We shall find the solution to this problem in terms of ψ_0 in a more systematic fashion using the Fourier transform. To take the "clutter" out of the equation, let us make a change of variable, writing

$$\psi(x,t) = u(\frac{\sqrt{2m}}{\hbar}x, \frac{t}{\hbar}).$$

The problem becomes

$$iu_t = -u_{xx}, \quad (x,t) \in R^2, \tag{7.31}$$

$$u(x,0) = f(x), \quad x \in R, \tag{7.32}$$

where $f(x) = \psi_0(\hbar x/\sqrt{2m})$. We take the Fourier transform in x of the equation $u_t = iu_{xx}$ and deduce that $\hat{u}(\xi,t)$ satisfies the ODE

$$\hat{u}_t(\xi,t) = -i\xi^2\hat{u}(\xi,t), \qquad \hat{u}(\xi,0) = \hat{f}(\xi).$$

Thus

$$\hat{u}(\xi,t) = \hat{u}(\xi,0)e^{-i\xi^2 t} = \hat{f}(\xi)e^{-i\xi^2 t}. \tag{7.33}$$

Using the convolution rule for the Fourier transform, we see that

$$u(x,t) = G(x,t) * f(x), \tag{7.34}$$

where $G(x,t)$ is the inverse Fourier transform of $\exp(-i\xi^2 t)$,

$$G(x,t) = \frac{1}{2\pi}\int_R e^{-i\xi^2 t}e^{ix\xi}\,d\xi. \tag{7.35}$$

However, $\exp(-i\xi^2 t)$ does not decay as $\xi \to \pm\infty$ so the integral (7.35) will not converge absolutely. On the other hand

$$e^{-i\xi^2 t} = \cos(\xi^2 t) - i\sin(\xi^2 t)$$

oscillates very rapidly as $\xi \to \pm\infty$, faster than $\xi \to \exp(ix\xi)$. This means that for large $|\xi|$, there will be a great deal of cancellation in the integral (7.35). It seems plausible that the integral converges conditionally and that

$$\lim_{a\to\infty} \int_{-a}^{a} e^{-i\xi^2 t} e^{ix\xi}\, d\xi$$

exists.

We can get $G(x, t)$ in closed form using the Fourier transform (6.30) which we write in terms of the inverse transform as

$$\frac{1}{2\pi} \int e^{-\xi^2/(4\mu)} e^{ix\xi}\, d\xi = \sqrt{\frac{\mu}{\pi}} e^{-\mu x^2}$$

for $Re\,\mu > 0$. Now we take

$$\mu_\varepsilon = \frac{1}{4(\varepsilon + i)t} \qquad \text{with } t > 0,$$

so that $-\xi^2/(4\mu_\varepsilon) = -(\varepsilon + i)\xi^2$. Then using the inverse transform rule above,

$$G_\varepsilon(x, t) \equiv \frac{1}{2\pi} \int_R e^{-i(\varepsilon+i)\xi^2 t} e^{ix\xi}\, d\xi$$

$$= \frac{1}{\sqrt{4\pi(\varepsilon + i)t}} e^{-\frac{x^2}{4(\varepsilon+i)t}}.$$

Now we let $\varepsilon \downarrow 0$ and find that

$$G(x, t) = \frac{1}{\sqrt{4\pi it}} e^{-\frac{x^2}{4it}} = \frac{1}{\sqrt{4\pi t}} e^{i(\frac{x^2}{4t} - \frac{\pi}{4})}. \tag{7.36}$$

It is easy to check that G solves (7.31) for $x \in R, t > 0$. With some care in taking the limit, we can show that

$$G * f = \lim_{\varepsilon\to 0} G_\varepsilon * f$$

is the solution of the IVP (7.31)-(7.32).

$G(x, t)$ given by (7.36) is the fundamental solution of the free Schrödinger equation. G is formally derived from the fundamental solution $S(x, t)$ of the heat equation by replacing k by i. However G has radically different properties.

First we observe that G may be defined for $t < 0$, because the evolution process governed by the Schrödinger is reversible in time, whereas the evolution governed by the heat equation is basically a one-way process. Note that if $u(x, t)$ solves (7.31), then so does $w(x, t) = \bar{u}(x, -t)$. Now $G(x, t)$ is defined for $t > 0$ by (7.36). We can define a fundamental solution $H(x, t)$ for $t < 0$ by the formula

$$H(x, t) = \overline{G(x, -t)} = \frac{1}{\sqrt{4\pi(-t)}}\left[e^{\frac{-x^2}{4i(-t)}}e^{-i\pi/4}\right]$$

$$= \frac{1}{\sqrt{4\pi(-t)}}e^{\frac{-x^2}{4it}}e^{i\pi/4}.$$

Thus we can extend the meaning of $G(x, t)$ to $t < 0$ with the same formula

$$G(x, t) = \frac{1}{\sqrt{4\pi it}}e^{\frac{-x^2}{4it}}$$

if, when we take $\sqrt{it}$, we always mean the root with positive real part:

$$\sqrt{it} = \begin{cases} \sqrt{t}e^{i\pi/4} & \text{for } t > 0 \\ \sqrt{-t}e^{-i\pi/4} & \text{for } t < 0 \end{cases}.$$

For fixed $t \neq 0$, $x \rightarrow G(x, t)$ does not decay as $x \rightarrow \pm\infty$, but oscillates rapidly. For fixed x as $t \rightarrow \pm\infty$, the amplitude decays as $|t|^{-1/2}$ and the oscillations slow down. On the other hand, for fixed x, as $t \rightarrow 0$, the amplitude blows up as $|t|^{-1/2}$ *and* the oscillations become more and more rapid. Thus $\lim_{t \rightarrow 0} G(x, t)$ does not exist.

We can summarize our results in

Theorem 7.2
Let f be complex-valued and piecewise continuous on R with $\int_R |f(x)|^2 dx < \infty$. Then there is a unique solution to the IVP (7.31)-(7.32) existing for all $t \in R$ given by the convolution

$$u(x, t) = G(x, t) * f(x) = \frac{1}{\sqrt{4\pi it}}\int_R e^{i(\frac{(x-y)^2}{4t})}f(y)dy. \tag{7.37}$$

Furthermore,

$$\int_R |u(x, t)|^2 dx = \int_R |f(x)|^2 dx \qquad \text{for all } t. \tag{7.38}$$

It remains to consider (7.38). In fact we used a multiplier technique to show that (7.38) is true for solutions of the Schrödinger equation. However it is interesting to verify (7.38) using properties of the Fourier transform. For the solution given by (7.37), (7.33) implies that $|\hat{u}(\xi, t)| = |\hat{f}(\xi)|$ for each t and all ξ. Then by the Plancherel theorem,

$$\int_R |u(x, t)|^2 dx = \frac{1}{2\pi} \int_R |\hat{u}(\xi, t)|^2 d\xi$$

$$= \frac{1}{2\pi} \int_R |\hat{f}(\xi)|^2 d\xi = \int_R |f(x)|^2 dx.$$

An excellent introductory treatment of quantum mechanics with historical motivation is given in [FT]. A more advanced treatment is the well known text [M].

Exercises 7.4

1. When the particle is moving at very high velocity, we must replace (7.27) by the relativistic form of the energy

    $$E^2 = p^2 c^2 + c^4 m^2,$$

    where c is the speed of light. Using the correspondence principle, derive the equation for the wave function which incorporates the relativistic form of the energy. This is the linear Klein-Gordon equation.

2. (a) Show that if u is a solution of $i u_t = -u_{xx} + V(x)u$, where V is a real valued potential, then $v(x, t) = \bar{u}(x, -t)$ is also a solution.

    (b) Let $f(x)$ be real, and let u be the solution of the IVP for $t > 0$. Show that for $t < 0$, $u(x, t) = \bar{u}(x, -t)$.

3. The fundamental solution of the free Schrödinger equation, $G(x, t)$, is given in equation (7.36). Make an mfile, bigg.m for $G(x, t)$. Plot on the interval $[-10, 10]$ for several values of $t > 0$. What happens to the oscillations as t gets smaller? As t gets larger?

The program unifm provides an approximation to the IVP (7.31), (7.32) by approximating the convolution $u = G * f$. Remember that the convolution integral

$$(G * f)(x) = \int G(x - y, t) f(y) dy$$

is approximated by the Riemann sums

$$\sum_{i=1}^{n} G(x - y_i, t) f(y_i) \Delta y.$$

We take the initial data $f(x) = 1/\sqrt{2}$ for $|x| < 1$, and $f = 0$ elsewhere. $|f(x)|^2$ is a uniform probability density on the interval $[-1, 1]$. Input parameters are the time t and the number of terms n. The solution is put into the vector u. The program plots $Re(u)$ in yellow, $Im(u)$ in magenta, and the probability density $\rho = |u|^2$ in turquoise (cyan) together on the same graph. The plots are done on $[0, 20]$ because the functions are even. You can also plot them separately. If you are running MATLAB5.0, you may enter the commmand `colordef black` before running the program to see the graphs more clearly.

4. Run program `unifm` for $t = .5$ and for $n = 4, 8, 12,$ and 20. After running for $n = 4$, save the results in the vector u_4 with the command `u4 = u;`. Then run again with $t = .5$, and $n = 8$, saving the results with the command `u8 = u;`. To compare the results for $t = .5$ and $n = 4, 8, 12, 20$, plot these vectors together pairwise, e.g., `plot(x,u4,x,u8)`. What is happening as the number of terms n increases? Now repeat the process for $t = .1$, and for $t = 2$. In each case, how large does n have to be to make the plots converge? Can you explain why the n should be different by looking at the integrand?

5. Compare the solution computed by program `unifm` with that computed by the program `heat1` for the heat equation. The initial data in both cases has step discontinuities. What similar properties and what different properties do you see? Which is more like the solution of the heat equation, $u(x, t)$, or the probability density $\rho(x, t) = |u(x, t)|^2$? For each $t > 0$, no matter how small, can you find points x near 20 where $\rho(x, t) > 0$? What is the physical significance of this fact?

7.5 The spectrum of the Schrödinger operator

7.5.1 Continuous spectrum

We return to the physical variables for the moment. The *Schrödinger operator* is the linear differential operator in the spatial variable

$$H\psi = -\frac{\hbar^2}{2m}\psi_{xx} + V(x)\psi. \tag{7.39}$$

We assume that the potential $V(x)$ is a real-valued function, which is bounded and piecewise continuous. Of course the definition of the operator is not complete until we specify its domain. We take the domain of the Schrödinger operator as the space W of complex-valued functions ψ, such that $\psi \in C^1(R)$, ψ'' is piecewise continuous on R and such that $\int_R |\psi(x)|^2 dx$, $\int_R |\psi'(x)|^2$, and $\int_R |\psi''(x)|^2 dx$ are all finite. In particular, if a function $\psi \in W$, then $\psi(x)$ and $\psi'(x) \to 0$ as $x \to \pm\infty$.

A function $\psi = \psi(x)$ is an eigenfunction of H if $\psi \in W$ and

$$H\psi = \lambda\psi \tag{7.40}$$

for some complex number λ. In fact the eigenvalues of H are real. Recall that the complex scalar product on W is given by

$$\langle f, g \rangle = \int_R f(x)\bar{g}(x)dx.$$

Using Green's second identity, it is easy to show that

$$\langle H\psi, \varphi \rangle = \langle \psi, H\varphi \rangle. \tag{7.41}$$

Then supposing that $\psi \in W$ is an eigenfunction of H with eigenvalue λ, we see that

$$\lambda\|\psi\|^2 = \langle H\psi, \psi \rangle = \langle \psi, H\psi \rangle = \langle \psi, \lambda\psi \rangle$$

$$= \bar{\lambda}\|\psi\|^2$$

whence $\lambda = \bar{\lambda}$, which implies that λ is real.

What is the physical meaning of an eigenvalue? If $\psi \in W$ and λ satisfy (7.40) then

$$e^{-i\lambda t/\hbar}\psi(x)$$

satisfies the time-dependent Schrödinger equation (7.30) and has angular frequency $\omega = \lambda/\hbar$. By the Einstein hypothesis, the energy E of such a matter wave is related to the frequency by $E = \hbar\omega$. Thus if ψ is an eigenfunction of H, then ψ has definite energy E which is the eigenvalue.

We start with the eigenvalue problem for the free Schrödinger equation which we write in nonphysical variables as

$$(Hu)(x) = -u''(x) = \lambda u(x). \tag{7.42}$$

Does this equation have any solutions in W? The answer is no. The general solution of (7.42) is

$$u(x) = Ae^{x\sqrt{\lambda}} + Be^{-x\sqrt{\lambda}}. \tag{7.43}$$

Now there is no choice of λ, or any choice of A or B, such that $u(x) \to 0$ as $x \to \infty$ and $x \to -\infty$.

However, there are functions in W which come arbitrarily close to being eigenfunctions of H. To see this, consider the action of H in the Fourier transform variable ξ:

$$\widehat{Hu}(\xi) = -(i\xi)^2\hat{u}(\xi) = \xi^2\hat{u}(\xi).$$

For any ξ_0 and $l > 0$, define the function

$$\varphi_l(x) = \frac{1}{\sqrt{l}}e^{ix\xi_0}e^{-(1/2)(x/l)^2}. \tag{7.44}$$

The Fourier transform of φ_l is given by

$$\hat{\varphi}_l(\xi) = \frac{1}{\sqrt{l}}\int_R e^{-(1/2)(x/l)^2}e^{-i(\xi-\xi_0)x}dx = (2\pi l)^{-1/2}e^{-l^2(\xi-\xi_0)^2/2},$$

where we have used the transform rule for the Gaussian (6.21) and the scaling rule (6.22) with $a = (1/l)$. Then the transform of $H\varphi_l$ is

$$\widehat{H\varphi_l}(\xi) = \xi^2\hat{\varphi}_l(\xi).$$

But for large l, $\hat{\varphi}_l$ is concentrated near ξ_0, so that

$$\widehat{H\varphi_l}(\xi) \approx \xi_0^2\hat{\varphi}_l(\xi).$$

Thus as $l \to \infty$, φ_l is closer and closer to being an eigenfunction of H with eigenvalue $\lambda = \xi_0^2$. More precisely, we can say that

$$\int_R |(H\varphi_l)(x) - \xi_0^2\varphi_l(x)|^2dx \to 0 \tag{7.45}$$

as $l \to \infty$. To see this we use the Plancherel theorem (6.23)

$$\int_R |(H\varphi_l)(x) - \xi_0^2\varphi_l(x)|^2dx = \frac{1}{2\pi}\int_R |\widehat{H\varphi_l}(\xi) - \xi_0^2\hat{\varphi}_l(\xi)|^2d\xi$$

$$= \frac{1}{2\pi}\int_R |\xi^2 - \xi_0^2|^2|\hat{\varphi}_l(\xi)|^2d\xi$$

$$= l \int_R |\xi^2 - \xi_0^2|^2 e^{-l^2(\xi - \xi_0)^2} d\xi$$

$$= \int_R |(\frac{\eta}{l})^2 + 2\xi_0(\frac{\eta}{l})|^2 e^{-\eta^2} d\eta$$

where we have used the change of variable $\eta = l(\xi - \xi_0)$. It is clear that this last integral tends to zero as $l \to \infty$. This establishes (7.45).

However, this is a trival result if $\varphi_l \to 0$ as $l \to \infty$. We must show that, in some sense, φ_l do not tend to zero as $l \to 0$. We take this to be the mean square sense. It is a simple calculation to show that

$$\int_R |\varphi_l(x)|^2 dx = \sqrt{\pi} \qquad \text{for all } l.$$

The values of λ for which there are "approximate" eigenfunctions, or "almost" eigenfunctions, constitute what is called the *continuous spectrum* of the free Schrödinger operator. We have seen that the continuous spectrum of the free Schrödinger operator consists of $\{\lambda \geq 0\}$ and that there are no eigenvalues. For $\lambda < 0$, there are no eigenvalues and no approximate eigenvalues.

7.5.2 Bound states of the square well potential

The situation can be quite different for the Schrödinger operator with a potential. We consider the case of a "square well" potential.' Let $Q > 0$ and take

$$V(x) = \begin{cases} 0 & -1 < x < 1 \\ Q, & |x| > 1 \end{cases}.$$

The Schrödinger operator is

$$(Hu)(x) = -u''(x) + V(x)u(x)$$

and the eigenvalue problem for H is

$$-\varphi''(x) + V(x)\varphi(x) = \lambda\varphi(x), \qquad (7.46)$$

where we require that $\varphi \in W$. First observe that if λ is an eigenvalue, then $\lambda \geq 0$. In fact, since $V \geq 0$,

$$\lambda \int_R |\varphi|^2 dx = \langle H\varphi, \varphi \rangle$$

$$= \int_R -\varphi''\bar{\varphi} + V|\varphi|^2 dx$$

$$= \int_R |\varphi'|^2 + V|\varphi|^2 dx \geq 0.$$

Because of the symmetry of $V(x)$, $\varphi(x)$ is a solution of (7.46) if and only if $\varphi(-x)$ is also a solution (with the same value λ). It can also be shown that the eigenspace associated with each eigenvalue is one-dimensional. Thus the eigenfunctions are either even functions or odd functions. Supposing that φ is even, (7.46) is equivalent to two ODE's

$$-\varphi'' = \lambda\varphi, \quad 0 < x < 1, \tag{7.47}$$

and

$$-\varphi'' + Q\varphi = \lambda\varphi, \quad x > 1, \tag{7.48}$$

with the boundary conditions $\varphi'(0) = 0$, $\varphi(x) \to 0$ as $x \to \infty$, and φ and φ' continuous at $x = 1$. We seek the solution in the form

$$\varphi(x) = \begin{cases} \cos(x\sqrt{\lambda}), & 0 < x < 1 \\ Ae^{-(x-1)\sqrt{Q-\lambda}}, & x > 1 \end{cases}.$$

We suppose that $\lambda < Q$ and take the decaying exponential solution of (7.48) to satisfy the requirement that $\varphi \to 0$ as $x \to \infty$. If $\lambda \geq Q$ there is no solution of (7.48) which decays as $x \to \infty$. Note also that if $\lambda = 0$, the solution $\varphi(x) \equiv 1$ in $0 < x < 1$. It cannot satisfy the requirement that φ' be continuous at $x = 1$. Thus the eigenvalues must lie in the open interval $0 < \lambda < Q$. The requirement that φ be continuous at $x = 1$ is satisfied if we take $A = \cos(\sqrt{\lambda})$. Finally, to make the derivative continuous at $x = 1$, we require that

$$-\sqrt{\lambda}\sin(\sqrt{\lambda}) = -\sqrt{Q-\lambda}\cos(\sqrt{\lambda})$$

or

$$\tan(\sqrt{\lambda}) = \frac{\sqrt{Q-\lambda}}{\sqrt{\lambda}}. \tag{7.49}$$

Now supposing that φ is an odd function, we want φ to satisfy (7.47) and (7.48) with the conditions $\varphi(0) = 0$, $\varphi \to 0$ as $x \to \infty$, and with φ and φ' continuous at $x = 1$. Accordingly we take φ in the form

$$\varphi(x) = \begin{cases} \sin(x\sqrt{\lambda}), & 0 < x < 1 \\ Be^{-(x-1)\sqrt{Q-\lambda}}, & x > 1 \end{cases}.$$

To make φ continuous at $x = 1$, we choose $B = \sin(\sqrt{\lambda})$. To make φ' continuous at $x = 1$, we require that λ be chosen so that

$$\sqrt{\lambda}\cos(\sqrt{\lambda}) = -\sqrt{Q - \lambda}\sin(\sqrt{\lambda}).$$

This means that λ must satisfy the equation

$$\tan(\sqrt{\lambda}) = -\frac{\sqrt{\lambda}}{\sqrt{Q - \lambda}}. \tag{7.50}$$

Setting $s = \sqrt{\lambda}$ in (7.49) and (7.50), we see that the eigenvalues correspond to the roots in the interval $0 < s < \sqrt{Q}$ of the two equations

$$\tan(s) = \frac{\sqrt{Q - s^2}}{s} \equiv p(s) \tag{7.51}$$

and

$$\tan(s) = -\frac{s}{\sqrt{Q - s^2}} \equiv q(s). \tag{7.52}$$

A rough estimate of the location of the roots can be made from Figure 7.2. The roots can be found by using a root finder, such as the routine `fzero` of MATLAB. The location of the roots s_k is determined as follows. Suppose that n is odd and $(n - 1)\pi/2 < \sqrt{Q} \leq n(\pi/2)$. Then there are n roots $s_k, k = 1, \ldots, n$ with

$$0 < s_1 < \frac{\pi}{2} < s_2 < \pi < \ldots < (n - 1)\pi/2 < s_n < \sqrt{Q} \leq n\frac{\pi}{2}.$$

If n is even, and $(n - 1)\pi/2 < \sqrt{Q} < n(\pi/2)$, then again there are n roots with the same location above. The spectrum of the Schrödinger operator in this case consists of the eigenvalues $0 < \lambda_1 < \ldots < \lambda_n < Q$ and continuous spectrum for $\lambda \geq Q$. If n is odd and $\sqrt{Q} = (n - 1)\pi/2$, then $s_n = (n - 1)\pi/2$ so that $\lambda_n = Q$. The roots s_k with k odd yield the even eigenfunctions, and those with k even yield the odd eigenfunctions. Since there are only a finite number of eigenfunctions, they cannot be complete. This is a fundamental difference between a boundary value problem on a finite interval, in which we expect an infinite number of eigenfunctions, and a problem on the whole line in which we may have no eigenfunctions (free Schrödinger equation) or only a finite number.

These eigenfunctions are called "bound states" because they represent the states in which the particle is captured by the potential. Notice however, that the eigenfunctions have an exponentially small "tail" for $|x| > 1$. Even though the particle is captured by the potential, there is still a small probability that it may be found outside the potential well. The eigenfunction corresponding to the lowest eigenvalue $\lambda_1 = s_1^2$ is called the "ground state." The graphs of the eigenfunctions

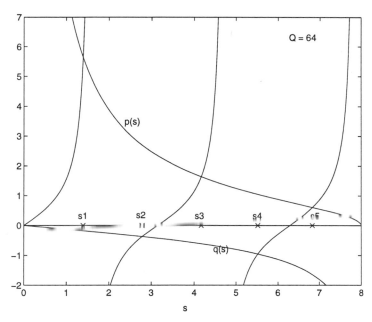

FIGURE 7.2
Graphs of $p(s)$, $q(s)$, and $\tan(s)$ for $Q = 64$.

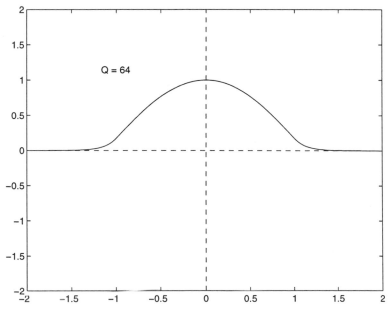

FIGURE 7.3
First bound state (ground state) for $Q = 64$, lowest eigenvalue $\lambda_1 = 1.947$.

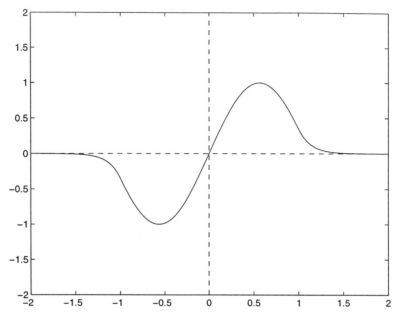

FIGURE 7.4
Second bound state for $Q = 64$, eigenvalue $\lambda_2 = 7.7613$.

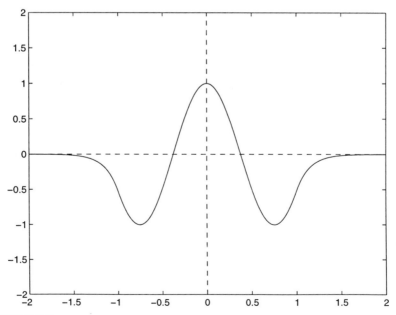

FIGURE 7.5
Third bound state for $Q = 64$, eigenvalue $\lambda_3 = 17.3458$.

corresponding to the first three bound states when $Q = 64$ are shown in Figures 7.3, 7.4, and 7.5.

Exercises 7.5

1. Normalize the eigenfunctions so that they have square integral over R equal to 1. Then calculate the probability that the particle lies outside the potential well for both even and odd bound states. How does the probability change as the eigenvalue gets closer to Q?

2. Let $Q = 100$ in the square well potential.

   (a) How many eigenvalues are there?

   (b) Use MATLAB to plot $p(s)$, $q(s)$ and $\tan(s)$ on $[0, 10]$ to give rough estimates for the first three roots s_1, s_2, s_3.

   (c) Use the MATLAB root finder `fzero` to find the first three roots and the first three eigenvalues.

   (d) Use MATLAB to plot the first three eigenfunctions on $[-2, 2]$.

3. How do the eigenvalues of the square well potential depend on Q?

   (a) Let $\lambda_k(Q)$ be the kth eigenvalue of (7.46). From the plots of the curves (7.51) and (7.52), find $\lim_{Q \to \infty} \lambda_k(Q)$.

   (b) What is the limit of the corresponding eigenfunctions when $Q \to \infty$?

   (c) When $Q = +\infty$, the potential well has infinitely high walls, and the probability of finding the particle outside the well is zero. This situation is called the "particle in a box." What problem for the vibrating string yields the same eigenvalues and eigenfunctions?

4. Let E_k be the kth eigenvalue of the eigenvalue problem in physical variables

   $$-\frac{\hbar^2}{2m}\psi_{xx} + V(x)\psi(x) = E\psi(x),$$

   where the potential is given by

   $$V(x) = \begin{cases} 0, & |x| < l \\ V, & |x| > l \end{cases}.$$

   (a) Write $\psi(x) = \varphi(x/l)$. Show that

   $$-\varphi''(x/l) + \frac{2ml^2}{\hbar^2}V(x)\varphi(x/l) = \frac{2ml^2}{\hbar^2}E\varphi(x/l).$$

(b)   In this equation set $y = x/l$ and conclude that now $\varphi(y)$ is a solution of the normalized eigenvalue problem (7.46) with potential

$$W(y) = \begin{cases} 0, & |y| < 1 \\ \frac{2ml^2 V}{\hbar^2}, & |y| > 1 \end{cases}.$$

(c)   Conclude that the kth energy level E_k of the problem in physical variables is given by

$$E_k(V, l, m) = \frac{\hbar^2}{2ml^2} \lambda_k \left(\frac{2ml^2 V}{\hbar^2} \right)$$

where $\lambda_k(Q)$ is the kth eigenvalue of (7.46).

5.   If l, V and m are fixed, how does $E_k(V, l, m)$ behave as $\hbar \to 0$? (Use the results of exercises 3 and 4). How do the corresponding eigenfunctions behave? Are they converging to nonzero functions as $\hbar \to \infty$? This is the classical limit of the discrete energy levels of the quantum theory.

7.6   Projects

1.   Find the eigenvalues and eigenfunctions for the time-independent Schrödinger equation with potential

$$V(x) = \begin{cases} 0, & |x| > 1 \\ -Q, & -1 < x < 0 \\ Q, & 0 < x < 1 \end{cases}.$$

Use `fzero` to find several of the eigenvalues, and graph the corresponding eigenfunctions.

Chapter 8

The Heat and Wave Equations in Higher Dimensions

8.1 Diffusion in higher dimensions

8.1.1 Derivation of the heat equation

We shall use $\mathbf{x}$ to denote points in R^2 or R^3 with components (x, y) or (x, y, z), and the Euclidean length of a vector is denoted $|\mathbf{x}| = \sqrt{x^2 + y^2}$ or $|\mathbf{x}| = \sqrt{x^2 + y^2 + z^2}$. The Laplace operator (Laplacian) in R^2 or R^3 is

$$\Delta u = u_{xx} + u_{yy} \qquad \text{or} \qquad \Delta u = u_{xx} + u_{yy} + u_{zz}.$$

The derivation of the heat equation in higher dimensions parallels the derivation in one dimension in Chapter 3. Let G be an open, bounded set in R^3 (for example, an open ball or cube), and let $u(\mathbf{x}, t)$ be the temperature at each point of G. Let B be a smaller ball contained entirely within G. The amount of heat energy contained in B is

$$\iiint_B c(\mathbf{x})\rho(\mathbf{x})u(\mathbf{x}, t) \, d\mathbf{x},$$

where c is the specific heat (calories/degree/volume) and ρ is the density (mass/volume). The flux of heat energy is given by a vector field $\mathbf{F}(\mathbf{x}, t)$. If A is the area of a flat surface element with unit normal $\mathbf{n}$, then the rate of heat flow through the surface element in the direction $\mathbf{n}$ is given by $(\mathbf{F} \cdot \mathbf{n})A$. Thus $\mathbf{F}$ has units calories/time/area. Now in the absence of any sources or sinks, the balance law for diffusion says that the rate of change of heat energy in B must equal the rate at which heat flows through the boundary surface ∂B, that is,

$$\frac{d}{dt} \iiint_B c\rho u \, d\mathbf{x} = -\iint_{\partial B} \mathbf{n} \cdot \mathbf{F} \, dS,$$

where $\mathbf{n}$ is the exterior unit normal to the surface ∂B. We see that if $\mathbf{F} \cdot \mathbf{n} > 0$ on ∂B, the amount of heat energy in B is decreasing. The divergence theorem (Section 1.1.3) says this surface integral may be rewritten as a volume integral, provided that the vector field $\mathbf{F}$ is C^1 in the interior of G:

$$\int\int_{\partial B} \mathbf{n} \cdot \mathbf{F} \, dS = \int\int\int_{B} div \mathbf{F} \, d\mathbf{x}.$$

Also assuming that u is C^1 in B, we may express the balance law for diffusion as

$$\int\int\int_{B} [c\rho u_t + div \mathbf{F}] \, d\mathbf{x} = 0.$$

Since this holds for all balls $B \subset G$, we conclude that the integrand must be zero in G,

$$c\rho u_t + div(\mathbf{F}) = 0 \quad \text{in } G.$$

However, Fourier's law of cooling says that heat flows in the direction opposite to the temperature gradient:

$$\mathbf{F} = -\kappa \nabla u.$$

Here, as in Chapter 3, κ is the diffusitivity of the material. Substituting in the differential form of the diffusion law, we conclude that

$$c\rho u_t - div(\kappa \nabla u) = 0.$$

If c, ρ and κ are constants, we set $k = \kappa/(c\rho)$, and we find that

$$u_t = k div(\nabla u) = k\Delta u,$$

which is the usual heat equation without sources.

8.1.2   The fundamental solution of the heat equation

The heat equation in higher dimensions has an interesting property that is useful in constructing solutions. We illustrate in the case of two space dimensions. If $u(x, t)$ and $v(y, t)$ each solve the homogeneous heat equation in one dimension, then $w(x, y, t) = u(x, t)v(y, t)$ solves the heat equation in two space dimensions. In fact,

$$w_t = u_t v + u v_t,$$

and

$$\Delta w = (\partial_x^2 + \partial_y^2)w = u_{xx}v + uv_{yy},$$

so that

$$w_t - k\Delta w = (u_t - ku_{xx})v + u(v_t - kv_{yy}) = 0.$$

It is not surprising then that we can construct the fundamental solution of the heat equation in higher dimensions by taking products of the fundamental solution in one dimension,

$$S_1(x, t) = \frac{1}{\sqrt{4\pi kt}} e^{-\frac{x^2}{4kt}}.$$

For example, we can show that the fundamental solution in two space dimensions is

$$S_2(x, y, t) = S_1(x, t)S_1(y, t) = \frac{1}{4\pi kt} \exp(-\frac{x^2 + y^2}{4kt}).$$

By the calculation above, this product solves the heat equation in two space dimensions. Furthermore, for $t > 0$,

$$\int \int_{R^2} S_2(x, y, t)\, dxdy = \int_R S_1(x, t)dx \int_R S_1(y, t)dy = 1.$$

Finally suppose that the initial data $f(x, y) = \varphi(x)\psi(y)$, where both φ and ψ are continuous and bounded on R. Then

$$u(x, y, t) = \int \int_{R^2} S_2(x - \xi, y - \eta, t)\varphi(\xi)\psi(\eta)d\xi d\eta.$$

solves $u_t = k(u_{xx} + u_{yy})$ and takes on the initial data as $t \to 0$. In fact

$$\lim_{t\to 0} u(x, y, t) = \lim_{t\to 0} \int_R S_1(x - \xi, t)\varphi(\xi)d\xi \lim_{t\to 0} \int_R S_1(y - \eta, t)\psi(\eta)d\eta$$

$$= \varphi(x)\psi(y).$$

By linearity, this limit also exits for any initial data which is a finite, linear combination of products:

$$f(x, y) = \sum_{i=1}^{N} \varphi_i(x)\psi_i(y).$$

Finally we can extend the limiting operation for any $f(x, y)$ bounded and continous on R^2.

The same argument works as well for the heat equation in any number of space variables. The fundamental solution in n space dimensions is given by

$$S_n(\mathbf{x}, t) = \frac{1}{(4\pi kt)^{n/2}} \exp(-\frac{|\mathbf{x}|^2}{4kt}).$$

As in Chapter 3, we can solve the IVP in n space dimensions in terms of a convolution. The solution of

$$u_t = k\Delta u, \quad \mathbf{x} \in R^n, t > 0, \qquad u(\mathbf{x}, 0) = f(\mathbf{x}), \quad \mathbf{x} \in R^n \qquad (8.1)$$

is given by

$$u(\mathbf{x}, t) = \int_{R^n} S_n(\mathbf{x} - \mathbf{y}, t) f(\mathbf{y}) d\mathbf{y}. \qquad (8.2)$$

It follows from the principle of Duhamel that the solution of the inhomogeneous equation

$$u_t = k\Delta u + q, \qquad u(\mathbf{x}, 0) = 0,$$

is given by

$$u(\mathbf{x}, t) = \int_0^t \int_{R^n} S_n(\mathbf{x} - \mathbf{y}, t - s) q(\mathbf{y}, s) d\mathbf{y} ds. \qquad (8.3)$$

The maximum principle holds as well in higher dimensions. Let G be an open, bounded set in R^n, and set $Q_T = G \times (0, T)$. Let Γ be that part of the boundary of Q given by

$$\Gamma = \{\partial G \times [0, T]\} \cup \bar{G} \times \{0\}.$$

Let $u \in C^2(Q_T)$ be continuous on $\bar{Q}_T$ and solve $u_t = k\Delta u$ in Q_T. Then

$$\max_{\bar{Q}_T} u = \max_{\Gamma} u.$$

Exercises 8.1

1. Let the sequence of functions $f_k(x, y) = k^2/\pi$ for $x^2 + y^2 < 1/k^2$ and $f_k = 0$ elsewhere. Let $u_k(x, y, t)$ be the solution of (8.1) with $u_k(x, y, 0) = f_k(x, y)$. Show that $u_k(x, y, t) \to S_2(x, y, t)$ as $k \to \infty$. Compare with exercise 8 of Section 3.3.

2. Let $u(x, y, t)$ be a solution of (8.1), such that $u, u_x, u_y, xu_x, yu_y \to 0$ rapidly as $r = \sqrt{x^2 + y^2} \to \infty$.

    (a) Show that $Q = \int \int u(x, y, t)dxdy$ is constant.

    (b) If $Q \neq 0$, show that the moments

    $$m_x = \frac{1}{Q} \int \int u(x, y, t)xdxdy \quad \text{and} \quad m_y = \frac{1}{Q} \int \int u(x, y, t)ydxdy$$

    are constant.

    (c) Let

    $$p(t) \equiv \frac{1}{Q} \int \int [(x - m_x)^2 + (y - m_y)^2]u(x, y, t)dxdy.$$

    Show that $p(t) = p(0) + 4kt$. Compare this result with exercise 6 of Section 3.3.

3. Let $u_1(x, y, t)$ be the solution of (8.1) with initial data $f_1(x, y)$, and let $u_2(x, y, t)$ be the solution with initial data $f_2(x, y)$. Let Q_1 and Q_2 be the integrals, respectively, of f_1 and f_2. Assume that $Q_1 \neq 0$, $Q_2 \neq 0$, and $Q_1 + Q_2 \neq 0$. Let m_x^1 and m_y^1 be the moments of f_1, and let m_x^2 and m_y^2 be the moments of f_2.

    If u is the sum solution $u(x, y, t) = u_1(x, y, t) + u_2(x, y, t)$ with total heat quantity Q and moments m_x, m_y, show that

    $$m_x = \frac{Q_1 m_x^1 + Q_2 m_x^2}{Q_1 + Q_2}$$

    and

    $$m_y = \frac{Q_1 m_y^1 + Q_2 m_y^2}{Q_1 + Q_2}.$$

4.   Let

$$u(x, y, t) = \frac{Q}{4\pi kt} e^{-\frac{(x-a)^2+(y-b)^2}{4kt}}.$$

(a)   Verify by direct computation that $\int\int u(x, y, t)dxdy = Q, m_x = a,$ $m_y = b$, and $p(t) = 4kt$ for $t > 0$.

(b)   What is the quantity of heat contained in the disk $(x-a)^2+(y-b)^2 < p(t)$? Compare with the result of exercise 6(e) of Section 3.3.

5.   Let u be the solution of (8.1) given by

$$u(x, y, t) = \int\int S_2(x - \xi, y - \eta, t)f(\xi, \eta)d\xi d\eta.$$

Let Q, m_x, m_y and $p_0 = p(0)$ be the quantities associated with f. Define

$$v(x, y, t) = Q(\frac{1 - p_0/4kt}{4\pi kt})e^{-\frac{(x-m_x)^2+(y-m_y)^2}{4kt}}$$

$$= Q(1 - p_0/4kt)S_2(x - m_x, y - m_y, t).$$

Expand $\xi, \eta \rightarrow S(x-\xi, y-\eta, t)$ in a Taylor series about $\xi = m_x, \eta = m_y$, and substitute in the integral for u. Show that

$$|u(x, y, t) - v(x, y, t)| \leq C/t^3$$

for x, y in bounded rectangles as $t \rightarrow \infty$.

The function mfile heat6.m computes the solution u of (8.1) with initial data

$$f(x, y) = \frac{Q_1}{4\pi s_1} e^{-\frac{(x-a_1)^2+(y-b_1)^2}{4s_1}} + \frac{Q_2}{4\pi s_2} e^{-\frac{(x-a_2)^2+(y-b_2)^2}{4s_2}}$$

The values $s_1 = .5$ and $s_2 = 2$ are set in the function. We group the remaining parameters into two vectors: centers $= [a_1, b_1, a_2, b_2]$ and $Q = [Q_1, Q_2]$. We declare these vectors global so that they may be accessed by the function heat6 and then assign values to them with the commands:

```
>> global centers Q
>> centers = [-3 -5 4 6 ]
>> Q = [1 4]
```

Now if we wish to calculate the value of $u(x, y, t)$ at, say, $x = 1, y = 4, t = 20$, we use the commmand u = heat6(1,4,20). We are more interested in plotting the solution. We shall do so on the square $[-10, 10] \times [-10, 10]$. To this end we set up a grid of mesh points in this square with the sequence of commands:

```
>> x = -10: .1 : 10;
>> y = x;
>> [X,Y] = meshgrid(x,y);
```

If we want to plot the graph of the function at time, say $t = 20$, we invoke the function heat6, now with matrix arguments, by the command U = heat6(X,Y,20); We can apply several plotting features with the following sequence of commands:

```
>> U = heat6(X,Y,20);
>> surf(X,Y,U); shading flat
>> pcolor(X,Y,U);shading flat
```

The command surf makes a 3-D plot, while the command pcolor makes a 2-D plot with colors indicating height. To see the level curves of the function, use the command contour(X,Y,U, 20) . The number 20 tells MATLAB to divide the range of the function into 21 equal intervals and plot 20 level curves, corresponding to the 20 interior values (excluding the maximum and minimum).

6. (a) Use the result of exercise 3 to calculate Q, m_x, m_y, and $p(t)$ for the solution of (8.1) computed by function heat6.

(b) Assign the values $a_1 = -3, b_1 = -5, a_2 = 4, b_2 = 6, Q_1 = 1, Q_2 = 4$. Note that the heights of the two humps are the same at $t = 0$. Choose times $t = 2, 5, 10, 20$. For each of these times make the pcolor plot and make the contour plot with 20 level curves. Estimate for what time t do you first get a temperature distribution with a single maximum? Where is it located (look at the contours)?

(c) Now plot the solution on the rectangle $[2, 3] \times [3.5, 4.5]$ by taking

```
>> x = 2:.02:3; y = 3.5:.02:4.5
>> [X,Y] = meshgrid(x,y);
```

Use the same quantities $a_1, b_1, a_2, b_2, Q_1, Q_2$ but with $t = 100, 500, 2000, 5000$. Check to see where the maximum occurs when there is single hump using the contour command. The asymptotic approximation of exercise 5 predicts that for large t, the solution should look like a single hump with its maximum at m_x, m_y. You can get explicit values for m_x, m_y using the assigned values for $a_1, b_1, a_2, b_2, Q_1, Q_2$ and the result of part (a).

For $t = 5000$, find where max u occurs, and compare with the predicted asymptotic values m_x, m_y.

7. Modify the function heat6 to yield the solution which is the sum of three Gaussian humps, and do plotting exercises as in exercise 6.

8. Let $u(x, y, t) = 4kt + x^2 + y^2$.

   (a) Verify that u solves (8.1)

   (b) Verify that the maximum (minimum) principle hold for u on any set $\{x^2 + y^2 \le a^2, 0 \le t \le T\}$.

9. Find an integral formula which gives the solution of the IBVP

$$u_t - k\Delta u = 0 \quad \text{in} \quad \{(x, y, t,) : x > 0, y \in R, t > 0\},$$

$$u(0, y, t) = 0 \quad \text{for} \quad y \in R, \ t \ge 0,$$

$$u(x, y, 0) = f(x, y) \quad \text{in} \quad \{(x, y) : x > 0, y \in R\}.$$

10. Show that if f has radial symmetry, i.e., $f(x, y) = f(r)$ where $r = \sqrt{x^2 + y^2}$, then so does the solution of the IVP (8.1). By putting in polar coordinates show, that the problem reduces to a PDE in variables r and t.

8.2 Boundary value problems for the heat equation

We illustrate several boundary value problems for the heat equation in two space dimensions. The formulation in three dimensions is exactly the same.

We think of a flat, thin plate of metal occupying the bounded, open set G of points in R^2. We assume that the plate is insulated on top and bottom so that no heat flows through these surfaces. The boundary ∂G is assumed to be piecewise C^1, that is, ∂G consists of a finite number of C^1 arcs with no sharp cusps. Examples would be a disk, rectangle, or any polygon. On each smooth piece of the boundary we can define a unit exterior normal vector $\mathbf{n}$. The normal derivative at the boundary of a function $u(x, y)$ defined on G is

$$\frac{\partial u}{\partial n} = \nabla u \cdot \mathbf{n}.$$

We assign boundary conditions of the first, second, or third kind to describe the heat flow through the boundary ∂G. The boundary value problems are

$$u_t = k\Delta u \quad \text{in } G \times (0, \infty), \qquad u(x, y, 0) = f(x, y), \qquad (8.4)$$

$$u = 0 \qquad \text{on } \partial G \times [0, \infty), \tag{8.5}$$

or

$$\frac{\partial u}{\partial n} = 0 \qquad \text{on } \partial G \times [0, \infty), \tag{8.6}$$

or

$$\frac{\partial u}{\partial n} + hu = 0 \qquad \text{on } \partial G \times [0, \infty), \tag{8.7}$$

where h is a continuous function on ∂G, $h > 0$. In each case we can make the boundary condition inhomogeneous by replacing the zero right-hand side by a function defined on ∂G. In the case of (8.5) we would be prescribing the temperature on the boundary, while in (8.6) we would be prescribing the flux.

In addition to (8.5) - (8.7), we can pose boundary value problems involving a combination of (8.5) and (8.6). For example, if ∂G is composed of two parts

$$\partial G = \Gamma_0 \cup \Gamma_1,$$

a mixed boundary condition is

$$u = 0 \quad \text{on } \Gamma_0, \qquad \frac{\partial u}{\partial n} = 0 \quad \text{on } \Gamma_1. \tag{8.8}$$

The method of separation of variables can be used to analyze the structure of these boundary value problems. Again we look for building-block solutions in the form

$$\varphi(x, y)\psi(t).$$

If we substitute this form in (8.4), we see that φ and ψ must satisfy

$$\frac{\psi'}{k\psi} = \frac{\Delta\varphi}{\varphi} = -\lambda,$$

where λ is constant. As before, this leads to an eigenvalue problem for the equation in the spatial variables:

$$-\Delta\varphi = \lambda\varphi \qquad \text{in } G, \tag{8.9}$$

with boundary conditions chosen from (8.5) -(8.8). The temporal equation is as before,

$$\psi' + k\lambda\psi = 0. \tag{8.10}$$

Now we set $X = C(\bar{G})$ and, for example,

$$W = \{u \in C^2(\bar{G}) : u = 0 \text{ on } \partial G\}$$

to implement the boundary condition (8.5). Note that W is a subspace of X. Then the eigenvalue problem can be expressed as that of finding eigenvalues and eigenfunctions of the linear operator $L : W \rightarrow X$ defined by $Lu = -\Delta u$, with domain W.

The boundary conditions (8.5)-(8.8) are all examples of symmetric boundary conditions for Δ. The appropriate scalar product for functions f, g on X is

$$\langle f, g \rangle = \int\int_G f(x, y)g(x, y)dxdy.$$

We will show that for each of the boundary conditions (8.5)-(8.8),

$$\langle Lu, v \rangle = \langle u, Lv \rangle$$

for all $u, v \in W$.

To see this, we shall need the higher dimensional form of the Green's identities. Let $u(x, y)$ and $v(x, y)$ be C^2 functions on G, and define the vector field $\mathbf{f}(x, y) = v(x, y)\nabla u(x, y)$. Observe that

$$div(\mathbf{f}) = div(v\nabla u) = \nabla u \cdot \nabla v + v\Delta u$$

because $div(\nabla u) = \Delta u$. Now apply the divergence theorem (Theorem 1.8) to $\mathbf{f}$. We find that

$$\int_{\partial G} v\frac{\partial u}{\partial n}\, ds = \int_{\partial G} \mathbf{n} \cdot \mathbf{f}\, ds = \int\int_G div(\mathbf{f})\, dxdy$$

$$= \int\int [\nabla u \cdot \nabla v + v\Delta u]\, dxdy.$$

This is *Green's first identity*, which can be rewritten

$$\int\int (-\Delta u)v\, dxdy = -\int_{\partial G} \frac{\partial u}{\partial n}\, ds + \int\int_G \nabla u \cdot \nabla v\, dxdy. \tag{8.11}$$

If we interchange the roles of u and v, we see that

$$\int\int_G u(-\Delta v)\,dxdy = -\int_{\partial G} u\frac{\partial v}{\partial n}\,ds + \int\int_G \nabla v \cdot \nabla u\,dxdy.$$

Subtract this last equation from (8.11) to deduce *Green's second identity*:

$$\int\int_G -\Delta uv\,dxdy = \int_{\partial G}[-\frac{\partial u}{\partial n}v + u\frac{\partial v}{\partial n}]ds + \int\int_G u(-\Delta v)dxdy. \quad (8.12)$$

In terms of the scalar product, we can express (8.12) as

$$\langle Lu, v\rangle = \langle u, Lv\rangle + \int_{\partial G}[-\frac{\partial u}{\partial n}v + u\frac{\partial v}{\partial n}]\,ds.$$

Now if $u, v \in W$, where W is determined by one of the boundary conditions (8.5) - (8.8), the boundary integral term

$$\int_{\partial G}[-\frac{\partial u}{\partial n}v + u\frac{\partial v}{\partial n}]ds = 0.$$

Thus for $u, v \in W$,

$$\langle Lu, v\rangle = \langle u, Lv\rangle. \quad (8.13)$$

This symmetry condition is also satisfied in the complex scalar product space, and implies that all the eigenvalues of a symmetric operator must be real. Using exactly the same argument as in Section 4.3, we can use (8.13) to show that the eigenfunctions corresponding to different eigenvalues are orthogonal in the scalar product. Note, however, that the eigenspaces corresponding to an eigenvalue λ are not always one-dimensional, that is, there are independent eigenfunctions with the same eigenvalue (to be seen in examples). The eigenspace for each eigenvalue is finite-dimensional, and we can always find an orthogonal basis of of eigenfunctions for each eigenspace.

For which of the boundary conditions (8.5) - (8.8) do we have all the eigenvalues $\lambda > 0$? Here we use Green's first identity. Let $u = v = \varphi$ be an eigenfunction with eigenvalue λ: $-\Delta\varphi = \lambda\varphi$. Then (8.11) becomes

$$\lambda \int\int_G \varphi^2 dxdy = \int\int_G (-\Delta\varphi)\varphi dxdy$$

$$= -\int_{\partial G} \frac{\partial \varphi}{\partial n}\varphi ds + \int\int_G |\nabla\varphi|^2 dxdy.$$

Now any of the boundary conditions (8.5) - (8.8) make the boundary integral nonnegative. Hence if φ is a eigenfunction (eigenfunctions cannot be identically zero) with eigenvalue λ,

$$\lambda \geq \frac{\int \int_G |\nabla \varphi|^2 dx dy}{\int \int_G \varphi^2 dx dy}. \tag{8.14}$$

The only way for $\lambda = 0$ to be an eigenvalue is to have the numerator vanish:

$$\int \int_G |\nabla \varphi|^2 dx dy = 0,$$

which implies $\varphi \equiv$ constant. Now going back to the boundary conditions, note that if $\varphi =$ constant, and satisfies either (8.5), (8.7), (8.8), then it follows that $\varphi \equiv 0$, so that φ is not an eigenfunction. Thus for each of the boundary conditions (8.5), (8.7), or (8.8) we conclude that

$$\lambda > 0. \tag{8.15}$$

The one exception is the boundary condition (8.6) for which $\varphi \equiv$ constant is an eigenfunction, with eigenvalue $\lambda_0 = 0$.

Now it can be shown (see [RR]) that for each of the boundary conditions (8.5) - (8.8), there is an infinite sequence of eigenvalues

$$\lambda_n \to \infty \quad \text{as } n \to \infty$$

and an infinite set of orthogonal eigenfunctions, which are complete. We shall label the eigenpairs (λ_n, φ_n) with the understanding that not all the λ_n are distinct because there may be several independent eigenfunctions $\varphi_{n_1}, \varphi_{n_2}, \ldots, \varphi_{n_k}$ with the same eigenvalue. We shall see this in the examples.

Then we can write the solution of (8.4) with one of the boundary conditions (8.5) - (8.8) as

$$u(x, y, t) = \sum_n A_n \exp(-\lambda_n k t) \varphi_n(x, y). \tag{8.16}$$

The coefficients A_n are found by the formula

$$A_n = \frac{\langle f, \varphi_n \rangle}{\langle \varphi_n, \varphi_n \rangle}, \tag{8.17}$$

where

$$\langle \varphi_n, \varphi_n \rangle = \int \int_G \varphi_n^2(x, y) dx dy.$$

As in Chapter 4, we can show that the series (8.16) converges rapidly for $t > 0$, and when $t = 0$, the series converges to f in the mean square sense in G:

$$\int \int_G |f(x, y) - \sum_1^N A_n \varphi_n(x, y)|^2 dxdy \to 0$$

as $N \to \infty$.

Exercises 8.2

1. Suppose that the region G is the upper semidisk centered at the origin of radius a. Suppose that the temperature is held at zero on the curved portion of the boundary, while the straight portion of the boundary is insulated.

   (a) Write down the IBVP for these boundary conditions.

   (b) Will all the eigenvalues λ be strictly positive? Why? What does this say about the behavior of the temperature as $t \to \infty$?

2. If the material in the plate is not uniform, the density, specific heat, and conductivity may vary from place to place. Then the heat equation becomes

   $$\sigma(x, y)u_t = div(\kappa(x, y)\nabla u)$$

   where $\sigma = c\rho$ is the product of the specific heat and the density.

   (a) Make a separation of variables $u(x, y, t) = \varphi(x, y)\psi(t)$. Show that the eigenvalue problem in the spatial variables now becomes

   $$-div(\kappa \nabla u) = \lambda \sigma u.$$

   (b) Define the weighted scalar product on G as

   $$\langle f, g \rangle = \int \int f(x, y)g(x, y)\sigma(x, y)dxdy.$$

   Let the operator
   $$Lu = -\frac{1}{\sigma}div(\kappa \nabla u)$$

   with W being the subspace of $C^2(G)$ defined by one of the boundary conditions (8.5) - (8.8). Show that

   $$\langle Lu, v \rangle = \langle u, Lv \rangle \quad \text{for all} \quad u, v \in W.$$

(c) Show that if φ is an eigenfunction of L with eigenvalue λ, then

$$\lambda \geq \frac{\int \int_G \kappa |\nabla \varphi|^2 dx dy}{\int \int_G \sigma |\varphi|^2 dx dy}.$$

8.3   Eigenfunctions for the rectangle

In this section we address the problem of finding the eigenfunctions in two space dimensions. We shall be able to determine the eigenfunctions explicitly by another application of separation of variables, but only for special geometries.

Let G be the rectangle

$$G = \{(x, y) : 0 < x < a, 0 < y < b\}.$$

We shall find the eigenfunctions for a mixed boundary value problem (8.8) in which the metal plate is insulated on the top and bottom edges

$$u_y(x, 0, t) = u_y(x, b, t) = 0 \qquad \text{for } 0 \leq x \leq a, \ t \geq 0, \qquad (8.18)$$

and is kept at temperature zero on the left and right edges:

$$u(0, y, t) = u(a, y, t) = 0 \qquad \text{for } 0 \leq y \leq b, \ t \geq 0. \qquad (8.19)$$

Look for eigenfunctions in the form of products

$$\varphi(x, y) = X(x)Y(y),$$

and substitute in equation (8.9). This yields

$$-\frac{X''}{X} - \frac{Y''}{Y} = \lambda.$$

Since X''/X depends only on x and Y''/Y depends only on y, we separate into two one-dimensional boundary value problems:

$$-Y'' = \mu Y, \qquad Y'(0) = Y'(b) = 0, \qquad (8.20)$$

and

$$-X'' = (\lambda - \mu)X, \qquad X(0) = X(a) = 0. \qquad (8.21)$$

The solutions of (8.20) are known:

$$Y_n(y) = \cos(\frac{n\pi y}{b}), \qquad \mu_n = (\frac{n\pi}{b})^2 \quad n = 0, 1, 2, \ldots.$$

Replacing μ by μ_n in (8.21) yields the solutions

$$X_m = \sin(\frac{m\pi x}{a})$$

with eigenvalues

$$\lambda_{m,n} - \mu_n = (\frac{m\pi}{a})^2, \quad m = 1, 2, 3, \ldots.$$

Thus the eigenvalues are

$$\lambda_{m,n} = (\frac{m\pi}{a})^2 + (\frac{n\pi}{b})^2$$

with eigenfunctions

$$\varphi_{m,n}(x, y) = \sin(\frac{m\pi x}{a})\cos(\frac{n\pi y}{b}) \quad \text{for } m = 1, 2, 3, \ldots, n = 0, 1, 2, \ldots.$$

Note here that if $a = b = \pi$, $\lambda_{m,n} = m^2 + n^2$. Thus, for example, when $m = 1$, $n = 2$, $\lambda_{1,2} = 5$ with eigenfunction $\varphi_{1,2} = \sin x \cos 2y$. However, $\varphi_{2,1} = \sin 2x \cos y$ also has eigenvalue 5, so that there are two independent eigenfunctions with the same eigenvalue. In fact there are eigenvalues with an arbitrarily large number of independent eigenfunctions. An example of an eigenvalue with four independent eigenfunctions is $\lambda = 85$ which can be written as

$$85 = 2^2 + 9^2 \qquad \text{and} \qquad 85 = 6^2 + 7^2.$$

The four independent eigenfunctions are $\varphi_{2,9}$, $\varphi_{9,2}$, $\varphi_{6,7}$, $\varphi_{7,6}$. An eigenvalue with six independent eigenfunctions is $\lambda = 850$ which can be written as a sum of squares in three distinct ways:

$$850 = 3^2 + 29^2 = 11^2 + 27^2 = 15^2 + 25^2.$$

Back to the case of the rectangle G with sides of length a and b, we see that the normalizing constants in (8.17) are

$$\langle \varphi_{m,n}, \varphi_{m,n} \rangle = \int \int_G \varphi_{m,n}^2(x, y) dx dy$$

$$= \int_0^a \sin^2(\frac{m\pi x}{a}) dx \int_0^b \cos^2(\frac{n\pi x}{b}) dy = \frac{ab}{4},$$

for $m, n \geq 1$, while

$$< \varphi_{m,0}, \varphi_{m,0} >= \frac{ab}{2}.$$

Hence

$$A_{m,n} = \frac{4}{ab} \int \int_G f(x, y) \varphi_{m,n}(x, y) dx dy$$

for $m, n \geq 1$, while

$$A_{m,0} = \frac{2}{ab} \int \int_G f(x, y) \varphi_{m,0} dx dy.$$

Example

If we take $f(x, y) = x + y^2$, the coefficients can be computed: For $m, n \geq 1$,

$$A_{m,n} = \frac{4}{ab} \int_0^a x \sin(\frac{m\pi x}{a}) dx \int_0^b \cos(\frac{n\pi y}{b}) dy$$

$$+ \frac{4}{ab} \int_0^a \sin(\frac{m\pi x}{a}) dx \int_0^b y^2 \cos(\frac{n\pi y}{b}) dy.$$

Now for $m \geq 1$,

$$\int_0^a \sin(\frac{m\pi x}{a}) dx = (\frac{a}{m\pi})[1 - (-1)^m],$$

and

$$\int_0^a x \sin(\frac{m\pi x}{a})dx = -(\frac{a^2}{m\pi})(-1)^m,$$

while for $n \geq 1$,

$$\int_0^b \cos(\frac{n\pi y}{b})dy = 0,$$

and

$$\int_0^b y^2 \cos(\frac{n\pi y}{b})dy = \frac{2b^3}{(n\pi)^2}(-1)^n.$$

Consequently for $m, n \geq 1$,

$$A_{m,n} = \frac{8b^2[1 - (-1)^m](-1)^n}{m\pi(n\pi)^2},$$

and for $m \geq 1$,

$$A_{m,0} = \frac{2}{ab}\int_0^a x \sin(\frac{m\pi x}{a})dx \int_0^b dy + \frac{2}{ab}\int_0^a \sin(\frac{m\pi x}{a})dx \int_0^b y^2 dy$$

$$= \frac{2}{m\pi}[\frac{b^3}{3} - (a + \frac{b^3}{3})(-1)^m].$$

Exercises 8.3

1. Find the eigenvalues and eigenfunctions for the rectangle G with the boundary conditions

    $$u(0, y, t) = u(a, y, t) = 0, \qquad 0 \leq y \leq b, \ t \geq 0,$$

    $$u(x, 0, t) = 0, \quad u_y(x, b, t) = 0, \qquad 0 \leq x \leq a, t \geq 0.$$

2. (a) In the boundary value problem of exercise 1, let $a = b = \pi$, and consider the two modes $\varphi_1(x, y) = \sin(x)\sin(y/2)$ and $\varphi_2(x, y) = \sin(2x)\sin(3y/2)$. Find the eigenvalues λ_1 and λ_2 associated with each mode.

(b)   Write a function mfile warm.m which will evaluate the solution

$$u(x, y, t) = .2\varphi_1(x, y)e^{-\lambda_1 t} + \varphi_2(x, y)e^{-\lambda_2 t}.$$

Make sure that it is array-smart. Then put a mesh over the square $[0, \pi] \times [0, \pi]$. Choose values of $t = 0, .2, .5, .75, 1$. For each value of t, make a pcolor plot using pcolor with flat shading, and super-impose the level curves. Use the following sequence of commands, which you can put in a script mfile with t as an input parameter.

```
>> U = warm(X,Y,t);
>> colormap(hot)
>> brighten(.5)
>> pcolor(X,Y,U); shading flat
>> hold on
>> contour(X,Y,U, 'k')
>> hold off
```

Notice how the contours change as the second mode makes a smaller contribution.

(c)   What angle do the contour lines make with the upper edge of the square? Why? What is the relation of the contour lines to the other three sides of the rectangle? Why? Phrase your answers in terms of the temperature gradient $\nabla u = (u_x(x, y, t), u_y(x, y, t))$.

3.   Consider the problem of heat conduction in an infinite slab, which is insu-lated on the top and bottom:

$$u_t - k\Delta u = 0, \qquad x \in R, 0 < y < b, t > 0,$$

$$u_y(x, 0, t) = u_y(x, b, t) = 0, \qquad x \in R, t \geq 0,$$

$$u(x, y, 0) = f(x, y), \qquad x \in R, 0 < y < b.$$

We shall impose the boundary condition that $u(x, y, t)$ remains bounded as $x \to \pm\infty$.

(a)   Writing $u = \varphi(x, y)\psi(t)$, what is the eigenvalue problem for φ?

(b)   Separate variables again, $\varphi = X(x)Y(y)$, and follow the procedure at the beginning of this section. What are the eigenvalues and eigen-functions for Y? What is the ODE for X? For which λ does this ODE have solutions which are bounded as $x \to \pm\infty$? Does this "eigenvalue problem" have discrete solutions ?

(c)   Since this is not our usual kind of eigenvalue problem, we shall use a different procedure. The eigenfunctions of part (b) are $Y_n(y) = \cos(n\pi y/b)$ with eigenvalues $(n\pi/b)^2$, $n = 0, 1, \ldots$. Assume we can expand the solution in terms of these eigenfunctions

$$u(x, y, t) = \sum_0^\infty u_n(x, t) Y_n(y)$$

with unknown functions $u_n(x, t)$. Substitute this expansion into the PDE. What PDE in (x, t) must each coefficient function u_n satisfy? Likewise, expand

$$f(x, y) = \sum_0^\infty f_n(x) Y_n(y).$$

Show that the initial condition for each u_n is $u_n(x, 0) = f_n(x)$.

(d)   Use the result of exercise 9 of Section 3.3 to solve the IVP for each u_n in terms of a convolution.

4.   Suppose that the rectangle is now insulated on three sides and the temperature is held at zero on the bottom side. Find the solution of the IBVP

$$u_t - \Delta u + \gamma u = 0 \quad \text{in } G, t > 0,$$

$$u(x, y, 0) = f(x, y) \quad \text{in } G,$$

with these boundary conditions.

5.   Suppose that the temperature is held at zero on the left, right, and bottom sides of the rectangle, while the boundary condition on the top $(y = b)$ is

$$u_y + hu = 0.$$

Solve the IBVP for the equation $u_t - k\Delta u = 0$ and these boundary conditions.

8.4   Eigenfunctions for the disk

In this section we study the eigenvalue problem (8.9) when G is the disk of radius a,

$$G = \{(x, y) \in R^2 : x^2 + y^2 < a^2\}.$$

We use polar coordinates to take advantage of the symmetry. We will be led to Bessel's equation and its solutions which are called Bessel functions. The Laplacian in polar coordinates (see Appendix A) is

$$\Delta\varphi = \varphi_{rr} + \frac{1}{r}\varphi_r + \frac{1}{r^2}\varphi_{\theta\theta}.$$

We separate variables again. This time we look for eigenfunctions which are products of functions of the radial variable r and the angle θ:

$$\varphi(r, \theta) = R(r)\Phi(\theta).$$

Substitution of this product in (8.9) yields

$$\frac{r^2 R''}{R} + \frac{r R'}{R} + \lambda r^2 = -\frac{\Phi''}{\Phi}.$$

The left side of this equation depends only on r while the right side depends only on θ. Hence we set both sides equal to a constant μ. This yields two ODE boundary value problems:

$$\Phi'' + \mu\Phi = 0, \tag{8.22}$$

where Φ must have period 2π, and

$$r^2 R'' + r R' + (\lambda r^2 - \mu)R = 0 \tag{8.23}$$

with boundary conditions that R must be bounded as $r \to 0$ and $R(a) = 0$. The 2π periodic solutions of (8.22) are obviously

$$\Phi_n(\theta) = A_n \cos(n\theta) + B_n \sin(n\theta), \quad \mu_n = n^2, \quad n = 0, 1, 2, \ldots,$$

where A_n and B_n are arbitrary constants.

Next we turn to the solution of (8.23) with μ replaced by $\mu_n = n^2$. We make the change of variables $\rho = r\sqrt{\lambda}$ and set $R(\lambda, r) = U(r\sqrt{\lambda})$. Then

$$R'(\lambda, r) = \sqrt{\lambda}U'(r\sqrt{\lambda}),$$

and

$$R''(\lambda, r) = \lambda U''(r\sqrt{\lambda}).$$

Hence (8.23) becomes

$$\rho^2 U''(\rho) + \rho U'(\rho) + (\rho^2 - n^2)U(\rho) = 0 \qquad (8.24)$$

with the boundary conditions that U must be bounded as $\rho \to 0$ and $U(a\sqrt{\lambda}) = 0$. Equation (8.24) is known as *Bessel's equation* of order n. For each n, there is a pair of independent solutions, $J_n(\rho)$ and $Y_n(\rho)$. J_n is called the *Bessel function of the first kind of order n*. Multiples of J_n are the only solutions which are bounded as $\rho \to 0$. All other solutions of (8.24) involve a component of Y_n which blows up as $\rho \to 0$. These are called the *Bessel functions of the second kind of order n*.

When $n = 0$, they are $J_0(\rho)$ and $Y_0(\rho)$ which have the properties

$$J_0(0) = 1, \qquad J_0(\rho) \approx \sqrt{\frac{2}{\pi\rho}} \cos(\rho - \pi/4) \qquad \text{as } \rho \to \infty$$

$$Y_0(\rho) \to -\infty \quad \text{as } \rho \to 0, \qquad Y_0(\rho) \approx \sqrt{\frac{2}{\pi\rho}} \sin(\rho - \pi/4) \quad \text{as } \rho \to \infty.$$

Their graphs are displayed here (Figure 8.1).

The condition that we require solutions U to be bounded as $\rho \to 0$ says that we must take U as a multiple of J_0. In fact, we take $U(\rho) = J_0(\rho)$. Then changing variables back again,

$$R(\lambda, r) = J_0(r\sqrt{\lambda}).$$

We see that there is a sequence $\lambda_{m,0} \to \infty$ such that $a\sqrt{\lambda_{m,0}} = \rho_{m,0}$, where $\rho_{m,0}$ is the mth zero of J_0. Similarly, there is a bounded solution of (8.24) for general n, $J_n(\rho)$, and a sequence $\lambda_{m,n} \to \infty$, such that $J_n(a\sqrt{\lambda_{m,n}}) = 0$. The $\lambda_{m,n}$ are the eigenvalues of this problem.

For $n = 0$, there is a single sequence of eigenfunctions, depending only on r:

$$\varphi_{m,0}(r) = J_0(r\sqrt{\lambda_{m,0}}).$$

For $n \geq 1$, there is a double sequence of two independent eigenfunctions for each pair m, n:

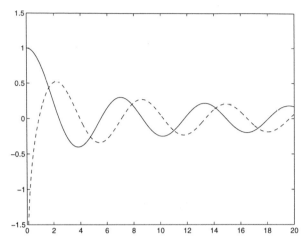

FIGURE 8.1
Graphs of Bessel functions $J_0(x)$ (solid curve) and $Y_0(x)$ (dashed curve).

$$\varphi_{m,n}^1(r, \theta) = J_n(r\sqrt{\lambda_{m,n}}) \cos(n\theta)$$

and

$$\varphi_{m,n}^2(r, \theta) = J_n(r\sqrt{\lambda_{m,n}}) \sin(n\theta).$$

When expressed in terms of x, y, these eigenfunctions are orthogonal in the scalar product (8.11), which in polar coordinates becomes

$$\langle f, g \rangle = \int_0^{2\pi} \int_0^a f(r, \theta) g(r, \theta) r \, dr \, d\theta.$$

The solution of the IBVP (8.4), (8.5), with G the disk of radius a, is given by

$$u(r, \theta, t) = \sum_{m=1}^{\infty} A_m \varphi_{m,0} \exp(-\lambda_{m,0} k t) \tag{8.25}$$

$$+ \sum_{m,n=1}^{\infty} \left[A_{m,n} \varphi_{m,n}^1(r, \theta) + B_{m,n} \varphi_{m,n}^2(r, \theta) \right] \exp(-\lambda_{m,n} k t)$$

where

$$A_m = \frac{\langle f, \varphi_{m,0} \rangle}{\langle \varphi_{m,0}, \varphi_{m,0} \rangle},$$

$$A_{m,n} = \frac{\langle f, \varphi_{m,n}^1 \rangle}{\langle \varphi_{m,n}, \varphi_{m,n} \rangle},$$

$$B_{m,n} = \frac{\langle f, \varphi_{m,n}^2 \rangle}{\langle \varphi_{m,n}^2, \varphi_{m,n}^2 \rangle}.$$

In particular if the initial data $f = f(r)$ is independent of θ, then $A_{m,n} = B_{m,n} = 0$ for all $m, n \geq 1$, and the solution reduces to the first sum in (8.25) with

$$A_m = \frac{\int_0^a f(r) \varphi_{m,0}(r) r \, dr}{\int_0^a \varphi_{m,0}^2(r) r \, dr}.$$

Since the Bessel functions are given in terms of power series, or some other integral representation, the coefficients A_m as a rule cannot be computed by hand. There are numerical means of evaluating the integrals. However, it may be easier to solve the IBVP by numerical means using the finite element method, as described in Chapter 10.

In many situations, for example exercise 6 of this section, after separating variables one arrives at equation (8.23) with μ not an integer. Let $s = \sqrt{\mu}$. Make the same change of variable $\rho = r\sqrt{\lambda}$ which yields

$$\rho^2 U'' + \rho U' + (\rho^2 - s^2)U = 0.$$

Solutions of this equation are called Bessel's functions of order s, denoted $J_s(\rho)$. More generally, by replacing n by a complex number α, one defines the Bessel functions J_α and Y_α. The solutions can be expressed in power series. There are also very useful integral representations of the Bessel functions. Bessel functions are so important that they have been computed to high accuracy in various handbooks, for example [AS]. However, now they are available at a key stroke in many software packages, including MATLAB. To find what is available in your version of MATLAB enter help bessel . Figure 8.1 was made with the MATLAB command

```
>> x = 0:.01:20;
>> plot(x, besselj(0,x), x,bessely(0,x))
```

The many important relationships between Bessel functions of different kinds and different orders, and asymptotic expansions can be found in [AS] and [Bo].

Exercises 8.4

1. An integral representation for the Bessel function $J_0(\rho)$. Let $v(x, y)$ be a solution of $\Delta v + v = 0$ in the disk G. Putting in polar coordinates, $v(\rho, \theta)$ must satisfy

    $$v_{\rho\rho} + \frac{1}{\rho} v_\rho + \frac{1}{\rho^2} v_{\theta\theta} + v = 0,$$

    and $\theta \to v(\rho, \theta)$ must have period 2π. Now let

    $$\varphi(\rho) = \int_{-\pi}^{\pi} v(\rho, \theta) d\theta.$$

    (a) Show that φ satisfies Bessel's equation of order zero.

    (b) Find a condition on α and β so that $\exp(\alpha x + \beta y)$ solves $\Delta v + v = 0$. Verify that $\exp(iy)$ is a solution of $\Delta v + v = 0$.

    (c) Change $\exp(iy)$ into polar coordinates, and deduce that a solution of Bessel's equation of order zero is given by

    $$\varphi(\rho) = \int_{-\pi}^{\pi} e^{i\rho \sin \theta} d\theta.$$

    Verify directly that $\varphi(\rho)$ is a solution of Bessel's equation of order zero.

    (d) Since $\varphi(0) = 2\pi$ (verify), we see that $J_0(\rho)$ is given by

    $$J_0(\rho) = \frac{1}{2\pi} \int_{-\pi}^{\pi} e^{i\rho \sin \theta} d\theta.$$

    Show that this representation can also be written

    $$J_0(\rho) = \frac{1}{\pi} \int_{0}^{\pi} \cos(\rho \sin \theta) d\theta.$$

2. (a) Use MATLAB to plot the Bessel function J_0 on the interval $[0, 20]$. Make a rough estimate of the location of the first three zeros ρ_1, ρ_2, ρ_3 of J_0.

    (b) We want to find the first three zeros of J_0 using the MATLAB root finder `fzero`. Since the Bessel function feature of MATLAB involves the order parameter α, and `fzero` accepts only functions of one argument, we must set up an mfile, called `g.m`, as follows:

```
function y = g(x)
        y = besselj(0,x)
```

Then apply `fzero` to this function. Using these values for the first three zeros of J_0, determine the first three (radial) eigenvalues for the disk of radius $r = a$.

(c) Taking $a = 1$, plot $r \rightarrow J_0(\rho_1 r)$, $r \rightarrow J_0(\rho_2 r)$, and $r \rightarrow J_0(\rho_3 r)$ on the same graph on the interval $[0, 1]$. These are the first three radially symmetric eigenfunctions of the problem $-\Delta \varphi = \lambda \varphi$ with the boundary condition $\varphi(1) = 0$.

3. Formulate and solve the problem of radial heat flow in the disk G of radius a with the temperature held at zero on the boundary $r = a$ and with initial data

$$f(x, y) = f(r) = \begin{cases} 1, & 0 \le r < a/2 \\ 0, & a/2 < r \le a \end{cases}.$$

4. (a) Show that if U satisfies Bessel's equation of order zero, then $V = U'$ satisfies the Bessel equation of order one.

(b) The Bessel function of order one $J_1(\rho)$ is taken to be

$$J_1(\rho) = -J_0'(\rho).$$

Find an integral representation for $J_1(\rho)$. Plot J_0 and J_1 on the same graph using MATLAB.

5. Formulate and solve the problem of radial heat flow in the disk G of radius of radius a, this time with the boundary $r = a$ being insulated. Show that the eigenvalues are related to the zeros of the Bessel function J_1.

6. We can also use separation of variables to solve boundary value problems in a sector. Let β be an angle with $0 < \beta < 2\pi$, and let G be the sector $\{(r, \theta) : 0 < r < a, 0 < \theta < \beta\}$. Let us suppose that the sides of the sector are insulated, and that the temperature is held at zero on the curved part the boundary, $r = a$. Then the eigenfunction problem is

$$-\Delta \varphi = \lambda \varphi \quad \text{in } G,$$

$$\varphi(a, \theta) = 0 \quad \text{for } 0 \le \theta \le \beta,$$

and
$$\frac{\partial u}{\partial \theta}(r, 0) = \frac{\partial u}{\partial \theta}(r, \alpha) = 0 \text{ for } 0 \le r < a.$$

As usual write $\varphi(r, \theta) = R(r)\Phi(\theta)$. The equation for R and the boundary conditions are the same as (8.23). The equation for Φ is the same as (8.22), but the boundary conditions are now

$$\Phi'(0) = \Phi'(\alpha) = 0.$$

Find the eigenfunctions and eigenvalues for this problem in the sector and solve the IBVP with initial condition

$$u(r, \theta) = f(r, \theta) \quad \text{in } G.$$

You will need to consider solutions of Bessel's equation of order $s = n\pi/\beta$.

8.5   Asymptotics and steady-state solutions

8.5.1   Approach to the steady state

In Chapter 4 we saw that under certain conditions, the solution of an inhomogeneous heat equation in one space dimension could tend to a nontrivial steady state as $t \to \infty$. We shall see similar asymptotic behavior in higher dimensions. Suppose given a time-independent source term $q = q(x, y)$, and consider the IBVP

$$u_t - k\Delta u = q \qquad \text{in } G \times (0, \infty), \tag{8.26}$$

$$u = 0 \qquad \text{on } \partial G \times [0, \infty),$$

$$u(x, y, 0) = f(x, y) \qquad \text{in } G.$$

We wish to solve this problem and see if we can determine the asymptotic behavior as $t \to \infty$. If the solution $u(x, y, t) \to U(x, y)$ as $t \to \infty$, we would expect the limiting solution $U(x, y)$ to solve the boundary value problem for the *Poisson* equation

$$-k\Delta U = q \qquad \text{in } G, \tag{8.27}$$

$$U = 0 \qquad \text{on } G.$$

We shall use a single index, n, for the eigenvalues and eigenfunctions, which we denote by $\varphi_n(x, y)$. Expand the source q in term of the eigenfunctions φ_n:

$$q(x, y) = \sum_1^\infty q_n \varphi_n(x, y),$$

and assume we can do the same for the desired solution,

$$u(x, y, t) = \sum_1^\infty u_n(t) \varphi_n(x, y).$$

The unknowns are now the sequence of functions $u_n(t)$. If we substitute these expansions into equation (8.26), on the left we will get

$$u_t - k\Delta u = \sum_1^\infty [u_n'(t) + k\lambda_n u_n(t)]\varphi_n(x, y),$$

and on the right, the expansion for q. Equating one expansion with another we obtain

$$\sum_1^\infty [u_n'(t) + k\lambda_n u_n(t) - q_n]\varphi_n(x, y) = 0.$$

This can be true only if each expression in square brackets is zero. Thus we can reduce the PDE to an infinite collection of ODE initial value problems,

$$u_n'(t) + k\lambda_n u_n(t) = q_n \tag{8.28}$$

$$u_n(0) = A_n = \frac{\langle f, \varphi_n \rangle}{\langle \varphi_n, \varphi_n \rangle}.$$

Now for each n, the solution of the IVP (8.28) is given by

$$u_n(t) = A_n e^{-\lambda_n k t} + \frac{q_n}{k\lambda_n}(1 - e^{-\lambda_n k t}).$$

Hence the solution of (8.26) can be written as the series

$$u(x, y, t) = \sum_1^\infty [A_n e^{-\lambda_n kt} + \frac{q_n}{\lambda_n k}(1 - e^{-\lambda_n kt})]\varphi_n(x, y). \qquad (8.29)$$

Let $U(x, y)$ be defined by

$$U(x, y) = \sum_1^\infty \frac{q_n}{\lambda_n k}\varphi_n(x, y).$$

Because of the decaying exponentials in (8.29),

$$u(x, y, t) \to U(x, y) \quad \text{as } t \to \infty.$$

$U(x, y)$ is the solution of the Poisson equation (8.27). We can check this directly.

$$-k\Delta U(x, y) = -k\Delta(\sum_1^\infty \frac{q_n}{\lambda_n k}\varphi_n(x, y))$$

$$= \sum_1^\infty \frac{q_n}{\lambda_n}(-\Delta)\varphi_n(x, y)$$

$$= \sum_1^\infty q_n\varphi_n(x, y) = q(x, y)$$

because $-\Delta\varphi_n(x, y) = \lambda_n\varphi_n(x, y)$.

Now suppose that the IBVP (8.26) is modified to be $u = g$ on ∂G. This means that we are going to specify the temperature on the boundary by a given function g which does not depend on t. Assume that the solution $\tilde{U}$ of the steady-state problem exists:

$$-k\Delta\tilde{U} = q \quad \text{in } G,$$

$$u = g \quad \text{on } \partial G.$$

We write the desired solution $u(x, y, t)$ as

$$u(x, y, t) = \tilde{U}(x, y) + v(x, y, t).$$

The function v will satisfy

$$v_t - k\Delta v = 0 \quad \text{in } G, \qquad u = 0 \quad \text{on } \partial G$$

$$v(x, y, 0) = f(x, y) - \tilde{U}(x, y).$$

It is easy to show that $v \to 0$ as $t \to \infty$, so that $u(x, y, t) \to \tilde{U}(x, y)$. We shall see how to find $\tilde{U}$ in Chapter 9.

8.5.2 Compatibility of source and boundary flux

In solving (8.28), note that we divided by λ_n, which was permissible, because all $\lambda_n > 0$ for boundary condition (8.5). But as we noted earlier, the boundary conditions (8.7) and (8.8) also yield only strictly positive eigenvalues. Thus this kind of problem can be solved in the same way and has the same kind of asymptotic behavior as $t \to \infty$.

However, the inhomogeneous version of (8.6) has an additional interesting feature because $\lambda = 0$ is an eigenvalue of the homogeneous problem

$$- \Delta\varphi = \lambda\varphi, \qquad \frac{\partial\varphi}{\partial n} = 0 \quad \text{on } \partial G. \tag{8.30}$$

We consider the IBVP

$$u_t - k\Delta u = q \qquad \text{in } G \times (0, \infty), \tag{8.31}$$

$$\frac{\partial u}{\partial n} = g \qquad \text{on } \partial G,$$

$$u(x, y, 0) = f(x, y).$$

We assume that both q and g do not depend on t. Can there exist a steady-state solution $U(x, y)$ to this problem? U would have to satisfy

$$-k\Delta U = q \quad \text{in } G, \qquad \frac{\partial U}{\partial n} = g \quad \text{on } \partial G.$$

Supposing that there existed a solution of this steady-state problem, integrate both sides of the PDE over G, and apply the divergence theorem to the integral on the left. We find that

$$-k\int_{\partial G} \frac{\partial U}{\partial n} ds = -k\int\int_G div(\nabla U)\, dxdy = -k\int\int_G \Delta U(x, y)\, dxdy$$

$$= \int\int_G q(x, y)\, dxdy.$$

But the flux condition on the boundary requires that $\partial U / \partial n = g$ on ∂G, where g is given. Thus we see that if a solution U exists to the steady-state problem, the given data q and g of the problem must satisfy a compatibility condition

$$\int\int_G q(x, y)\, dxdy + k \int_{\partial G} g\, ds = 0. \tag{8.32}$$

In words, the net heat creation or absorption produced by the source q in the region G must balance the heat flux g through the boundary. This condition is the higher dimensional analogue of the one-dimensional condition (4.50).

Note further that the solutions of the steady-state problem are not unique. If $U(x, y)$ is a solution, then so is $U(x, y) + C$ for any constant C. We shall see that when the time-dependent solutions of (8.31) tend to a steady state, a unique limiting steady state is determined by the initial data f.

We shall attempt to find a solution u to (8.31) in the special case that $g = 0$. Assume that there is a complete, orthogonal set of eigenfunctions of the eigenvalue problem (8.30). We expand the desired solution u in terms of the eigenfunctions $\varphi_n(x, y)$:

$$u(x, y, t) = u_0(t) + \sum_1^\infty u_n(t)\varphi_n(x, y),$$

where

$$u_0(t) = \frac{\int\int_G u(x, y, t)dxdy}{area(G)},$$

and

$$u_n(t) = \frac{\langle u(x, y, t), \varphi_n(x, y)\rangle}{\langle \varphi_n, \varphi_n \rangle},$$

for $n \geq 1$. We also expand the source term q:

$$q(x, y) = q_0 + \sum_1^\infty q_n\varphi_n(x, y),$$

where

$$q_0 = \frac{\int\int_G q(x, y)\, dxdy}{area(G)}, \qquad q_n = \frac{\langle q, \varphi_n\rangle}{\langle \varphi_n, \varphi_n\rangle}, n \geq 1.$$

We substitute in the PDE (8.31) for u to deduce the ODE's for u_n. They are

$$u_0'(t) = q_0, \qquad u_0(0) = A_0 = \frac{\int\int_G f(x, y)\, dxdy}{area(G)}$$

and

$$u'_n(t) + k\lambda_n u_n(t) = q_n, \qquad u_n(0) = A_n = \frac{\langle f, \varphi_n \rangle}{\langle \varphi_n, \varphi_n \rangle}, \qquad n \geq 1.$$

These equations have solutions

$$u_0(t) = A_0 + q_0 t$$

and

$$u_n(t) = A_n \exp(-\lambda_n kt) + \frac{q_n}{k\lambda_n} \left[1 - \exp(-\lambda_n kt) \right], \qquad n \geq 1.$$

Then subsituting in the expansion for u, we find that

$$u(x, y, t) = A_0 + q_0 t$$

$$+ \sum_{n \geq 1} \left[A_n \exp(-\lambda_n kt) + \frac{q_n}{k\lambda_n} \left[1 - \exp(-\lambda_n kt) \right] \right] \varphi_n(x, y).$$

We can see that u will remain bounded as $t \to \infty$ if and only if the coefficient $q_0 = 0$. This is precisely the balance condition (8.32) of the internal source with the flux through the boundary when $g = 0$. When this condition is satisfied, $u(x, y, t) \to U(x, y)$ as $t \to \infty$ where

$$U(x, y) = A_0 + \sum_{n \geq 1} \frac{q_n}{k\lambda_n} \varphi_n(x, y).$$

The case when g is not zero can be treated in a similar fashion by first subtracting off a function $v(x, y)$ such that $\partial v / \partial n = g$ on ∂G. Then u is replaced by $w(x, y, t) = u(x, y, t) - v(x, y)$, and q is replaced by $\tilde{q} = q + k\Delta v$. Now it follows that u will converge to a steady state $U(x, y)$ if and only if the compatibility condition (8.32) is satisfied.

How is the particular solution of the steady-state problem determined by the initial data? If we integrate a solution of (8.31) over G and use the divergence theorem, we find that

$$\frac{d}{dt} \int \int_G u(x, y, t) \, dx dy = \int \int_G [k\Delta u(x, y, t) + q(x, y)] \, dx dy$$

$$= k \int_{\partial G} \frac{\partial u}{\partial n} \, ds + \int \int_G q(x, y) \, dx dy = k \int_{\partial G} g \, ds + \int \int_G q(x, y) \, dx dy.$$

Thus if u solves (8.31) and the data q and g satisfy the compatibility condition (8.32), the average temperature $\int \int_G u(x, y, t)\, dxdy$ is constant. In this case the limiting steady state $U(x, y)$ satisfies

$$\int \int_G U(x, y)\, dxdy = \lim_{t \to \infty} \int \int_G u(x, y, t)\, dxdy = \int \int_G f(x, y)\, dxdy.$$

We see that when a solution of the steady-state problem exists for given q and g, the solution of the time-dependent problem (8.31) tends to that steady state U having the same average temperature as the initial temperature f.

Exercises 8.5

1. Consider the IBVP in the (open) square $G = \{(0, \pi) \times (0, \pi)\}$

$$u_t - \Delta u = 0 \quad \text{in } G \times \{t > 0\}$$

with boundary conditions

$$u(0, y, t) = \sin(y), \quad u(\pi, y, t) = 0, \quad 0 < y < \pi, \quad t \geq 0,$$

$$u(x, 0, t) = u(x, \pi, t) = 0, \quad 0 < x < \pi, \quad t \geq 0,$$

and initial condition

$$u(x, y, 0) = 0, \quad \text{for } (x, y) \in G.$$

Note that the combination of initial condition and boundary conditions makes $(x, y) \to u(x, y, 0)$ discontinuous at the left edge of the square.

(a) Explain why we cannot apply the maximum principle directly to the solution u of this IBVP. Make an argument using continuous approximations to show that nevertheless, $u(x, y, t) \geq 0$ for all $(x, y) \in \bar{G}, t \geq 0$.

(b) Look for the steady state solution $U(x, y)$ of this boundary value problem in the form $U(x, y) = X(x) \sin(y)$. Find the ODE that X must solve and the appropriate boundary conditions. Find X.

(c) Write the solution u of the time-dependent problem as

$$u(x, y, t) = v(x, y, t) + U(x, y).$$

What IBVP must v satisfy? Find v using the appropriate eigenfunction expansion. Calculate the coefficients.

  (d) If we truncate the infinite series at M terms, we have only an approximate solution of the given IBVP. In what sense is it approximate?
What condition is not satisfied exactly?

The function `heat7` calculates an approximate solution to this boundary value
problem by summing up M terms of the eigenfunction expansion. To call the
function `heat7`, summing $M = 5$ terms, and to plot the approximate solution on
the line $y = \pi/2$, $0 \le x \le \pi$, at time $t = .5$, use the sequence of commands

```
>> global M
>> delx = pi/50;
>> x = 0: delx : pi;
>> M = 5;
>> t = .5;
>> u = heat7(x, pi/2, t);
>> plot(x,u)
```

To make a `pcolor` plot on the square G with contour lines, we must create a mesh
and proceed as follows:

```
>> global M
>> delx = pi/50;
>> x = 0:delx:pi;
>> y = x;
>> [X,Y] = meshgrid(x,y);
>> M = 5;
>> t = .5;
>> u = heat7(X,Y,t);
>> colormap(hot); brighten(.5)
>> pcolor(X,Y,u);shading flat
>> hold on
>> contour(X,Y,u,20,'k')
>> hold off
```

Since we will want to repeat this plotting procedure several times, you may wish
to make a script mfile of this sequence of commands. To be even more efficient,
you can include statements which ask for M and t as input. Of course you can also
make 3-D plots using the `surf` command.

 2. (a) Make the `pcolor` plots with contour lines of the approximate solution
with $M = 20$ for $t = .05, .1, .2, .5, 1$. Describe the heat flow that
you see in these plots. Do you see the temperature tending to a steady
state? Compare the `pcolor` plot at, say, $t = 5$, with the plot at $t = 1$.
You will need to open another MATLAB window to do this.

(b) Plot the trace of the approximate solution, i.e., its restriction to the line $y = \pi/2$, for $t = 0$, and $M = 5, 10, 20$. This gives an idea of how well the eigenfunction expansion is approximating the initial temperature, which is everywhere zero. What phenomenon do you see happening near $x = 0$?

(c) Now plot the trace on the line $y = \pi/2$ with $t = .05$ and $M = 5$. Note that the function takes on negative values. Explain how this observation fits in with part (d) of exercise 1. Try $M = 10$. Now use the zoom feature to see the same behavior, but on a smaller scale.

(d) Next plot the trace on the same line, this time with $M = 20$ and with $t = .1, .2, .5, 1$. Put the curves all on the same graph along with the trace of the steady-state solution $x \rightarrow U(x, \pi/2)$. Do the traces of the time-dependent problem approach that of the steady-state solution? With $M = 20$ how large must we choose t to ensure that the approximate solution is within 5% of the steady state solution?

3. Keeping the same initial condition, modify the boundary conditions of exercise 1 to be

$$u(0, y, t) = \sin(y), \quad u(\pi, y, t) = 2\sin(y), \qquad 0 < y < \pi,$$

$$u(x, 0, t) = u(x, \pi, t) = 0, \qquad 0 < x < \pi.$$

Repeat the analysis of exercises 1 and 2. Write a function mfile to evaluate the approximate solution by summing M terms of the eigenfunction expansion. Use the mfile heat7.m as a guide.

4. Consider the IBVP in the rectangle $G = \{0 < x < a, 0 < y < b\}$

$$u_t - k\Delta u = q = \sin(\pi x/a)\cos(\pi y/b)\sin t$$

with boundary conditions

$$u(0, y, t) = u(a, y, t) = 0, \quad 0 < y < b,$$

$$u_y(x, 0, t) = u_y(x, b, t) = 0, \quad 0 < x < a,$$

and the initial condition

$$u(x, y, 0) = (x - a)^2/2 \quad \text{for } (x, y) \in G.$$

(a) What are the eigenvalues $\lambda_{m,n}$ and the eigenfunctions $\varphi_{m,n}$ for this problem?

(b) Write $u = v + w$, where v satisfies $v_t - k\Delta v = 0$, $v(x, y, 0) = (x - a)^2/2$ and the boundary conditions; and w satisfies $w_t - k\Delta w = q$, $w(x, y, 0) = 0$ and the boundary conditions. Solve the problem for v, and show that $v \to 0$ as $t \to \infty$.

(c) Solve the problem for w. Show that there is a function $z(x, y, t)$ with $t \to z(x, y, t)$ having period 2π and such that $w - z \to 0$ as $t \to \infty$.

8.6 The wave equation

8.6.1 The initial-value problem

The linear wave equation in two or three space dimensions is

$$u_{tt} - c^2 \Delta u = 0, \qquad (8.33)$$

where $u = u(\mathbf{x}, t)$ with $\mathbf{x} = (x, y) \in R^2$ or $\mathbf{x} = (x, y, z) \in R^3$. It arises in acoustical theory as the linearization of the equations of gas dynamics (Chapter 5), in linear elasticity theory to describe the vibrations of elastic solids, and in the theory of electricity and magnetism, Section 8.10.

Plane and spherical waves

Before we solve the IVP for the wave equation, we look for some special solutions in closed form. Plane waves are a class of solutions of (8.33) in both two and three space dimensions. For instance in $R^3 \times R$, we let $\mathbf{k} = (k_1, k_2, k_3)$ and let $k = \sqrt{k_1^2 + k_2^2 + k_3^2}$. Then

$$u(\mathbf{x}, t) = \exp(i(\omega t - \mathbf{k} \cdot \mathbf{x}))$$

is a solution of (8.33) when $\omega = ck$. ω is the angular frequency and k is the wave number. The phase of the plane wave is given by

$$\omega t - \mathbf{k} \cdot \mathbf{x} = k(ct - \theta \cdot \mathbf{x}),$$

where $\theta = \mathbf{k}/k$ is a unit vector pointing in the direction of propagation. The planes of constant phase are

$$\{\mathbf{x} : \theta \cdot \mathbf{x} = ct\}$$

and all move with speed c, independent of the wave number k. The real and imaginary parts of u are also solutions of (8.33). They are

$$\cos(\omega t - \mathbf{k} \cdot \mathbf{x}) \quad \text{and} \quad \sin(\omega t - \mathbf{k} \cdot \mathbf{x}).$$

Plane waves do not have finite energy (see section 8.7).

In three space dimensions, we can find spherically symmetric solutions of (8.33) in closed form. If $u(\mathbf{x}, t)$ is a spherically symmetric solution, i.e., $u(\mathbf{x}, t) = u(r, t)$ with $r = \sqrt{x^2 + y^2 + z^2}$, equation (8.33) reduces to

$$u_{tt} - c^2(u_{rr} + \frac{2}{r}u_r) = 0.$$

It is not hard to show (see exercise 1 of this section) that $v(r, t) = ru(r, t)$ satisfies the one-dimensional wave equation

$$v_{tt} - c^2 v_{rr} = 0 \text{ in } r > 0.$$

Since the general solution of this equation is

$$v(r, t) = F(r - ct) + G(r + ct),$$

the general spherically symmetric solution of (8.33) is given by

$$u(r, t) = \frac{F(r - ct) + G(r + ct)}{r}. \tag{8.34}$$

The first term represents a spherical *outgoing* wave whose amplitude decreases as $1/r$. The second term is a spherical *incoming* wave which converges at the origin. For important physical reasons, there is not a similar simple form for radially symmetric solutions of (8.33) in R^2.

Now we turn to the problem (8.33) on $R^3 \times R$ with initial data

$$u(\mathbf{x}, 0) = f(\mathbf{x}), \qquad u_t(\mathbf{x}, 0) = g(\mathbf{x}). \tag{8.35}$$

Recall that in Section 8.1.1 we found that solutions of the heat equation in higher dimensions could be constructed by taking products of solutions in one dimension. Because the wave equation has a second derivative in the t variable, this idea does not work for its solutions. However, we can reduce the dimension of the problem for the wave equation by taking averages of the solution over spheres. This results in spherically symmetric solutions which have the form (8.34).

First we state an invariance property (8.36) of the Laplace operator, which will be derived in Chapter 9. Let u be a C^2 function on R^3, and fix $\mathbf{x}_0 \in R^3$. Define

$$M(r) = M(r, \mathbf{x}_0, u) = \frac{1}{4\pi r^2} \int \int_{|\mathbf{x} - \mathbf{x}_0| = r} u(\mathbf{x}) dS(\mathbf{x}).$$

$4\pi r^2$ is the surface area of the sphere of radius r so that M is the *mean* (average) of u over the sphere of radius r, centered at $\mathbf{x}_0$. We shall often suppress $\mathbf{x}_0$ and u to make the notation less cumbersome.

The Laplace operator commutes with the spherical mean operator in the following sense (for a proof see Section 9.1.2):

$$\Delta M(r, u) = M(r, \Delta u). \tag{8.36}$$

On the left side of the equation, we are taking the Laplacian of a spherically symmetric function:

$$\Delta M(r, u) = M_{rr} + \frac{2}{r} M_r.$$

Now let $u(\mathbf{x}, t)$ be a C^2 solution of (8.33), and define

$$M(r, t, \mathbf{x}_0, u) = \frac{1}{4\pi r^2} \int \int_{|\mathbf{x}-\mathbf{x}_0|=r} u(\mathbf{x}, t) dS(\mathbf{x}).$$

M is the spherical mean of $\mathbf{x} \to u(\mathbf{x}, t)$ over the sphere of radius r centered at $\mathbf{x}_0$. For the moment we suppress $\mathbf{x}_0$. We can differentiate under the integral sign and use (8.33) to find that

$$M(r, t, u)_{tt} = \frac{1}{4\pi r^2} \int \int_{|\mathbf{x}-\mathbf{x}_0|=r} u_{tt} dS = \frac{c^2}{4\pi r^2} \int \int_{|\mathbf{x}-\mathbf{x}_0|=r} \Delta u dS = c^2 M(r, t, \Delta u).$$

Combine this equation with (8.36) to deduce that

$$M_{tt} = c^2 \Delta M = c^2 \left(M_{rr} + \frac{2}{r} M_r \right)$$

or

$$(rM)_{tt} = c^2(r M_{rr} + 2M_r) = c^2 \partial_r^2(rM).$$

Hence $v(r, t) \equiv r M(r, t, u)$ satisfies the one-dimensional wave equation

$$v_{tt} = c^2 v_{rr} \qquad \text{for} \quad 0 \le r < \infty, \ t \in R.$$

The initial data for v is

$$v(r, 0) = v_0(r) \equiv \frac{1}{4\pi r} \int \int_{|\mathbf{x}-\mathbf{x}_0|=r} f(\mathbf{x}) dS(\mathbf{x}) \tag{8.37}$$

and

$$v_t(r, 0) = v_1(r) \equiv \frac{1}{4\pi r} \int \int_{|\mathbf{x}-\mathbf{x}_0|=r} g(\mathbf{x}) dS(\mathbf{x}). \tag{8.38}$$

In addition we see that

$$v(0, t) = 0 \qquad \text{for all } t.$$

Thus we have a half-line problem for the one-dimensional wave equation of the type solved in Chapter 5. For $0 \le r \le ct$,

$$v(r, t) = [v_0(ct + r) - v_0(ct - r)]/2 + (1/2c) \int_{ct-r}^{ct+r} v_1(s)ds \qquad (8.39)$$

$$= (1/2c)\partial_t \int_{ct-r}^{ct+r} v_0(s)ds + (1/2c) \int_{ct-r}^{ct+r} v_1(s)ds.$$

Since u is assumed continuous at $\mathbf{x}_0$, we conclude that

$$u(\mathbf{x}_0, t) = \lim_{r \downarrow 0} M(r, t, \mathbf{x}_0)$$

$$= \lim_{r \downarrow 0} \frac{v(r, t) - v(0, t)}{r} = v_r(0, t).$$

We can calculate $v_r(0, t)$ from (8.39):

$$v_r(r, t) = (1/2c)\partial_t[v_0(ct + r) + v_0(ct - r)] + (1/2c)[v_1(ct + r) + v_1(ct - r)].$$

Hence

$$v_r(0, t) = \partial_t[(1/c)v_0(ct)] + (1/c)v_1(ct). \qquad (8.40)$$

Going back to (8.37) and (8.38) to express v_0 and v_1 in terms of f and g, we find that

$$v_0(ct) = \frac{1}{4\pi ct} \int\int_{|\mathbf{x}-\mathbf{x}_0|=ct} f(\mathbf{x})dS(\mathbf{x})$$

and

$$v_1(ct) = \frac{1}{4\pi ct} \int\int_{|\mathbf{x}-\mathbf{x}_0|=ct} g(\mathbf{x})dS(\mathbf{x}).$$

Finally substituting these expressions in (8.40) we arrive at *Kirchoff's formula* for the solution of the IVP (8.33), (8.35):

$$u(\mathbf{x}_0, t) = \partial_t \left[\frac{1}{4\pi c^2 t} \int_{|\mathbf{x}-\mathbf{x}_0|=ct} f(\mathbf{x}) dS(\mathbf{x}) \right] + \frac{1}{4\pi c^2 t} \int_{|\mathbf{x}-\mathbf{x}_0|=ct} g(\mathbf{x}) dS(\mathbf{x}).$$

$$(8.41)$$

8.6.2 The method of descent

We shall derive the formula for the solution in $R^2 \times R$ from (8.41) by the *method of descent*. Now we seek the solution of (8.33), (8.35) with $\mathbf{x} = (x, y)$. Suppose that we had a solution $u(x, y, t)$; we consider it a solution in $R^3 \times R$ by assuming it constant in z, and we do the same with the initial functions $f(x, y)$ and $g(x, y)$. Then we must examine the integrals in (8.41) and see how they can be simplified when f and g depend only on (x, y). For ease of notation, let us take $\mathbf{x}_0 = 0$ and consider the integral for f:

$$\int_{|\mathbf{x}|=ct} f(\mathbf{x}) dS(\mathbf{x}) = \int_{S^+} f(\mathbf{x}) dS(\mathbf{x}) + \int_{S^-} f(\mathbf{x}) dS(\mathbf{x}),$$

where S^+ is the hemisphere in $z \geq 0$ and S^- is the hemisphere in $z \leq 0$. For both hemispheres,

$$dS(\mathbf{x}) = \frac{ct}{\sqrt{c^2 t^2 - x^2 - y^2}} dx dy.$$

Using the fact that f does not depend on z,

$$\int_{S^\pm} f(\mathbf{x}) dS(\mathbf{x}) = ct \int \int_{x^2+y^2 \leq c^2 t^2} f(x, y) \frac{dx dy}{\sqrt{c^2 t^2 - x^2 - y^2}}.$$

Thus

$$\frac{1}{4\pi c^2 t} \int_{|\mathbf{x}|=ct} f(\mathbf{x}) dS(\mathbf{x}) = \frac{1}{2\pi c} \int \int_{x^2+y^2 \leq c^2 t^2} f(x, y) \frac{dx dy}{\sqrt{c^2 t^2 - x^2 - y^2}}.$$

If we make the same calculation for the integral of g and rewrite the result in vector notation, $\mathbf{x} = (x, y)$, we arrive at the formula for the solution in $R^2 \times R$:

$$u(\mathbf{x}_0, t) = \partial_t \frac{1}{2\pi c} \left[\int \int_{|\mathbf{x}-\mathbf{x}_0| \leq ct} \frac{f(\mathbf{x}) d\mathbf{x}}{\sqrt{c^2 t^2 - |\mathbf{x} - \mathbf{x}_0|^2}} \right] \qquad (8.42)$$

$$+ \frac{1}{2\pi c} \int \int_{|\mathbf{x}-\mathbf{x}_0| \leq ct} \frac{g(\mathbf{x}) d\mathbf{x}}{\sqrt{c^2 t^2 - |\mathbf{x} - \mathbf{x}_0|^2}}.$$

The integrals here are two-dimensional integrals over the disk $|\mathbf{x} - \mathbf{x_0}| \leq ct$ in R^2, while the integrals in (8.41) are taken over the sphere $|\mathbf{x} - \mathbf{x_0}| = ct$ in R^3.

In both formulas we see that information can travel no faster than speed c. Indeed if $f = g = 0$ for $|\mathbf{x}| \geq a$, then the solution $u(\mathbf{x}, t) = 0$ for $|\mathbf{x}| \geq a + ct$ (draw a picture). Thus at time $t > 0$, the domain of influence of the ball in R^3 of radius $a > 0$, is the ball $\{|\mathbf{x}| \leq a + ct\}$ in R^3, and the domain of influence of the disk of radius $a > 0$, is the disk $\{|\mathbf{x}| \leq a + ct\}$ in R^2.

However there is an important difference between the domains of dependence of a point $(\mathbf{x_0}, t_0)$ in two and three space dimensions. In three space dimensions, the domain of dependence is given by

$$\{|\mathbf{x} - \mathbf{x_0}| = ct_0\},$$

which is a *spherical surface* centered at $\mathbf{x_0}$. Thus if $f = g = 0$ for $|\mathbf{x}| \geq a$, the solution $u(\mathbf{x}, t) = 0$ in the forward light cone $\{|\mathbf{x}| \leq ct - a\}$ for $ct \geq a$. This feature occurs because, if $(\mathbf{x}, t)$ is a point in this cone, the sphere of radius ct centered at $\mathbf{x}$ will lie outside the ball $\{|\mathbf{x}| \leq a\}$ and hence meets only points where f and g are both zero. This sharp propagation of information, in which all signals travel exactly at speed c, is called *Huyghens' principle*.

In the two-dimensional formula (8.42), the integral is taken over the full disk $\{|\mathbf{x} - \mathbf{x_0}| \leq ct\}$ in R^2. This means that for some initial data, there will always be some residual signal in the cone $\{|\mathbf{x}| \leq ct - a\}$ in two space dimensions. Huyghens' principle does not hold in two space dimensions.

The integrals in (8.41) and (8.42) are not easy to evaluate. This representation is important for our analytic understanding of the solutions, but does not yield a practical method of producing numbers.

One method of solving PDE's is to use plane waves and spherical means to decompose the solutions and arrive at solution formulas, as we did for the wave equation. This method is explored in [John2].

Exercises 8.6

1. Suppose that $u(\mathbf{x}, t)$ solves the wave equation $u_{tt} - c^2 \Delta u = 0$ in $R^3 \times R$, and that $u(\mathbf{x}, t) = u(r, t)$ where $r = |\mathbf{x}|$. u is spherically symmetric.

   (a) Use the formula for the Laplace operator in spherical coordinates (Appendix A) to verify that

   $$u_{tt} - c^2 (u_{rr} + \frac{2}{r} u_r) = 0.$$

   (b) Show that $v = ru$ satisfies the one-dimensional wave equation

   $$v_{tt} - c^2 v_{rr} = 0.$$

2. Consider the IBVP

$$u_{tt} - c^2 \Delta u = 0 \quad \text{in } R^3 \times R$$

with

$$u(\mathbf{x}, 0) = f(\mathbf{x}), \qquad u_t(\mathbf{x}, 0) = g(\mathbf{x}) \quad \text{on } R^3,$$

where f and g are spherically symmetric.

(a) Reduce to an IBVP for the one-dimensional wave equation as in exercise 1.

(b) Extend f and g as even functions of r, and solve using d'Alembert's formula. Show that when $r > 0$, the formula for $u(r, t)$ is

$$u(r, t) = \frac{1}{2r}[(r + t)f(r + t) + (r - t)f(r - t) + \int_{r-t}^{r+t} sg(s)ds].$$

Show that this formula may be put in the form (8.34). These two formulas are therefore equivalent representations of spherically symmetric solutions.

(c) Show that

$$\lim_{r \to 0} u(r, t) = f(t) + f'(t) + tg(t).$$

3. Let $u(r, t) = F(r - ct)/r$ be a spherically symmetric outgoing solution of (8.33) in $R^3 \times R$. Assume that $F(s) = 0$ for s outside some finite interval $[a, b]$, $0 < a < b$. Calculate the energy (see Section 8.7) of u directly. Show that it is conserved with

$$e(t) = 4\pi c^2 \int_a^b [F'(s)]^2 ds.$$

Note how the factor of r^2 in the volume element of spherical coordinates exactly balances the factor of r^{-2} which appears in the expression $u_t^2 + c^2 |\nabla u|^2$.

4. We cannot plot the graphs of solutions of the wave equation in three space dimensions and time; it would require four dimensions. However we can attempt to visualize some spherically symmetric solutions.

(a) Use the formula of exercise 2 with $f(r) = \exp(-r^2)$ and $g(r) = 0$. Plot the solution $u(r, t)$ as a function of r on the interval $.01 \le r \le 5$

for $t = 1, 2, 3, 4, 5$. Put the graph on the same axes so that you can compare the heights. Describe how the initial disturbance propagates. How does its amplitude change? Is there is a residue signal left behind?

(b) Repeat part (a) with initial data $f(r) = 0$ and $g(r) = \exp(-r^2)$. You can do the integral involving g analytically. Compare the graphs of this solution (at the same times) with those of part (a). How are they the same? How do they differ?

5. In this exercise, we try another method of visualizing solutions of the wave equation in three space dimensions. We can plot $(x, y) \to u(x, y, 0, t)$ for fixed t. Let us do this for spherically symmetric solutions. In (8.34) take $G = 0$ and $F(s) = \exp(-.3s^2) \cos(2s)$. Write an mfile bigf.m for this F.

(a) To see what the profile of the wave looks like, plot $r \to F(r - t)/r$ on $.1 \le r \le 20$ for $t = 4, 8, 12, 16$.

(b) Write an mfile to plot $(x, y) \to F(r - t)/r$, where $r = \sqrt{x^2 + y^2}$. Note that this function will *not* satisfy the 2-D wave equation (verify!). Put a meshgrid on the square $[.2, 20] \times [.2, 20]$ with $\Delta x = \Delta y = .2$. Put the following sequence of commands in a mfile, wave.m, with t as input parameter:

```
t = input('enter the value of t   ')
x = .2:.2:20; y = x;
[X,Y] = meshgrid(x,y);
R = sqrt(X.^2 + Y.^2);
w = bigf(R-t)./R;
colormap(cool)
surf(X,Y,w); shading flat
view(45,45)
axis([0 20 0 20 -.3 .3])
```

You may change the colormap, the viewpoint, or the vertical scale on the graph.

(c) Do the plots for part b) for $t = 4, 8, 12, 16$. What should be the appearance of the solution surface ahead of the wave and *behind* the wave? What principle does this illustrate?

6. Modify the plotting program you wrote in exercise 5 to plot a real-valued plane-wave solution of the wave equation (with $c = 1$) in two space dimensions. Plot

$$(x, y) \to u(x, y, t) = \cos(t - \mathbf{k} \cdot (x, y))$$

for several choices of **k**. **k** must be a unit vector.

7. Suppose that **x** is in R^2 or in R^3, and that $f(\mathbf{x}) = g(\mathbf{x}) = 0$ for $\mathbf{x} \cdot \alpha \geq 0$ for some unit vector α. Show that the solution of (8.33), (8.35) satisfies $u = 0$ for $\mathbf{x} \cdot \alpha \geq ct$ for all $t \geq 0$.

8. Change to polar coordinates in R^2, and show that a radially symmetric solution $u(r, t)$ of the wave equation on $R^2 \times R$ solves

$$u_{tt} - c^2(u_{rr} + \frac{1}{r}u_r) = 0.$$

9. Recall from Section 8.3 that $\psi(r) = J_0(r\sqrt{\lambda})$ solves $-\Delta\varphi = \lambda\varphi$ for all r and for each $\lambda > 0$.

   (a) Find the proper value of ω (in terms of λ), so that $u(r, t, \lambda) = e^{i\omega t}\varphi(r)$ solves (8.33) on $R^2 \times R$. This provides a family of radially symmetric solutions of (8.33) on $R^2 \times R$ in closed form (modulo calculation of the Bessel function). These are called *standing waves*.

   (b) For an integrable function $g(\lambda)$, show that

$$v(r, t) = \int_0^\infty u(r, t, \lambda)g(\lambda)d\lambda$$

   provides another radially symmetric solution of (8.33).

   (c) Show that the other eigenfunctions found in Section 8.3, which involve the higher order Bessel functions also give rise to solutions of (8.33) on $R^2 \times R$. These solutions will now have a θ-dependence.

10. Once again modify the plotting program of exercise 5 to plot radially symmetric standing waves on R^2.

8.7 Energy

The *energy* of a solution $u(\mathbf{x}, t)$ contained in the ball of radius a, or disk of radius a, centered at the origin, at time t, is defined as

$$e_a(t) = \frac{1}{2}\int_{|\mathbf{x}|\leq a}[u_t^2 + c^2|\nabla u|^2]d\mathbf{x}.$$

Here the integral is either two or three-dimensional. When we omit the subscript, writing $e(t)$, we shall mean the integral to be taken over all of R^2 or R^3.

Conservation of energy can be shown as in the one-dimensional case, but here we prefer to give a proof using the divergence theorem. We describe the proof in two space dimensions. The same proof works in three space dimensions as well. For any C^2 function $u(\mathbf{x}, t)$, verify that we can write

$$(u_{tt} - c^2 \Delta u)u_t = div\ \mathbf{f}(\mathbf{x}, t),$$

where

$$\mathbf{f}(\mathbf{x}, t) = (-c^2 u_t \nabla u, \ \frac{1}{2}(u_t^2 + c^2|\nabla u|^2)).$$

The vector $\mathbf{f}(\mathbf{x}, t)$ has four components when there are three space dimensions, and three components when there are two space dimensions. Now assume that u satisfies the wave equation so that $div\ \mathbf{f} = 0$. Apply the divergence theorem over the truncated cone (Figure 8.2)

$$K_T = \{(\mathbf{x}, t) : |\mathbf{x}| \le a + c(T - t), 0 \le t \le T\}.$$

Let $v = (v_\mathbf{x}, v_t)$ be the exterior unit normal to the surface ∂K_T of K_T. We find that

$$0 = \int \int \int_{K_T} div\ \mathbf{f}(\mathbf{x}, t)\ dxdt = \int \int_{\partial K_T} v \cdot \mathbf{f}\ dS.$$

The surface ∂K_T of K_T consists of three pieces:

$$\begin{aligned}
\text{top} \quad &= \{(\mathbf{x}, t) : |\mathbf{x}| \le a, t = T\} \\
\text{bottom} &= \{(\mathbf{x}, t) : |\mathbf{x}| \le a + cT, t = 0\} \\
\text{side} \quad &= \{(\mathbf{x}, t) : |\mathbf{x}| = a + c(T - t), 0 \le t \le T\}
\end{aligned}$$

On top, $v = (0, 1)$ and on the bottom, $v = (0, -1)$. Hence

$$0 = \int \int_{\partial K_T} v \cdot \mathbf{f}\ dS = \int \int_{\partial K_T} [(v_t/2)(u_t^2 + c^2|\nabla u|^2) - c^2(v_\mathbf{x} \cdot \nabla u)u_t]dS(\mathbf{x})$$

$$= \frac{1}{2} \int \int_{\text{top}} (u_t^2 + c^2|\nabla u|^2)dx - \frac{1}{2} \int \int_{\text{bottom}} (u_t^2 + c^2|\nabla u|^2)dx$$

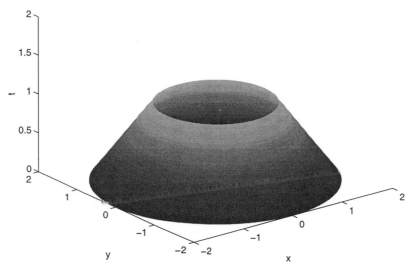

FIGURE 8.2
The truncated cone K_T with $a = c = T = 1$.

$$+ \int \int_{\text{side}} [(v_t/2)(u_t^2 + c^2|\nabla u|^2) - c^2(v_{\mathbf{x}} \cdot \nabla u)u_t \, dS(\mathbf{x}, t).$$

On the sloping side of the cone,

$$v = (\frac{\mathbf{x}}{|\mathbf{x}|}, c)/\sqrt{1 + c^2}$$

so that

$$v_t = c|v_{\mathbf{x}}|.$$

This implies that the integral over the sloping side of the cone is nonnegative. Indeed, by the Schwarz inequality,

$$c^2|(v_{\mathbf{x}} \cdot \nabla u)u_t| \leq (c|v_{\mathbf{x}}|)c|\nabla u||u_t| \leq (1/2)c|v_{\mathbf{x}}|(u_t^2 + c^2|\nabla u|^2),$$

so that the integrand on the sloping side is greater than or equal to

$$\frac{1}{2}(v_t - c|v_{\mathbf{x}}|)(u_t^2 + c^2|\nabla u|^2) \geq 0.$$

Hence the energy at the top of the cone is less than or equal to the energy at the bottom:

$$e_a(T) \le e_{a+cT}(0).$$

Now let $a \to \infty$. It follows that if the initial data f, g has *finite energy*, that is,

$$\int \int_{R^2} [g^2(\mathbf{x}) + c^2|\nabla f(\mathbf{x})|^2] d\mathbf{x}$$

is finite, then the solution has finite energy for $t = T$ and

$$e(T) = \frac{1}{2} \int \int_{R^2} [u_t^2 + c^2|\nabla u|^2] d\mathbf{x} \le \frac{1}{2} \int \int_{R^2} [g^2 + c^2|\nabla f|^2] d\mathbf{x}. \qquad (8.43)$$

Since $T > 0$ is arbitrary, this inequality implies that for all t,

$$e(t) \le e(0).$$

Now reverse the argument by turning the cone upside down. In the same way, we can deduce that $e(0) \le e(t)$ for all t. Consequently we have conservation of energy $e(t) = e(0)$ for all t.

Exercises 8.7

1. Consider the modified wave equation with dissipation d:

    $$u_{tt} + 2du_t - c^2\Delta u + d^2u = 0$$

    for $\mathbf{x}$ in R^2 or R^3. Show that $v = \exp(dt)u$ solves the equation $v_{tt} - c^2\Delta v = 0$. Show that the energy of u decays like $\exp(-dt)$.

2. Let G be the half plane $\{(x, y) : x > 0\}$. Show that energy is conserved for solutions of $u_{tt} - c^2\Delta u = 0$ in $G \times R$ with either of the boundary conditions $u(0, y, t) = 0$ or $u_x(0, y, t) = 0$.

3. An approximate radially symmetric solution to the wave solution in two space dimensions is given by

    $$\tilde{u}(r, t) = \frac{\varphi(r - ct)}{r^{1/2}}.$$

Assume that $\varphi(s) = 0$ for s outside a finite interval $[a, b]$, where $0 < a < b$. This approximation gets better for large r and t.

(a)   Show that

$$\tilde{u}_{tt} - c^2 \Delta \tilde{u} = -\frac{c^2}{4r^{5/2}} \varphi(r - ct).$$

(b)   Compute the energy of $\tilde{u}$. Show that it converges to

$$2\pi \int_R [\varphi'(s)]^2 ds$$

as $t \to \infty$.

8.8   Sources

Finally we obtain a representation formula for the solution of the inhomogeneous IVP

$$u_{tt} - c^2 \Delta u = q \tag{8.44}$$

$$u(\mathbf{x}, 0) = u_t(\mathbf{x}, 0) = 0 \qquad \text{for } \mathbf{x} \in R^2 \text{ or } R^3.$$

We write the solution as an integral using the principle of Duhamel

$$u(\mathbf{x}, t) = \int_0^t z(\mathbf{x}, t, s) ds,$$

where for each s, $0 \le s \le t$, z solves the IVP

$$z_{tt} - c^2 \Delta z = 0 \tag{8.45}$$

$$z(\mathbf{x}, s, s) = 0, \qquad z_t(\mathbf{x}, s, s) = q(\mathbf{x}, s).$$

Now in three space dimensions (use (8.41) with $f = 0$, $g = q$, and replace t by $t - s$) we obtain

$$z(\mathbf{x}, t, s) = \frac{1}{4\pi c^2(t - s)} \int \int_{|\mathbf{y}-\mathbf{x}|=c(t-s)} q(\mathbf{y}, s) dS(\mathbf{y}),$$

so that

$$u(\mathbf{x}, t) = \frac{1}{4\pi c^2} \int_0^t \frac{1}{(t - s)} \int \int_{|\mathbf{y}-\mathbf{x}|=c(t-s)} q(\mathbf{y}, s) dS(\mathbf{y}) ds. \tag{8.46}$$

How can we interpret the integral (8.46)? We shall put it in geometric terms. Let $K_{\mathbf{x},t}$ be the solid cone in four-dimensional space-time (see Figure 8.3):

$$K_{\mathbf{x},t} = \{(\mathbf{y}, t) : |\mathbf{y} - \mathbf{x}| \le c(t - s), 0 \le s \le t\}.$$

The vertex of the cone $K_{\mathbf{x},t}$ is at the point $(\mathbf{x}, t)$, and its base is the ball in three space $\{\mathbf{y} \in R^3 : |\mathbf{y} - \mathbf{x}| \le ct\}$. Let $\Sigma_{\mathbf{x},t}$ be the three-dimensional lateral surface of $K_{\mathbf{x},t}$:

$$\Sigma_{\mathbf{x},t} = \{(\mathbf{y}, s) : |\mathbf{y} - \mathbf{x}| = c(t - s), 0 \le s \le t\}.$$

Now the domain of dependence of $z(\mathbf{x}, t, s)$ is the sphere in R^3 which is the intersection of $\Sigma_{\mathbf{x},t}$ with the plane $t = s$. $\Sigma_{\mathbf{x},t}$ is the union of these spheres, and hence, is the domain of dependence of u on the source q. The value of the solution u at the point $(\mathbf{x}, t)$ in four-dimensional space-time is determined by the values of q over this conical surface.

To make the dependence of u on q clearer, we shall transform the integral (8.46) into a three-dimensional volume integral over the base of the cone $K_{\mathbf{x},t}$. Let $\rho = c(t - s), 0 \le s \le t$. Then $cds = -d\rho$ and

$$u(\mathbf{x}, t) = \int_{ct}^0 \frac{1}{4\pi c\rho} \int_{|\mathbf{y}-\mathbf{x}|=\rho} q(\mathbf{y}, t - \rho/c) dS(\mathbf{y})(-d\rho/c)$$

$$= \frac{1}{4\pi c^2} \int_0^{ct} \frac{q(\mathbf{y}, t - \rho/c)}{\rho} dS(\mathbf{y}) d\rho.$$

$dS(\mathbf{y})$ in (8.46) is the element of surface area on the sphere of radius $\rho = c(t - s)$, and $d\mathbf{y} = dS(\mathbf{y})d\rho$ is the ordinary three-dimensional volume element. Thus

$$u(\mathbf{x}, t) = \frac{1}{4\pi c^2} \int_{|\mathbf{y}-\mathbf{x}|\le ct} \frac{q(\mathbf{y}, t - |\mathbf{x} - \mathbf{y}|/c)}{|\mathbf{x} - \mathbf{y}|} d\mathbf{y}. \tag{8.47}$$

In particular we see that, if $q(\mathbf{x}, t) = 0$ for $|\mathbf{x}| \geq a > 0$ for all t, then the influence of the source is zero outside the expanding region $|\mathbf{x}| \leq a + ct$, $t > 0$. See Figure 8.4a below.

An important special case is that of a source oscillating with frequency ω,

$$q(\mathbf{x}, t) = e^{i\omega t} q_0(\mathbf{x}).$$

We use the complex exponential $\exp(i\omega t)$ to simplify the computations. If we need real solutions, we simply take the real and imaginary parts of u. Now substitute this form of q in (8.47). We see that

$$q(\mathbf{y}, t - |\mathbf{x} - \mathbf{y}|/c) = e^{i\omega t} e^{-ik|\mathbf{x} - \mathbf{y}|} q_0(\mathbf{y}),$$

where we have introduced the wave number $k = \omega/c$. Then (8.47) becomes

$$u(\mathbf{x}, t) = \frac{e^{i\omega t}}{4\pi c^2} \int_{|\mathbf{y} - \mathbf{x}| \leq ct} \frac{e^{-ik|\mathbf{x} - \mathbf{y}|}}{|\mathbf{x} - \mathbf{y}|} q_0(\mathbf{y}) \, d\mathbf{y}.$$

Now suppose that the oscillating source q is localized in space,

$$q_0(\mathbf{x}) = 0 \quad \text{for } |\mathbf{x}| \geq a > 0.$$

We would expect the solution to describe radiation from this source, and that it would have the same frequency as the source. In fact this is true on any fixed ball in space, if we wait long enough. For each fixed $\mathbf{x}$, the integral over the region $|\mathbf{x} - \mathbf{y}| \leq ct$ reduces to the integral over the ball $|\mathbf{y}| \leq a$ for $t > |\mathbf{x}| + a/c$. See Figure 8.4b. Hence for each fixed $\mathbf{x}$, $u(\mathbf{x}, t)$ becomes (for $t > |\mathbf{x}| + a/c$)

$$u(\mathbf{x}, t) = e^{i\omega t} v(\mathbf{x}),$$

where

$$v(\mathbf{x}) = \frac{1}{4\pi} \int \frac{e^{-ik|\mathbf{x} - \mathbf{y}|} \tilde{q}_0(\mathbf{y})}{|\mathbf{x} - \mathbf{y}|} \, d\mathbf{y} \tag{8.48}$$

with $\tilde{q}_0 = q_0/c^2$. Thus we see that the solution $u(\mathbf{x}, t)$ converges for each $\mathbf{x}$ to another solution which is a product of a time factor $\exp(i\omega t)$ and a spatial factor $v(\mathbf{x})$. This is known as the *principle of limiting amplitude*. For a function of the form $\exp(i\omega t)v(\mathbf{x})$ to solve the PDE

$$u_{tt} - c^2 \Delta u = c^2 \tilde{q}_0(\mathbf{x}) e^{i\omega t},$$

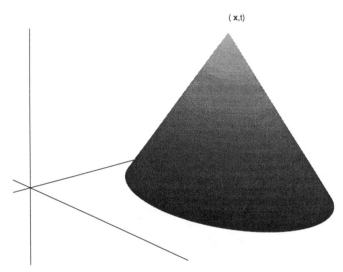

FIGURE 8.3
Schematic representation of the backward cone $K_{\mathbf{x},t}$. Vertical axis is the t axis,
horizontal plane represents three space dimensions.

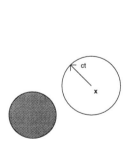

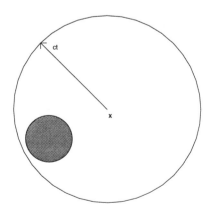

FIGURE 8.4a
Source $q = 0$ outside the ball B_a
(shaded area). When $ct < |\mathbf{x}| - a$,
the ball of radius ct, center $\mathbf{x}$, does
not meet B_a.

FIGURE 8.4b
Source $q = 0$ outside the ball B_a (shaded
area). When $ct > |\mathbf{x}| - a$, the ball of radius
ct, center $\mathbf{x}$, now includes B_a.

v must satisfy the *reduced wave equation*, often called the *Helmholtz equation*

$$-(\Delta v + k^2 v) = \tilde{q}_0(x).$$

Of course, this can also be seen directly from (8.48). Notice that this solution of the PDE does not satisfy the initial conditions $u(\mathbf{x}, 0) = u_t(\mathbf{x}, 0) = 0$. It is more akin to a steady-state solution of the heat equation. We shall see this solution of the Helmholtz equation again in Chapter 9, when we treat solutions of the Poisson equation.

Exercises 8.8

1. Consider (8.44) in R^3, and take the source term q to depend only on $r = \sqrt{x^2 + y^2 + z^2}$ and t, $q = q(r, t)$.

   (a) Use a geometric argument to show that the solution given by (8.46) is also spherically symmetric.

   (b) Show that $w(r, t) = ru(r, t)$ satisfies the one-dimensional wave equation for $r > 0$ with right hand side $rq(r, t)$ and boundary condition $w(0, t) = 0$. Think of $q(r, t)$ as defined as an even function of r. Then $r \to rq(r, t)$ is odd. Solve the problem $w_{tt} - c^2 w_{rr} = rq(r, t)$ on the whole line using (5.29). Show that w found this way is odd, and hence satisfies $w(0, t) = 0$.

2. Again in three space dimensions, assume the source is of the form $q(r, t) = \exp(i\omega t)q_0(r)$, with $q_0(r) = 0$ for $r > a$. Use the results of exercise 1 to show directly that the solution $u(r, t)$ of (8.44) converges as $t \to \infty$ to a product $\exp(i\omega t)v(r)$ where v satisfies the Helmholtz equation with an appropriate right-hand side. In particular, find $u(r, t)$ and $v(r)$ when $q_0(r) = 1$ for $0 < r < a$ and $q_0(r) = 0$ for $r > a$. Use the results of exercise 13, Section 5.3.

8.9 Boundary value problems for the wave equation

8.9.1 Eigenfunction expansions

We shall consider problems in two space dimensions, but the ideas are easily extended to higher dimensions. Let G be an open, bounded set in R^2 with ∂G piecewise C^1. The initial-boundary-value problems are

$$u_{tt} - c^2 \Delta u = 0 \qquad \text{in } G \times R, \tag{8.49}$$

$$u(x, y, 0) = f(x, y), \qquad u_t(x, y, 0) = g(x, y),$$

$$u = 0 \qquad \text{on } \partial G, \tag{8.50}$$

$$\frac{\partial u}{\partial n} = 0 \qquad \text{on } \partial G, \tag{8.51}$$

and

$$\frac{\partial u}{\partial n} + hu = 0 \qquad \text{on } \partial G. \tag{8.52}$$

Here $h = h(x, y)$ is a nonnegative function defined on ∂G. The mixed boundary condition is given by

$$u = 0 \quad \text{on } \Gamma_0, \qquad \frac{\partial u}{\partial n} = 0 \quad \text{on } \Gamma_1, \tag{8.53}$$

where $G = \Gamma_1 \cup \Gamma_2$.

We follow the same procedure used earlier in this chapter for the heat equation. First we look for solutions in the form

$$\varphi(x, y)\psi(t).$$

This leads to the temporal equation

$$\psi'' + c^2 \lambda \psi = 0,$$

and the eigenvalue problem

$$L\varphi = \lambda\varphi,$$

where $L\varphi = -\Delta\varphi$ is the linear operator with domain W being the set of $\varphi \in C^2(\bar{G})$, which satisfy one of the boundary conditions (8.50) - (8.53). As we saw before, there is a sequence of eigenfunctions φ_n with eigenvalues λ_n. The solution of (8.49) with one of these boundary conditions can be written

$$u(x, y, t) = \sum_n [A_n \cos(\omega_n t) + B_n \sin(\omega_n t)]\varphi_n(x, y),$$

where $\omega_n = c\sqrt{\lambda_n}$ is the angular frequency,

$$A_n = \frac{\langle f, \varphi_n \rangle}{\langle \varphi_n, \varphi_n \rangle},$$

and

$$B_n = \frac{1}{\omega_n} \frac{\langle g, \varphi_n \rangle}{\langle \varphi_n, \varphi_n \rangle}.$$

Example

Let G be the rectangle $(0, a) \times (0, b)$ with the Dirichlet boundary conditions (8.50). Then we can think of $u(x, y, t)$ as the vertical displacement of a rectangular drum head. The eigenfunctions will be the modes of vibration of the drumhead. They are found to be

$$\varphi_{m,n}(x, y) = \sin(\frac{m\pi x}{a}) \sin(\frac{n\pi y}{b}), \qquad m, n = 1, 2, 3, \ldots,$$

with eigenvalues

$$\lambda_{m,n} = (\frac{m\pi}{a})^2 + (\frac{n\pi}{b})^2.$$

The nodal lines or curves of each mode are the lines where $\varphi_{m,n} = 0$.

8.9.2 Nodal curves

More complex nodal line and curve structures can be found by taking sums and differences of the eigenfunctions. To illustrate this idea, let us suppose that G is a square with $a = b = \pi$. Then the eigenfunctions $\varphi_{m,n}$ and $\varphi_{n,m}$ both have the eigenvalue

$$\lambda = \lambda_{m,n} = \lambda_{n,m} = m^2 + n^2,$$

and both vibrate with the angular frequency

$$\omega = \omega_{m,n} = \omega_{n,m} = c\sqrt{\lambda}.$$

Hence for each α,

$$\varphi = \alpha\varphi_{m,n} + (1 - \alpha)\varphi_{n,m}$$

is a mode of vibration which also vibrates with angular frequency ω. The nodal lines or curves are now those of φ. As α varies from one down to zero, the nodal lines of $\varphi_{m,n}$ are deformed into the nodal lines of $\varphi_{n,m}$. Contour maps of the mode $\varphi = \alpha\varphi_{2,3} + (1 - \alpha)\varphi_{3,2}$ on the square $a = b = \pi$ are shown for various values of α in Figures 8.5 - 8.8. The nodal curves are the solid curves. They separate the square into several different regions which vibrate independently. The number of regions changes from 6 to 4 and back to 6 as α ranges from 1 to 0.

8.9.3 Conservation of energy

The energy of a solution u of (8.49) with boundary conditions (8.50), (8.51). or (8.53) is

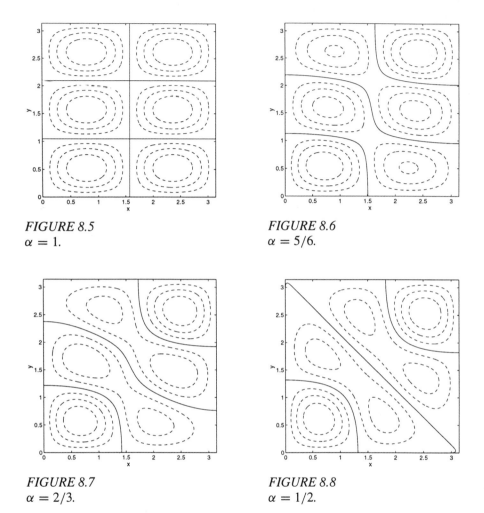

FIGURE 8.5
$\alpha = 1$.

FIGURE 8.6
$\alpha = 5/6$.

FIGURE 8.7
$\alpha = 2/3$.

FIGURE 8.8
$\alpha = 1/2$.

$$e(t) = \frac{1}{2}\int\int_G [u_t^2 + c^2|\nabla u|^2]dxdy.$$

We will show that this energy is conserved by a slightly different method than we used in the one-dimensional case. Let u be a solution of (8.49) and multiply (8.49) by u_t, and then integrate over G. We find that

$$\int\int_G u_{tt}u_t\,dxdy - \int\int_G c^2(\Delta u)u_t\,dxdy = 0.$$

Now

$$\int\int_G u_{tt}u_t dx dy = \frac{1}{2}\frac{d}{dt}\int\int_G u_t^2 dx dy,$$

and, using Green's first identity on the second integral, we find that

$$-\int\int_G \Delta u u_t dx dy = -\int_{\partial G}\frac{\partial u}{\partial n}u_t ds + \int\int_G \nabla u \cdot \nabla u_t dx dy.$$

Thus

$$\frac{de(t)}{dt} = \frac{1}{2}\frac{d}{dt}\int\int_G [u_t^2 + |\nabla u|^2]dx dy = \int_{\partial G}\frac{\partial u}{\partial n}u_t ds.$$

In each of the boundary conditions (8.50), (8.51), and (8.53), the integral over ∂G vanishes so that

$$\frac{de(t)}{dt} = 0,$$

which shows that energy is conserved. In the case of the Robin condition (8.52), notice that the boundary term becomes

$$\int_{\partial G}\frac{\partial u}{\partial n}u_t ds = -\int_{\partial G}huu_t ds = -\frac{1}{2}\frac{d}{dt}\int_{\partial G}hu^2 ds.$$

Hence for (8.52), a different energy quantity is conserved, namely

$$\frac{1}{2}\frac{d}{dt}\left[\int\int_G (u_t^2 + c^2|\nabla u|^2)dx dy + \int_{\partial G}hu^2 ds\right] = 0.$$

8.9.4 Inhomogeneous problems

The inhomogeneous equation with a right-hand side forcing term

$$u_{tt} - c^2\Delta u = q \tag{8.54}$$

$$u(x, y, 0) = 0, \qquad u_t(x, y, 0) = 0,$$

with one of the boundary conditions (8.50) - (8.53) is solved using the eigenfunction expansions. We expand q as

$$q(x, y, t) = \sum_n q_n(t)\varphi_n(x, y)$$

and u as

$$u(x, y, t) = \sum_n u_n(t)\varphi_n(x, y).$$

The coefficients $u_n(t)$ must satisfy the ODE's

$$u_n'' + \omega_n^2 u_n = q_n$$

$$u_n(0) = u_n'(0) = 0.$$

It is easy to show that

$$u_n(t) = \frac{1}{\omega_n} \int_0^t \sin(\omega_n(t-s))q_n(s)ds. \tag{8.55}$$

Of course we can treat inhomogeneous boundary conditions following the methods of Chapters 4 or 5.

Control theory provides a setting in which to consider these inhomogeneous boundary conditions from a different point of view. Suppose that we are given initial conditions $f(x, y)$ and $g(x, y)$ and the inhomogeneous boundary condition

$$u(x, y, t) = l(x, y, t) \qquad \text{for } (x, y) \in \partial G.$$

Our objective is to choose the function (control) l so as to make the solution u converge to zero as rapidly as possible. By applying an optimal boundary control we hope to damp out vibrations as quickly as possible. Furthermore, there may be practical considerations which allow us to control only a portion of the boundary. The control l will be zero for $(x, y) \notin \Gamma_0$, where Γ_0 is a subset of ∂G.

Exercises 8.9

1. Solve the IVP (8.49) with boundary conditions (8.50), where G is the square $\{0 < x < \pi, 0 < y < \pi\}$ and the initial data is

$$f(x, y) = \sin x \sin(2y) + \frac{1}{2}\sin(4x)\sin(2y), \qquad g(x, y) = 0.$$

   (a) What are the angular frequencies of the modes involved? Is the motion periodic? If so, what is the period?

   (b) Same questions for the initial data

$$f(x, y) = \sin x \sin(2y) + \frac{1}{2}\sin(3x)\sin(y), \qquad g(x, y) = 0.$$

2. Consider the inhomogeneous problem in the same square G as in exercise 1,

$$u_{tt} - c^2 \Delta u = q,$$

where $q = \sin(\omega t) \sin(3x) \sin(4y)$ with Dirichlet boundary conditions, and with initial conditions

$$u(x, y, 0) = 0, \qquad u_t(x, y, 0) = 0.$$

Find the solution as an integral in the form (8.55). For what value(s) of ω will resonance occur, that is, for what values of ω will the solution grow without bound as $t \to \infty$? Describe the behavior of the solution as $t \to \infty$ in the nonresonant case.

3. Let G be the same square. Consider (8.49) with boundary conditions (8.50) and initial data

$$f(x, y) = 0, \qquad g(x, y) = \begin{cases} \frac{1}{\varepsilon\sqrt{2}}, & |x - \pi/2|, |y - \pi/2| \leq \varepsilon \\ 0, & \text{elsewhere} \end{cases}.$$

(a) Verify that $(1/2) \int \int_G g(x, y)^2 dx dy = 1$, so that the solution will have finite, conserved, energy $= 1$.

This initial condition corresponds to the drum being hit with a unit amount of energy concentrated in the small square of side 2ε centered at $(\pi/2, \pi/2)$.

(b) Find the solution in terms of the eigenfunction expansion.

(c) Now move the small square where the drum is struck to the corner of the drum, so that it occupies the set $(\pi - 2\varepsilon, \pi) \times (\pi - 2\varepsilon, \pi)$. Find the solution again in terms of the eigenfunction expansion.

(d) Compare the $\omega_{1,1}$ frequency component for small ε for both cases. What happens when you hit the drum in corner vs. hitting it in the middle?

Now we use MATLAB to analyze the nodal structure of various modes of vibration, and to view some time-dependent solutions which represent the vibrations of a drum head. To do this efficiently, we must write an mfile of a sequence of MATLAB commands. As usual the mfile must begin with commands to make a meshgrid of the square $G = \{[0, \pi] \times [0, \pi]\}$ with spacing $\Delta x = \Delta y = \pi/100$. Make the pair of integers m, n and the parameter α input data with the commands:

```
M = input('Enter the integer pair [m,n]    ')
m = M(1); n = M(2);
alpha = input('Enter the value of alpha    ')
```

Next write a line that defines the mode

$$z = \alpha \sin(mx)\sin(ny) + (1-\alpha)\sin(nx)sin(my).$$

Remember to use the meshgrid variables X, Y. Next write the plotting commands:

```
pcolor(X,Y,z);shading flat
hold on
contour(X,Y,z,1,'k')
hold off
```

In the `contour` command, we set the number of desired contour levels at one. Because min $z = -$ max z for these modes, the single level set will be the curve(s) where $z = 0$, i.e., the nodal curves. You can add more level sets and still keep the nodal curve if you ask for an odd number of level sets. You can also omit the `pcolor` command to produce a plot of just the nodal curves, which is easier to print out. However, you should change the color to `'c'` which is cyan.

4. To check your program, run it first with $m = 1, n = 2$, and $\alpha = 1$. The square should divide into two rectangles separated by the horizontal nodal line $y = \pi/2$. Use the command `surf(X,Y,z);shading flat` to see a 3-D plot of the surface. You will be able to see which parts of the graph are above the x, y plane and which are below.

   (a) Now run your program with $m = 1, n = 3$, and the parameter values $\alpha = 1, .75, .5, .25, 0$. Count the number of pieces of the square separated by nodal curves for each value of α. At about what value of α does the number change from 3 to 2?

   (b) Run again with $m = 1, n = 7$, and repeat the count of part a). Now make a conjecture about the number of separate pieces there are when $\alpha = .5$ for any pair 1, n, where n is odd. Try it out with $n = 9$.

5. Suppose that the mode is

$$\varphi = \alpha \sin(x)\sin(ny) + (1-\alpha)\sin(nx)\sin(y)$$

   with n even and $\alpha = .5$.

   (a) Use paper and pencil to show that the line $y = \pi - x$ is always a nodal line.

(b) Now use your program to view this situation, say, for n = 6. Into how many separate pieces do the nodal curves divide the square? Run the program again for $\alpha = .5$ and $n = 8$. Same question. What is your conjecture about the number of pieces for $\alpha = .5$ and general n even?

(c) To show that a small change in α can make a dramatic change in the nodal structure, again try $n = 6$, or $n = 8$, but perturb α a bit to $\alpha = .51$. How many distinct nodal curves are there? How many separate pieces of the square?

Now modify your program to give snapshots of the time-dependent solution

$$u(x, y, t) = \cos(\omega t)[\alpha \sin(mx) \sin(ny) + (1 - \alpha) \sin(nx) \sin(my)],$$

where $\omega = \sqrt{m^2 + n^2}$. Add lines to calulate ω and the period $T = 2\pi/\omega$. Make t an input datum. Change the plotting commands to the commands

```
>> surf(X,Y,z);shading flat
>> axis([0 pi 0 pi -1 1])
```

This last command fixes the vertical scale.

6. (a) Run your program with $\alpha = 1, m = 2, n = 3$ and $t = 0, T/8, T/4, T/2, T$. You may wish to fix the values of α, m, n in your program instead of making them input data. Verify that there are six vibrating pieces of the square separated by the nodal lines.

(b) Now change α to .5, same m, n. Run for the same values of t. Verify that the nodal curves separate vibrating pieces of the square. You can check that the nodal curves remain the same at each time by plotting with the command contour(X,Y,z,1, 'c'). They will appear in cyan (turquoise) on the black background. If you are running MATLAB5.0, enter the command colordef black before running the program, to give a black background.

7. Consider the two modes

$$\varphi_1 = \sin(2x) \sin(4y) \quad \text{and} \quad \varphi_2 = \sin(6x) \sin(3y).$$

(a) Calculate the frequency and period for each mode.

(b) Show that the two modes have a common period T by finding integers k and l and a frequency ω, such that $\omega_1 = k\omega$ and $\omega_2 = l\omega$. Then $T = 2\pi/\omega$. This means that the solution

$$u(x, y, t) = \cos(\omega_1 t)\varphi_1(x, y) + \cos(\omega_2 t)\varphi_2(x, y)$$

will be a periodic motion with period T.

(c)   Modify the program used in exercises 5 and 6 to plot snapshots of this solution. Find times t_* and t_{**}, such that when $t = t_*$, φ_2 is not visible, and when $t = t_{**}$, φ_1 is not visible.

(d)   Are there any lines or curves which are nodal lines for this motion, that is, are there any lines or curves inside the square where u is always zero?

8.   Consider a circular drum of radius a. Use the eigenfunctions found in Section 8.4 for the disk to solve the IVP (8.49) with boundary condition (8.50).

8.10   The Maxwell equations

8.10.1   The electric and magnetic fields

The Maxwell equations are a linear system of partial differential equations that describe the fluctuations of four vector quantities:

E,	the electric field,
D,	the displacement field,
B,	the magnetic induction, and
H,	the magnetic field.

Each of these vector quantities has three components and they are functions of $(\mathbf{x}, t) = (x, y, z, t)$. The basic equations in the rationalized mks system of units are

$$\frac{\partial \mathbf{D}}{\partial t} - \nabla \times \mathbf{H} = -\mathbf{J} \tag{8.56}$$

$$\frac{\partial \mathbf{B}}{\partial t} + \nabla \times \mathbf{E} = 0 \tag{8.57}$$

$$\nabla \cdot \mathbf{D} = \rho \tag{8.58}$$

$$\nabla \cdot \mathbf{B} = 0. \tag{8.59}$$

Here $\mathbf{J}$ is the current density, and ρ is the charge density. Each of the equations expresses a physical law. For example, (8.57) is a reformulation of Faraday's law of induction which implies that a changing magnetic field can create an electric field. Equation (8.58) is Coulomb's law which states that an electric field is created by a charge. (8.59) states that there are no magnetic poles. However, the key equation, which led to the theory of electromagnetic waves, is (8.56). For time-independent phenomena, the first equation is

$$\nabla \times \mathbf{H} = \mathbf{J}$$

which yields Ampere's law that $\nabla \cdot \mathbf{J} = 0$ because $\nabla \cdot (\nabla \times \mathbf{H}) = 0$. It was Maxwell who suggested that the magnetic field could also be excited by a changing electric field, and added the term $\partial \mathbf{D}/\partial t$ to make equation (8.56).

The four field quantities are related by the constitutive relations

$$\mathbf{D} = \epsilon \mathbf{E}$$

and

$$\mathbf{B} = \mu \mathbf{H}.$$

ϵ and μ may be matrices, but we shall assume that they are scalars called the *dielectric constant* and the *magnetic permeability* respectively. They are properties of the medium. With these relations, (8.56) - (8.59) is a system of eight equations in the six unknowns $\mathbf{E} = (E_1, E_2, E_3)$ and $\mathbf{H} = (H_1, H_2, H_3)$. In the absence of any free charge or current, the system has a beautiful symmetry.

$$\epsilon \frac{\partial \mathbf{E}}{\partial t} - \nabla \times \mathbf{H} = 0 \tag{8.60}$$

$$\mu \frac{\partial \mathbf{H}}{\partial t} + \nabla \times \mathbf{E} = 0 \tag{8.61}$$

$$\nabla \cdot \mathbf{E} = 0 \tag{8.62}$$

$$\nabla \cdot \mathbf{H} = 0. \tag{8.63}$$

We shall consider this system on all of R^3, with no boundary value conditions. The initial-value problem is to find a solution of (8.60)-(8.63) with initial conditions

$$\mathbf{E}(\mathbf{x}, 0) = \mathbf{E}^0(\mathbf{x}), \tag{8.64}$$

and

$$\mathbf{H}(\mathbf{x}, 0) = \mathbf{H}^0(\mathbf{x}). \tag{8.65}$$

We assume that $\mathbf{E}^0$ and $\mathbf{H}^0$ satisfy (8.62) and (8.63). Now it is a fact that a solution $\mathbf{E}, \mathbf{H}$ of (8.60) and (8.61), which satisfies (8.62) and (8.63) at $t = 0$ continues to satisfy (8.62) and (8.63) for all t. To see this, take the divergence of each term in (8.60). Interchanging the order of differentiation, we see that

$$\epsilon \frac{\partial}{\partial t}(\nabla \cdot \mathbf{E}) = \nabla \cdot (\nabla \times \mathbf{H}) = 0.$$

Thus $\nabla \cdot \mathbf{E}$ is independent of t, and if $\nabla \cdot \mathbf{E} = 0$ at $t = 0$, then $\nabla \cdot \mathbf{E} = 0$ for all t. The same argument, using (8.61) and (8.63), applies to $\nabla \cdot \mathbf{H}$.

Now if $\mathbf{E}$ and $\mathbf{H}$ are C^2 functions satisfying (8.60) - (8.63), then each component of $\mathbf{E}$ and $\mathbf{H}$ satisfies the wave equation

$$u_{tt} - c^2 \Delta u = 0, \tag{8.66}$$

where

$$c = \frac{1}{\sqrt{\epsilon \mu}}$$

is the speed of propagation in the medium. In vacuum, c is the speed of light with value $c \approx 3 \times 10^8$ meters/second. Let us verify this for the components of $\mathbf{E}$. Differentiate (8.60) with respect to t:

$$\epsilon \frac{\partial^2 \mathbf{E}}{\partial^2 t} = \frac{\partial}{\partial t}(\nabla \times \mathbf{H}) = \nabla \times \frac{\partial \mathbf{H}}{\partial t},$$

and substitute for $\partial \mathbf{H}/\partial t$ using (8.61). We find that

$$\frac{\partial^2 \mathbf{E}}{\partial^2 t} = -\frac{1}{\epsilon \mu}[\nabla \times (\nabla \times \mathbf{E})]. \tag{8.67}$$

However for any vector field $\mathbf{v}$, we have the identity

$$\nabla \times (\nabla \times \mathbf{v}) = \nabla(\nabla \cdot \mathbf{v}) - \Delta \mathbf{v}$$

where $\Delta \mathbf{v}$ means apply Δ to each component of $\mathbf{v}$. Applying this identity to the right side of (8.67) and using the fact that $\nabla \cdot \mathbf{E} = 0$, we see that

$$\frac{\partial^2 \mathbf{E}}{\partial^2 t} = \frac{1}{\epsilon \mu} \Delta \mathbf{E},$$

which is to say that each component of $\mathbf{E}$ satisfies (8.64) with $c^2 = (\epsilon \mu)^{-1}$.

8.10.2 The initial-value problem

Next we turn our attention to the solution of the initial-value problem (8.60) - (8.63) with initial data (8.64), (8.65). Since each component of **E** and **H** solves the wave equation (8.66), we may use the Kirchoff formulas (8.41). Now from (8.60) and (8.61), we see that

$$\left.\frac{d\mathbf{E}}{dt}\right|_{t=0} = -\frac{1}{\epsilon}\nabla \times \mathbf{H}^0(\mathbf{x})$$

and

$$\left.\frac{d\mathbf{H}}{dt}\right|_{t=0} = \frac{1}{\mu}\nabla \times \mathbf{E}^0(\mathbf{x}).$$

Hence

$$\mathbf{E}(\mathbf{x}, t) = -\frac{1}{4\pi c^2 t} \int\int_{|\mathbf{y}-\mathbf{x}|=ct} \frac{1}{\epsilon}(\nabla \times \mathbf{H}^0)(\mathbf{y})dS(\mathbf{y})$$

$$+ \frac{\partial}{\partial t}\left(\frac{1}{4\pi c^2 t} \int\int_{|\mathbf{y}-\mathbf{x}|=ct} \mathbf{E}^0(\mathbf{y})dS(\mathbf{y})\right)$$

and

$$\mathbf{H}(\mathbf{x}, t) = \frac{1}{4\pi c^2 t} \int\int_{|\mathbf{y}-\mathbf{x}|=ct} \frac{1}{\mu}(\nabla \times \mathbf{E}^0)(\mathbf{y})dS(\mathbf{y}) \tag{8.68}$$

$$+ \frac{\partial}{\partial t}\left(\frac{1}{4\pi c^2 t} \int\int_{|\mathbf{y}-\mathbf{x}|=ct} \mathbf{H}^0(\mathbf{y})dS(\mathbf{y})\right).$$

8.10.3 Plane waves

An important class of solutions of the Maxwell equations (8.60)- (8.63) is the class of plane waves. They are of the form

$$\mathbf{E}(\mathbf{x}, t) = \mathbf{E}^0 e^{i(\omega t - \mathbf{k}\cdot\mathbf{x})}$$

$$\mathbf{H}(\mathbf{x}, t) = \mathbf{H}^0 e^{i(\omega t - \mathbf{k}\cdot\mathbf{x})}.$$

$\mathbf{E}^0$ and $\mathbf{H}^0$ are constant vectors. How should $\mathbf{k}$, $\mathbf{E}^0$, and $\mathbf{H}^0$ be related to yield a solution of (8.60) - (8.63)? First, since each component of **E** and **H** must satisfy (8.66), we require that $|\mathbf{k}| = \omega/c$. Next we impose the divergence-free conditions (8.62), (8.63). They will be satisfied provided that

$$\mathbf{E}^0 \cdot \mathbf{k} = \mathbf{H}^0 \cdot \mathbf{k} = 0.$$

Thus both $\mathbf{E}^0$ and $\mathbf{H}^0$ must be orthogonal to the direction of propagation $\mathbf{k}$. Such a wave is called a transverse wave.

Next we want to substitute the formulas for $\mathbf{E}$ and $\mathbf{H}$ in (8.60) and (8.61). We will use the fact that for a constant vector $\mathbf{v}$ and scalar function $f(\mathbf{x})$,

$$\nabla \times (f\mathbf{v}) = \nabla f \times \mathbf{v}.$$

Hence

$$\nabla \times \mathbf{H} = \nabla(e^{i(\omega t - \mathbf{k} \cdot \mathbf{x})}) \times \mathbf{H}^0$$

$$= (-i\mathbf{k} \times \mathbf{H}^0)e^{i(\omega t - \mathbf{k} \cdot \mathbf{x})}$$

and

$$\epsilon \frac{\partial \mathbf{E}}{\partial t} = i\omega\epsilon \mathbf{E}^0 e^{i(\omega t - \mathbf{k} \cdot \mathbf{x})}.$$

Thus to satisfy (8.60), we require that

$$\epsilon\omega\mathbf{E}^0 = -\mathbf{k} \times \mathbf{H}^0 = \mathbf{H}^0 \times \mathbf{k}.$$

Similarly to satisfy (8.61), we require that

$$\mu\omega\mathbf{H}^0 = -\mathbf{E}^0 \times \mathbf{k}.$$

Thus if $\mathbf{k}$ is given, and $\mathbf{E}^0$ is chosen so that $\mathbf{k} \cdot \mathbf{E}^0 = 0$, then we define $\mathbf{H}^0$ by

$$\mathbf{H}^0 = \frac{1}{\mu\omega}(\mathbf{k} \times \mathbf{E}^0).$$

Since $|\mathbf{k}|^2 = (\omega/c)^2$, it follows that $\mathbf{H}^0 \times \mathbf{k} = \epsilon\omega\mathbf{E}^0$. The three vectors $\mathbf{k}, \mathbf{E}^0, \mathbf{H}^0$ form an orthogonal set.

8.10.4   Electrostatics

Electrostatics deals with steady-state solutions of the Maxwell equations. Assume the charge density $\rho = \rho(\mathbf{x})$ and that $\mathbf{E}$ and $\mathbf{H}$ are time-independent. Then the system (8.56) - (8.59) decouples. The equations for $\mathbf{E}$ become

$$\nabla \times \mathbf{E} = 0, \qquad \epsilon \nabla \cdot \mathbf{E} = \rho.$$

Now assume that these equations are satisfied in a simply connected region $G \subset R^3$. Then, because $\nabla \times \mathbf{E} = 0$, there exists a scalar potential function $V(\mathbf{x})$, such that $\mathbf{E} = -\nabla V$ in G. Hence V solves the equation

$$-\Delta V = -\nabla \cdot \nabla V = \nabla \cdot \mathbf{E} = \frac{\rho}{\epsilon}.$$

This is the Poisson equation. If the boundary ∂G of the region is "grounded", the potential is zero on ∂G, and we have the boundary value problem for the Poisson equation

$$-\Delta V = \frac{\rho}{\epsilon} \quad \text{in } G, \qquad V = 0 \quad \text{on } \partial G.$$

We shall study this boundary value problem in Chapter 9.

8.10.5 Conservation of energy

First, since each component of $\mathbf{E}$ and $\mathbf{H}$ satisfies (8.66), and solutions of (8.66) have a finite speed of propagation, it follows that the same must be true for $\mathbf{E}$ and $\mathbf{H}$.

Secondly, since solutions of (8.66) have a conserved energy, we should expect that solutions of (8.60) - (8.63) also have some conserved energy quantity. We shall show that the energy

$$e(t) = \frac{1}{2}[\epsilon \int_{R^3} |\mathbf{E}(\mathbf{x}, t)|^2 dx + \mu \int_{R^3} |\mathbf{H}(\mathbf{x}, t)|^2 dx] \tag{8.69}$$

is conserved. If we are dealing with the energy contained in a subset G of R^3, then we use

$$e_G(t) = \frac{1}{2}[\epsilon \int_G |\mathbf{E}(\mathbf{x}, t)|^2 dx + \mu \int_G |\mathbf{H}(\mathbf{x}, t)|^2 dx].$$

When certain boundary conditions are imposed $e_G(t)$ is also a conserved quantity. Let us calculate the rate of change of $e_G(t)$. First note that

$$\frac{d|\mathbf{E}|^2}{dt} = 2\mathbf{E} \cdot \mathbf{E}_t \quad \text{and} \quad \frac{d|\mathbf{H}|^2}{dt} = 2\mathbf{H} \cdot \mathbf{H}_t.$$

Then

$$\frac{d}{dt} e_G(t) = \int_G \epsilon \mathbf{E}_t \cdot \mathbf{E} + \mu \mathbf{H}_t \cdot \mathbf{H} dx.$$

Now take scalar the product of (8.60) with $\mathbf{E}$, and the scalar product of (8.61) with $\mathbf{H}$, and add. This shows that the integrand of this last integral is

$$\epsilon \mathbf{E}_t \cdot \mathbf{E} + \mu \mathbf{H}_t \cdot \mathbf{H} = (\nabla \times \mathbf{H}) \cdot \mathbf{E} - (\nabla \times \mathbf{E}) \cdot \mathbf{H} = -\nabla \cdot (\mathbf{E} \times \mathbf{H}),$$

where we have used the fact that for any vector fields $\mathbf{M}$ and $\mathbf{N}$,

$$\nabla \cdot (\mathbf{M} \times \mathbf{N}) = \mathbf{N} \cdot (\nabla \times \mathbf{M}) - \mathbf{M} \cdot (\nabla \times \mathbf{N}).$$

Now by the divergence theorem,

$$\int_G \nabla \cdot (\mathbf{E} \times \mathbf{H}) dx = \int_{\partial G} \mathbf{n} \cdot (\mathbf{E} \times \mathbf{H}) dS(\mathbf{x}).$$

Here $\mathbf{n}$ is the exterior unit normal to ∂G. Thus we conclude that

$$\frac{d}{dt} e_G(t) = -\int_{\partial G} \mathbf{n} \cdot (\mathbf{E} \times \mathbf{H}) dS(\mathbf{x}). \tag{8.70}$$

The vector $\mathbf{E} \times \mathbf{H}$ is called the *Poynting* vector. It is the flux of the electromagnetic energy. The quantity $\mathbf{n} \cdot (\mathbf{E} \times \mathbf{H})$ has units of energy/time/area and thus is the flow of power/unit area across the surface ∂G in the outward dirction $\mathbf{n}$. If the integral of $\mathbf{n} \cdot (\mathbf{E} \times \mathbf{H})$ over ∂G is positive, then from (8.71) we see that the energy contained in G is decreasing.

As a special case, take G be the ball $B_a = \{|\mathbf{x}| \le a\}$, and denote $e_G(t) = e_a(t)$. We show that the energy (8.70) is conserved under the assumption that the initial data $\mathbf{E}(\mathbf{x}, 0) = 0$ and $\mathbf{H}(\mathbf{x}, 0) = 0$ outside some ball $B_b = \{|\mathbf{x}| < b\}$. Then the set of points where $(\mathbf{E}, \mathbf{H})(\mathbf{x}, t) \ne 0$, is contained in the ball $\{|\mathbf{x}| \le b + ct\}$, which is growing with speed c. Given a $T > 0$, choose a so large that $a > b + cT$. Then $e(t) = e_a(t)$ for $|t| \le T$. Hence

$$\frac{d}{dt} e(t) = \frac{d}{dt} e_a(t) = -\int_{|\mathbf{x}|=a} \mathbf{n} \cdot (\mathbf{E} \times \mathbf{H}) dS(\mathbf{x})$$

$$= 0$$

because $\mathbf{E} = \mathbf{H} = 0$ on the sphere $\{|\mathbf{x}| = a\}$ for $|t| \le T$. Since T is arbitrary, we see that $e(t)$ is constant. In fact this argument can be modified to show that if the integral

$$\int_{R^3} [|\mathbf{E}(\mathbf{x}, 0)|^2 + |\mathbf{H}(\mathbf{x}, 0)|^2] d\mathbf{x}$$

is finite, then $e(t)$ is finite for all t, and $e(t)$ is constant.

A standard reference for electromagnetism is [Ja].

Exercises 8.10

1. Working with the basic equations (8.56) - (8.59), deduce the continuity equation

$$\frac{\partial \rho}{\partial t} + \nabla \cdot \mathbf{J} = 0.$$

2. In conducting material, of conductivity γ, the current density $\mathbf{J} = \gamma \mathbf{E}$. Assuming no charge is present, and the constitutive relations $\mathbf{D} = \epsilon \mathbf{E}$, and $\mathbf{B} = \mu \mathbf{H}$, equation (8.56) becomes

$$\epsilon \frac{\partial \mathbf{E}}{\partial t} - \nabla \times \mathbf{H} = -\gamma \mathbf{E}. \tag{8.71}$$

Join this equation together with (8.61) - (8.63). Further assume time-harmonic dependence

$$\mathbf{E}(\mathbf{x}, t) = \mathbf{U}(\mathbf{x}) e^{i\omega t} \quad \text{and} \quad \mathbf{H}(\mathbf{x}, t) = \mathbf{V}(\mathbf{x}) e^{i\omega t}.$$

(a) Substitute this form of $\mathbf{E}$ and $\mathbf{H}$ into equations (8.61) - (8.63), and (8.72). What equations must $\mathbf{U}$ and $\mathbf{V}$ satisfy?

(b) Using the identity $\nabla \times (\nabla \times \mathbf{M}) = -\Delta \mathbf{M} + \nabla(div\ M)$, deduce that $\mathbf{U}$ satisfies

$$\Delta \mathbf{U} + \Gamma^2 \mathbf{U} = 0,$$

where $\Gamma^2 = -i\omega\mu(\gamma + i\omega\epsilon) = k^2 - i\gamma\omega\mu$. Remember that $k = \omega/c$ is the wave number. When the conductivity $\gamma = 0$, each component of $\mathbf{U}$ (and of $\mathbf{V}$) satisfies the homogeneous Helmholtz equation.

3. Look for solutions of (8.60) - (8.63) in the form

$$\mathbf{E} = (0, u(x, t), 0) \quad \text{and} \quad \mathbf{H} = (0, 0, v(x, t)).$$

(a) Verify that $\mathbf{E}$ and $\mathbf{H}$ in this form satisfy (8.62), (8.63). Then (8.60), (8.61) reduce to a system of two PDE's for u, v. Find this system.

(b) Verify that both u and v must satisfy the one-dimensional wave equation $u_{tt} - c^2 u_{xx} = 0$.

(c) Suppose that $u(x, t) = F(x - ct) + G(x + ct)$. From the system of equations of part (a), find $v(x, t)$.

4. When the boundary of G is a perfect conductor, the boundary conditions for $\mathbf{E}$ and $\mathbf{H}$ are that $\mathbf{E}$ is normal to ∂G and $\mathbf{H}$ is tangential to ∂G. This means that there is some function $\beta(\mathbf{x})$, defined on ∂G, such that

$$\mathbf{E} = \beta(\mathbf{x})\mathbf{n} \quad \text{on } \partial G$$

and

$$\mathbf{H} \cdot \mathbf{n} = 0.$$

Using (8.71), show that with these boundary conditions, the energy $e_G(t)$ is conserved, that is, show that the energy flux through the boundary is zero.

5. Let H be the right half-space, $H = \{(x, y, z) : x > 0\}$, with boundary ∂H (the y, z plane) being a perfect conductor. Suppose a solution in H of the form discussed in exercise 3.

(a) What boundary conditions must u and v satisfy on the plane $x = 0$? (see exercise 4).

(b) Suppose that $u(x, t) = G(x + ct)$ for $t \le 0$, where $G(s)$ is defined for all s and $G(s) = 0$ for s outside some interval $[a, b]$, $0 < a < b$. Find $u(x, t)$ for $t \ge 0$, and then construct $v(x, t)$ using the results of exercise 3.

6. Let the Maxwell equations (8.60) - (8.63) be satisfied in the region $0 < x < L$, $y, z \in R$, and assume that the infinite planes $x = 0$ and $x = L$ are both perfect conductors.

(a) What is the IBVP that $\mathbf{E}$ and $\mathbf{H}$ must satisfy?

(b) Again assume that $\mathbf{E} = (0, u(x, t), 0)$ and $\mathbf{H} = (0, 0, v(x, t))$. What is the IBVP that u and v must satisfy? Use the system of exercise 3.

(c) Find the solution of this IBVP by substituting expansions

$$u(x, t) = \sum u_n(t) \sin(n\pi/L)$$

and

$$v(x, t) = \sum v_n(t) \cos(n\pi/L)$$

in the system for u and v. Derive a system of ODE's for each pair u_n and v_n. Solve the system for each n.

8.11 Projects

1. The Bessel functions $J_n(z)$ arise as the complex Fourier coefficients of the function $f(\theta) = \exp(iz \sin \theta)$:

$$e^{iz \sin \theta} = \sum_{-\infty}^{\infty} J_n(z) e^{in\theta}.$$

$f(\theta)$ is said to be a *generating function* for the Bessel functions. It follows that

$$J_n(z) = \frac{1}{2\pi} \int_0^{2\pi} e^{i(z \sin \theta - n\theta)} d\theta.$$

Now we can use the FFT to compute approximations to the Fourier coefficients of $f(\theta)$. Write a program that uses the MATLAB `fft` to compute values for $J_n(z)$ for $0 \le n \le 30$ and $|z| < 30$. The integrand is highly oscillatory for large z, and so must be sampled frequently in the interval $[0, 2\pi]$. Try $N = 128$ sample points, and compare the values you compute this way with the values produced by the MATLAB function `besselj`.

2. Let G be the square of side 2π, and consider the boundary value problem for the Poisson equation

$$-\Delta u(x, y) = q(x, y) \quad \text{in } G, \qquad u = 0 \quad \text{on } \partial G.$$

Much as in project 2 of Chapter 6, we may expand both u and q in terms of the eigenfunctions $\varphi_{m,n} = \sin(mx)\sin(ny)$:

$$u(x, y) = \sum_{m,n=1} u_{m,n} \varphi_{m,n}(x, y), \quad q(x, y) = \sum_{m,n} q_{m,n} \varphi_{m,n}(x, y),$$

where

$$u_{m,n} = \frac{q_{m,n}}{m^2 + n^2}.$$

Use the double FFT of MATLAB to find approximate values for the $q_{m,n}$, divide by $m^2 + n^2$, and use the inverse double FFT to compute an approximate solution u. Write a program that accepts input data q, computes u, and graphs the result.

3. Use the double FFT to compute solutions of the heat equation in the square G of side 2π with the boundary condition that $u = 0$ on ∂G.

Chapter 9

Equilibrium

In Chapter 8, we have seen how the Laplace and Poisson equations occur in steady-state heat flow. In this chapter we investigate properties of solutions of these equations and representations of the solutions in terms of integrals. We characterize these solutions as minimizers of certain variational problems, and again arrive at the notion of weak solutions.

9.1 Harmonic functions

9.1.1 Examples

We begin by looking at the Laplace equation in an open set $G \subset R^2$ or R^3. Suppose that $u \in C^2(G)$ and

$$\Delta u = 0 \qquad \text{in } G.$$

Solutions of this equation are called *harmonic* functions. u is said to be harmonic in G. In one dimension, a harmonic function $u(x)$ solves $u''(x) = 0$ and hence $u(x) = ax + b$, and its graph is a straight line. Harmonic functions in several variables are much more varied. These are some examples of harmonic functions in two variables:

(1) $u(x, y) = x^2 - y^2$
(2) $u(x, y) = 2xy$
(3) $u(x, y) = x^3 - 3xy^2$
(4) $u(x, y) = 3x^2y - y^3$.

These examples arise by taking the real and imaginary parts of analytic functions of $z = x + iy$. Examples (1) and (2) are the real and imaginary parts of z^2 while (3) and (4) are the real and imaginary parts of z^3. In general the real and imaginary parts of any analytic function are harmonic functions of two variables. They have a special relationship in that their level curves are orthogonal (see exercise 9). Many of the properties of harmonic functions of two variables can be seen in a different form in any treatment of analytic functions. Here are some examples in three variables:

(1) $u(x, y, z) = (x^2 + y^2 + z^2)^{-1/2}$ for $(x, y, z) \neq (0, 0, 0)$.
(2) $u(x, y, z) = x^2 + y^2 - 2z^2$.

9.1.2 The mean value property

A linear function in one variable $u(x) = ax + b$ has the property that

$$u(\frac{x+y}{2}) = \frac{1}{2}(u(x) + u(y))$$

for any $x, y \in R$, that is, the value of u at $(x + y)/2$ is the average of the values of u at x and at y. Harmonic functions in two and three variables have an analogous property called the *mean value property*. First we define the analogue of the average $(u(x) + u(y))/2$.

For $r \geq 0$, and $x_0 \in R^2$, let C_r be the circle and D_r be the disk with center x_0 and radius r. For $x_0 \in R^3$, let S_r be the sphere and B_r the ball with center x_0 and radius r. Recall from Chapter 8 that the *spherical mean* of a function $u(x)$, $x \in R^3$, is

$$M(r, x_0, u) = \frac{1}{4\pi r^2} \int\int_{S_r} u(x)\, dS.$$

For a function $u(x)$, $x \in R^2$, of two variables, the *circular mean* is

$$M(r, x_0, u) = \frac{1}{2\pi r} \int_{C_r} u(x)\, ds.$$

The mean value property will be a consequence of the following property of the circular or spherical means. Let $u(x)$ be a C^2 function on an open set G contained in either R^2 or R^3. Let $x_0 \in G$. Then for all r such that C_r or S_r is contained in G, in R^2,

$$\frac{d}{dr}M(r, x_0, u) = \frac{1}{2\pi r} \int\int_{D_r} \Delta u(x)\, dx, \tag{9.1}$$

and in R^3,

$$\frac{d}{dr}M(r, x_0, u) = \frac{1}{4\pi r^2} \int\int\int_{B_r} \Delta u(x)\, dx. \tag{9.2}$$

We verify (9.1). Note that on the circle C_r, the exterior normal derivative $\partial u/\partial n = u_r$. Apply the divergence theorem to the vector field $\mathbf{f}(\mathbf{x}) = \nabla u(\mathbf{x})$ on the disk D_r:

$$\int\int_{D_r} \Delta u(\mathbf{x})\, dx = \int\int_{D_r} div(\nabla u)(\mathbf{x})\, dx = \int_{C_r} u_r\, ds.$$

Now using polar coordinates centered at $\mathbf{x}_0 \in D_r$, the element of arc length on the circle C_r is $ds = rd\theta$, $0 \leq \theta \leq 2\pi$. Hence

$$\frac{1}{2\pi r}\int\int_{D_r} \Delta u(\mathbf{x})\, dx = \frac{1}{2\pi r}\int_{C_r} u_r\, ds = \frac{1}{2\pi}\int_0^{2\pi} u_r(r,\theta)\, d\theta$$

$$= \frac{d}{dr}\Big(\frac{1}{2\pi}\int_0^{2\pi} u(r,\theta)\, d\theta\Big) = \frac{d}{dr}\Big(\frac{1}{2\pi r}\int_{C_r} u\, ds\Big),$$

which establishes (9.1). Equation (9.2) is established the same way using the divergence theorem on B_r and the fact that the element of surface area on S_r is $dS = r^2 d\xi$ where $d\xi$ is the element of surface area on the unit sphere.

Now we state the *mean value property*. We say that a continuous function $u(\mathbf{x})$ of two variables $\mathbf{x} = (x,y)$ has the mean value property in G if

$$u(\mathbf{x}_0) = M(r, \mathbf{x}_0, u) = \frac{1}{2\pi r}\int_{C_r} u(\mathbf{x})\, ds, \tag{9.3}$$

for all circles C_r with center $\mathbf{x}_0$ such that $D_r \subset G$. For a function of three variables $\mathbf{x} = (x, y, z)$, the mean value property is

$$u(\mathbf{x}_0) = M(r, u, \mathbf{x}_0) = \frac{1}{4\pi r^2}\int\int_{S_r} u(\mathbf{x})\, dS, \tag{9.4}$$

for all spheres S_r with center $\mathbf{x}_0$ such that $B_r \subset G$.

Harmonic functions have the mean value property. This is a consequence of (9.1) and (9.2). In two dimensions, if $\Delta u = 0$, the right side of (9.1) is zero which says that $r \to M(r, u, \mathbf{x}_0)$ is constant. Now for any continuous function,

$$\lim_{r\to 0} M(r, u, \mathbf{x}_0) = u(\mathbf{x}_0).$$

Hence $M(r, u, \mathbf{x}_0) \equiv u(\mathbf{x}_0)$ for all r such $D_r \subset G$. Exactly the same argument applies in three dimensions using (9.2).

In fact, u having the mean value property in G is equivalent to u being harmonic in G. Let $\mathbf{x}_0 \in G$. Suppose that $u \in C^2(G)$ and has the mean value property. This means that $u(\mathbf{x}_0) = M(r, \mathbf{x}_0, u)$ for all r, $0 < r \leq r_0$, some $r_0 > 0$. Then the left side of (9.1) is zero implying that for $0 < r < r_0$,

$$\frac{1}{\pi r^2} \int \int_{D_r} \Delta u(\mathbf{x}) \, d\mathbf{x} = 0.$$

Consequently,

$$(\Delta u)(\mathbf{x}_0) = \lim_{r \to 0} \frac{1}{\pi r^2} \int \int_{D_r} \Delta u(\mathbf{x}) \, d\mathbf{x} = 0.$$

Since this is true for any $\mathbf{x}_0 \in G$, we see that u is harmonic in G. The same argument hold in three dimensions using (9.2).

A second property equivalent to the mean value property over spheres or circles is the *solid mean value property*:

$$u(\mathbf{x}_0) = \frac{1}{\pi r^2} \int \int_{D_r} u(\mathbf{x}) d\mathbf{x} \qquad (9.5)$$

for all r such that $D_r \subset G$. The three-dimensional version of the solid mean value property is

$$u(\mathbf{x}_0) = \frac{1}{\frac{4}{3}\pi r^3} \int \int \int_{B_r} u(\mathbf{x}) \, d\mathbf{x}. \qquad (9.6)$$

We claim that (9.3) is true if and only (9.5) is true. In fact, suppose that (9.3) is true. Then for all $0 < \rho \leq r$,

$$2\pi\rho \, u(\mathbf{x}_0) = \int_{C_\rho} u \, ds.$$

Integrating on $[0, r]$, we obtain

$$\pi r^2 u(\mathbf{x}_0) = \int_0^r \int_{C_\rho} u \, ds d\rho = \int \int_{D_r} u(\mathbf{x}) \, d\mathbf{x}.$$

This is (9.5). On the other hand, if (9.5) is true then,

$$\pi r^2 u(\mathbf{x}_0) = \int \int_{D_r} u(\mathbf{x}) \, d\mathbf{x}$$

for $0 < r < r_0$, some $r_0 > 0$. If we differentiate this expression we obtain (9.3). In the same way, (9.4) is seen to be true if and only if (9.6) is true.

We summarize these results in

Theorem 9.1
Suppose that $u \in C^2(G)$, G an open set of R^2. Then the following are equivalent.
  (1) u satisfies (9.3) for each $\mathbf{x}_0 \in G$ and each circle $C_r \subset G$.
  (2) u satisfies (9.5) for each $\mathbf{x}_0 \in G$ and each disk $D_r \subset G$.
  (3) $\Delta u = 0$ in G.

The same theorem holds true in three dimensions substituting (9.4) and (9.6).

The Laplace operator and spherical means

Before discussing further properties of harmonic functions, we record a useful fact about the Laplace operator and spherical means. The Laplace operator is invariant under a rotation of coordinates (see the exercises of this section). As a consequence it is also invariant under spherical means in the following sense. Let $u(\mathbf{x})$ be any C^2 function in an open set $G \subset R^3$. Fix $\mathbf{x}_0 \in G$, and introduce spherical coordinates centered at $\mathbf{x}_0$, with $r = |\mathbf{x} - \mathbf{x}_0|$ the distance from $\mathbf{x}_0$, θ the polar angle, and ϕ the angle in the x, y plane (see Figure 9.1). Cartesian coordinates are expressed in terms of the spherical coordinates by

$$\mathbf{x} = \mathbf{x}_0 + r(\sin\theta\cos\phi, \ \sin\theta\sin\phi, \ \cos\theta).$$

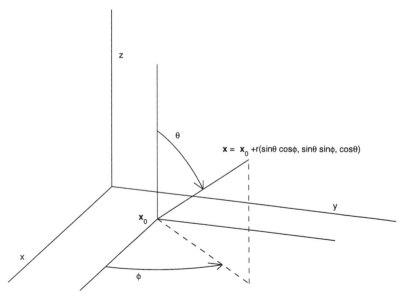

FIGURE 9.1
Spherical coordinates centered at $\mathbf{x}_0$.

Let $M(r, u)$ be the spherical mean over S_r (we suppress the $\mathbf{x}_0$). Then

$$\Delta M(r, u) = M(r, \Delta u). \tag{9.7}$$

The left side of (9.7) is the Laplace operator applied to the purely radial function $r \to M(r, u)$. In spherical coordinates, the Laplace operator is expressed by

$$\Delta u = u_{rr} + \frac{2}{r} u_r + \frac{1}{r^2} \left[u_{\theta\theta} + (\cot \theta) u_\theta + \frac{1}{\sin^2 \theta} u_{\phi\phi} \right]$$

(see Appendix A, section A.4). Since M does not depend on θ or ϕ, the left side of (9.7) reduces to

$$M_{rr} + \frac{2}{r} M_r = \frac{1}{r^2} \frac{d}{dr} (r^2 M_r).$$

On the other hand, if we multiply (9.2) by $4\pi r^2$ and differentiate again, we see that

$$4\pi \frac{d}{dr}(r^2 M_r) = \frac{d}{dr} \int \int \int_{B_r} \Delta u(\mathbf{x}) \, dx = \frac{d}{dr} \int_0^r \int \int_{S_\rho} \Delta u \, dS d\rho$$

$$= \int \int_{S_r} \Delta u \, dS.$$

Finally, dividing by $4\pi r^2$, we see that

$$\frac{1}{r^2} \frac{d}{dr} (r^2 M_r) = \frac{1}{4\pi r^2} \int \int_{S_r} \Delta u \, dS = M(r, \Delta u).$$

This is exactly (9.7).

9.1.3   The maximum principle

A second important property of harmonic functions is that they obey a maximum principle. This property follows directly from the mean value property. Let G be a bounded, open set in R^2 or R^3. We further assume that G is *pathwise-connected*. This means that any two points in G can be connected by a C^1 curve which lies entirely within G. For example, $G = \{x^2 + y^2 < 1\} \cup \{x^2 + y^2 > 3\}$ is *not* pathwise-connected.

Theorem 9.2
(Maximum principle). Let $u \in C^2(G)$, $u \in C(\bar{G})$ and suppose that $\Delta u = 0$ in G. Then

(1) (Weak form) $\max_{\bar{G}} u = \max_{\partial G} u$.
(2) (Strong form) If $\max_{\bar{G}} u$ *is attained at some interior point* $(x_0, y_0) \in G$, *then* $u(x, y) = u(x_0, y_0)$ *for all* $(x, y) \in G$.
Similar statements hold for the minimum.

Statement (2) is called the strong form of the maximum principle because it makes a more dramatic assertion about harmonic functions. Note that the weak form allows that u could attain its maximum inside G as well as on ∂G. The strong form implies the weak form and says this can only happen when u is a constant function.

We give the proof in two dimensions; the same argument works in three dimensions. We know that $\max_{\bar{G}} u$ exists because $\bar{G}$ is a compact set and u is a continuous function on $\bar{G}$ (Theorem 1.1). We shall use the solid mean value property to prove the strong form. Suppose that $(x_0, y_0) \in G$ and that

$$M = \max_{\bar{G}} u(x, y) = u(x_0, y_0).$$

Let $D_\rho = \{(x, y) : (x - x_0)^2 + (y - y_0)^2 < \rho^2\}$ be the largest open disk with center at (x_0, y_0) such that $\bar{D}_\rho \subset G$. We claim that $u(x, y) \equiv M$ on D_ρ. Suppose not. Then there must exist a point $(\hat{x}, \hat{y}) \in D_\rho$, where $u(\hat{x}, \hat{y}) < M$. Because u is continuous, there must exist another smaller disk $\hat{D} \subset D_\rho$, centered at $(\hat{x}, \hat{y})$, such that $u \leq \hat{M} \equiv (1/2)(M + u(\hat{x}, \hat{y})) < M$. But then the solid mean value property (9.5) forces us into a contradiction because

$$M = u(x_0, y_0) = \frac{1}{\pi \rho^2} \int\!\!\int_{D_\rho} u\, dx dy = \frac{1}{\pi \rho^2} \left[\int\!\!\int_{\hat{D}} u\, dx dy + \int\!\!\int_{D_\rho - \hat{D}} u\, dx dy \right]$$

$$\leq \frac{1}{\pi \rho^2} \left[\hat{M} a(\hat{D}) + M a(D_\rho - \hat{D}) \right] < M,$$

where $a(\hat{D})$ and $a(D_\rho - \hat{D})$ are the areas of these sets. Thus at any point $(x, y) \in G$ where $u(x, y) = M$, we can show that there is a disk D, centered at (x, y), such that $u \equiv M$ on D. Now let (x_*, y_*) be any other point in G. Because we assume that G is pathwise-connected, there is a C^1 path γ lying entirely within G, running from (x_0, y_0) to (x_*, y_*). We may cover γ by a finite sequence of overlapping disks $D_0 = D_\rho, D_1, D_2, \ldots, D_n$, such that $(x_*, y_*) \in D_n$ and the center of D_k lies in D_{k-1}. Repeating the argument for D_0 on each disk D_k, we find that $u = M$ all along γ, and hence $u(x_*, y_*) = M$. as well. This shows that $u(x, y) \equiv M$ on G and completes the proof of the maximum principle. $\square$

We shall use the maximum principle to prove uniqueness of solutions of boundary problems.

Exercises 9.1

1. Verify that the following are harmonic functions.

   (a) $u(r) = \ln r$ in $\{(x, y) : r = \sqrt{x^2 + y^2} > 0\}$.

   (b) $u(r) = 1/r$ in $\{(x, y, z) : r = \sqrt{x^2 + y^2 + z^2} > 0\}$.

2. Verify (9.2).

3. Verify (9.7) for the case of two dimensions. Put the expression for Δu in polar coordinates (see Appendix A) into $M(r, \Delta u)$ and integrate the term involving $u_{\theta\theta}$.

4. The Laplace operator Δ is invariant under translations and certain linear transformations of the independent variables.

   (a) In R^2, a translation is a mapping $x = a + \xi$, $y = b + \eta$. Let $v(\xi, \eta) = u(a + \xi, b + \eta)$. Using the chain rule, show that

   $$v_{\xi\xi} + v_{\eta\eta} = u_{xx} + u_{yy}.$$

   (b) A rotation in R^2 is a mapping

   $$x = (\cos\theta)\xi - (\sin\theta)\eta, \qquad y = (\sin\theta)\xi + (\cos\theta)\eta.$$

   Let $v(\xi, \eta) = u(x, y)$. Use the chain rule to show that

   $$v_{\xi\xi} + v_{\eta\eta} = u_{xx} + u_{yy}.$$

   (c) In R^3, an orthogonal linear transformation is of the form $\mathbf{x} = A\xi$, where $A = [a_{i,j}]$ is 3×3 matrix, such that the columns of A form an orthonormal set:

   $$\sum_{i=1}^{3} a_{i,j} a_{i,k} = \begin{cases} 0, & j \neq k \\ 1, & j = k \end{cases}.$$

   Let $v(\xi) = u(\mathbf{x})$. Show that $\Delta_\xi v = \Delta_{\mathbf{x}} u$.

5. A *conformal* mapping in R^2 is a mapping

$$x = f(\xi, \eta), \qquad y = g(\xi, \eta),$$

such that $f_\xi = g_\eta$ and $f_\eta = -g_\xi$. The rotations are special cases of conformal mappings. If $v(\xi, \eta) = u(f(\xi, \eta), g(\xi, \eta))$ show that

$$(f_\xi^2 + g_\eta^2)(v_{\xi\xi} + v_{\eta\eta}) = u_{xx} + u_{yy}.$$

6. Let

$$u(x, y) = \frac{1 - x^2 - y^2}{1 - 2x + r^2 + y^2}.$$

(a) Verify that u is harmonic in the open disk $D_1 = \{x^2 + y^2 < 1\}$.

(b) Use the maximum principle to find the maximum of u on the closed disk $\bar{D}_\rho$ for each $\rho < 1$. Where does it occur?

(c) What are the values of u on the boundary of D_1? Is the maximum principle satisfied on $\bar{D}_1$?

7. Define a function $u(x, y)$ as *subharmonic* in an open set $G \subset R^2$ if $-\Delta u \leq 0$.

(a) Follow the derivation of the mean value property, and show that u is subharmonic in G if and only if, for each point $p \in G$,

$$u(p) \leq \frac{1}{2\pi r} \int_{C_r} u \, ds$$

for all circles C_r with center p, such that C_r and its interior are contained in G.

(b) Let $G \subset R^2$ be an open, bounded set (so that $\bar{G}$ is compact). Show that u subharmonic in G and continuous on $\bar{G}$ implies that

$$max_{\bar{G}} u = max_{\partial G} u.$$

(c) Consider the solution of the inhomogeneous problem

$$-\Delta u = q \quad \text{in } G, \qquad u = g \quad \text{on } \partial G,$$

where $q < 0$ in G. Let v be the solution of

$$\Delta v = 0 \quad \text{in } G, \qquad v = g \quad \text{on } \partial G.$$

Show that $u \leq v$ in G.

The program mvp (mean value property) computes the values of three different functions $u(x, y)$, $v(x, y)$, and $w(x, y)$ in the square $G = \{0 < x < 1, 0 < y < 1\}$ and writes the numbers to the matrix Z. The user first chooses the function by entering 1,2,or 3. Next the user enters the coordinates of the center of a circle, and the radius of the circle. The center and radius should be chosen so that the circle is contained in G. The program computes the average of the function around the circle. The average and the value at the center of the circle are printed on the screen. The values of the function around the circle are put into the vector trace. The program plots the function. The circle is shown in red in the plane $z = 0$. The trace on the surface above is shown in white. You can plot the values of the trace around the circle and compare with the value at the center with the command plot(theta, trace, theta, valctr).

8. Run program mvp several times for each choice of function with different circles and radii.

   (a) Why is the average close to the value in the center when the radius is small?

   (b) Which of the functions is harmonic? Which is subharmonic? It is possible that one of them is neither.

9. Let $f(z)$ be an analytic function of the complex variable $z = x + iy$. As a consequence of being differentiable in z, it follows that the $u = Re(f)$ and $v = Im(f)$ satisfy the Cauchy Riemann equations :

$$\frac{\partial u}{\partial x} = \frac{\partial v}{\partial y}$$

$$\frac{\partial u}{\partial y} = -\frac{\partial v}{\partial x}.$$

   (a) Assuming that u and v are C^2 and satisfy the Cauchy Riemann equations, show that both u and v are harmonic. u and v are called *conjugate* harmonic functions.

   (b) Show that if u and v are conjugate harmonic functions, then $\nabla u \cdot \nabla v = 0$ for all (x, y). Verify that this is true for $u(x, y) = x^2 - y^2$ and $v(x, y) = xy$.

(c) To see this graphically using MATLAB, put a meshgrid on the unit square. Plot the contours of u and v on the same graph with the commands

```
>> u = X.^2 - Y.^2;
>> v = X.*Y;
>> contour(X,Y,u,'r')
>> hold on
>> contour(X,Y,v,'b')
>> hold off
```

Indicate directions of the gradients on your plots. Why must they be orthogonal? Do this with some other pairs of conjugate harmonic functions, for example, $Re(z^3)$ and $Im(z^3)$.

10. This is another version of the maximum principle for harmonic functions. It is the analogue of the second version of the maximum principle for solutions of the heat equation given in Theorem 3.2.

Let $u(x, y)$ be C^2 in the half-plane $H = \{(x, y) : x > 0\}$ and continuous on $\bar{H} = \{(x, y) : x \geq 0\}$. Suppose that $\Delta u = 0$ in H. Assume that there is a constant C such that

$$u(x, y) \leq \frac{C}{r} \quad \text{as } r \to \infty \text{ in } H,$$

and that there is a constant M, such that $u(0, y) \leq M$ for all $y \in R$. Then

$$u(x, y) \leq M \quad \text{for all } (x, y) \in \bar{H}.$$

In less precise terms,

$$\max_{\bar{H}} u(x, y) = \max_{R} u(0, y).$$

To prove this assertion, apply the maximum principle of Theorem 9.2 to u on each half-disk of radius ρ, $\{(x, y) \in \bar{H} : x^2 + y^2 \leq \rho^2\}$.

9.2 The Dirichlet problem

9.2.1 Fourier series solution in the disk

What kind of boundary value problems will be well posed for the Laplace equation $\Delta u = 0$? Our experience with steady-state heat flow suggests that the following boundary value problems are appropriate for this equation:

$$\Delta u = 0 \quad \text{in } G, \qquad u = f \quad \text{on } \partial G, \tag{9.8}$$

or

$$\Delta u = 0 \quad \text{in } G, \qquad \frac{\partial u}{\partial n} = g \quad \text{on } \partial G. \tag{9.9}$$

The first boundary value problem is often referred to as the Dirichlet problem, and the second as the Neumann problem.

We shall begin by solving (9.8) in the open disk $D_\rho \subset R^2$ of radius ρ centered at the origin. Following the procedure used before in the discussion of the heat equation, we look for certain building-block solutions. First we write (9.8) in polar coordinates (see Appendix A):

$$u_{rr} + \frac{1}{r}u_r + \frac{1}{r^2}u_{\theta\theta} = 0 \quad \text{for} \quad 0 \le r < \rho, \quad 0 \le \theta \le 2\pi, \tag{9.10}$$

$$u(\rho, \theta) = f(\theta) \quad \text{for} \quad 0 \le \theta \le 2\pi.$$

Then we seek solutions in the form of a product

$$u(r, \theta) = R(r)\Phi(\theta),$$

and substitute in (9.10) to arrive at

$$\frac{r^2 R''}{R} + \frac{r R'}{R} = -\frac{\Phi''}{\Phi}. \tag{9.11}$$

Again, the left side of (9.11) depends only on r while the right side depends only on θ. Hence we set both sides equal to a constant μ and deduce the following two ODE problems:

$$\Phi'' + \mu\Phi = 0, \qquad \Phi \text{ has period } 2\pi, \tag{9.12}$$

$$r^2 R'' + r R' - \mu R = 0, \qquad R \text{ bounded as } r \downarrow 0. \tag{9.13}$$

This is the same as equation (8.23), but with $\lambda = 0$. The solutions of (9.12) are clearly $\mu = n^2$ and

$$\Phi(\theta) = \cos n\theta, \quad \sin n\theta, \quad n = 0, 1, 2, \ldots.$$

Now look for solutions of (9.13) with $\mu = n^2$ in the form r^α. Substituting this form in (9.13) yields

$$r^2\alpha(\alpha - 1)r^{\alpha-2} + r\alpha r^{\alpha-1} - n^2 r^\alpha = 0$$

or

$$r^\alpha(\alpha^2 - n^2) = 0.$$

Hence the general solution of (9.13) is given by

$$A_0 + B_0 \log r \qquad \text{for } n = 0,$$

and

$$A_n r^n + B_n r^{-n} \qquad \text{for } n \geq 1.$$

Since we can admit only solutions of (9.13) which are bounded as $r \downarrow 0$, our building-block solutions are

$$1 \quad \text{for } n = 0; \quad r^n \cos n\theta, \quad r^n \sin n\theta \quad \text{for } n \geq 1.$$

We shall seek a solution of (9.8) in the form

$$u(r, \theta) = \frac{a_0}{2} + \sum_1^\infty r^n (a_n \cos n\theta + b_n \sin n\theta).$$

To satisfy the boundary condition $u(\rho, \theta) = f(\theta)$, we must choose the coefficients a_n, b_n such that

$$f(\theta) = u(\rho, \theta) = \frac{a_0}{2} + \sum_1^\infty \rho^n (a_n \cos n\theta + b_n \sin n\theta). \qquad (9.14)$$

Now $f(\theta)$, which has period 2π, has the Fourier series

$$f(\theta) = \frac{A_0}{2} + \sum_1^\infty A_n \cos n\theta + B_n \sin n\theta,$$

where A_n and B_n are given by (6.6) and (6.7). Then (9.14) will be satisfied if we choose

$$a_0 = A_0, \quad a_n = \rho^{-n} A_n \quad \text{and} \quad b_n = \rho^{-n} B_n \quad \text{for} \quad n \geq 1.$$

Theorem 9.3
Let $D_\rho \subset R^2$ be the open disk of radius ρ centered at the origin. Let $f(\theta)$ be a continuous function of period 2π. Then there is a unique solution $u \in C^\infty(D_\rho)$ of (9.8) which is continuous on $\bar{D}_\rho$. In polar coordinates

$$u(r, \theta) = \frac{A_0}{2} + \sum_{n=1}^{\infty} (\frac{r}{\rho})^n [A_n \cos n\theta + B_n \sin n\theta], \tag{9.15}$$

where A_n and B_n are the Fourier coefficients of f.

We have derived the formula (9.15) without worrying about the manner in which the series converges. We shall not prove everything in the statement of the theorem. However, we note that inside D_ρ, we have $r/\rho < 1$. Hence the terms of the series are bounded by those of a geometric series $A \sum (r/\rho)^n$ which converges. Actually, the series converges uniformly on each smaller disk $0 \le r \le \hat{\rho}$ with $\hat{\rho} < \rho$, so that the sum of the series must be continuous on D_ρ. If we formally compute u_r from (9.15), we obtain

$$u_r(r, \theta) = \sum_{1}^{\infty} \frac{n}{\rho} \frac{r}{n}(\frac{r}{\rho})^{n-1} [A_n \cos n\theta + B_n \sin n\theta]. \tag{9.16}$$

The terms in this series can be bounded by

$$|\frac{n}{\rho}(\frac{r}{\rho})^{n-1} [A_n \cos n\theta + B_n \sin n\theta]| \le Cn(\frac{r}{\rho})^{n-1},$$

where C is constant. $\sum n(r/\rho)^{n-1}$ converges by the ratio test as long as $r < \rho$. This means that the differentiated series (9.16) converges uniformly for $0 \le \theta \le 2\pi$ and $0 \le r \le \hat{\rho}$ for each $\hat{\rho} < \rho$. In fact we can differentiate to any order in either θ or r, and the resulting series will converge uniformly in the same manner. Thus we see that actually $u \in C^\infty(D_\rho)$.

We shall prove that u is continuous on the closed disk $\bar{D}_\rho$ under the stronger assumption on f that f is continuously differentiable. Then we know from Chapter 6 that the Fourier coefficients A_n and B_n satisfy

$$\sum_{1}^{\infty} |A_n| < \infty \qquad \text{and} \qquad \sum_{1}^{\infty} |B_n| < \infty.$$

This implies that the partial sums in the series (9.15) converge to the solution u, uniformly on the closed disk $0 \le r \le \rho$. Since each of the partial sums in the series (9.15) is continuous we see that u must be continuous on the closed disk. To show that the conclusion holds true even when f is only assumed continous requires a more technical argument involving the maximum principle.

We can use the maximum principle to prove uniqueness. Suppose that v is another solution of $\Delta v = 0$ in D_ρ, $v = f$ on ∂D_ρ, $v \in C^2(D_\rho) \cap C(\bar{D}_\rho)$. Then $w = u - v$ solves $\Delta w = 0$ in D_ρ, $w = 0$ on ∂D_ρ. Hence

$$\max_{\bar{D}_\rho} w = \max_{\partial D_\rho} w = 0,$$

so that $w \leq 0$ on $\bar{D}_\rho$. But we can also apply the maximum principle to $-w$, whence $-w \leq 0$ on $\bar{D}_\rho$ as well. We conclude that $w = 0$. □

9.2.2 The Poisson kernel

We can find another way of representing the solution of the Dirichlet problem for the disk. Each of the coefficients A_n, B_n in (9.15) is the value of an integral. We shall combine all of these integrals into a single integral of f against a kernel $P_\rho(r, \theta, \varphi)$ called the Poisson kernel (for the disk D_ρ). This integral representation will give us more physical insight into the properties of the solution u given by (9.15). We write out the integral formulas for the coefficients A_n, B_n for $n \geq 1$.

$$A_n \cos n\theta + B_n \sin n\theta = \frac{1}{\pi} \int_{-\pi}^{\pi} f(\varphi) \cos n\varphi \, d\varphi \cos n\theta + \frac{1}{\pi} \int_{-\pi}^{\pi} f(\varphi) \sin n\varphi \, d\varphi \sin n\theta$$

$$= \frac{1}{\pi} \int_{-\pi}^{\pi} f(\varphi)[\cos n\varphi \cos n\theta + \sin n\varphi \sin n\theta] d\varphi$$

$$= \frac{1}{\pi} \int_{-\pi}^{\pi} f(\varphi) \cos n(\theta - \varphi) d\varphi.$$

The leading term

$$\frac{A_0}{2} = \frac{1}{2\pi} \int_{-\pi}^{\pi} f(\varphi) d\varphi.$$

Formally interchanging the integration and the summation of the series (9.15), we see that

$$u(r, \theta) = \frac{1}{\pi} \int_{-\pi}^{\pi} \left[\frac{1}{2} + \sum_{1}^{\infty} \left(\frac{r}{\rho} \right)^n \cos n(\theta - \varphi) \right] f(\varphi) d\varphi. \tag{9.17}$$

The sum under the integral sign looks complicated, but in fact, we can sum this series in closed form for $r < \rho$. We can write

$$\left(\frac{r}{\rho} \right)^n \cos n(\theta - \varphi) = Re[z^n],$$

where $z = (r/\rho)\exp(i(\theta - \varphi))$. Then the sum under the integral is

$$\frac{1}{2} + \sum_{1}^{\infty}(\frac{r}{\rho})^n \cos n(\theta - \varphi) = \sum_{0}^{\infty} Re[z^n] - \frac{1}{2} = Re\left[\frac{1}{1-z}\right] - \frac{1}{2}$$

$$= \frac{Re[1-z]}{(Re[1-z])^2 + (Im[z])^2} - \frac{1}{2}.$$

Here $Im[z] = (r/\rho)\sin(\theta - \varphi)$ so that the series sums to

$$\frac{1 - (r/\rho)\cos(\theta - \varphi)}{(1 - (r/\rho)\cos(\theta - \varphi))^2 + ((r/\rho)\sin(\theta - \varphi))^2} - \frac{1}{2}$$

$$= \frac{1}{2}\frac{\rho^2 - r^2}{\rho^2 - 2r\rho\cos(\theta - \varphi) + r^2}.$$

Hence (9.15) becomes

$$u(r, \theta) = \frac{1}{2\pi}\int_{-\pi}^{\pi}\frac{\rho^2 - r^2}{\rho^2 - 2r\rho\cos(\theta - \varphi) + r^2}f(\varphi)d\varphi. \qquad (9.18)$$

This equation is called the *Poisson formula* for the disk of radius ρ.

Now for the promised physical insight. We shall rewrite the solution using vector notation for points in the plane. Let $\mathbf{x} = (x, y) = (r\cos\theta, r\sin\theta)$ and $\mathbf{y} = (\rho\cos\varphi, \rho\sin\varphi)$. Then the denominator in (9.18) is $|\mathbf{x} - \mathbf{y}|^2$ while the numerator is $|\mathbf{y}|^2 - |\mathbf{x}|^2$.

Theorem 9.4
The unique solution of the Dirichlet problem (9.8) can be represented by

$$u(\mathbf{x}) = \int_{|\mathbf{y}|=\rho} P_\rho(\mathbf{x}, \mathbf{y})f(\mathbf{y})ds \qquad (9.19)$$

for any continuous function f given on the boundary. The function

$$P_\rho(\mathbf{x}, \mathbf{y}) = \frac{1}{2\pi\rho}\frac{|\mathbf{y}|^2 - |\mathbf{x}|^2}{|\mathbf{y} - \mathbf{x}|^2} \qquad (9.20)$$

is called the Poisson kernel *for the disk of radius ρ. The additional factor of ρ in the denominator is needed because the the element of arc length of the circle of radius ρ is $ds = \rho d\varphi$. A solution of the Dirichlet problem is displayed in Figure 9.2.*

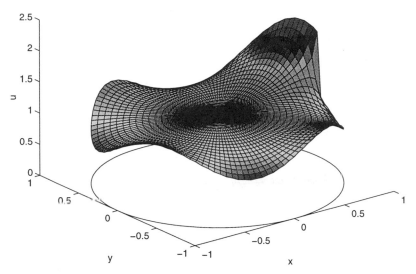

FIGURE 9.2
Solution of the Dirichlet problem in the unit disk for boundary data
$f(\theta) = \sin(4\theta)/4 + \exp(-2\theta^2) + 1.$

Note that the integral

$$\frac{1}{2\pi\rho} \int_{|\mathbf{y}|=\rho} f(\mathbf{y})ds$$

is the average value of f over the circle $|\mathbf{y}| = \rho$. When $\mathbf{x} = 0$, (9.19) reduces to this average in agreement with the mean value property seen in Section 9.1. Another interesting property of the Poisson kernel becomes evident if we substitute the function $u(\mathbf{x}) \equiv 1$ into (9.19). u is the unique solution of $\Delta u = 0$ with $u = 1$ on the circle of radius ρ. Hence (9.19) tells us that for any $\mathbf{x} \in D_\rho$, we have

$$1 = \int_{|\mathbf{y}|=\rho} P_\rho(\mathbf{x}, \mathbf{y})ds(\mathbf{y}). \tag{9.21}$$

We see that the solution (9.19) of the Dirichlet problem (9.8) is a superposition of solutions of Laplace's equation. In the context of heat flow, $P_\rho(\mathbf{x}, \mathbf{y})$ is the steady-state heat flow produced by an infinitely hot source concentrated at the point $\mathbf{y} \in C_\rho$ with the temperature held at zero at all other points of C_ρ. The level curves of $\mathbf{x} \to P_\rho(\mathbf{x}, \mathbf{y})$ are circles, all of which are tangent at the point $\mathbf{y} \in C_\rho$.

We note here that the Poisson kernel for the ball of radius ρ in R^3 is given by

$$P_\rho(\mathbf{x}, \mathbf{y}) = \frac{1}{4\pi\rho} \frac{\rho^2 - |\mathbf{x}|^2}{|\mathbf{y} - \mathbf{x}|^3}. \tag{9.22}$$

The Poisson kernel has several properties in common with the heat kernel.

(1) For each fixed $\mathbf{y} \in C_\rho$, $\mathbf{x} \to P_\rho(\mathbf{x}, \mathbf{y})$ is harmonic on D_ρ.
(2) $P_\rho(\mathbf{x}, \mathbf{y}) > 0$ for $\mathbf{y} \in C_\rho$, $\mathbf{x} \in D_\rho$.
(3) For $|\mathbf{x}| < \rho$, $\int_{C_\rho} P_\rho(\mathbf{x}, \mathbf{y})ds = 1$.
(4) If $|\mathbf{y}| = |\mathbf{y}'| = \rho$, $\mathbf{y} \neq \mathbf{y}'$, then $P_\rho(\mathbf{x}, \mathbf{y}) \to 0$ as $\mathbf{x} \to \mathbf{y}'$, but $P_\rho(\mathbf{x}, \mathbf{y})$ blows up as $\mathbf{x} \to \mathbf{y}$.

The condition $|\mathbf{x}| < \rho$ for the Poisson kernel corresponds to the condition $t > 0$ for the heat kernel $S(x - y, t)$. The corresponding properties are

(1) For each $y \in R$, $(x, t) \to S(x - y, t)$ solves the heat equation for $t > 0$.
(2) $S(x - y, t) > 0$ for $x, y \in R$ and $t > 0$.
(3) For $t > 0$ and $x \in R$, $\int_R S(x - y, t)dy = 1$.
(4) For $x \neq y$, $S(x - y, t) \to 0$ and $t \downarrow 0$, while $S(0, t)$ blows up as $t \downarrow 0$.

One can find the Poisson kernel in closed form for some other special geometries by separation of variables e.g., the annulus $\rho_1 < r < \rho_2$; a sector $0 < r < \rho, 0 < \theta < \theta_0$; a half-plane or a quarter-plane. We can show the existence of the Poisson kernel for more general shapes of G, but will not be able to find it in closed form.

The Neumann problem (9.9) can be solved in terms of a series like (9.15), provided that g satisfies the compatibility condition

$$\int_{-\pi}^{\pi} g(\theta)d\theta = 0.$$

Solutions are unique up to a constant. These questions will be discussed in the exercises.

9.2.2 Liouville's theorem

We can also use the Poisson kernel to deduce another important property of harmonic functions.

Theorem 9.5 (Liouville).
Let $u(\mathbf{x})$ be harmonic on R^2 or R^3, and suppose that there is a constant M, such that either $u(\mathbf{x}) \geq M$ for all $\mathbf{x}$ (u is bounded below), or $u(\mathbf{x}) \leq M$ for all $\mathbf{x}$ (u is bounded above). Then u is a constant function.

We give the proof for R^2. If we assume that u is bounded below, then by adding a constant we can assume $u(x) \geq 0$ for all x. We shall show that $u(x) = u(0)$ for all x. Fix x and let $\rho > |x|$. Then by (9.19) and (9.20),

$$u(x) = \frac{1}{2\pi\rho} \int_{C_\rho} \frac{|y|^2 - |x|^2}{|y - x|^2} u(y) ds.$$

Here we are using the restriction of u to the circle C_ρ as the boundary values f in (9.19), thus reproducing u. Now we make upper and lower estimates on the kernel. Since $|x| < \rho$, it follows that $|y - x| \geq |y| - |x|$, whence

$$\frac{|y|^2 - |x|^2}{|y - x|^2} \leq \frac{(|y| + |x|)(|y| - |x|)}{(|y| - |x|)^2} = \frac{\rho + |x|}{\rho - |x|}.$$

On the other hand, $|y - x| \leq |y| + |x|$, so that

$$\frac{|y|^2 - |x|^2}{|y - x|^2} \geq \frac{(|y| - |x|)(|y| + |x|)}{(|y| + |x|)^2} = \frac{\rho - |x|}{\rho + |x|}.$$

Because $u(x) \geq 0$,

$$\left(\frac{\rho - |x|}{\rho + |x|}\right)\frac{1}{2\pi\rho} \int_{C_\rho} u(y) ds \leq \frac{1}{2\pi\rho} \int_{C_\rho} \frac{|y|^2 - |x|^2}{|y - x|^2} u(y) ds \leq \left(\frac{\rho + |x|}{\rho + |x|}\right)\frac{1}{2\pi\rho} \int_{C_\rho} u(y) ds.$$

The middle term is obviously $u(x)$, and by the mean·value property,

$$\frac{1}{2\pi\rho} \int_{C_\rho} u(y) ds = u(0).$$

Hence we have the *Harnack inequality* for nonnegative harmonic functions on the disc $D_\rho \subset R^2$:

$$\left(\frac{\rho - |x|}{\rho + |x|}\right) u(0) \leq u(x) \leq \left(\frac{\rho + |x|}{\rho - |x|}\right) u(0). \tag{9.23}$$

Now (9.23) holds for any disk D_ρ that includes the point x. Thus we may take ρ arbitrarily large in (9.23). It follows that $u(x) = u(0)$.

In the case that $u(x) \leq M$, we may apply the argument to $-u$ which is bounded below. $\square$

In the same way, using the expression (9.22) for the Poisson kernel for the ball of radius ρ in R^3, we can derive the Harnack inequality for nonnegative harmonic functions on the ball of radius ρ:

$$\frac{\rho(\rho - |\mathbf{x}|)}{(\rho + |\mathbf{x}|)^2} u(\mathbf{0}) \leq u(\mathbf{x}) \leq \frac{\rho(\rho + |\mathbf{x}|)}{(\rho - |\mathbf{x}|)^2} u(\mathbf{0}). \tag{9.24}$$

Liouville's theorem for harmonic functions on R^3 follows in the same way from this inequality.

Since the real and imaginary parts of analytic functions are harmonic functions of two variables, much of the theory of harmonic functions is intertwined with analytic function theory. In particular the theory of conformal maps provides solutions of boundary value problems in ideal fluid flow. For a discussion of these and other questions see [Fi].

Exercises 9.2

1. Show analytically that the level curves of $P_\rho(x, y)$ are circles.

We want to view the Poisson kernel in the unit disk ($\rho = 1$), and to do so, we must introduce polar coordinates. This can be done using the following sequence of MATLAB commands:

```
r = 0: .05 :1;
theta = 0: 2*pi/100: 2*pi;
X = r'*cos(theta);
Y = r'*sin(theta);
```

The third command puts the values of r in a column vector and then takes the matrix product of this column vector with the row vector cos(theta) to make the matrix X with X(i,j) = r(i)*cos(theta(j)) . Similarly for the fourth command. Now if a function f is expressed $f = f(x, y)$, and there are symmetries which make it natural to consider f as function of polar coordinates, we use the MATLAB expression f(X,Y). For example, if we have an mfile for f, say f.m, then surf(X,Y,f(X,Y)) will plot f on the unit disk.

2. Write a script mfile using the above sequence of commands which will plot the Poisson kernel for the unit disk. In formula (9.20), we will take $\mathbf{y} = (a, b)$ and $\mathbf{x} = (x, y)$, so that (9.20) becomes

$$(\frac{1}{2\pi}) \frac{1 - x^2 - y^2}{(x - a)^2 + (y - b)^2}.$$

The point (a, b), where the Poisson kernel becomes infinite, is on the unit circle so that $a^2 + b^2 = 1$. Make the pair (a, b) input data for your program. Write one line in your program for the numerator, one for the denominator and one for the quotient. Write your formulas in terms of the variables X and Y.

```
num   = . . . . . ;
denom = . . . . . ;
pk    = num./denom;
```

Plot for several values of (a, b), using the command surf(X,Y,pk); shading flat. You should restrict the scale on the z axis to the interval $[0, 5]$ with the command axis([-1 1 -1 1 0 5]).

Now suppose that we wish to plot a function h expressed in polar coordinates, $h = h(r, \theta)$. After the row vectors r and theta are defined, and the polar meshgrid [X,Y] is defined as above, we define matrices R and TH as follows:

```
R  = r'*ones(size(theta));
TH = ones(size(r'))*theta;
```

Then you can plot h on the unit disk with the commands surf(X,Y,h(R,TH)) or mesh(X,Y,h(R,TH)) . The plots produced by the mesh command are easier to print out.

3. Plot several of the harmonic functions $r^n \cos(n\theta)$ and $r^n \sin(n\theta)$ in the unit disk. Note that $r^n \cos(n\theta) = Re((x + iy)^n)$ and $r^n \sin(n\theta) = Im((x + iy)^n)$.

Program dirch solves the Dirichlet problem in the unit disk $D_1 \subset R^2$

$$\Delta u = 0 \quad \text{in } D_1 \qquad u = f \quad \text{on } r = 1.$$

You must write an mfile f.m for the boundary data f, letting theta be the independent variable. The function f should have period 2π. The program computes the solution using the Fourier series expansion and puts the result in the matrix U. It plots it with the command surf(X,Y,U) which makes a pretty picture. However, such a picture takes a long time to print on paper, and may even have a file too long to print. For printing purposes, it is better to use the command mesh(X,Y,U). The view is from the first quadrant. The arrows indicate the positive x and y axes. Of course, you can change the viewpoint with the command view.

4. (a) Solve the Dirichlet problem in the disk D_1 analytically for $f(\theta) = 1 + \sin\theta$. Verify that the solution satisfies the mean value property for circles centered at the origin.

    (b) Run program dirch for this data choice. Using MATLAB or by analytic means, find the maximum and minimum values of the solution u over $\bar{D}_1$ and where they are attained.

5. (a) Solve the Dirichlet problem analytically for

$$f(\theta) = 2 + .5\sin\theta - .1\cos\theta + \sin^2\theta + \cos 4\theta.$$

Use the identity $\sin^2 x = (1 - \cos(2x))/2$. Verify that the solution satisfies the mean value property for circles centered at the origin.

(b) Run program dirch with this data choice. Where are the maximum and minimum values over $\bar{D}_1$ attained and what are they ?

6. (a) Solve the Dirichlet problem analytically for $f(\theta) = \cos\theta + 4\sin^3\theta$. Use trig identities to find the Fourier series for f. Verify that the solution satisfies the mean value property for circles centered at the origin.

(b) Run program dirch with this choice of data. Where are the maximum and minimum attained? What are they?

7. (a) Run program dirch with $f(\theta) = \exp(-(\theta - \pi)^2) - 3\exp(-2(\theta - 3\pi/2)^2)$. This data is essentially periodic because $f(0), f(2\pi) \approx 0$. Where are the max and min attained? What are they?

(b) Find the value of $u(0,0)$ by computing the boundary integral

$$\frac{1}{2\pi}\int_0^{2\pi} f(\theta)d\theta.$$

To do this, express the integral in terms of the error function, and evaluate using the error function of MATLAB. Compare with the value of u at $(0,0)$ as computed by the program dirch. To see what this is, type U(1,1). The indices $i = 1, j = 1$, correspond to the point $(0,0)$.

8. Recall that the Neumann problem for the disk D_ρ is

$$\Delta u = 0 \quad \text{in } D_\rho, \qquad \frac{\partial u}{\partial r}(\rho,\theta) = g(\theta), \quad 0 \le \theta \le 2\pi.$$

We look for solutions in the form

$$\frac{a_0}{2} + \sum_1^\infty r^n[a_n\cos(n\theta) + b_n\sin(n\theta)].$$

Then

$$\frac{\partial u}{\partial r} = \sum_1^\infty nr^{n-1}[a_n\cos(n\theta) + b_n\sin(n\theta)].$$

Assuming that g has the Fourier expansion

$$g(\theta) = \frac{A_0}{2} + \sum_1^\infty [A_n \cos(n\theta) + B_n \sin(n\theta)],$$

find the coefficients a_n and b_n so that the boundary condition is satisfied. What compatibility condition must g satisfy? Will the solution be unique?

9. Suppose that u and v are two bounded solutions of the Poisson equation $-\Delta u = q = -\Delta v$ on R^2 or R^3. Use Liouville's theorem to show that $u = v + \text{constant}$.

9.3   The Dirichlet problem in a rectangle

We shall use the technique of separation of variables to solve the Dirichlet problem in the rectangle

$$G = \{(x, y) : 0 < x < a, 0 < y < b\}.$$

Let f, g, h, k be given continuous functions. We want to solve

$$\Delta u = 0 \quad \text{in } G$$

$$u(0, y) = f(y) \quad u(a, y) = g(y), \quad 0 \le y \le b,$$

$$u(x, 0) = h(x) \quad u(x, b) = k(x), \quad 0 \le x \le a.$$

First we split u into a sum $u = v + w$, where v and w are to solve the problems

$$\Delta v = 0 \quad \text{in } G,$$

$$v(0, y) = f(y) \quad v(a, y) = g(y), \quad 0 \le y \le b,$$

$$v(x, 0) = v(x, b) = 0, \quad 0 \le x \le a,$$

and

$$\Delta w = 0 \quad \text{in } G,$$

$$w(0, y) = w(a, y) = 0, \quad 0 \le y \le b,$$

$$w(x, 0) = h(x) \quad w(x, b) = k(x), \quad 0 \le x \le a.$$

We use separation of variables to find building-block solutions of the problem for v. If we substitute a product $X(x)Y(y)$ into the problem for v, we obtain

$$\frac{X''(x)}{X(x)} = -\frac{Y''(y)}{Y(y)} = \mu,$$

so that X and Y must solve the ODE's

$$Y'' + \mu Y = 0, \quad Y(0) = Y(b) = 0,$$

and

$$X'' - \mu X = 0.$$

The first equation has the solutions

$$Y_n(y) = \sin(\frac{n\pi y}{b}), \quad \mu_n = (\frac{n\pi}{b})^2, \quad n = 1, 2, \ldots$$

The general solution of the ODE for X can be written conveniently in the form

$$X(x) = A \sinh(\sqrt{\mu}(a - x)) + B \sinh(\sqrt{\mu}x).$$

Then taking $\mu = \mu_n$, we get a family of solutions

$$X_n(x) = A_n \sinh(\sqrt{\mu_n}(a - x)) + B_n \sinh(\sqrt{\mu_n}x).$$

We seek v in the form of a series

$$v(x, y) = \sum_1^\infty [A_n \sinh(\sqrt{\mu_n}(a - x)) + B_n \sinh(\sqrt{\mu_n}x)] \sin(\frac{n\pi y}{b}).$$

We note that v in this form solves $v(x, 0) = v(x, b) = 0$. It remains to choose the coefficients A_n and B_n so that the other boundary conditions for v are satisfied. This will be the case if

$$f(y) = v(0, y) = \sum_1^\infty A_n \sinh(a\sqrt{\mu_n}) \sin(\frac{n\pi y}{b}),$$

and

$$g(y) = v(a, y) = \sum_{1}^{\infty} B_n \sinh(a\sqrt{\mu_n}) \sin(\frac{n\pi y}{b}).$$

It follows that the coefficients must be

$$A_n = \frac{2}{b \sinh(a\sqrt{\mu_n})} \int_{0}^{b} f(y) \sin(\frac{n\pi y}{b}) dy$$

$$B_n = \frac{2}{b \sinh(a\sqrt{\mu_n})} \int_{0}^{b} g(y) \sin(\frac{n\pi y}{b}) dy.$$

Thus the boundary value problem for v is solved.

In like manner, the boundary problem for w can be solved in the form

$$w = \sum_{1}^{\infty} [C_m \sinh(\sqrt{\lambda_m}(b - y)) + D_m \sinh(\sqrt{\lambda_m}y)] \sin(\frac{m\pi x}{a}).$$

Here $\lambda_m = (m\pi/a)^2$, and

$$C_m = \frac{2}{a \sinh(b\sqrt{\lambda_m})} \int_{0}^{a} h(x) \sin(\frac{m\pi x}{a}) dx,$$

$$D_m = \frac{2}{a \sinh(b\sqrt{\lambda_m})} \int_{0}^{a} k(x) \sin(\frac{m\pi x}{a}) dx.$$

Then $u = v + w$ solves the Dirichlet boundary value problem.

We can also use this same procedure to solve a mixed Dirichlet-Neumann problem for $\Delta u = 0$ in G. We assign the boundary conditions

$$u(0, y) = f(y) \quad u(a, y) = g(y), \quad 0 \le y \le b,$$

$$u_y(x, 0) = h(x) \quad u_y(x, b) = k(x), \quad 0 \le x \le a.$$

We proceed as before, writing $u = v + w$ where $\Delta v = \Delta w = 0$ in G, with boundary conditions

$$v(0, y) = f(y) \quad v(a, y) = g(y), \quad 0 \le y \le b,$$

$$v_y(x, 0) = v_y(x, b) = 0, \qquad 0 \le x \le a,$$

and

$$w(0, y) = w(a, y) = 0, \qquad 0 \le y \le b,$$

$$w_y(x, 0) = h(x) \quad w_y(x, b) = k(x), \qquad 0 \le x \le a.$$

Now we separate variables for the problem involving v and look for building-block solutions of the form $X(x)Y(y)$. We find that X and Y must satisfy the ODE's

$$Y'' + \mu Y = 0, \qquad Y'(0) = Y'(b) = 0$$

and

$$X'' - \mu X = 0.$$

Thus there is a sequence of solutions

$$Y_n(y) = \cos(\frac{n\pi y}{b}), \qquad \mu_n = (\frac{n\pi}{b})^2, \qquad n = 0, 1, 2, \ldots.$$

For convenience we take the solutions of the ODE for X (with $\mu = \mu_n$) as

$$X_0(x) = \frac{1}{2}(A_0(a - x) + B_0 x)$$

and for $n \ge 1$,

$$X_n(x) = A_n \sinh(\sqrt{\mu_n}(a - x)) + B_n \sinh(\sqrt{\mu_n}x).$$

Then we seek v in the form

$$v(x, y) = \frac{1}{2}(A_0(a - x) + B_0 x) + \sum_1^\infty [A_n \sinh(\sqrt{\mu_n}(a - x))$$

$$+ B_n \sinh(\sqrt{\mu_n}x)] \cos(\frac{n\pi y}{b}).$$

Now applying the boundary conditions for v, we must choose the coefficients A_n and B_n, so that

$$f(y) = v(0, y) = \frac{a A_0}{2} + \sum_1^\infty A_n \sinh(a\sqrt{\mu_n}) \cos(\frac{n\pi y}{b})$$

and

$$g(y) = v(b, y) = \frac{a B_0}{2} + \sum_1^\infty B_n \sinh(a\sqrt{\mu_n}) \cos(\frac{n\pi y}{b}).$$

This means that A_n and B_n must be given by

$$A_n = \frac{2}{b \sinh(a\sqrt{\mu_n})} \int_0^b f(y) \cos(\frac{n\pi}{b}) dy$$

for $n \geq 1$ with

$$A_0 = \frac{2}{ab} \int_0^b f(y) dy$$

and

$$B_n = \frac{2}{b \sinh(a\sqrt{\mu_n})} \int_0^b g(y) \cos(\frac{n\pi y}{b}) dy$$

for $n \geq 1$ with

$$B_0 = \frac{2}{ab} \int_0^b g(y) dy.$$

Similarly, we find that w is given by the series

$$w(x, y) = \sum_1^\infty [C_m \cosh(\sqrt{\lambda_m}(b - y)) + D_m \cosh(\sqrt{\lambda_m} y)] \sin(\frac{m\pi x}{a}),$$

where

$$C_m = -\frac{2}{a\sqrt{\lambda_m} \sinh(b\sqrt{\lambda_m})} \int_0^a h(x) \sin(\frac{m\pi x}{a}) dx$$

and

$$D_m = \frac{2}{a\sqrt{\lambda_m} \sinh(b\sqrt{\lambda_m})} \int_0^a k(x) \sin(\frac{m\pi x}{a}) dx.$$

Exercises 9.3

1. In the same rectangle G, solve the boundary value problem $\Delta u = 0$ in G

$$u(0, y) = f(y) \qquad u(a, y) = g(y), \quad 0 \le y \le b,$$

$$u(x, 0) = h(x) \qquad u_y(x, b) = k(x), \quad 0 \le x \le a.$$

9.4 The Poisson equation

9.4.1 The Poisson equation without boundaries

We are proceeding in a different manner than we did for the heat and wave equations in that we considered boundary value problems for the Laplace equation before discussing solutions without boundary conditions. Now we want to consider a fundamental solution of Laplace's equation, like the fundamental solution of the heat equation. We will use this fundamental solution to solve the Poisson equation in R^2 and R^3.

We set

$$s(\mathbf{x}) = \frac{1}{2\pi} \log(\frac{1}{|\mathbf{x}|}) \quad \text{when } n = 2 \tag{9.25}$$

and

$$s(\mathbf{x}) = \frac{1}{4\pi} \frac{1}{|\mathbf{x}|} \quad \text{when } n = 3. \tag{9.26}$$

$s(\mathbf{x})$ is the *fundamental solution* of the Laplace equation. In exercise 1 of Section 9.1 we already verified that $\Delta s = 0$ for $|\mathbf{x}| \ne 0$

In the same way that we used the fundamental solution of the heat equation to solve the pure initial value-value problem by a convolution, we can use the fundamental solution of Laplace's equation to solve the equation $-\Delta u = q$. To get a uniqueness result for these solutions, we need to impose some kind condition at infinity (instead of a boundary condition).

A function $u(\mathbf{x})$ on R^2 or R^3 *vanishes* (uniformly) *at infinity,* if there is a function $\hat{u}(r) \ge 0$, such that

$$|u(\mathbf{x})| \le \hat{u}(|\mathbf{x}|)$$

and $\hat{u}(r) \to 0$ as $r \to \infty$.

We remark that if u is continuous and vanishes at infinity, then u is bounded on R^2 or R^3. In fact, by the definition we know that there is a $\rho > 0$ such that $|u(\mathbf{x})| \le \hat{u}(r) \le 1$ for $r \ge \rho$. Letting M be the maximum of $|u(\mathbf{x})|$ over $|\mathbf{x}| \le \rho$, we see that

$$|u(\mathbf{x})| \le \max\{M, 1\}.$$

Now we are ready to state a theorem about the Poisson equation.

Theorem 9.6
*Suppose that $q(\mathbf{x})$ is a C^1 function such that $q = 0$ for $|\mathbf{x}| \ge \rho$ for some $\rho > 0$.
Then*

$$u(\mathbf{x}) = \int s(\mathbf{x} - \mathbf{y})q(\mathbf{y})dy \tag{9.27}$$

solves the Poisson equation

$$-\Delta u = q. \tag{9.28}$$

In R^3, this solution is unique in the class of functions that vanish at infinity. In R^2, this solution is unique, up to a constant, in the class of functions u such that ∇u vanishes at infinity.

In (9.27) we use the single integral sign to denote integration over R^2 or R^3. The integral (9.27) is also a convolution and may be abbreviated

$$u = s * q.$$

First we verify that the formula (9.27) provides a solution of equation (9.28) for the case $n = 3$. The calculation for $n = 2$ is similar. The integrand of (9.27) is singular at $\mathbf{y} = \mathbf{x}$ because

$$s(\mathbf{x} - \mathbf{y}) = \frac{1}{4\pi} \frac{1}{|\mathbf{x} - \mathbf{y}|}$$

blows up when $\mathbf{x} = \mathbf{y}$. However the integral still converges because the integral of $|\mathbf{x}|^{-p}$ over any ball $B = \{\mathbf{x} \in R^3 : |\mathbf{x}| < a\}$ is finite for $p < 3$ (see Chapter 1). We can differentiate once under the integral sign in (9.27) and still have a convergent integral,

$$\nabla u(\mathbf{x}) = \int \nabla_\mathbf{x} s(\mathbf{x} - \mathbf{y})q(\mathbf{y})dy, \tag{9.29}$$

because

$$\nabla_{\mathbf{x}}s(\mathbf{x} - \mathbf{y}) = \frac{1}{4\pi}\nabla_{\mathbf{x}}(\frac{1}{|\mathbf{x} - \mathbf{y}|}) \qquad (9.30)$$

$$= -\frac{1}{4\pi}\frac{1}{|\mathbf{x} - \mathbf{y}|^2}\nabla_{\mathbf{x}}|\mathbf{x} - \mathbf{y}| = -\frac{1}{4\pi}\frac{\mathbf{x} - \mathbf{y}}{|\mathbf{x} - \mathbf{y}|^3},$$

so that

$$|\nabla_{\mathbf{x}}s(\mathbf{x} - \mathbf{y})| \le \frac{1}{4\pi}\frac{1}{|\mathbf{x} - \mathbf{y}|^2}.$$

Use the fact the $\nabla_{\mathbf{x}}s(\mathbf{x} - \mathbf{y}) = -\nabla_{\mathbf{y}}s(\mathbf{x} - \mathbf{y})$, and integrate by parts in $\mathbf{y}$:

$$\nabla u(\mathbf{x}) = -\int \nabla_{\mathbf{y}}s(\mathbf{x} - \mathbf{y})q(\mathbf{y})dy = +\int s(\mathbf{x} - \mathbf{y})\nabla q(\mathbf{y})dy. \qquad (9.31)$$

There are no boundary terms because $q = 0$ for $|\mathbf{x}| \ge \rho$. We differentiate again in (9.31):

$$-\Delta u(\mathbf{x}) = -div(\nabla u(\mathbf{x})) = -\int \nabla_{\mathbf{x}}s(\mathbf{x} - \mathbf{y}) \cdot \nabla_{\mathbf{y}}q(\mathbf{y})dy$$

$$= \int \nabla_{\mathbf{y}}s(\mathbf{x} - \mathbf{y}) \cdot \nabla_{\mathbf{y}}q(\mathbf{y})dy.$$

We cannot put two derivatives on s because the second derivatives individually are not integrable. However, we can cut out a small ball of radius $\varepsilon > 0$ about $\mathbf{y} = \mathbf{x}$ and observe that

$$-\Delta u(\mathbf{x}) = \lim_{\varepsilon\downarrow 0}\int_{|\mathbf{x}-\mathbf{y}|\ge\varepsilon}\nabla_{\mathbf{y}}s(\mathbf{x} - \mathbf{y}) \cdot \nabla_{\mathbf{y}}q(\mathbf{y})dy.$$

Now integrating by parts back onto s over this region, we see that

$$\int_{|\mathbf{x}-\mathbf{y}|\ge\varepsilon}\nabla_{\mathbf{y}}s((\mathbf{x} - \mathbf{y}) \cdot \nabla q(\mathbf{y})dy \qquad (9.32)$$

$$= \int_{|\mathbf{x}-\mathbf{y}|=\varepsilon}\frac{\partial s}{\partial n_{\mathbf{y}}}(\mathbf{x} - \mathbf{y})q(\mathbf{y})dS(\mathbf{y}) - \int_{|\mathbf{x}-\mathbf{y}|\ge\varepsilon}\Delta_{\mathbf{y}}s(\mathbf{x} - \mathbf{y})q(\mathbf{y})dy$$

$$= \int_{|\mathbf{x}-\mathbf{y}|=\varepsilon}\frac{\partial s}{\partial n_{\mathbf{y}}}(\mathbf{x} - \mathbf{y})q(\mathbf{y})dS(\mathbf{y})$$

because $\Delta_\mathbf{y} s(\mathbf{x} - \mathbf{y}) = 0$ for $|\mathbf{x} - \mathbf{y}| > 0$. To evaluate this last integral, we introduce spherical coordinates $\mathbf{y} = \mathbf{x} + r\xi$, where

$$\xi = (\sin\theta \sin\phi, \sin\theta \sin\phi, \cos\theta).$$

Then $dS(\mathbf{y}) = \varepsilon^2 d\xi$ on the sphere of radius $\varepsilon > 0$, and $d\xi = \sin\theta\,d\theta\,d\phi, 0 \le \theta \le \pi, 0 \le \phi \le 2\pi$, is the element of area on the unit sphere. The exterior normal $\mathbf{n}$ in (9.32) is pointing *into* the sphere of radius ε, centered at $\mathbf{x}$. Hence

$$\left.\frac{\partial s}{\partial n\,\mathbf{y}}(\mathbf{x} - \mathbf{y})\right|_{|\mathbf{x}-\mathbf{y}|=\varepsilon} = -\left.\frac{\partial}{\partial r}s(r\xi)\right|_{r=\varepsilon} = \frac{1}{4\pi\varepsilon^2}.$$

It follows that

$$\int_{|\mathbf{x}-\mathbf{y}|=\varepsilon} \frac{\partial s}{\partial n\,\mathbf{y}} s(\mathbf{x} - \mathbf{y})q(\mathbf{y})dS(\mathbf{y}) = \frac{1}{4\pi\varepsilon^2} \int_{|\xi|=1} q(\mathbf{x} + \varepsilon\xi)\varepsilon^2 d\xi$$

is the average of $\mathbf{y} \to q(\mathbf{y})$ over the sphere of radius ε, center $\mathbf{x}$. Consequently

$$-\Delta u(\mathbf{x}) = \lim_{\varepsilon\downarrow 0} \frac{1}{4\pi\varepsilon^2} \int_{|\xi|=1} q(\mathbf{x} + \varepsilon\xi)\varepsilon^2 d\xi = q(\mathbf{x}).$$

We have established (9.28).

Finally we turn to the uniqueness questions. We do the case $n = 3$. We must show that the solution given by (9.27) vanishes at infinity and that it is the only solution of (9.28) in that class. Let Q be the maximum of $|q|$. Q is finite because $q(\mathbf{y}) = 0$ for $|\mathbf{y}| \ge \rho$. Now when $n = 3$, the solution given by (9.27) can be estimated:

$$|u(\mathbf{x})| \le \frac{Q}{4\pi} \int_{|\mathbf{y}|\le\rho} \frac{d\mathbf{y}}{|\mathbf{x} - \mathbf{y}|}.$$

Let us assume that $|\mathbf{x}| > \rho$. Then the denominator $|\mathbf{x} - \mathbf{y}| \ge |\mathbf{x}| - \rho$, so that for $|\mathbf{x}| > \rho$,

$$|u(\mathbf{x})| \le \frac{Q}{4\pi} \int_{|\mathbf{y}|\le\rho} \frac{d\mathbf{y}}{|\mathbf{x}| - \rho} = \frac{Q\rho^3}{3}\left(\frac{1}{|\mathbf{x}| - \rho}\right).$$

We take the right-hand side of this inequality to be $\hat{u}$. It is clear that $\hat{u}(r) = \hat{u}(|\mathbf{x}|) \to 0$ as $r \to \infty$. Thus u given by (9.27) vanishes at infinity.

   To show that the solution is unique in this class, suppose that v is another solution
of (9.28), also vanishing at infinity. Then because the Laplace operator is linear,
$u - v$ is harmonic. Furthermore,

$$|u(\mathbf{x}) - v(\mathbf{x})| \leq |u(\mathbf{x})| + |v(\mathbf{x})| \leq \hat{u}(r) + \hat{v}(r) \to 0$$

as $r \to \infty$. In particular, this implies that $u - v$ is bounded on R^3. But then
Liouville's Theorem says that $u - v$ must be a constant function. Since $u - v$
vanishes at infinity, $u - v$ must be identically zero, or $u \equiv v$. We have finished
the proof of Theorem 9.6 in the case $n = 3$. We leave the uniqueness part when
$n = 2$ to the exercises. $\square$

Remark   The δ function in one dimension is defined in Chapter 1 as that generalized function such that for any continuous function $f(x)$,

$$\int \delta(x) f(x) dx = f(0)$$

and more generally

$$\int \delta(x - y) f(y) dy = f(x).$$

The analogous generalized function is also defined in higher dimensions. The
generalized function $\mathbf{x} \to \delta(\mathbf{x})$ is a limit of functions sharply peaking at $\mathbf{x} = 0$ and
has the property that, for $f(\mathbf{x})$ continuous,

$$\int_{R^n} \delta(\mathbf{x} - \mathbf{y}) f(\mathbf{y}) dy = f(\mathbf{x}).$$

The δ function in R^2 or R^3 can be thought of as products of the one-dimensional
δ functions.

   The calculation we made to prove Theorem 9.6 can be summarized symbolically
as

$$-\Delta_{\mathbf{x}} \int s(\mathbf{x} - \mathbf{y}) q(\mathbf{y}) dy = -\int \Delta_{\mathbf{x}} s(\mathbf{x} - \mathbf{y}) q(\mathbf{y}) dy = q(\mathbf{x}).$$

In short hand notation we write

$$-\Delta_{\mathbf{x}} s(\mathbf{x} - \mathbf{y}) = \delta(\mathbf{x} - \mathbf{y}).$$

The Helmholtz equation

In Section 8.8 we saw that solutions of the wave equation in three dimensions with a localized, time-harmonic, source converge as $t \to \infty$ to a solution of the form $v(\mathbf{x}) \exp(i\omega t)$. The spatial factor v satisfies the Helmholtz equation

$$-\Delta v - k^2 v = q.$$

In fact we saw that (equation (8.48))

$$v(\mathbf{x}) = \frac{1}{4\pi} \int_{R^3} \frac{e^{-ik|\mathbf{x}-\mathbf{y}|}}{|\mathbf{x}-\mathbf{y}|} q(\mathbf{y}) \, d\mathbf{y}.$$

When $k = 0$ this formula reduces to that for the solution of the Poisson equation, (9.27). Thus we see that the fundamental solution $s(\mathbf{x})$ for the Laplace equation in R^3 belongs to the family of fundamental solutions $s_{\pm k}(\mathbf{x})$ for the Helmholtz equation,

$$s_{\pm k}(\mathbf{x}) = \frac{e^{\pm ik|\mathbf{x}|}}{4\pi |\mathbf{x}|}.$$

It is easy to verify that $s_{\pm k}$ satisfies $(\Delta + k^2)s_{\pm k} = 0$ for $\mathbf{x} \neq 0$. s_{-k} is used to find the solution u of the wave equation which represents outgoing radiation produced by the source $q(\mathbf{x}) \exp(i\omega t)$, and s_k is used for incoming radiation (or outgoing with time reversed). Note that

$$Re(s_{\pm k}(\mathbf{x})) = \frac{\cos k|\mathbf{x}|}{4\pi |\mathbf{x}|}$$

has the same singularity as does $s(\mathbf{x})$ at $\mathbf{x} = 0$. On the other hand,

$$Im(s_{\pm k}(\mathbf{x})) = \frac{\pm \sin k|\mathbf{x}|}{4\pi |\mathbf{x}|}$$

is perfectly regular at $\mathbf{x} = 0$.

9.4.2 The Green's function

The Poisson equation with zero boundary conditions in the bounded open set G is

$$- \Delta u = q \quad \text{in } G, \qquad u = 0 \quad \text{on } \partial G. \tag{9.33}$$

In Chapter 8 we saw that, when there existed a complete set of eigenfunctions for the problem $-\Delta\varphi = \lambda\varphi$ in G, $\varphi = 0$ on ∂G, we could solve (9.33) by expanding q in terms of the eigenfunctions. Now we find a different way to represent the solution in terms of an integral.

The one-dimensional case

By way of motivation, we first consider a two-point, boundary value problem for an ODE. Let q(x) be a given continuous function on the interval $[0, L]$. We seek the solution of the problem

$$-u''(x) = q(x), \quad 0 < x < L, \qquad u(0) = u(L) = 0.$$

The general solution to the ODE, involving two arbitrary constants, can be found by integrating twice:

$$u(x) = -\int_0^x \int_0^y q(s)ds\,dy + Ax + B.$$

To satisfy the boundary conditions we must set $B = 0$, and

$$A = \frac{1}{L}\int_0^L \int_0^y q(s)ds\,dy.$$

The formula for u can then be rewritten as a single integral expression

$$u(x) = \int_0^L \gamma(x, y)q(y)dy,$$

where

$$\gamma(x, y) = \begin{cases} (y/L)(L - x), & y < x \\ (x/L)(L - y), & y > x \end{cases}.$$

$\gamma(x, y)$ is called the *Green's function* for this problem. In the formalism of generalized functions, we write

$$-\frac{d^2}{dx^2}\gamma(x, y) = \delta(x - y).$$

The integral defines a linear operator

$$q \to Kq = \int_0^L \gamma(x, y)q(y)dy$$

on $X = C[0, L]$ into itself. Because $\gamma(x, y) = \gamma(y, x)$, it is easy to verify that in the usual inner product on X,

$$\langle Ku, v \rangle = \langle u, Kv \rangle$$

for all $u, v \in X$. Thus K is a *symmetric* linear integral operator. There is a well developed theory of such operators as regards their eigenvalues and eigenfunctions because they have in much in common with symmetric matrix operators on R^n. K is the inverse of the linear operator M defined by

$$Mu = -u''$$

with domain the subspace $W \subset X$ given by

$$W = \{u \in C^2[0, L] : u(0) = u(L) = 0\}.$$

We see that as operators, $K \circ M = I$ on W and $M \circ K = I$ on X. The integral representation of the solution provides useful information about the eigenvalue problem $M\varphi = \lambda\varphi$. If λ is an eigenvalue of M, then $1/\lambda$ is an eigenvalue of K, and the two operators have the same eigenfunctions.

We shall derive a similar integral expression for the solution of the Poisson equation with zero boundary conditions in two or three dimensions. We assume that we can solve the Dirichlet problem (9.8) in G for any continuous function f given on ∂G. There are certain kinds of regions with sharply spiked boundaries for which this is not true, so that this is not an empty assumption. However it is certainly true in many cases of importance in applications. For each $y \in G$, let $x \to v(x, y)$ be the solution of

$$\Delta_x v = 0 \quad \text{in } G, \qquad v(\mathbf{x}, \mathbf{y}) = -s(\mathbf{x} - \mathbf{y}) \quad \text{on } \partial G.$$

Then the *Green's function* for G is given by

$$\gamma(\mathbf{x}, \mathbf{y}) = s(\mathbf{x} - \mathbf{y}) + v(\mathbf{x}, \mathbf{y}). \tag{9.34}$$

Theorem 9.7
Let G be an open, bounded set in R^2 or R^3 with piecewise C^1 boundary ∂G, such that we can always solve the Dirichlet problem in G for continuous data on ∂G. Let $q \in C^1(\bar{G})$. Then there is a unique solution to (9.33) given by

$$u(\mathbf{x}) = \int_G \gamma(\mathbf{x}, \mathbf{y}) q(\mathbf{y}) d\mathbf{y}. \tag{9.35}$$

To verify that (9.35) provides a solution of (9.33), we use (9.34) and our previous calculation with the fundamental solution $s(\mathbf{x} - \mathbf{y})$:

$$-\Delta u(\mathbf{x}) = -\Delta \int_G [s(\mathbf{x} - \mathbf{y}) + v(\mathbf{x}, \mathbf{y})] q(\mathbf{y}) d\mathbf{y}$$

$$= q(\mathbf{x}) + \int_G -\Delta_{\mathbf{x}} v(\mathbf{x}, \mathbf{y}) q(\mathbf{y}) d\mathbf{y} = q(\mathbf{x})$$

because $\Delta_{\mathbf{x}} v(\mathbf{x}, \mathbf{y}) = 0$. In addition, for $\mathbf{x} \in \partial G$,

$$u(\mathbf{x}) = \int_G \gamma(\mathbf{x}, \mathbf{y}) q(\mathbf{y}) d\mathbf{y} = 0.$$

Uniqueness of the solution follow from the linearity of the Laplace operator and the maximum principle. $\square$

For reference we collect some important

Properties of the Green's function.

(1) $\Delta_{\mathbf{x}} \gamma(\mathbf{x}, \mathbf{y}) = 0$ for $\mathbf{x} \neq \mathbf{y}$ and in the sense of generalized functions,

$$-\Delta_{\mathbf{x}} \gamma(\mathbf{x}, \mathbf{y}) = \delta(\mathbf{x} - \mathbf{y}).$$

(2) $\gamma(\mathbf{x}, \mathbf{y}) = 0$ for $\mathbf{y} \in G$, $\mathbf{x} \in \partial G$.
(3) Symmetry: $\gamma(\mathbf{x}, \mathbf{y}) = \gamma(\mathbf{y}, \mathbf{x})$
(4) $\gamma(\mathbf{x}, \mathbf{y}) > 0$.

The first property follows from the definition of γ because $\Delta_{\mathbf{x}} v(\mathbf{x}, \mathbf{y}) = 0$ for all $\mathbf{x} \in G$. The second is a consequence of the choice of v.

Here is a quick derivation of (3), using the language of generalized functions. Let $\mathbf{a}, \mathbf{b} \in G$ and set $u(\mathbf{x}) = \gamma(\mathbf{x}, \mathbf{a})$ and $w(\mathbf{x}) = \gamma(\mathbf{x}, \mathbf{b})$. We shall apply Green's second identity (8.12), rather loosely, to the generalized functions $-\Delta u(\mathbf{x}) = \delta(\mathbf{x} - \mathbf{a})$ and $-\Delta w(\mathbf{x}) = \delta(\mathbf{x} - \mathbf{b})$. Because $u = w = 0$ on ∂G,

$$\gamma(\mathbf{a}, \mathbf{b}) = \int_G \delta(\mathbf{x} - \mathbf{a}) w(\mathbf{x}) d\mathbf{x} = \int_G (-\Delta u) w d\mathbf{x}$$

$$= \int_G u(-\Delta w) d\mathbf{x} = \int_G u(\mathbf{x}) \delta(\mathbf{x} - \mathbf{b}) d\mathbf{x} = \gamma(\mathbf{b}, \mathbf{a}).$$

This result can be made rigorous by a repetition of the arguments used to prove Theorem 9.6. Finally, (4) is proved in the exercises using the maximum principle.

We can give a physical interpretation of the Green's function in the context of electrostatics. Let G be an open set in R^3, and let $\mathbf{E}$ be the electric field produced by the charge density q. Then $div\mathbf{E} = q$. Furthermore, $\mathbf{E} = -\nabla u$, where u is the electrostatic potential produced by the charge. Combining these two equations, we see that

$$-\Delta u = div\mathbf{E} = q,$$

which is the Poisson equation. Now define a sequence of charge densities

$$q_n(\mathbf{x}) = \begin{cases} 3n^3/4\pi & \text{for } |\mathbf{x}| < 1/n \\ 0 & \text{for } |\mathbf{x}| > 1/n \end{cases}.$$

For fixed $\mathbf{y} \in G$, $\mathbf{x} \to q_n(\mathbf{x} - \mathbf{y})$ is a charge density concentrated on the ball $\{|\mathbf{x} - \mathbf{y}| < 1/n\}$ and

$$\int q_n(\mathbf{x})d\mathbf{x} = \int q_n(\mathbf{x} - \mathbf{y})d\mathbf{x} = 1$$

for all n. We set

$$\gamma_n(\mathbf{x}, \mathbf{y}) = \int_G \gamma(\mathbf{x}, \mathbf{z})q_n(\mathbf{z} - \mathbf{y})d\mathbf{z}$$

$$= \frac{3n^3}{4\pi} \int_{|\mathbf{y}-\mathbf{z}|<1/n} \gamma(\mathbf{x}, \mathbf{z})d\mathbf{z}.$$

$\mathbf{x} \to \gamma_n(\mathbf{x}, \mathbf{y})$ is the potential associated with the charge density $q_n(\mathbf{x} - \mathbf{y})$. For fixed $\mathbf{x}$, the integral is the average of $\gamma(\mathbf{x}, \mathbf{y})$ over the ball of radius $1/n$ centered at $\mathbf{y}$. Hence as $n \to \infty$, $\gamma_n(\mathbf{x}, \mathbf{y}) \to \gamma(\mathbf{x}, \mathbf{y})$. Thus we can think of $\gamma(\mathbf{x}, \mathbf{y})$ as the electrostatic potential produced by a unit amount of charge placed over an infinitely small ball centered at $\mathbf{y}$. The boundary is "grounded", so that the potential is zero there.

In the context of electrostatics, the symmetry property (3) of the Green's function is called the *principle of reciprocity*, which states that the potential at point $\mathbf{x}$ due to a unit charge at point $\mathbf{y}$ is the same as the potential at point $\mathbf{y}$ due to a unit charge at point $\mathbf{x}$.

There is an important connection between the Green's function $\gamma(\mathbf{x}, \mathbf{y})$ and the Poisson kernel used to represent solutions of the Dirichlet problem. Let $u(\mathbf{x})$ be C^2 on $\bar{G}$. We can employ a limiting argument, as we did to verify (9.28), to see that

$$\int_G (-\Delta u)(\mathbf{y}) \gamma(\mathbf{x}, \mathbf{y}) dy$$

$$= \int_{\partial G} \left[-\frac{\partial u}{\partial n_{\mathbf{y}}}(\mathbf{y}) \gamma(\mathbf{x}, \mathbf{y}) + u(\mathbf{y}) \frac{\partial \gamma}{\partial n_{\mathbf{y}}}(\mathbf{x}, \mathbf{y}) \right] dS(\mathbf{y}) + \int_G u(\mathbf{y})(-\Delta_{\mathbf{y}}) \gamma(\mathbf{x}, \mathbf{y}) dy$$

$$= \int_{\partial G} u(\mathbf{y}) \frac{\partial \gamma}{\partial n_{\mathbf{y}}}(\mathbf{x}, \mathbf{y}) dS(\mathbf{y}) + u(\mathbf{x})$$

because $\gamma(\mathbf{x}, \mathbf{y}) = 0$ for $\mathbf{y} \in \partial G$ and $\Delta_{\mathbf{y}} \gamma(\mathbf{x}, \mathbf{y}) = \Delta_{\mathbf{y}} \gamma(\mathbf{y}, \mathbf{x}) = \delta(\mathbf{x} - \mathbf{y})$. Now if u is the solution of the Dirichlet problem

$$\Delta u = 0 \quad \text{in } G, \qquad u = f \quad \text{on } \partial G$$

then, from the formula above, we see that

$$u(\mathbf{x}) = -\int_{\partial G} u(\mathbf{y}) \frac{\partial \gamma}{\partial n_{\mathbf{y}}}(\mathbf{x}, \mathbf{y}) dS(\mathbf{y}).$$

Thus the Poisson kernel for the general open set G is

$$P(\mathbf{x}, \mathbf{y}) = -\frac{\partial \gamma}{\partial n_{\mathbf{y}}}(\mathbf{x}, \mathbf{y}), \quad \mathbf{y} \in \partial G, \ \mathbf{x} \in G. \tag{9.36}$$

Formulas (9.35) and (9.36) show how important it is to be able to calculate the Green's function for a region G. Unfortunately this can only be done in closed form for sets G with simple geometries, such as the disk, rectangle, half-plane, quarter-plane in R^2, and the corresponding geometries in R^3.

Example A

We calculate the Green's function for the half-plane $H = \{(x_1, x_2) : x_1 > 0\}$ using the method of balancing charges. If a unit charge is concentrated at the point $\mathbf{y} = (y_1, y_2) \in H$, the resulting potential over all of R^2 is given by the fundamental solution

$$s(\mathbf{x}, \mathbf{y}) = -\frac{1}{2\pi} \log |\mathbf{x} - \mathbf{y}|.$$

To achieve the desired boundary value of zero on the x_2 axis, we place another unit charge, with opposite sign, at the symmetric point $\hat{\mathbf{y}} = (-y_1, y_2)$. Then we add the contributions from both potentials to yield the Green's function for H,

$$\gamma(\mathbf{x}, \mathbf{y}) = \frac{1}{2\pi}(\log(|\mathbf{x} - \hat{\mathbf{y}}| - \log|\mathbf{x} - \mathbf{y}|)$$

$$= \frac{1}{2\pi} \log\left(\frac{|\mathbf{x} - \hat{\mathbf{y}}|}{|\mathbf{x} - \mathbf{y}|}\right) \qquad .$$

Next we use (9.36) to find the Poisson kernel for the half-plane H. The exterior normal vector on the boundary of H is $\mathbf{n} = (-1, 0)$ which implies that the normal derivative to the boundary $x_1 = 0$ is given by

$$\partial/\partial n = -\partial/\partial x_1.$$

We wish to use the formula (9.36), so we must calculate

$$\frac{\partial}{\partial y_1}\gamma(\mathbf{x}, \mathbf{y}) = \frac{1}{2\pi}\frac{\partial}{\partial y_1}[\log|\mathbf{x} - \hat{\mathbf{y}}| - \log|\mathbf{x} - \mathbf{y}|]$$

$$= \frac{1}{2\pi}\left[\frac{x_1 + y_1}{|\mathbf{x} - \hat{\mathbf{y}}|} + \frac{x_1 - y_1}{|\mathbf{x} - \mathbf{y}|}\right].$$

When $y_1 = 0$, $\mathbf{y} = (0, y_2) = \hat{\mathbf{y}}$. Using (9.36) we see that the Poisson kernel for the half-plane H is

$$P(\mathbf{x}, \mathbf{y}) = -\frac{\partial \gamma}{\partial n_y}(\mathbf{x}, \mathbf{y}) = \left(\frac{x_1}{\pi}\right)\frac{1}{|\mathbf{x} - \mathbf{y}|^2}.$$

The Dirichlet problem in H for f given on the boundary $x_1 = 0$ is

$$\Delta u(\mathbf{x}) = 0 \quad \text{in } H, \qquad u = f \quad \text{on } x_1 = 0.$$

If f is continuous and bounded on R, a solution of the Dirichlet problem in H is given (in vector notation) by

$$u(\mathbf{x}) = \frac{x_1}{\pi}\int_{y_1=0}\frac{f(\mathbf{y})dS}{|\mathbf{x} - \mathbf{y}|^2}.$$

If we use the notation $\mathbf{x} = (x, y)$, $\mathbf{y} = (\xi, \eta)$, we can write the formula as

$$u(x, y) = \frac{x}{\pi} \int_R \frac{f(\eta)\,d\eta}{x^2 + (y - \eta)^2}.$$

Example B

The Green's function for the disk D_ρ centered at the origin, will take the form

$$\gamma(\mathbf{x}, \mathbf{y}) = s(\mathbf{x} - \mathbf{y}) + v(\mathbf{x}, \mathbf{y})$$

$$= \frac{1}{2\pi} \log(\frac{1}{|\mathbf{x} - \mathbf{y}|}) + v(\mathbf{x}, \mathbf{y})$$

where, for fixed $\mathbf{y} \in D_\rho$,

$$\Delta_{\mathbf{x}} v(\mathbf{x}, \mathbf{y}) = 0 \ \text{ for } \mathbf{x} \in D_\rho, \ \mathbf{x} \neq \mathbf{y},$$

$$v(\mathbf{x}, \mathbf{y}) = -\frac{1}{2\pi} \log(\frac{1}{|\mathbf{x} - \mathbf{y}|}) \quad \text{for } |\mathbf{x}| = \rho.$$

Think of $s(\mathbf{x} - \mathbf{y})$ as the potential from a unit charge placed at $\mathbf{y}$. We shall construct v by placing another unit charge outside the disk which cancels out the potential $s(\mathbf{x} - \mathbf{y})$ on $|\mathbf{x}| = \rho$. Let $\hat{\mathbf{y}} = \rho^2 \mathbf{y}/|\mathbf{y}|^2$. $\hat{\mathbf{y}}$ is a multiple of $\mathbf{y}$ which lies outside of D_ρ:

$$|\hat{\mathbf{y}}| = \frac{\rho^2}{|\mathbf{y}|} > \rho.$$

We say that $\hat{\mathbf{y}}$ is the *inversion* of $\mathbf{y}$ with respect to the circle C_ρ (see Figure 9.3). We want to compare the distances $|\mathbf{x} - \mathbf{y}|$ and $|\mathbf{x} - \hat{\mathbf{y}}|$ for $|\mathbf{x}| = \rho$. When $|\mathbf{x}| = \rho$, we calculate

$$|\mathbf{x} - \hat{\mathbf{y}}|^2 = |\mathbf{x} - \frac{\rho^2 \mathbf{y}}{|\mathbf{y}|^2}|^2 = |\mathbf{x}|^2 - \frac{2\rho^2}{|\mathbf{y}|^2} < \mathbf{x}, \mathbf{y} > + \frac{\rho^4 |\mathbf{y}|^2}{|\mathbf{y}|^4}$$

$$= \frac{\rho^2}{|\mathbf{y}|^2} [|\mathbf{y}|^2 - 2 < \mathbf{x}, \mathbf{y} > + \rho^2] = \frac{\rho^2}{|\mathbf{y}|^2} |\mathbf{x} - \mathbf{y}|^2.$$

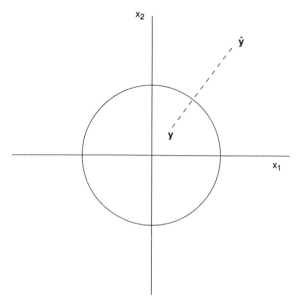

FIGURE 9.3
Inversion with respect to the circle of radius ρ.

Thus

$$\frac{1}{|\mathbf{x} - \mathbf{y}|} = \frac{\rho}{|\mathbf{y}|}(\frac{1}{|\mathbf{x} - \hat{\mathbf{y}}|}).$$

We take

$$v(\mathbf{x}, \mathbf{y}) = -\frac{1}{2\pi} \log\left[(\frac{\rho}{|\mathbf{y}|})\frac{1}{|\mathbf{x} - \hat{\mathbf{y}}|}\right]. \tag{9.37}$$

With this choice of v, (9.34) yields the Green's function for D_ρ. According to (9.36), the Poisson kernel for D_ρ is given by

$$-\frac{\partial\gamma}{\partial n_\mathbf{y}}(\mathbf{x}, \mathbf{y})\bigg|_{|\mathbf{y}|=\rho} = -\frac{\mathbf{y}}{|\mathbf{y}|}\cdot\nabla_\mathbf{y}\gamma(\mathbf{x}, \mathbf{y})\bigg|_{|\mathbf{y}|=\rho}.$$

If we finish this calculation, we get back exactly formula (9.20).

The Green's function for the ball of radius ρ in R^3 is found in exactly the same manner:

$$\gamma(\mathbf{x}, \mathbf{y}) = \frac{1}{4\pi}\left[\frac{1}{|\mathbf{x} - \mathbf{y}|} - \frac{\rho}{|\mathbf{y}|}\frac{1}{|\mathbf{x} - \hat{\mathbf{y}}|}\right] \tag{9.38}$$

with $\hat{\mathbf{y}} = \rho^2\mathbf{y}/|\mathbf{y}|^2$.

Exercises 9.4

1. Prove the uniqueness part of Theorem 9.6 in the case $n = 2$. Show that, when $q = 0$ for $|\mathbf{x}| \geq \rho$, the solution u given by (9.27) satisfies $|\nabla u| \to 0$ as $|\mathbf{x}| \to \infty$. Then apply Liouville's theorem to the derivatives of $u - v$, when v is another solution of (9.28).

2. Solve analytically the boundary value problem in the unit disk $D_1 \subset R^2$:

$$-\Delta u \equiv 1 \text{ in } D_1, \quad u = 0 \text{ on } \partial D_1.$$

Look for a radially symmetric solution $u = u(r)$. In polar coordinates, the equation becomes

$$-\frac{1}{r}(ru_r)_r = -u_{rr} - \frac{1}{r}u_r = 1, \ 0 < r < 1, \ u'(0) = 0, \ u(1) = 0.$$

3. Solve the boundary value problem in the unit disk D_1 with a piecewise constant right-hand side q:

$$-\Delta u = q \quad \text{in } D_1, \qquad u = 0 \quad \text{on } \partial D_1,$$

where

$$q(x, y) = \begin{cases} q_0, & x^2 + y^2 < \rho^2 \\ 0, & \rho^2 < x^2 + y^2 < 1 \end{cases}$$

and $0 < \rho < 1$. In polar coordinates, the problem becomes

$$-\frac{1}{r}(ru_r)_r = -(u_{rr} + \frac{1}{r}u_r) = q(r), \quad 0 < r < 1, \quad u'(0) = 0, \quad u(1) = 0.$$

4. Solve the problem $\Delta u = q$ in R^3 with the same q. In spherical coordinates, the equation becomes

$$-\frac{1}{r^2}\partial_r(r^2 u_r) = q, \quad 0 < r < 1, \qquad u'(0) = 0, \ u(r) \to 0 \text{ as } r \to \infty.$$

Here $q(r) = q_0$ for $0 < r < \rho < 1$, and $q(r) = 0$ for $r > \rho$.

   (a) Solve the ODE with the boundary conditions at $r = 0$ and $r = \infty$.

   (b) Solve the problem using the formula (9.27). Compare the results.

The function green plots the Green's function for the unit disk. The function call is green(x,y,xi,eta). Make a polar coordinate meshgrid for the unit disk as in exercise 2, Section 9.2. To view the Green's function $\gamma(x, y, \xi, \eta)$ with singularity located at $\xi = a, \eta = b$, you can use the commands

```
>> Z = green(X,Y,a,b);
>> surf(X,Y,Z)
```

You can also use mesh(X,Y,Z).

5. In this exercise we illustrate the principle of superposition for solutions of the Poisson equation in the unit disk with zero boundary conditions. Choose points $(a_1, b_1), (a_2, b_2), (a_3, b_3)$ in the unit disk, and weights q_1, q_1, q_3. Each of the functions

$$z_1 = \gamma(x, y, a_1, b_1)$$

$$z_2 = \gamma(x, y, a_2, b_2)$$

$$z_3 = \gamma(x, y, a_3, b_3)$$

is a singular solution of the Poisson equation, and so is

$$z = q_1 z_1 + q_2 z_2 + q_3 z_3.$$

Plot this function in the unit disk for several choices of the points and weights.

Program deflect solves the boundary value problem in the unit disk $D_1 \subset R^2$

$$-\Delta u = q \quad \text{in } D_1, \qquad u = 0 \quad \text{on } \partial D_1,$$

where

$$q(x, y) = \begin{cases} q_0, & (x - a)^2 + (y - b)^2 \le \rho_0^2 \\ 0, & \text{elsewhere} \end{cases}.$$

In this case, the integral (9.35) reduces to

$$u(x, y) = q_0 \int_{\hat{D}} \gamma(x, y, \xi, \eta) d\xi d\eta,$$

where $\hat{D}$ is the small disk with center (a, b) and radius ρ_0. The solution is computed by extending the process used in exercise 5 to sum over a larger sum to approximate the integral.

The solution surface can be thought of as the deflection caused by pushing your finger up on a drum head from underneath at the point (a, b). The user must enter the values a, b, ρ_0, and q_0. Make sure that $\sqrt{a^2 + b^2} + \rho_0 \leq 1$. The solution is written into the matrix U. The trace $x \rightarrow u(x, 0)$, $-1 \leq x \leq 1$, is written into the vector trace. To view the trace, use the command plot(x,trace) .

6. Run program deflect with $q_0 = 5$, $\rho_0 = .2$ and $b = 0$ and values $a = 0, .2, .4, .6, .8$. Plot the trace of each solution. For what value of a is the deflection the greatest? Can you guess a function $d(a)$ which gives the maximum deflection of the surface as a ranges from 0 to .8?

7. Prove property (4) of the Green's function. Fix $\mathbf{y} \in G$. We know that $\mathbf{x} \rightarrow \gamma(\mathbf{x}, \mathbf{y})$ blows up as $\mathbf{x} \rightarrow \mathbf{y}$. Hence we can find an $\varepsilon > 0$ such that $\gamma(\mathbf{x}, \mathbf{y}) \geq 1$ on $|\mathbf{x} - \mathbf{y}| \geq \varepsilon$. On the other hand, $\mathbf{x} \rightarrow \gamma(\mathbf{x}, \mathbf{y})$ is harmonic on $G_\varepsilon = \{\mathbf{x} \in G : |\mathbf{x} - \mathbf{y}| > \varepsilon\}$. Use the maximum principle.

8. (a) Find the Green's function for the quarter-plane in R^2, $Q = \{(x_1, x_2) : x_1, x_2 > 0\}$. Use the method of balancing charges.

   (b) Find the Poisson kernel for the quarter-plane using the results of part (a) and (9.36).

9. Find the Green's function for the half-ball in R^3, $B_\rho^+ = \{\mathbf{x} \in B_\rho : x_3 > 0\}$. Use a reflection of the Green's function for the whole ball.

10. Find the Green's function for the exterior of the ball of radius ρ in R^3. That is, find the Green's function for the problem

$$-\Delta u = q \text{ in } |\mathbf{x}| > \rho,$$

$$u = 0 \text{ on } |\mathbf{x}| = \rho, \quad u \rightarrow 0 \text{ as } |\mathbf{x}| \rightarrow \infty.$$

11. From Example A, recall that $H = \{(x, y) : x > 0\}$ is the half-plane in R^2 and that the Poisson kernel for H is given by

$$P(\mathbf{x}, \mathbf{y}) = P(x, y, \eta) = \frac{x}{\pi} \frac{1}{x^2 + (y - \eta)^2}.$$

Verify the following properties of the Poisson kernel for H.

   (a) For fixed η, the level curves of $(x, y) \rightarrow P(x, y, \eta)$ are circles in H tangent to the y axis.

   (b) $\Delta_{x,y} P = 0$ in H.

(c) $\int_R P(x, y, \eta) dy = 1$ for each $x > 0$, $\eta \in R$.

(d)   If $-a \le \eta \le a$ and $|y| > a$,

$$|P(x, y, \eta)| \le \frac{r}{\pi} \frac{1}{(r-a)^2}, \quad r = \sqrt{x^2 + y^2}.$$

12.   Let $f(\eta) = 1$ for $|\eta| < 1$, $f = 0$ elsewhere.

(a)   Find the solution $u(x, y)$ of the Dirichlet problem in H with f given on the y axis. Use the Poisson kernel representation. The integral can be evaluated in closed form.

(b)   Show that $\lim_{x \downarrow 0} u(x, y) = f(y)$ for $y \ne \pm 1$. What happens when $y = \pm 1$?

(c)   Use MATLAB to make pcolor plots of the solution u in the rectangle $0 \le x \le 10$, $-10 \le y \le 10$. Use the command `colormap(hsv)`, and then superimpose the contour lines in black.

13.   Let $f(\eta)$ be continuous on R with $f = 0$ for $|\eta| > a$. Let $u(x, y)$ be the solution given by the Poisson formula of the Dirichlet problem in H with boundary data f.

(a)   Use part (d) of exercise 11 to show that there is a constant C, such that $|u(x, y)| \le C/r$ as $r \to \infty$.

(b)   Use the version of the maximum principle in exercise 10 of Section 9.1 to show that u, given by the Poisson formula, is the unique solution of the Dirichlet problem in H, which decays like $1/r$ as $r \to \infty$.

14.   Recall the compatibility conditions which must be satisfied for the existence of solutions in the case of the Poisson equation with homogeneous Neumann boundary conditions,

(i) $$-\Delta u = q \text{ in } G, \quad \frac{\partial u}{\partial n} = 0 \text{ on } \partial G$$

requires $\int_G q \, d\mathbf{x} = 0$; and the Neumann problem

(ii) $$\Delta u = 0 \text{ in } G, \quad \frac{\partial u}{\partial n} = g \text{ on } \partial G$$

requires $\int_{\partial G} g \, dS = 0$.

We want to construct a *Neumann* function $N(\mathbf{x}, \mathbf{y})$ for problem (i) which is the analogue of the Green's function. If we followed the procedure used to construct $\gamma(\mathbf{x}, \mathbf{y})$, we would look for N in the form

$$N(\mathbf{x}, \mathbf{y}) = s(\mathbf{x} - \mathbf{y}) + v(\mathbf{x}, \mathbf{y}).$$

v must be chosen so that for $\mathbf{y} \in G$, $\Delta_{\mathbf{x}} v = 0$, and

$$\frac{\partial N}{\partial n_{\mathbf{x}}} = \frac{\partial s}{\partial n_{\mathbf{x}}} + \frac{\partial v}{\partial n_{\mathbf{x}}} = 0 \quad \text{on } \partial G.$$

(a) Can v be found such that this equation on ∂G is satisfied? Keep in mind the compatibility condition for (ii).

(b) Instead we modify the procedure. Seek N in the form

$$N(\mathbf{x}, \mathbf{y}) = s(\mathbf{x} - \mathbf{y}) + \frac{|\mathbf{x} - \mathbf{y}|^2}{2nA} + v(\mathbf{x}, \mathbf{y})$$

where $A = \text{area}(G)$. Now v must satisfy

$$\Delta_{\mathbf{x}} v = 0 \quad \text{in } G,$$

$$\frac{\partial v}{\partial n_{\mathbf{x}}} = -\frac{\partial}{\partial n_{\mathbf{x}}} [s(\mathbf{x} - \mathbf{y}) + \frac{|\mathbf{x} - \mathbf{y}|^2}{2nA}] \quad \text{on } \partial G.$$

Show that the compatibility condition for (ii) is now satisfied. Is N constructed this way unique?

(c) Show that if q is given on G, and N is constructed as in part (b), then $u(\mathbf{x}) = \int_G N(\mathbf{x}, \mathbf{y}) q(\mathbf{y}) d\mathbf{y}$ solves

$$-\Delta u = q - \frac{1}{A} \int q(\mathbf{x}) d\mathbf{x} \quad \text{in } G, \quad \frac{\partial u}{\partial n} = 0 \quad \text{on } \partial G.$$

Thus if $\int_G q d\mathbf{x} = 0$, then $-\Delta u = q$. Is the solution constructed this way unique?

15. Construct a Neumann function $N(x, y)$ for the one-dimensional problem

$$-u''(x) = q(x), \quad 0 < x < L,$$

$$u'(0) = u'(L) = 0.$$

N must satisfy

$$-N_{xx}(x, y) = \delta(x - y) - \frac{1}{L}, \quad N_x(0, y) = N_x(L, y) = 0.$$

Note that once you have found a Neumann function $N(x, y)$, $N(x, y) + g(y)$ is another one for an arbitrary C^2 function $g(y)$. Find a choice of N such that $N(x, y) = N(y, x)$.

16. In some cases a Neumann function can be constructed using symmetry arguments. Use the method of balancing charges to construct the Neumann function for the half-space $\{x \in R^3 : x_1 > 0\}$.

17. Let $D_\rho^+ \subset R^2$ be the half-disk $\{x \in D_\rho : x_1 > 0\}$. Use the Green's function $\gamma(x, y)$ for the disk D_ρ to construct a kernel function $\beta(x, y)$, so that $u(x) = \int \beta(x, y)q(y)dy$ solves the problem

$$-\Delta u = q \quad \text{in } G, \quad u = 0 \quad \text{on } |x| = \rho, \quad \frac{\partial u}{\partial n} = 0 \quad \text{on } x_1 = 0.$$

18. Using a single index n, let $\varphi_n(x)$ and λ_n be the eigenfunctions and eigenvalues of the problem $-\Delta\varphi = \lambda\varphi, \quad \varphi = 0$ on ∂G. The eigenfunction expansion of the solution of the problem $-\Delta u = q$ in G, $u = 0$ on ∂G is given by

$$u(x) = \sum_1^\infty \frac{q_n}{\lambda_n}\varphi_n(x),$$

where $q_n = \int_G q(x)\varphi_n(x)dx/\langle\varphi_n, \varphi_n\rangle$. Starting from this point, show that the Green's function has the expansion

$$\gamma(x, y) = \sum_1^\infty \frac{1}{\lambda_n} \frac{\varphi_n(x)\varphi_n(y)}{\langle\varphi_n, \varphi_n\rangle}.$$

Incidentally, this expansion also shows the symmetry of γ.

19. Find the eigenfunction expansion for a Neumann function $N(x, y)$. This particular Neumann function will satisfy $N(x, y) = N(y, x)$.

20. Conformal mappings (exercise 5, Section 9.1) can be used to solve the Dirichlet problem in two dimensions. The mapping $(x, y) \rightarrow (x^2 - y^2, 2xy)$ maps the first quadrant Q onto the upper half-plane.

(a) Verify that this mapping is conformal.

(b) Solve the problem

$$\Delta u = 0 \quad \text{in } Q, \quad u(x, 0) = A, \quad u(0, y) = B$$

by first solving the problem in the half-plane $y > 0$,

$$\Delta v = 0 \quad \text{in } y > 0, \quad v(x, 0) = B \quad \text{for } x < 0, \quad v(x, 0) = A \quad \text{for } x > 0.$$

9.5   Variational methods and weak solutions

9.5.1   Problems in variational form

In many cases the solutions of the Laplace and Poisson equations represent an equilibrium state. Steady-state heat flow, steady-state ideal fluid flow, and electrostatics are examples. The steady state or equilibrium state of a system is often characterized as having a minimal energy. In this section we investigate how the solutions of the Dirichlet problem and the solutions of the Poisson equation can be seen as minimizing certain quadratic functionals. The notion of a weak solution arises quite naturally in this context.

We begin with the Dirichlet problem which we restate here for convenience.

$$\Delta u = 0 \quad \text{in } G, \qquad u = f \quad \text{on } \partial G, \tag{9.39}$$

where f is a given continuous function on ∂G, and ∂G is assumed to be piecewise C^1. Define the sets of *admissible functions*

$$\mathcal{A} = \{v \in C^1(G) : v \in C(\bar{G}), v = f \text{ on } \partial G\}$$

and

$$\mathcal{A}_0 = \{\varphi \in C^1(G) : \varphi \in C(\bar{G}), \varphi = 0 \text{ on } \partial G\}.$$

Notice that if $v \in \mathcal{A}$, then $v + \varphi \in \mathcal{A}$ for all $\varphi \in \mathcal{A}_0$. $\mathcal{A}$ is not a linear subspace but is a translate (depending on f) of $\mathcal{A}_0$. Now we define the quadratic functional on $\mathcal{A}$

$$F(v) = \int_G |\nabla v|^2 dx, \quad v \in \mathcal{A}. \tag{9.40}$$

Since $F(v) \geq 0$ for all $v \in \mathcal{A}$, it is reasonable to ask if there is a function $u \in \mathcal{A}$ such that

$$F(v) \geq F(u) \quad \text{for all } v \in \mathcal{A}.$$

We call u a *minimizer* for F on $\mathcal{A}$.

Let us suppose that $u \in \mathcal{A}$ is such a minimizer for F. Then for any $\varphi \in \mathcal{A}_0$, and all $\varepsilon \in R$, $u + \varepsilon\varphi \in \mathcal{A}$ and

$$F(u + \varepsilon\varphi) \geq F(u).$$

This means that for each $\varphi \in \mathcal{A}_0$, the real-valued function

$$\varepsilon \to F(u + \varepsilon\varphi) = \int_G |\nabla(u + \varepsilon\varphi)|^2 d\mathbf{x}$$

has a minimum at $\varepsilon = 0$. The *variation* of F at u is given by

$$\delta F(u)\varphi = \frac{d}{d\varepsilon} F(u + \varepsilon\varphi)\bigg|_{\varepsilon=0}.$$

We can calculate this to be

$$\frac{d}{d\varepsilon} \int_G \left[|\nabla u|^2 + 2\varepsilon\nabla u \cdot \nabla\varphi + \varepsilon^2|\nabla\varphi|^2\right]d\mathbf{x}\bigg|_{\varepsilon=0} = 2\int_G \nabla u \cdot \nabla\varphi d\mathbf{x}.$$

Because it is assumed that u is a minimizer for F on $\mathcal{A}$,

$$\delta F(u)\varphi = 2\int_G \nabla u \cdot \nabla\varphi d\mathbf{x} = 0 \tag{9.41}$$

for all $\varphi \in \mathcal{A}_0$. Suppose we knew in addition that $u \in C^2(G)$. Then we could integrate by parts in (9.41), using the fact that $\varphi = 0$ on ∂G, to deduce that

$$0 = \int_G \nabla u \cdot \nabla\varphi d\mathbf{x} = -\int_G \Delta u \varphi d\mathbf{x} + \int_{\partial G} \frac{\partial u}{\partial n}\varphi dS = -\int_G \Delta u \varphi d\mathbf{x}.$$

Since this equation holds for all $\varphi \in \mathcal{A}_0$, we deduce that $\Delta u = 0$, and $u = f$ on ∂G because $u \in \mathcal{A}$. Thus a C^2 minimizer must solve the Dirichlet problem (9.39).

The PDE associated with minimizing the functional F is called the *Euler-Lagrange* equation for F. In this case the Euler-Lagrange equation is just $\Delta u = 0$.

The characterization of the solution of the Dirichlet problem (9.39) as a minimizer for F suggests that we might construct solutions of (9.39) by finding a sequence of functions $u_n \in \mathcal{A}$ such that

$$F(u_n) \geq F(u_{n+1}) \geq F(u_{n+2}) \geq \ldots.$$

The u_n would be a minimizing sequence. However we would have to prove that such a sequence converged to some function u, and then prove u had two derivatives so that it would be a strict solution of (9.39). If the the limiting function u were at least in $\mathcal{A}$, we would know that it would satisfy (9.41). In this case we would say that u is a *weak* solution of (9.39). We use the same terminology as we did when describing waves. However, the difference with Laplace's equation is that one can prove that a weak solution of $\Delta u = 0$ is in fact C^∞ and thus is a strict solution.

For a second example of a variational problem, we consider the minimal surface problem. Let the set $\mathcal{A}$ be as before where now we take G as a bounded, open set of R^2. We can think of f (given on ∂G) as the height of a wire over ∂G. Let $v \in \mathcal{A}$. Then the surface area of the graph of v is

$$J(v) = \int\int_G \sqrt{1 + v_x^2 + v_y^2}\, dx dy. \tag{9.42}$$

We want to find that function u which minimizes J. The graph of u would be like a soap bubble supported by the wire defined by the graph of f. To find the Euler-Lagrange equation of a minimizer u of J, we must find the variation of J. Again let $\varphi \in \mathcal{A}_0$, and differentiate with respect to ε the function

$$\varepsilon \to J(u + \varepsilon\varphi) = \int\int_G \sqrt{1 + (u_x + \varepsilon\varphi_x)^2 + (u_y + \varepsilon\varphi_y)^2}\, dx dy.$$

Now

$$\frac{d}{d\varepsilon} J(u + \varepsilon\varphi) = \int\int_G \frac{(u_x + \varepsilon\varphi_x)\varphi_x + (u_y + \varepsilon\varphi_y)\varphi_y}{\sqrt{1 + (u_x + \varepsilon\varphi_x)^2 + (u_y + \varepsilon\varphi_y)^2}}\, dx dy$$

so that

$$\delta J(u)\varphi = \frac{d}{d\varepsilon} J(u + \varepsilon\varphi)\Big|_{\varepsilon=0} \tag{9.43}$$

$$= \int\int_G \frac{u_x\varphi_x + u_y\varphi_y}{\sqrt{1 + u_x^2 + u_y^2}}\, dx dy = 0.$$

Assuming that $u \in C^2(G)$, we can integrate by parts in (9.43), putting all the derivatives on u, to arrive at

$$0 = \int\int_G \left[\partial_x\left(\frac{u_x}{\sqrt{1 + u_x^2 + u_y^2}}\right) + \partial_y\left(\frac{u_y}{\sqrt{1 + u_x^2 + u_y^2}}\right) \right]\varphi\, dx dy,$$

which holds for all $\varphi \in \mathcal{A}_0$. Hence the Euler-Lagrange equation for the minimal surface area problem is a boundary value problem for the nonlinear PDE

$$\partial_x\left(\frac{u_x}{\sqrt{1+u_x^2+u_y^2}}\right) + \partial_y\left(\frac{u_y}{\sqrt{1+u_x^2+u_y^2}}\right) = 0 \quad \text{in } G, \qquad (9.44)$$

$$u = f \quad \text{on } \partial G.$$

This nonlinear PDE is very difficult to analyze.

A third problem which can be put in variational form is the Poisson equation with zero boundary condition:

$$-\Delta u = q \quad \text{in } G, \qquad u = 0 \quad \text{on } \partial G. \qquad (9.45)$$

G is a bounded, open subset of either R^2 or R^3. Here we take the functional on $\mathcal{A}_0$ as

$$H(v) = \int_G \left[|\nabla v|^2 - 2vq \right] d\mathbf{x}.$$

If $u \in \mathcal{A}_0$ is a minimizer of H on $\mathcal{A}_0$, then

$$0 = \delta H(u)\varphi = 2 \int_G \left[\nabla u \cdot \nabla \varphi - q\varphi \right] d\mathbf{x}, \qquad (9.46)$$

holding for all $\varphi \in \mathcal{A}_0$. Now again, if $u \in C^2(G)$, we can integrate by parts in (9.46) to deduce that

$$\int_G [-\Delta u - q]\varphi d\mathbf{x} = 0$$

for all $\varphi \in \mathcal{A}_0$, which implies that u solves (9.45). If u is only C^1 and satisfies (9.46) for all $\varphi \in \mathcal{A}_0$, then we say that u is a weak solution of (9.45).

Remarks

The process in which one constructs a minimizing sequence and shows that it converges to a minimizer is called the *direct method of the calculus of variations*. Carrying out this procedure can be very difficult. There are variational problems in which the minimizing sequences do not converge, and in which there is no minimizing function. For an example, see exercise 1 of this section.

9.5.2 The Rayleigh-Ritz procedure

Finally we look at the eigenvalue problem

$$-\Delta \varphi = \lambda \varphi \quad \text{in } G, \qquad \varphi = 0 \quad \text{on } \partial G \qquad (9.47)$$

where G is a bounded, open set of R^2 or R^3. For geometries other than a disk, a ball, or a rectangle, it is usually not possible to find the eigenvalues and eigenfunctions explicitly. Nevertheless, it is still possible to get information about the eigenvalues in a general region G using variational methods. Recall from Chapter 8 that if λ is an eigenvalue and φ a corresponding eigenfunction satisfying (9.47), then

$$\lambda = \frac{\int_G |\nabla\varphi|^2 d\mathbf{x}}{\int_G \varphi^2 d\mathbf{x}}. \tag{9.48}$$

The expression (9.48) is called the *Rayleigh quotient*.

Theorem 9.8
Let λ_1 be the lowest eigenvalue of (9.47). Then

$$\lambda_1 = \min_{v \in \mathcal{A}_0} \frac{\int_G |\nabla v|^2 d\mathbf{x}}{\int_G v^2 d\mathbf{x}}. \tag{9.49}$$

Here is the proof. Let $v \in C^2(G) \cap C^1(\bar{G})$ with $v = 0$ on ∂G. Then we can expand v in terms of the eigenfunctions of (9.47). We shall use a single index n and label the eigenfunctions and eigenvalues φ_n and λ_n. Then

$$v(\mathbf{x}) = \sum_{n=1}^{\infty} c_n \varphi_n(\mathbf{x}), \qquad c_n = \frac{\langle v, \varphi_n \rangle}{\langle \varphi_n, \varphi_n \rangle}.$$

Now using Green's first identity and the fact that $v = 0$ on ∂G, we see that

$$\int_G |\nabla v|^2 d\mathbf{x} = \int_G (-\Delta v) v d\mathbf{x} = \langle -\Delta v, v \rangle, \tag{9.50}$$

and

$$-\Delta v = \sum_{n=1}^{\infty} c_n(-\Delta \varphi_n) = \sum_{n=1}^{\infty} \lambda_n c_n \varphi_n.$$

Since the φ_n are orthogonal,

$$\langle -\Delta v, v \rangle = \langle \sum_1^{\infty} \lambda_n c_n \varphi_n, \sum_1^{\infty} c_m \varphi_m \rangle = \sum_1^{\infty} \lambda_n c_n^2 \langle \varphi_n, \varphi_n \rangle$$

$$\geq \lambda_1 \sum_1^{\infty} c_n^2 \langle \varphi_n, \varphi_n \rangle = \lambda_1 \langle v, v \rangle = \lambda_1 \int_G v^2 d\mathbf{x}.$$

Combining this with (9.50) yields

$$\int_G |\nabla v|^2 dx \geq \lambda_1 \int_G v^2 dx$$

for all $v \in C^2(G) \cap C^1(\bar{G})$ with $v = 0$ on ∂G. Equality is actually attained when $v = \varphi_1$. Thus (9.49) is proved for the class of $v \in C^2(G) \cap C^1(\bar{G})$. More technical arguments are needed to demonstrate the validity of (9.49) for $v \in \mathcal{A}_0$. $\square$

Now let us see how (9.49) can be used to get some numerical approximation to λ_1. We can experiment with various trial functions in (9.49), trying to make the Rayleigh quotient as small as possible. For instance, in one dimension, the lowest eigenvalue of the boundary value problem

$$- u''(x) = \lambda u(x) \quad \text{for } 0 < x < 1, \tag{9.51}$$

$$u(0) = u(1) = 0,$$

is $\lambda_1 = \pi^2$, and the eigenfunction is $\varphi_1(x) = \sin \pi x$. If we did not know the eigenfunction φ_1, but had some idea from physical reasoning that it would not have any zeros in the interior of the interval, we could try $v(x) = x(1 - x)$, which is the parabola such that $v(0) = v(1) = 0$. Now $v'(x) = 1 - 2x$ so that the Rayleigh quotient is

$$\frac{\int_0^1 (v')^2(x)dx}{\int_0^1 v^2(x)dx} = \frac{\int_0^1 (1 - 2x)^2 dx}{\int_0^1 x^2(1 - x)^2 dx} = \frac{1/3}{1/30} = 10$$

while $\pi^2 \approx 9.98696$. Not too bad for a first guess !

The minimum characterization of the lowest eigenvalue of problem (9.47) can be extended to higher eigenvalues. Let $\mathcal{A}_{n-1}$ be the subspace of functions $v \in \mathcal{A}_0$ which are orthogonal to the first $n - 1$ eigenfunctions:

$$\langle v, \varphi_1 \rangle = \langle v, \varphi_2 \rangle = \cdots \langle v, \varphi_{n-1} \rangle = 0.$$

Theorem 9.9
Let λ_n be the nth eigenvalue in the sequence of eigenvalues $\lambda_1 \leq \lambda_2 \cdots$ of the problem (9.47). Then

$$\lambda_n = \min_{\mathcal{A}_{n-1}} \frac{\int_G |\nabla v|^2 dx}{\int_G v^2 dx}. \tag{9.52}$$

This result is proved in the same manner as Theorem 9.8.

The Rayleigh-Ritz procedure is a systematic way of using (9.52). Suppose we choose a basis of trial functions $v_1, \ldots, v_n \in \mathcal{A}_0$. Let $\mathcal{V}$ be the linear subspace of $\mathcal{A}_0$ of linear combinations $v = c_1 v_1 + \ldots + c_n v_n$. The trial functions v_j are probably not pairwise orthogonal so that we must calculate

$$b_{j,k} = \langle v_j, v_k \rangle = \int_G v_j(\mathbf{x}) v_k(\mathbf{x}) d\mathbf{x}.$$

Hence

$$\int_G v^2 d\mathbf{x} = \langle v, v \rangle = \sum_{j,k} c_j c_k \langle v_j, v_k \rangle = B\mathbf{c} \cdot \mathbf{c},$$

where $B = [b_{j,k}]$ is an $n \times n$ symmetric positive definite matrix, and $\mathbf{c} = (c_1, \ldots c_n)$ is the vector of coefficients. Let

$$a_{j,k} = \langle \nabla v_j, \nabla v_k \rangle = \int_G \nabla v_j(\mathbf{x}) \cdot \nabla v_k(\mathbf{x}) d\mathbf{x},$$

so that

$$\int_G |\nabla v|^2 d\mathbf{x} = \langle \nabla v, \nabla v \rangle = \sum_{j,k} a_{j,k} c_j c_k$$

$$= A\mathbf{c} \cdot \mathbf{c}$$

where again $A = [a_{j,k}]$ is a symmetric, positive-definite, $n \times n$ matrix.

If an eigenfunction φ of (9.47), with eigenvalue λ, could be written as a linear combination of the trial functions v_j,

$$\varphi = c_1 v_1 + \cdots c_n v_n,$$

then for each $j = 1, \ldots n$,

$$\langle -\Delta \varphi, v_j \rangle = \lambda \langle \varphi, v_j \rangle. \tag{9.53}$$

Substitution of the expansion for φ into (9.53) would yield

$$\sum_j a_{i,j} c_j = \lambda \sum_j b_{i,j} c_j$$

or, more succinctly,

$$A\mathbf{c} = \lambda B\mathbf{c}. \tag{9.54}$$

Thus λ would be an eigenvalue of the matrix equation (9.54). It is highly unlikely that we could find trial functions v_j so that we could write φ as a linear combination of the v_j. However, we might able to approximate φ by a linear combination of the v_j, in which case the eigenvalues of (9.54) might approximate those of (9.47). Since both A and B are positive-definite, symmetric, $n \times n$ matrices, there is a set of n independent eigenvectors with real eigenvalues $\tilde{\lambda}_1, \ldots, \tilde{\lambda}_n$ of (9.54). It is not hard to show that

$$\tilde{\lambda}_n = \max_{\mathbf{c} \neq 0} \frac{A\mathbf{c} \cdot \mathbf{c}}{B\mathbf{c} \cdot \mathbf{c}} = \max_v \frac{\int_G |\nabla v|^2 dx}{\int_G v^2 dx}. \tag{9.55}$$

We can vary the choice of trial functions, thereby getting better or worse approximations of the eigenvalues of (9.47). The following result gives a one-sided estimate.

Theorem 9.10 (Minimax).

$$\lambda_n = \min_V \tilde{\lambda}_n = \min_V \max_{v \in V} \frac{\int_G |\nabla v|^2 \, dx}{\int_G v^2 \, dx}.$$

First we show that for any choice of the trial function v_j, we have $\tilde{\lambda}_n \geq \lambda_n$. Among the functions in V, we can find at least one, v_*, that also lies in A_{n-1}. This will be the case if we can find coefficients $c_1, \ldots, c_n$, such that

$$\langle \varphi_j, v_* \rangle = c_1 \langle \varphi_j, v_1 \rangle + \cdots + c_n \langle \varphi_j, v_n \rangle = 0$$

for $j = 1, \ldots, n-1$. Since this is a linear homogeneous system of $(n-1)$ equations in n unknowns, we know that there must exist a nontrivial solution c_*. Now using (9.52) and (9.55),

$$\lambda_n = \min_{A_{n-1}} \frac{\int_G |\nabla v|^2 dx}{\int_G v^2 dx} \leq \frac{\int_G |\nabla v_*|^2 dx}{\int_G v_*^2 dx} \leq \max_{v \in V} \frac{\int_G |\nabla v|^2 dx}{\int_G v^2 dx} = \tilde{\lambda}_n.$$

To prove the theorem, it will suffice to find one choice of trial functions v_j, such that $\tilde{\lambda}_n = \lambda_n$. But this is easy. Take $v_j = \varphi_j$ for $j = 1, \ldots, n$. Then the matrix A is diagonal with the eigenvalues λ_j on the diagonal, and the matrix B is the identity. Hence in this case,

$$\tilde{\lambda}_n = \max_{v \in V} \frac{\int_G |\nabla v|^2 dx}{\int_G v^2 dx} = \max_{\mathbf{c} \neq 0} \frac{A\mathbf{c} \cdot \mathbf{c}}{B\mathbf{c} \cdot \mathbf{c}} = \lambda_n.$$

The theorem is proved. $\square$

Note that this result gives a way of estimating the nth eigenvalue, without knowing the lower eigenvalues.

Let us go back to the one-dimensional eigenvalue problem (9.51). We have already taken $v(x) = x(1-x)$ as an approximation of the first eigenfunction. Now take $w(x) = x(1/2 - x)(1 - x)$ as an approximation to the second eigenfunction. The matrix A is

$$A = \begin{bmatrix} \int (v')^2 & \int v'w' \\ \int v'w' & \int (w')^2 \end{bmatrix} = \begin{bmatrix} 1/3 & 0 \\ 0 & 1/20 \end{bmatrix}$$

and the matrix B is

$$B = \begin{bmatrix} \int v^2 & \int vw \\ \int vw & \int w^2 \end{bmatrix} = \begin{bmatrix} 1/30 & 0 \\ 0 & 1/840 \end{bmatrix}.$$

This means that the lowest eigenvalue of the system (9.54) is $\tilde{\lambda}_1 = (1/3)(30) = 10$, and the next eigenvalue of (9.54) is $\tilde{\lambda}_2 = (1/20)(840) = 42$. As we saw before, the lowest eigenvalue of (9.51) is $\lambda_1 = \pi^2 \approx 9.98696$ and the second eigenvalue of (9.51) is $\lambda_2 = 4\pi^2 \approx 39.47842$. Thus $\tilde{\lambda}_1 \geq \lambda_1$, and $\tilde{\lambda}_2 \geq \lambda_2$.

The classic reference for variational problems is [GF]. See also [CourHilb]. A more recent treatment at the undergraduate level is [T].

Exercises 9.5

1. This is an example where there is no minimizing function in the space of admissible functions. Let $\mathcal{A} = \{u \in C[0, 1] : f(0) = f(1) = 1\}$, and let the functional F be

$$F(u) = \int_0^1 u^2(x)dx.$$

    Show that there is a sequence of functions $u_n(x) \in \mathcal{A}$, such that $F(u_n) \to 0$, but that for any $u \in \mathcal{A}$, $F(u) > 0$.

2. A second example. Let $\mathcal{A} = \{u \in C^1[0, 1] : u(0) = u(1) = 0\}$. Convergence of a sequence of functions u_n in $\mathcal{A}$ is defined to mean that u_n and u_n' converge uniformly. Now let the functional F be

$$F(u) = \int_0^1 [u^2 + 1 - (u')^2]dx.$$

    Construct a sequence u_n of "sawtooth" functions such that $F(u_n) \to 0$, with $u_n \to 0$ uniformly on $[0, 1]$, but such that the derivatives u_n' do not converge uniformly.

3. Let the space of admissible functions be $\mathcal{A} = \{u \in C^1[0, L] : u(0) = a, u(1) = b\}$, and let the functional

$$F(u) = \int_0^1 \sqrt{1 + u'(x)^2}\, dx.$$

   F is the arc length of the graph of u. Using the Euler-Lagrange equations, show that the minimizing function is the straight line $u(x) = a + (b - a)x$.

4. Find the Euler-Lagrange equation for the following functionals.

   (a) $F(u) = \int \int [x^2 u_x^2 + y^2 u_y^2]\, dx\, dy, \ u = u(x, y)$.

   (b) $F(u) = \int \int [u_t^2 - c^2 u_x^2]\, dx\, dt, \ u = u(x, t)$.

5. Let G be the ellipse in the x, y plane,

$$(\frac{x}{a})^2 + (\frac{y}{b})^2 \leq 1.$$

   Calculate the Rayleigh quotient for the trial function

$$v(x, y) = 1 - (\frac{x}{a})^2 - (\frac{y}{b})^2.$$

   What is the estimated value for the first eigenvalue of (9.47)? When $a = b = 1$, how does this estimated value compare with the first eigenvalue as found in exercise 2 of Section 8.4 ?

9.6 Projects

1. Modify program dirch so that you can solve the Neumann problem in the unit disk D_1:

$$\Delta u = 0 \quad \text{in } D_1, \qquad \frac{\partial u}{\partial n} = g \quad \text{on } r = 1.$$

   What condition must be placed on g to ensure existence of a solution? What other condition can you place on the solution u to make it unique when it exits?

Chapter 10

Numerical Methods for Higher Dimensions

In this chapter we investigate a finite difference scheme for the Laplace equation. Then we turn our attention to the finite element method for the Poisson equation. Finally we look at the Galerkin method for a time-dependent problem.

10.1 Finite differences

Let G be a rectangle in R^2, $0 < x < a$, $0 < y < b$, and consider the boundary value problem

$$\Delta u = 0 \quad \text{in } G, \qquad u = f \quad \text{on } \partial G. \tag{10.1}$$

We wish to discretize this equation, as we have before with the heat and wave equations, by replacing the derivatives in the equation with difference quotients. Introduce a grid in G of points (x_i, y_j) with

$$x_i = i\Delta x, \quad i = 0, \dots, I, \qquad y_j = j\Delta y, \quad j = 0, \dots, J,$$

where $\Delta x = a/I$, and $\Delta y = b/J$. Let $u_{i,j}$ denote the approximation of $u(x_i, y_j)$. We approximate the derivatives in (10.1) with centered differences:

$$u_{xx}(x_i, y_j) \approx \frac{u_{i+1,j} - 2u_{i,j} + u_{i-1,j}}{(\Delta x)^2},$$

$$u_{yy}(x_i, y_j) \approx \frac{u_{i,j+1} - 2u_{i,j} + u_{i,j-1}}{(\Delta y)^2}.$$

Substituting these expressions into (10.1), we arrive at

$$-\theta_x[u_{i+1,j} + u_{i-1,j}] + u_{i,j} - \theta_y[u_{i,j+1} + u_{i,j-1}] = 0,$$

where

$$\theta_x = \frac{(\Delta y)^2}{2((\Delta x)^2 + (\Delta y)^2)}, \qquad \theta_y = \frac{(\Delta x)^2}{2((\Delta x)^2 + (\Delta y)^2)}.$$

Thus

$$u_{i,j} = \theta_x u_{i+1,j} + \theta_x u_{i-1,j} + \theta_y u_{i,j+1} + \theta_y u_{i,j-1},$$

and $2(\theta_x + \theta_y) = 1$, so that $u_{i,j}$ is a weighted average of its four nearest neighbors. This is a discrete version of the mean value property of harmonic functions. In particular if $\Delta x = \Delta y$, the equation becomes

$$u_{i,j} = (1/4)[u_{i+1,j} + u_{i-1,j} + u_{i,j+1} + u_{i,j-1}]. \tag{10.2}$$

Now let us write down the full set of equations in the case $I = J = 4$ and $\Delta x = \Delta y$. The grid points are displayed in Figure 10.1. Note that because of the boundary condition $u = f$ on ∂G, values of $u_{i,j}$ with either $i = 0$, or $j = 0$, or $i = 4$, or $j = 4$ are already assigned. Thus the unknowns are $u_{i,j}$ at the nine interior grid points (x_i, y_j), $i = 1, 2, 3$, $j = 1, 2, 3$.

We write all the unknowns as a nine vector as follows:

$$\mathbf{u} = (u_{1,1}, u_{1,2}, u_{1,3}, u_{2,1}, u_{2,2}, u_{2,3}, u_{3,1}, u_{3,2}, u_{3,3}).$$

We shall group the boundary data in the nine vector:

$$\mathbf{f} = (\frac{u_{0,1} + u_{1,0}}{4}, \frac{u_{0,2}}{4}, \frac{u_{0,3} + u_{1,4}}{4}, \frac{u_{2,0}}{4}, 0, \frac{u_{2,4}}{4}, \frac{u_{3,0} + u_{4,1}}{4}, \frac{u_{4,2}}{4}, \frac{u_{3,4} + u_{4,3}}{4}).$$

Finally the system of linear equations is $A\mathbf{u} = \mathbf{f}$ where A is the 9×9 matrix

$$A = \begin{bmatrix} 1 & -\frac{1}{4} & 0 & -\frac{1}{4} & 0 & 0 & 0 & 0 & 0 \\ -\frac{1}{4} & 1 & -\frac{1}{4} & 0 & -\frac{1}{4} & 0 & 0 & 0 & 0 \\ 0 & -\frac{1}{4} & 1 & 0 & 0 & -\frac{1}{4} & 0 & 0 & 0 \\ -\frac{1}{4} & 0 & 0 & 1 & -\frac{1}{4} & 0 & -\frac{1}{4} & 0 & 0 \\ 0 & -\frac{1}{4} & 0 & -\frac{1}{4} & 1 & -\frac{1}{4} & 0 & -\frac{1}{4} & 0 \\ 0 & 0 & -\frac{1}{4} & 0 & -\frac{1}{4} & 1 & 0 & 0 & -\frac{1}{4} \\ 0 & 0 & 0 & -\frac{1}{4} & 0 & 0 & 1 & -\frac{1}{4} & 0 \\ 0 & 0 & 0 & 0 & -\frac{1}{4} & 0 & -\frac{1}{4} & 1 & -\frac{1}{4} \\ 0 & 0 & 0 & 0 & 0 & -\frac{1}{4} & 0 & -\frac{1}{4} & 1 \end{bmatrix}.$$

This is an example of a sparse matrix. It is a *banded* matrix in that all the entries are zero, except those lying in a band of width $I - 1$ about the diagonal. No row has more than five nonzero elements.

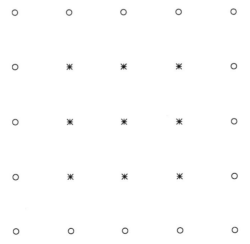

FIGURE 10.1
Grid points in the mesh when $I = J = 4$. Interior mesh points are indicated by asterisks.

For a modest mesh size of $\Delta x = \Delta y = .01$ on the unit square we see that we must solve a linear system involving $n = 10^4$ unknowns. For a full $n \times n$ matrix, Gaussian elimination requires on the order of n^3 operations, so that our problem would require 10^{12} operations. Fortunately there are other methods which exploit the sparseness of the matrix A that results from the discretization of the Laplace equation.

We shall discuss (very briefly) a class of methods called *iterative methods* which exploit the structure of sparse, banded matrices. The first of these is the *Jacobi method.* Consider the equation

$$A\mathbf{u} = \mathbf{f}, \tag{10.3}$$

where A is an $n \times n$ matrix and $\mathbf{f}$ is an n-vector. Assume that the diagonal elements $a_{i,i} \neq 0$. We write

$$A = D + N,$$

where N is the matrix of off-diagonal elements of A. Then (10.3) can be written

$$D\mathbf{u} = -N\mathbf{u} + \mathbf{f},$$

whence the solution $\mathbf{u}$ must satisfy

$$\mathbf{u} = B_J\mathbf{u} + \mathbf{g}_J. \tag{10.4}$$

Here $B_J = -D^{-1}N$, and $\mathbf{g}_J = D^{-1}\mathbf{f}$. We use the subscript J to remind us that we are using the Jacobi method. Now we make a first guess for the solution, say $\mathbf{u}^0$, and substitute in the right side of (10.4). We find that

$$\mathbf{u}^1 = B_J\mathbf{u}^0 + \mathbf{g}_J.$$

We hope that $\mathbf{u}^1$ lies closer to the true solution $\mathbf{u}$ than does $\mathbf{u}^0$. In this way we generate a sequence of approximate solutions by iterating formula (10.4) over and over again:

$$\mathbf{u}^k = B_J\mathbf{u}^{k-1} + \mathbf{g}_J.$$

In terms of components, the Jacobi method is given by

$$u_i^k = -\frac{1}{a_{i,i}} \sum_{j \neq i} a_{i,j} u_j^{k-1} + \frac{f_i}{a_{i,i}}.$$

If the off-diagonal elements of A are "small" relative to the diagonal elements, the iterates converge to the exact solution.

In the case of the matrix A which arises from the discretization of Laplace's equation, $D = I$, and the matrix N is mostly zeros. The Jacobi method is very easy to implement using (10.2). Make a first guess $u^0_{i,j}$ at all the interior grid points. Then use (10.2) to calculate the next iterate. In general,

$$u^k_{i,j} = \frac{1}{4}[u^{k-1}_{i+1,j} + u^{k-1}_{i-1,j} + u^{k-1}_{i,j+1} + u^{k-1}_{i,j-1}].$$

A finite difference scheme for the linear heat equation in the rectangle G provides another way to view the Jacobi method in this situation. If we approximate u_t by a forward difference, and the derivatives u_{xx} and u_{yy} by centered differences, we arrive at a scheme which reduces to the Jacobi iteration when we choose $k\Delta t/\Delta x^2 = 1/4$. Thus the Jacobi iterative method amounts to solving the heat equation for large times when the solution converges to the steady state.

Unfortunately the Jacobi method converges rather slowly and is not practical. The *Gauss-Seidel* method is another iterative method which converges about twice as fast as the Jacobi method. Gauss-Seidel is derived by splitting A differently. Write instead

$$A = L + U,$$

where is L is lower triangular and contains the diagonal elements of A and U is upper triangular with zeros on diagonal. Then (10.3) can be rewritten

$$\mathbf{u} = B_G\mathbf{u} + \mathbf{g}_G,$$

where now

$$B_G = -L^{-1}U \qquad \text{and} \qquad \mathbf{g}_G = L^{-1}\mathbf{f}.$$

We use the subscript G to denote the Gauss-Seidel method. The iterative scheme is

$$\mathbf{u}^k = B_G\mathbf{u}^{k-1} + \mathbf{g}_G. \tag{10.5}$$

In terms of components of $\mathbf{u}$, (10.5) can be expressed

$$u^k_i = -\frac{1}{a_{i,i}}[\sum_{j<i} a_{i,j}u^k_j + \sum_{j>i} a_{i,j}u^{k-1}_j] + \frac{f_i}{a_{i,i}}.$$

Thus in computing u_i^k, we use the updated value u_j^k for $j < i$, and use the older value u_j^{k-1} for $j > i$.

The Gauss-Seidel method can be accelerated in the following way. Having arrived at the iterate $\mathbf{u}^{k-1}$, we calculate

$$\mathbf{v}^k = B_G \mathbf{u}^{k-1} + \mathbf{g}_G.$$

But we do not use $\mathbf{v}^k$ as the next iterate. Rather we take $\mathbf{u}^k$ as a convex combination

$$\mathbf{u}^k = \omega \mathbf{v}^k + (1 - \omega)\mathbf{u}^{k-1}.$$

If $0 < \omega < 1$, then $\mathbf{u}^k$ lies on the line between $\mathbf{u}^{k-1}$ and $\mathbf{v}^k$ and we say that $\mathbf{u}^k$ is *underrelaxed*. If $\omega > 1$ then $\mathbf{u}^k$ is an extrapolation of $\mathbf{v}^k$ and $\mathbf{u}^{k-1}$, and we say that $\mathbf{u}^k$ is *overrelaxed*. It is found that for the right choice of ω, the overrelaxed scheme converges significantly faster. This is called the method of *successive overrelaxations* or SOR.

A more complete discussion of iterative methods is given in [KC].

Exercises 10.1

1. To compare the various methods, we consider the boundary value problem

    $$-u''(x) = f(x), \quad 0 < x < L,$$

    $$u(0) = u(L) = 0.$$

    Let J be a postive integer, and set $\Delta x = L/J$. Use the $(J+1)$ mesh points $x_j = j\Delta x$, $j = 0, 1, 2, \ldots, J$. Approximate u'' by the centered difference (2.32).

    (a) The unknowns are the $(J+1)$ quantities $u_0, \ldots, u_J$. Use the boundary conditions $u_0 = u_J = 0$ to reduce the system to $(J-1)$ equations in $(J-1)$ unknowns.

    (b) Write out the Jacobi iteration scheme to solve this scheme. Write a short MATLAB program to solve the system this way. Use $L = 10$. Let J and the number of iterations N be input variables. Use an mfile f.m to define the right-hand side. To begin, use the function $f(x) \equiv 1$ with exact solution $u(x) = x(10-x)/2$. The values of this u at the node points x_j are, in fact, exact solutions of the difference equations. You can use the values $u(x_j)$ to check your computed solution.

(c) With $J = 100$, how many iterations does it take for the the computed solution to come uniformly within .1 of the exact solution? How much time does this require? Use the commands `tic` and `toc`.

(d) For the same problem, write out the Gauss-Seidel iteration scheme. Write a MATLAB program to solve the system this way. Same questions as in part (b).

2. The system of exercise 1, part (a) can be written

$$T\mathbf{u} = \mathbf{f},$$

where $T = S/(\Delta x)^2$ and S is the $(J - 1) \times (J - 1)$ tridiagonal matrix

$$S = \begin{bmatrix} 2 & -1 & 0 & 0 & \ldots & 0 & 0 \\ -1 & 2 & -1 & 0 & \ldots & 0 & 0 \\ 0 & -1 & 2 & -1 & \ldots & 0 & 0 \\ . & . & . & & \ldots & 0 & 0 \\ . & . & . & & \ldots & 0 & 0 \\ 0 & 0 & 0 & 0 & 0 & -1 & 2 \end{bmatrix}.$$

MATLAB solves this system using a version of Gaussian elimination which exploits the sparse nature of the matrix. *It does not find the inverse* . To do this we must enter the matrix in the sparse format. For this matrix S, we can use the commands

```
D = sparse([1:J-1], [1:J-1], 2);
E = sparse([1:J-2], [2:J-1],-1,J-1, J-1);
S = E + D + E';
```

For more information on the sparse format, enter `help sparse`.

(a) Use the right-hand side $f(x) = \sin(\pi x/10)$. The exact solution is $u(x) = (10/\pi)^2 f(x)$. Now solve the system for $J = 100$, and compare the time with that used for the Jacobi and Gauss-Seidel methods.

(b) Solve the system for $J = 5, 10, 20, 40$. For each J compute the maximum error at the mesh points, $\max_j |u(x_j) - u_j|$. What power law governs the rate of decrease of the error as J increases?

3. Next consider the boundary value problem with Neumann boundary conditions
$$-u'' + cu = f, \quad 0 < x < L,$$
$$u'(0) = u'(L) = 0.$$

(a) What are the eigenvalues and eigenfunctions of the related problem
$-\varphi'' + c\varphi = \lambda\varphi$, $\varphi'(0) = \varphi'(L) = 0$. ?

(b) Put a mesh of points on the interval $[0, L]$ as in exercise 1. Approximate u'' with a centered difference. What is the system of equations to be solved, independent of the boundary conditions? By (2.32), the truncation error is $O(\Delta x^2)$.

(c) If we approximate the boundary conditions by a forward difference on the left and a backward difference on the right, we add the two equations

$$\frac{u_1 - u_0}{\Delta x} = \frac{u_J - u_{J-1}}{\Delta x} = 0.$$

What is the resulting system of $(J-1)$ equations in $(J-1)$ unknowns? Setting $f(x) \equiv 0$, does this system have a nontrivial solution when $c = 0$? Keep part (a) in mind.

(d) The truncation error introduced by approximating the boundary conditions by forward and backward differences is $O(\Delta x)$, which is a loss of accuracy from the $O(\Delta x^2)$ of part (b). To make a better approximation for the boundary conditions, we use a centered difference approximation (2.31), and introduce the "ghost points" (see Section 3.6) x_{-1} and x_{J+1} outside the interval. Then the boundary conditions become

$$\frac{u_{-1} - u_1}{2\Delta x} = \frac{u_{J+1} - u_{J_1}}{2\Delta x} = 0.$$

Using these two equations and the centered differences for the PDE at the points $x_0, \ldots, x_J$, construct a system of $(J + 1)$ equations in $(J + 1)$ unknowns. Can this system be solved when $c = 0$?

(e) Take $L = 10$. Write a MATLAB program using the sparse matrix format to solve this system. Make J and c input parameters. Use the right-hand side $f(x) = x + \exp(-2x^2)$. Set $J = 100$, and experiment with different values of c ranging from $c = .1$ to $c = 10$. Describe how the solution changes as c changes.

4. (a) Set up the equations for the finite difference scheme for the Poisson equation $-\Delta u = q$ in G, $u = 0$ on ∂G.

(b) Do the same for the equation with variable coefficients

$$-(a(x, y)u_x)_x - (b(x, y)u_y)_y = q(x, y) \quad \text{in } G, \qquad u = 0 \quad \text{on } \partial G.$$

5. Prove that the maximum principle holds for the discretized version (10.2) of (10.1).

6. Consider the Dirichlet problem (10.1) with G being the unit square $G = \{0 < x < 1, 0 < y < 1\}$. Let the boundary values be $f = 0$ on $x = 0$ and on $x = 1$, $f = \sin(\pi x)$ on $y = 0$ and $f = e^{\pi} \sin(\pi x)$ on $y = 1$.

    (a) Use a mesh $\Delta x = \Delta y = 1/3$ so that $I = J = 2$. Write down the 4×4 system for the difference scheme.

    (b) Taking $u_{i,j} = 0$ at all of the interior grid points as an initial guess, make four iterations using the Jacobi method.

    (c) Use the same initial, guess and make four iterations using the Gauss-Seidel method.

    (d) Find the exact solution of the 4×4 system using the MATLAB operation A\b. Compare the results of parts a) and b) with the exact solution.

    (e) Compare the values of the exact solution of the 4×4 system with the values of the exact solution $u(x, y) = \sin(\pi x)e^{\pi y}$ at the grid points.

7. Write out the explicit, finite difference scheme for the heat equation $u_t = k(u_{xx} + u_{yy})$ in the rectangle G, with $\Delta x = \Delta y$. Verify that when $k\Delta t/\Delta x^2 = 1/4$, the successive time steps for the heat equation are the same as the Jacobi iterations.

10.2 Finite elements

In section 10.1, we saw that it was possible to discretize the problem (10.1) using finite differences and arrive at a system of linear equations which approximated the problem (10.1). However, this is not easily done when the set G is not a rectangle, which happens often in practice. In this section we present another method which reduces a partial differential equation to a system of linear equations, and which is easy to adapt to quite generally shaped regions G. We shall develop the method for the Poisson equation with zero boundary condition:

$$ - \Delta u = q \quad \text{in } G, \qquad u = 0 \quad \text{on } \partial G. \qquad (10.6) $$

To illustrate the method in a simple context, we begin with a one-dimensional boundary value problem:

$$ - u'' = q \quad \text{for} \quad 0 < x < L, \qquad u(0) = u(L) = 0. \qquad (10.7) $$

Recall from Chapter 9 that a weak solution u of (10.6) satisfies the following integral relation

$$\int_0^L u'(x)v'(x)dx = \int_0^L q(x)v(x)dx \tag{10.8}$$

for all v which are piecewise C^1 with $v(0) = v(L) = 0$. Now introduce a mesh on the interval $[0, L]$ with mesh points $x_j = jh$, $j = 0, \ldots, N + 1$ where $h = L/(N + 1)$. In (10.8) we are going to substitute continuous piecewise linear functions which are joined at the mesh points. To put some structure into the problem, we define a basis of piecewise linear functions called *hat functions*. A typical hat function is shown in Figure 10.2. For $j = 1, \ldots, N$ set,

$$\varphi_j(x) = \begin{cases} 0 & \text{for } 0 \le x \le x_{j-1} \\ \frac{1}{h}(x - x_{j-1}) & \text{for } x_{j-1} \le x \le x_j \\ 1 - \frac{1}{h}(x - x_j) & \text{for } x_j \le x \le x_{j+1} \\ 0 & \text{for } x \ge x_{j+1} \end{cases}.$$

With this construction, $\varphi_j(x_i) = 0$ for $i \ne j$, and $\varphi_j(x_j) = 1$.
In the intervals where φ is linear, we can compute the derivative:

$$\varphi_j'(x) = \begin{cases} 0 & \text{for } 0 < x < x_j - 1 \\ \frac{1}{h} & \text{for } x_{j-1} < x < x_j \\ -\frac{1}{h} & \text{for } x_j < x < x_{j+1} \\ 0 & \text{for } x > x_{j+1} \end{cases}. \tag{10.9}$$

Let $v(x)$ be a linear combination of these hat functions:

$$v(x) = \sum_1^N a_j\varphi_j(x).$$

Note that $v(x_j) = a_j$ for $j = 1, \ldots, N$ and $v(0) = \varphi_1(0) = 0$, and $v(L) = \varphi_N(L) = 0$. The graph of v is generated by connecting the points (x_j, a_j) with straight lines, and hence is continuous and piecewise linear. In fact any continuous piecewise linear function with "joints" at the x_j and vanishing at the end points can be represented in this way (see Figure 10.3).

We shall seek a continuous piecewise linear approximate solution to (10.8) in the form

$$u_h(x) = \sum_1^N u_j\varphi_j(x)$$

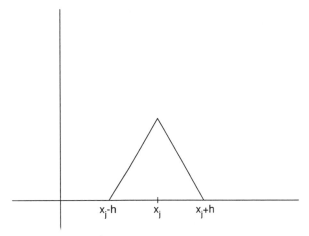

FIGURE 10.2
Typical hat function.

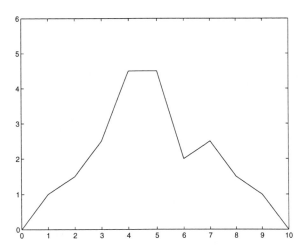

FIGURE 10.3
Piecewise linear function.

and require that (10.8) be satisfied for $v = \varphi_i$, $i = 1, \ldots, N$, that is,

$$\int_0^L u_h'(x)\varphi_i'(x)dx = \int_0^L q(x)\varphi_i(x)dx$$

for $i = 1, \ldots, N$. But

$$\int_0^L u_h'(x)\varphi_i'(x)dx = \int_0^L \sum_{j=1}^N u_j\varphi_j'(x)\varphi_i'(x)dx$$

$$= \sum_{j=1}^N a_{i,j}u_j,$$

where

$$a_{i,j} = \int_0^L \varphi_i'(x)\varphi_j'(x)dx.$$

Then the coefficients u_j must satisfy the N linear equations

$$\sum_{j=1}^N a_{i,j}u_j = \int_0^L q(x)\varphi_i(x)dx = q_i \tag{10.10}$$

for $i = 1, \ldots, N$, where $q_i = \int_0^L q\varphi_i dx$.

The matrix A with elements $a_{i,j}$ is called the *stiffness* matrix from its origin in elasticity theory. The elements of A are easily calculated using (10.9) as

$$a_{i,j} = \frac{1}{h}\begin{cases} 0 & i < j-1 \text{ or } i > j+1 \\ -1 & i = j-1 \text{ or } i = j+1. \\ 2 & i = j \end{cases}$$

The matrix A is tridiagonal and, up to a factor of h, is the same matrix we would get if we approximated $-u''$ by the centered difference

$$-u''(x) \approx \frac{-u(x+h) + 2u(x) - u(x-h)}{h^2}.$$

The resulting set of equations for the values u_j is almost the same as the one we obtained using finite differences. If we approximate q_i by

$$q_i = \int_0^L q\varphi_i dx \approx q(x_i)\int_0^L \varphi_i dx = hq(x_i),$$

we get exactly the same set of equations as in the finite difference treatment. In the one-dimensional context, finite difference methods and finite elements are equivalent.

Note that we could have chosen other basis functions φ_j, looked for an approximate solution $U_N = \sum u_j \varphi_j$, and substituted in (10.8). The advantage of using the hat functions is that when $|i - j| > 1$, the set of points where φ_i is nonzero does not overlap the set of points where φ_j is nonzero. This makes the matrix A tridiagonal so that the system (10.10) is easily solved by an LU factorization.

Now we turn our attention to the finite element method in two dimensions. We return to the boundary value problem (10.1), but we do not assume that G is a rectangle. G may be any open, bounded set in R^2 with ∂G consisting of a finite number of C^1 curves. The first step is to approximate G by a union of triangles. This step, which was not necessary in one dimension, is called *triangulation*. In this triangulation, it is assumed that if two triangles have points in common, they have either a vertex or an entire side in common. The points where the vertices of the triangles meet are called nodes (see Figure 10.4).

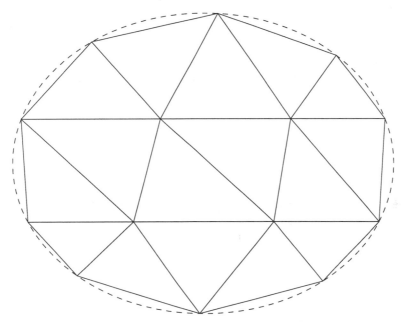

FIGURE 10.4
Triangulation of a convex region.

In the example of Figure 10.4, there are six interior nodes. More generally, we will have N interior nodes at points (x_j, y_j), $j = 1, \ldots, N$. Let S_j be the union of all the triangles which have a vertex at the jth node (x_j, y_j). Now construct a continuous piecewise linear function φ_j on each S_j as follows. We take $\varphi_j(x, y) = ax + by + c$ on each of the triangles in S_j. The coefficients a, b, c will vary from triangle to triangle and are chosen so that φ_j is continuous, $\varphi(x_j, y_j) = 1$, and $\varphi_j = 0$ on the boundary of S_j. For example, consider the region S_j shown in Figure 10.5 below.

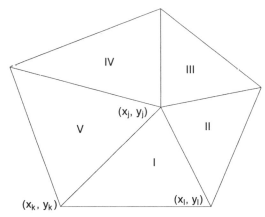

FIGURE 10.5
S_j, *the union of triangles with vertex at* (x_j, y_j).

The coefficients a, b, c of $\varphi_j(x, y)$ in the triangle denoted I in figure 10.5 are determined by the system of three linear equations

$$ax_j + by_j + c = 1$$
$$ax_k + by_k + c = 0.$$
$$ax_l + by_l + c = 0$$

A similar set equations must be solved for each of the triangles II, III, IV, and V. We can think of the graph of φ_j as a tent, pinned down around the boundary of S_j and rising to height 1 at (x_j, y_j).

A linear combination

$$v(x, y) = \sum_{j=1}^{N} c_j \varphi_j(x, y)$$

is a function on the union $S = \cup S_j$, which is continuous and linear on each of the triangles. Furthermore $v(x_j, y_j) = c_j$ and $v = 0$ on the boundary of S. We let V denote the vector space of all such linear combinations. By picking the triangulation carefully, ∂S will be a good approximation of ∂G. In particular, if ∂G has sharp curves or corners, we will need smaller triangles there because we expect the solution to vary rapidly in these areas. We measure the fineness of a particular triangulation S by letting h be largest side of all the triangles in S. To indicate the dependence on h, we shall write V_h for V.

In Chapter 9 we saw that the weak form of (10.6) is the integral relation

$$\int\int_G \nabla u \cdot \nabla v \, dx dy = \int\int_G q v \, dx dy,$$

which holds for all $v \in C^1(\bar{G})$, such that $v = 0$ on ∂G. As in one dimension, we shall seek a piecewise linear approximate solution in the form

$$u_h(x, y) = \sum_{j=1}^{N} u_j \varphi_j(x, y) \in V_h.$$

We note that $u_h(x_j, y_j) = u_j$ and $u_h = 0$ on ∂S. If we substitute this form in the integral relation and do the integration over S instead of over G, we can require that the relation hold for $v = \varphi_1, \varphi_2, \ldots, \varphi_N$. This yields the approximating problem

$$\int\int_S \nabla u_h \cdot \nabla \varphi_i \, dx dy = \int\int_S q \varphi_i \, dx dy$$

for $i = 1, \ldots, N$. Now the left-hand side is

$$\int\int_S \nabla u_h \cdot \nabla \varphi_i \, dx dy = \sum_{j=1}^{N} a_{i,j} u_j,$$

where

$$a_{i,j} = \int\int_S \nabla \varphi_j \cdot \nabla \varphi_i \, dx dy,$$

so that the approximating problem is a system of linear equations

$$\sum_{j=1}^{N} a_{i,j} u_j = q_i \qquad i = 1, \cdots, N, \tag{10.11}$$

where

$$q_i = \int\int_S q \varphi_i \, dx dy.$$

The matrix $A = [a_{i,j}]$ is again called the *stiffness matrix*. A is symmetric and positive-definite, so that the system (10.11) has a unique solution. Furthermore A is a sparse matrix because the set of points where $\varphi_j > 0$, i.e., S_j, overlaps only S_i

when the node (x_i, y_i) is a nearest neighbor of the node (x_j, y_j) and there are only a small number of these. Iterative techniques are often used to solve these sparse systems.

In what sense do the approximate solutions u_h converge to the true solution u as the triangulation gets finer and finer, i. e., as $h \to 0$? To state a precise result, we must assume that the boundary of G is a polygon, and that as $h \to 0$, the triangles do not become too skinny (the angles do not become too small).

Theorem 10.1
There are positive constants C_0 and C_1, independent of h and of the solution u of (10.6), such that

$$\|u - u_h\| \le C_0 h^2 \|u\|_2 \tag{10.12}$$

and

$$\|u - u_h\|_1 \le C_1 h \|u\|_2. \tag{10.13}$$

The quantities involved here are norms. For a function $f(x, y)$ on an open set G, they are defined by

$$\|f\|^2 = \int\int_G |f(x, y)|^2 dxdy,$$

$$\|f\|_1^2 = \|f\|^2 + \int\int_G |\nabla u|^2 dxdy,$$

and

$$\|f\|_2^2 = \|f\|_1^2 + \int\int_G [|f_{xx}|^2 + |f_{xy}|^2 + |f_{yy}|^2]dxdy.$$

Each of these norms, called Sobolev norms, is a way of measuring the magnitude of a function. $\| \ \|$ without a subscript is the same norm used in Section 4.2 to discuss the convergence of the eigenfunction expansions. The norms $\| \ \|_1$ and $\| \ \|_2$ have the same properties (1), (2), and (3) following equation (4.15), but take into account the behavior of the derivatives of the function as well. The Sobolev norms are widely used in the study of partial differential equations.

A good introducton to the finite element method is given in [Johnson]. More advanced treatments are found in [BaS] and [BS].

Finally we remark that MATLAB has a quite sophisticated toolbox which employs the finite element method and Galerkin method (described in the next section) to solve BVP's and IBVP's in quite arbitrary domains. The PDE toolbox has a very easy to use graphical user interface which allows one to specify the region in which the equation is to be solved, the equation, and the boundary conditions.

Exercises 10.2

1. Consider the BVP in one dimension

$$-u'' = q, \quad 0 < x < L, \quad u'(0) = u'(L) = 0.$$

   (a) Show that if u is a strict solution of this BVP and $v \in C^1[0, L]$, then

   $$\int_0^L u'(x)v'(x)dx = \int_0^L q(x)v(x)dx.$$

   (b) Show that if u is a C^2 function and satisfies this relation for all $v \in C^1[0, L]$, then u is a solution of the BVP.

   (c) Derive the system of linear equations for the coefficients in a finite element approximation.

2. Verify that the stiffness matrix A is positive-definite.

3. Find a formula for the area of the triangle T with vertices (x_1, y_1), (x_2, y_2), and (x_3, y_3).

4. Solve the set of three equations for the coefficients a, b, c for one sloping plane of the tent function φ_j (see Figure 10.5). The formula should express a, b, c in terms of the coordinates (x_j, y_j), (x_k, y_k), and (x_l, y_l) of the vertices.

5. Let G be the unit square $\{0 \le x \le 1, 0 \le y \le 1\}$. Create a triangulation of G with the three lines

$$y = 1/2, \quad y = 2x, \quad y = 2 - 2x.$$

   There are ten triangles and two interior nodes at $(1/4, 1/2)$ and $(3/4, 1/2)$. Let φ_1 be the tent function such that $\varphi_1(1/4, 1/2) = 1$, and let φ_2 be the tent function such that $\varphi_2(3/4, 1/2) = 1$.

   (a) Find formulas for φ_1 and φ_2. Calculate $\int \int_G \varphi_j dx dy$.

   (b) Calculate the 2×2 stiffness matrix A for this pair of tent functions.

   (c) Consider the boundary value problem

   $$-\Delta u = 1 \text{ in } G, \quad u = 0 \text{ on } \partial G.$$

Using the results of parts (a) and (b), write out the system of two
equations (10.11) for this problem. Solve the system for u_1 and u_2.
The linear combination $u_1\varphi_1(x, y) + u_2\varphi_2(x, y)$ is the finite element
approximation to the solution of this boundary value problem.

10.3   Galerkin methods

The term Galerkin method refers to a wide class of methods in which one seeks
to approximate solutions of differential or integral equations by finite linear combinations of basis functions with desirable properties. This was done using the
finite element method in section 10.2 for solutions of the Poisson equation. However the term also refers to methods for time-dependent equations in which the
finite-dimensional approximation is done in the spatial variable, thus reducing the
PDE to a system of ODE's. This is the sense in which we shall use the term.

Let us consider the heat equation in two space variables. For definiteness we
consider the following IBVP for $u = u(x, y, t)$.

$$u_t - \Delta u = q \quad \text{in } G \times (0, \infty), \qquad u = 0 \text{ on } \partial G \times [0, \infty), \qquad (10.14)$$

$$u(x, y, 0) = f(x, y) \quad \text{in } G,$$

where G is an open, bounded set in R^2 with piecewise C^1 boundary ∂G.

If G is a rectangle, we could introduce a rectangular mesh and approximate
all the derivatives in (10.14) by difference quotients. We would arrive at a finite
difference scheme of the kind discussed in Chapter 3. Instead of pursuing the
finite difference approach, we shall see how the the finite element method can be
employed to solve (10.14) numerically in more general domains G.

We shall formulate a notion of weak solution for the problem (10.14) in a manner
similar to that used in Section 10.2 for the Poisson equation. Let $v \in C^1(\bar{G})$ with
$v = 0$ on ∂G. Multiply the PDE in (10.14) by v, and integrate over G:

$$\int\int_G u_t v \, dx \, dy - \int\int_G \Delta u v \, dx \, dy = \int\int_G q v \, dx \, dy.$$

Using Green's theorem and the fact that $v = 0$ on ∂G, we find that

$$-\int\int_G \Delta u v \, dx \, dy = \int\int_G \nabla u \cdot \nabla v \, dx \, dy.$$

Thus we see that if u is a solution of (10.14), then

$$\int\int_G u_t v \, dx dy + \int\int_G \nabla u \cdot \nabla v \, dx dy = \int\int_G q v \, dx dy \qquad (10.15)$$

for all $v \in C^1(\bar{G})$ with $v = 0$ on ∂G. Notice that a function u that satisfies (10.15) does not have to possess two continuous derivatives because we have put one derivative on the function v. This will allow us to construct approximations to u that are linear combinations of tent functions, which are only piecewise C^1.

Now we shall discretize the problem in the spatial variables. Suppose that $S = \cup S_j$ is a triangulation of G with a corresponding basis of tent functions $\varphi_j(x, y)$. Let N be the number of interior nodes, and let h be the maximum length of the side of a triangle in S. We look for an approximate solution in the form

$$u_h(x, y, t) = \sum_{j=1}^{N} u_j(t) \varphi_j(x, y).$$

Substituting this expression in (10.15) and taking $v = \varphi_i$, we generate the set of equations

$$\int\int_G \sum_j u_j'(t) \varphi_j \varphi_i + \int\int_G \sum_j u_j(t) \nabla \varphi_j \cdot \nabla \varphi_i \, dx dy = \int\int_G q \varphi_i \, dx dy \equiv q_i$$

for $i = 1, \ldots, N$. We see that the N vector $\mathbf{u}(t) = (u_1(t), \ldots, u_N(t))$ satisfies the linear system of ODE's

$$M \frac{d\mathbf{u}}{dt} + A\mathbf{u} = \mathbf{q}, \qquad (10.16)$$

where $\mathbf{q}(t) = (q_1(t), \ldots, q_N(t))$. As before, the matrix A has elements

$$a_{i,j} = \int\int_G \nabla \varphi_i \cdot \nabla \varphi_j \, dx dy.$$

The matrix M has elements

$$m_{i,j} = \int\int_G \varphi_i \varphi_j \, dx dy.$$

Both A and M are symmetric, positive-definite matrices. Even in this context of diffusion, A is called the stiffness matrix, and M is called the *mass matrix*.

We must choose initial conditions for (10.16) which are consistent with the initial condition $u(x, y, 0) = f(x, y)$ and the form of our approximation for u. For each $i = 1, \ldots, N$,

$$\int\int_G u(x, y, 0)\varphi_i(x, y)\,dxdy = \int\int_G f(x, y)\varphi_i(x, y)\,dxdy \equiv f_i.$$

Then substitute the form of the approximation

$$u_h(x, y, 0) = \sum_j u_j(0)\varphi_j(x, y)$$

into the left side of the previous equation, and keep the equality. This yields

$$M\mathbf{u}(0) = \mathbf{f}, \tag{10.17}$$

where $\mathbf{f} = (f_1, \ldots, f_N)$.

We have used the finite element in the spatial variables to reduce the original PDE to a system of ODE's. Notice that we can think of this procedure as a generalization of the technique of separation of variables that we saw in Chapter 8. Instead of using a basis of eigenfunctions of the Laplace operator, we are using the finite element basis. When we use the eigenfunction expansions of Chapter 8, the resulting system of ODE's is completely decoupled, and the solution of each equation is a decaying exponential. The difficult part of the problem, computationally, is to find the eigenfunctions. The finite element approach avoids this difficulty but leads to the coupled system (10.16), which may not be easy to solve accurately.

What can we say analytically about the system (10.16) when $\mathbf{q} = 0$? We can look for special solutions of the form

$$\mathbf{u}(t) = \mathbf{w}e^{-\lambda t}. \tag{10.18}$$

We find that a function of the form (10.18) is a solution of (10.16) if

$$(A - \lambda M)\mathbf{w} = 0. \tag{10.19}$$

Because A and M are both symmetric and positive-definite, there will be a basis in R^N of eigenvectors $\mathbf{w}_j$, and the general solution of (10.16) will be given by

$$\sum_{j=1}^{N} \alpha_j \mathbf{w}_j e^{-\lambda_j t},$$

where $(A - \lambda_j M)\mathbf{w}_j = 0$. We must chose the coefficients α_j, so that the initial condition (10.17) is satisfied.

Thus we have reduced the original PDE, first to a system of ODE's, and then to an eigenvalue problem for $N \times N$ matrices. However we must bear in mind that N

may be quite large, so that finding the eigenvectors and eigenvalues of (10.19) is no small task. It is also true that the eigenvalues will be spread over a wide range of values. In fact the ratio of the largest eigenvalue to the smallest eigenvalue of the system (10.19) is on the order of h^{-2} for a typical triangulation. This means that the basic solutions (10.18) decay on very different time scales. Again using terminology from mechanics, the system (10.16) is said to be a *stiff* system.

Instead of trying to find the eigenvectors and eigenvalues of the system (10.16), we can use numerical methods to integrate (10.16). Because the system is stiff, explicit methods, such as the Euler method, can require a very small time step to be accurate. For an introductory discussion of the numerical problems associated with stiff systems, see [KC].

We will use an implicit method which has better stability properties. We continue to assume $\mathbf{q} = 0$ in (10.16). Let $\mathbf{u}_n$ denote the approximate value of $\mathbf{u}$ at time $t = t_n$. If we replace $\mathbf{u}'(t)$ in (10.16) by a backward difference,

$$\mathbf{u}'(t) \approx \frac{\mathbf{u}(t_n) - \mathbf{u}(t_{n-1})}{\Delta t},$$

this leads to a backward Euler scheme for (10.16):

$$M\mathbf{u}_n - M\mathbf{u}_{n-1} + \Delta t\, A\mathbf{u}_n = 0.$$

A more accurate implicit method is obtained by averaging the forward and backward Euler methods to obtain the Crank-Nicolson (trapezoid) scheme:

$$M\mathbf{u}_n - M\mathbf{u}_{n-1} + \frac{\Delta t}{2} A(\mathbf{u}_n + \mathbf{u}_{n-1}) = 0.$$

Now for each time step, we must solve for $\mathbf{u}_n$ in the system

$$(M + \frac{\Delta t}{2} A)\mathbf{u}_n = (M - \frac{\Delta t}{2} A)\mathbf{u}_{n-1}.$$

Many sophisticated methods are available to integrate stiff systems, e.g., the Gear methods.

Remark: Spectral Methods

The Galerkin method, as developed here, does not have to use the finite element basis functions. Any other suitable choice of basis functions φ_j that satisfy the boundary conditions yields the system of ODE's (10.16), although the mass matrix and the stiffness matrix may be different.

In particular, if the domain G is a rectangle, one may choose a family of orthogonal polynomials as the basis functions. The mass matrix and the stiffness matrix will no longer be sparse, but the rapid convergence allows one to take fewer terms in the expansion. The orthogonal polynomials may be thought of as extending the notion of eigenfunctions. For this reason, the Galerkin method in this context is called a *spectral method*. An introduction to spectral methods is given in [GO]. A very interesting introduction to Galerkin methods and mathematical modelling is given in the text [EEHJ].

Exercises 10.3

1. Reduce the discussion of the Galerkin method for the heat equation to the one dimensional setting by considering the IBVP

    $$u_t - u_{xx} = q, \ 0 < x < L, \quad u(x,0) = f(x),$$

    $$u(0,t) = u(L,t) = 0.$$

    Use the piecewise linear hat functions of Section 10.2. The stiffness matrix A has already been computed there.

    (a) Compute the mass matrix M for the basis of hat functions.

    (b) Write out the system (10.16), (10.17) of ODE's.

    (c) Take $L = 10$, $N = 9$. Write a MATLAB program to solve the system of nine ODE's using the Crank-Nicolson scheme. Make the time step Δt an input variable as well as the number of time steps. Use initial data $f(x) = \sin(\pi x/10)$ and $q = 0$ which has an easily found exact solution. Now increase N to get a better approximation.

2. A Galerkin approximation to solutions of a wave equation. Consider the IBVP in one dimension

    $$u_{tt} - u_{xx} + r(t)u = 0, \ 0 < x < L, \quad u(0,t) = u(0,L) = 0,$$

    $$u(x,0) = f(x), \ u_t(x,0) = g(x), \ 0 < x < L.$$

    Here $r(t)$ is a given continuous function. An interesting choice for $r(t)$ would be a periodic function. Let $\varphi_j(x)$ be a sequence of basis functions (such as the piecewise linear hat functions), such that $\varphi_j(0) = \varphi_j(L) = 0$ for all j.

    (a) Show that if u is a solution of the IBVP, then for each i,

    $$\int_0^L u_{tt}(x,t)\varphi_i(x)dx + \int_0^L u_x(x,t)\varphi_i'(x)dx$$

$$+ r(x) \int_0^L u(x, t)\varphi_i(x)dx = 0.$$

(b) Now approximate u by a linear combination of the basis functions

$$u(x, t) \approx u_h(x, t) = \sum_{j=1}^N u_j(t)\varphi_j(x).$$

Substitute this approximation for u in the left side of the previous equation and set the result to zero. Show that the vector valued function $\mathbf{u}(t) = (u_1, \ldots, u_N)$ satisfies the system of ODE's

$$M\mathbf{u}''(t) + A\mathbf{u}(t) + r(t)M\mathbf{u}(t) = 0,$$

where the stiffness matrix A and the mass matrix M are as before.

(c) Show that a consistent choice of initial conditions for $\mathbf{u}$ is

$$M\mathbf{u}(0) = \mathbf{f}, \quad M\mathbf{u}'(0) = \mathbf{g},$$

where the ith component of $\mathbf{f}$ is $\int_0^L f\varphi_i dx$ and the ith component of $\mathbf{g}$ is $\int_0^L g\varphi_i dx$.

(d) Use the stiffness matrix and mass matrix for the basis of piecewise linear hat functions as computed in exercise 1. Write out the second-order system of equations for this problem. Make this into a first-order system of $2N$ equations in $2N$ unknowns by adding the new variables $v_i(t) = u_i'(t)$.

A pseudospectral method for the KdV equation

The Korteweg-deVries equation (KdV) arises in the study of shallow water waves, in particular, long waves moving down a channel. It combines the "shocking up" effect of a nonlinear conservation law with a third derivative dispersive term,

$$u_t + uu_x + \gamma u_{xxx} = 0. \tag{10.20}$$

The KdV equation has been studied extensively (see [Wh]). Linearization about the constant solution $u \equiv 1$ yields the linear KdV equation studied in Chapter 7. Here we present a numerical technique to treat this equation. We shall compute approximate numerical solutions that have period 2π in x, using a spectral method. We assign initial data

$$u(x, 0) = f(x),$$

where f has period 2π. Since we seek periodic functions, it is natural to expand the desired solution in terms of complex Fourier series. We write the nonlinear term as $v = u u_x$, and expand

$$u(x,t) = \sum_{-M}^{M} u_k(t)e^{ikx}, \qquad v(x,t) = \sum_{-M}^{M} v_k(t)e^{ikx}.$$

Then

$$u_t(x,t) = \sum u_k'(t)e^{ikx},$$

and

$$\gamma u_{xxx} = \gamma \sum (ik)^3 u_k(t)e^{ikx}.$$

Substituting these expressions in the PDE (10.20), and equating the Fourier coefficients, we arrive at the $(2M+1)$ ODE's

$$u_k'(t) + v_k(t) + \gamma(ik)^3 u_k(t) = 0, \tag{10.21}$$

$$u_k(0) = f_k,$$

where f_k is the kth Fourier coefficient of the initial data f. This is really a coupled system of equations because v_k depends on u and u_x and hence on all the Fourier coefficients, $l = -M, \ldots, M$. Forgetting for moment the nonlinear term, the linear system

$$u_k'(t) + \gamma(ik)^3 u_k(t) = 0$$

can be solved exactly

$$u_k(t) = e^{-\gamma(ik)^3 t} f_k, \quad k = -M, \ldots, M.$$

Over a wide range of k, the speed of oscillation varies greatly. This suggests that the nonlinear system (10.21) would be difficult to integrate numerically. For this reason, we "precondition" the system by making a change of dependent variable which factors out the rapid oscillations. We set

$$w_k(t) = e^{\gamma(ik)^3 t} u_k(t).$$

Then

$$w_k'(t) = F(w(t)), \quad w_k(0) = f_k, \tag{10.22}$$

where $w = (w_k)$ is the vector of coefficients. The function $F(w)$ is now fairly complicated, described by the sequence of operations

$$w_k \to u_k = e^{-\gamma(ik)^3 t} w_k, \qquad (10.23)$$

$$u_k \to u = \sum u_k e^{ikx}, \quad iku_k \to u_x = \sum iku_k e^{ikx},$$

$$uu_k = v \to F_k = -e^{\gamma(ik)^3} v_k.$$

If the coefficients u_k, v_k, w_k are the usual Fourier coefficients determined by integrals $\int u \exp(-ikx)dx$, this would be called a Galerkin spectral method. However, to exploit the speed of the FFT, we shall instead use the DFT coefficients. The method is now called a *pseudospectral method*. For an interesting application of the pseudospectral method, see the article [HLS].

Program kdv prepares the data, solves the system of ODE's (10.21), and then reconstructs the solution at the desired viewing time T. The function mfile wwprime.m implements the sequence of operations (10.23) using the FFT of MATLAB. Initial data is given in the mfile f.m. The file kdv.m must be modified according to which version of MATLAB you are using, MATLAB4.2 or MATLAB5.0. Enter help kdv

3. (a) What is the dispersion relation for the linear equation $u_t + \gamma u_{xxx} = 0$? Which wavelengths travel faster, longer or shorter?

   (b) When $\gamma = 0$, the equation becomes the nonlinear conservation law (2.17). The solutions of (2.17) developed shock waves when the initial data $f(x)$ was decreasing. Why will the dispersion term u_{xxx} prevent that?

   (c) Choose the initial data for kdv to be $f(x) = \cos x$. Run program kdv with $\gamma = 1$, $N = 64$, and $T = 1$. With the student edition you may have to use $N = 32$ to avoid the limitation on matrix size. Now run again with smaller values for γ, eventually getting down to $\gamma = 0$. Describe what you see. What happens to the solution when $\gamma = 0$? When $\gamma = 0$, to what discontinuous wave form should the exact solution converge as t increases?

   (d) Change the initial data to $f(x) = \exp(-2(x - \pi)^2)$. Again run with $\gamma = 1$, $N = 64$, and $T = 1$. How do you explain the small ripples that develop at the left (tail) side of the wave as t increases?

10.4   A reaction-diffusion equation

In this section we show how the Galerkin method can be applied to the diffusion equation with a nonlinear source term. We begin in the one-dimensional setting. Suppose that we have a cell culture in solution in a tube of liquid. We assume that the cell population is uniform over a cross section of the tube. Let $u(x, t)$ denote the linear density of cells at location x at time t so that u has units of cells/length. Recalling the discussion at the beginning of Chapter 3, we see that

$$\int_b^a u(x, t)dx$$

is the number of cells in the tube in the interval $[a, b]$. We shall write the flux through a cross section at x at time t as $F(x, t)$. F has units cells/time. Then our fundamental balance law is expressed as

$$\frac{d}{dt} \int_a^b u(x, t)dx = F(a, t) - F(b, t) + \text{ source }.$$

As usual we adopt the convention that $F \geq 0$ when the flow of cells is from left to right, and $F \leq 0$ otherwise. Under the usual smoothness assumptions, we derive a differential equation

$$u_t + F_x = q(x, t, u), \tag{10.24}$$

where the source has been written in the form of a function q. We are allowing q to depend on the cell density u and on x and t. To get a PDE in the single function u, we assume the Fickian law of diffusion

$$F = -Du_x, \tag{10.25}$$

where $D > 0$ is the diffusion coefficient. This equation says that cells will diffuse from high concentration to low concentration. With this relation the PDE (10.24) becomes

$$u_t - Du_{xx} = q(x, t, u). \tag{10.26}$$

This is an example of a *reaction-diffusion* equation. In other settings, such as combustion, the right-hand side q models the production and absorption of material due to chemical reaction. In our case of cell growth, we will take q in the form

$$q(u) = ru(1 - u/K).$$

Here r is the growth rate of the cells, and K is the carrying capacity of the environment (in our case the solution in the tube).

A basic ODE of population growth is the *logistics* equation

$$\psi'(t) = r\psi(t)(1 - \psi(t)/K), \quad \psi(0) = \psi_0, \tag{10.27}$$

which has solution

$$\psi(t) = \frac{K\psi_0}{\psi_0 + (K - \psi_0)e^{-rt}}.$$

When the initial data ψ_0 is taken in the realistic range $0 < \psi_0 < K$, the solution tends to the carrying capacity as $t \to \infty$. $\psi(t) \equiv 0$ is an unstable solution, and $\psi(t) \equiv K$ is an asymptotically stable solution.

The reaction-diffusion equation

$$u_t - Du_{xx} = ru(u - u/K) \tag{10.28}$$

combines the diffusive properties of the heat equation with the growth properties of the logistics equation. Equation (10.28) is known as Fisher's equation. It is possible to find travelling wave solutions of (10.28) as in Chapter 3, Section 5. We shall instead consider an IBVP by studying (10.28) on the interval $(0, L)$, where we impose the conditions

$$u(0, t) = u(L, t) = 0, \quad t \geq 0, \tag{10.29}$$

$$u(x, 0) = f(x).$$

We know that all solutions of the heat equation with the boundary conditions (10.29) decay exponentially to zero as $t \to \infty$. However, now we add on the source term $q = ru(u - u/K)$, and we also know that all solutions of the ODE (10.27) grow exponentially when ψ is small. Which of these effects will dominate in the solutions of (10.28), (10.29)? In fact for L sufficiently large, there are non-trivial steady-state solutions, but they are not all stable.

To show how the Galerkin method can be applied to a reaction-diffusion equation, we consider a higher dimensional analogue of (10.28), (10.29). Let G be an open, bounded set of R^2 with boundary a polygon. Let $D(x, y) = [d_{i,j}(x, y)]$ be a matrix such that at each $(x, y) \in G$, D is symmetric and positive-definite. D is called the diffusion matrix. We shall study the IBVP in G

$$u_t - \nabla \cdot (D\nabla u) = q \quad \text{in } G, \tag{10.30}$$

$$u = 0 \quad \text{on } \partial G,$$

$$u(x, y, 0) = f(x, y) \quad \text{in } G.$$

We assume only that the reaction term q has the form

$$q = q(x, y, t, u).$$

How can we use the finite element method to generate numerical solutions of (10.30)? The weak form of (10.30) is

$$\int\int_G u_t v \, dx \, dy + \int\int_G (D\nabla u) \cdot \nabla v \, dx \, dy = \int\int_G q v \, dx \, dy \qquad (10.31)$$

which holds for all $v \in C^1(\bar{G})$ with $v = 0$ on ∂G. Now let $S = \cup S_j$ be a triangulation of G, and φ_j a corresponding basis of tent functions. As we did in Section 10.3, let us seek an approximate solution in the form

$$u_h(x, y, t) = \sum u_j(t)\varphi(x, y).$$

Substituting this form in (10.31) and taking $v = \varphi_i, i = 1, \ldots, N$, we arrive at the nonlinear system of ODE's

$$\frac{d}{dt} M\mathbf{u} + A\mathbf{u} = \mathbf{q}. \qquad (10.32)$$

The elements of the matrix A are

$$a_{i,j} = \int\int_G (D\nabla\varphi_i) \cdot \nabla\varphi_j \, dx \, dy.$$

As before the elements of the matrix M are

$$m_{i,j} = \int\int_G \varphi_i\varphi_j \, dx \, dy.$$

The components of the vector function $\mathbf{q} = (q_1(t, \mathbf{u}), \ldots, q_N(t, \mathbf{u}))$ are obtained from the integrations

$$q_i(t, \mathbf{u}) = \int\int_G q(x, y, t, \sum_j u_j(t)\varphi_j(x, y))\varphi_i(x, y) \, dx \, dy.$$

The matrices A and M are again symmetric and positive-definite.

A new question arises now that we did not have to consider when the system of equations was linear. The system of ODE's (10.32) is nonlinear, and one does not know a priori that solutions exist on a given time interval. It is possible that solutions blow up in finite time, depending on the initial data. This question can usually be settled by going back to the original PDE and making estimates on the solutions which imply that they cannot blow up in finite time. These estimates in turn depend crucially on the nature of the nonlinearity.

Reaction-diffusion equations have been studied extensively, see [Sm]. For a discussion of Fisher's equation in the biological context, see [E].

Exercises 10.4

1. Show that the matrix A is positive-definite using the properties of the diffusion matrix D.

2. What is the Crank-Nicolson method for (10.28)? What set of nonlinear equations must be solved at each time step?

3. Use a finite element basis and the Galerkin method to derive a system of second-order ODE's in t which approximate solutions of the nonlinear wave equation

$$u_{tt} - \Delta u + u^3 = 0 \quad \text{in } G, \tag{10.33}$$

$$u = 0 \qquad \text{on } \partial G,$$

$$u(x, y, 0) = f(x, y), \quad u_t(x, y, 0) = g(x, y) \quad \text{in } G.$$

4. Show that solutions of the nonlinear system of ODE's that corresponds to (10.33) exist for all time, regardless of the initial data, by making energy estimates for solutions of (10.33).

Chapter 11

Epilogue: Classification

In Section 2.1 we indicated that there is a way of grouping second order PDE's into classes, such that the solutions of PDE's in the same class have similar qualitative properties. We shall discuss this classification now and relate it to our physical classification into equations which describe diffusion, wave propagation, and equilibrium situations. We let

$$Lu = au_{xx} + 2bu_{xy} + cu_{yy} + du_x + eu_y + fu \qquad (11.1)$$

be the general second-order operator in two variables with constant coefficients. For convenience in this discussion, we shall assume that $a > 0$. The characteristic polynomial associated with L is

$$q(x, y) = ax^2 + 2bxy + cy^2 + dx + ey + f. \qquad (11.2)$$

We know from the theory of quadratic forms that by making a linear change of coordinates,

$$\xi = \alpha x + \beta y$$

$$\eta = \gamma x + \delta y,$$

we can bring the polynomial q into the following form, depending on the the sign of the quantity $ac - b^2$.

(1) $ac - b^2 > 0$ ellipse
$$\tilde{q}(\xi, \eta) = \xi^2 + \eta^2 + \text{lower order}$$

(2) $ac - b^2 = 0$ parabola
$$\tilde{q}(\xi, \eta) = \xi^2 + \text{lower order}$$

(3) $ac - b^2 < 0$ hyperbola
$$\tilde{q}(\xi, \eta) = \xi^2 - \eta^2 + \text{lower order}$$

The same change of coordinates, when applied to the differential operator L, leads to a similar classification. Set $u(x, y) = v(\alpha x + \beta y, \gamma x + \delta y)$. Then using the chain rule, we see that the differential operator L in the (ξ, η) coordinates becomes

$$\tilde{L}v = \tilde{a}v_{\xi\xi} + 2\tilde{b}v_{\xi\eta} + \tilde{c}v_{\eta\eta} + \text{lower order} , \qquad (11.3)$$

where

$$\tilde{a} = a\alpha^2 + 2b\alpha\beta + c^2\beta^2$$

$$\tilde{b} = a\alpha\gamma + b(\alpha\delta + \beta\gamma) + c\beta\delta$$

$$\tilde{c} = a\gamma^2 + 2b\gamma\delta + c\delta^2.$$

Let A and $\tilde{A}$ be the matrices

$$A = \begin{bmatrix} a & b \\ b & c \end{bmatrix} \qquad \tilde{A} = \begin{bmatrix} \tilde{a} & \tilde{b} \\ \tilde{b} & \tilde{c} \end{bmatrix}.$$

By studying the formulas for $\tilde{a}, \tilde{b}, \tilde{c}$, we see that

$$\tilde{A} = B^t A B,$$

where

$$B = \begin{bmatrix} \alpha & \gamma \\ \beta & \delta \end{bmatrix} \qquad \text{and} \qquad B^t = \begin{bmatrix} \alpha & \beta \\ \gamma & \delta \end{bmatrix}.$$

We note that the sign of $ac - b^2$ is preserved under this change of coordinates. In fact, from the rule for determinants,

$$\det(AB) = (\det A)(\det B),$$

valid for any two square matrices A, B of the same size, we see that

$$\tilde{a}\tilde{c} - \tilde{b}^2 = \det\tilde{A} = \det(B^t A B) = (\det B)^2 \det A$$

$$= (\alpha\delta - \beta\gamma)^2(ac - b^2)$$

Of course, we assume that the change of variable maps the (x, y) plane onto all of the (ξ, η) plane so that the determinant of B is nonzero.

Now we want to choose the change of variable so that $\tilde{A}$ is diagonal. However, A is real and symmetric, and hence has two orthogonal eigenvectors. If we choose the column vector $[\alpha, \beta]^t$ as one of the eigenvectors, with eigenvalue λ_1, and $[\gamma, \delta]^t$ as the other, with eigenvalue λ_2, then $\tilde{b} = 0$ and

$$\tilde{A} = \begin{bmatrix} \lambda_1(\alpha^2 + \beta^2) & 0 \\ 0 & \lambda_2(\gamma^2 + \delta^2) \end{bmatrix}.$$

The eigenvalues of A are given by the formulas

$$\lambda_1 = \frac{a + c + \sqrt{(a + c)^2 - 4(ac - b)^2}}{2a}$$

$$\lambda_2 = \frac{a + c - \sqrt{(a + c)2 - 4(ac - b^2)}}{2a}.$$

At least one of λ_1, λ_2 is nonzero, for otherwise A would be zero. Since we have assumed that $a > 0$, $ac - b^2 > 0$ if and only if $c > 0$ as well. Hence, from the formulas for λ_1 and λ_2, it follows that

$$ac - b^2 > 0 \qquad \text{if and only if } \lambda_1, \lambda_2 > 0,$$

$$ac - b^2 = 0 \qquad \text{if and only if } \lambda_1 > 0, \lambda_2 = 0,$$

$$ac - b^2 < 0 \qquad \text{if and only if } \lambda_2 < 0 < \lambda_1.$$

This means that in the first case $\tilde{a}, \tilde{c} > 0$, in the second case $\tilde{a} > 0$ and $\tilde{c} = 0$, and in the third case, $\tilde{c} < 0 < \tilde{a}$. We can make a further scaling of the independent variables to bring the operator into the following form;

(1) $ac - b^2 > 0$ \qquad $v_{\xi\xi} + v_{\eta\eta} +$ lower order

(2) $ac - b^2 = 0$ \qquad $v_{\xi\xi} +$ lower order

(3) $ac - b^2 < 0$ \qquad $v_{\xi\xi} - v_{\eta\eta} +$ lower order .

Borrowing the terminology from the classification of quadratic polynomials, we say that a differential operator (11.1) is

elliptic \qquad when $ac - b^2 > 0$

parabolic \qquad when $ac - b^2 = 0$

hyperbolic \qquad when $ac - b^2 < 0$.

In the second case, it may happen that there is no term v_η in the operator $\tilde{L}$, which means that $\tilde{L}v = 0$ is simply an ODE. When this happens, we say that $Lu = 0$ is degenerate.

The Laplace equation $\Delta u = 0$ is, of course, the model example of an elliptic equation. Elliptic equations usually model equilibrium situations. Diffusion processes are usually modelled by a parabolic equation like the heat equation $u_t - ku_{xx} = 0$. In the elliptic and parabolic case, the solutions of $Lu = 0$ obey a maximum principle and they are infinitely differentiable. This terminology also extends to certain systems of equations.

The wave equation $u_{tt} - c^2 u_{xx} = 0$ is the standard example of a hyperbolic equation. This type of equation models waves or signals which have a finite speed of propagation. Singularities in the solution are propagated: they are not smoothed out. The term hyperbolic is also applied to the first-order differential equations we studied in Chapter 2, and to systems of first-order equations. In this context hyperbolic means that information is propagated with a finite speed.

For an equation whose coefficients $a, b, c \ldots$ depend on the variables x, y, the classification is done at each point (x, y). There are equations of changing type which are hyperbolic at some points and elliptic at other points. This occurs in the study of supersonic flow past an aircraft. See the reference [CF].

Many important equations do not fit this classification. An example is the Schrödinger equation studied in Chapter 7 which combines some properties of parabolic and hyperbolic equations.

Appendix A

Recipes and Formulas

For easy reference this appendix provides a summary of the technique of separation of variables and formulas for the fundamental solutions of important equations. Details of the calculations in specific examples are found in earlier chapters.

A.1 Separation of variables in space-time problems

Separation of variables is a technique used to reduce the problem of solving a PDE to solving problems of lower dimension, ultimately to one-dimensional problems. The applicability of the technique depends on the form of the equation and the geometry of the region. Often it is possible to separate the variables of space and time. This is the case if the problem has the form

$$N(t, \partial_t)u + M(\mathbf{x}, \partial_{\mathbf{x}})u = 0 \qquad \text{in } G, \tag{A.1}$$

$$B(\mathbf{x}, \partial_{\mathbf{x}})u = 0 \qquad \text{on } \partial G.$$

Here $\mathbf{x}$ lies in G where G is an open, bounded set in R, R^2, or R^3. $N(t, \partial t)$ is an operator involving $\partial/\partial t$ with coefficients depending possibly on t, and $M(\mathbf{x}, \partial_{\mathbf{x}})$ is an operator involving the spatial derivatives $\partial/\partial x_i$, with coefficients depending possibly on $\mathbf{x}$. B is a boundary operator that may involve coefficients depending on $\mathbf{x}$ and derivatives $\partial/\partial x_i$ on the boundary ∂G. We shall assume that equations (A.1) are linear and homogeneous. For example, we might have

$$N(t, \partial_t)u = a(t)u_{tt} + b(t)u_t + c(t)u,$$

$$M(\mathbf{x}, \partial_{\mathbf{x}})u = -div(c(\mathbf{x})\Delta u),$$

and

$$Bu = \frac{\partial u}{\partial n} + c(\mathbf{x})u,$$

where $\partial/\partial n$ is the exterior normal derivative at the boundary.

A solution to (A.1) can be sought in the form $u(x,t) = \varphi(\mathbf{x})\psi(t)$. If we substitute in (A.1) we see that

$$[N(t, \partial_t)\psi]\varphi + [M(\mathbf{x}, \partial_\mathbf{x})\varphi]\psi = 0,$$

whence

$$\frac{M(\mathbf{x}, \partial_\mathbf{x})\varphi}{\varphi} = -\frac{N(t, \partial_t)\psi}{\psi} = \lambda.$$

This yields the two equations

$$N(t, \partial_t)\psi = -\lambda\psi \tag{A.2}$$

$$M(\mathbf{x}, \partial_\mathbf{x})\varphi = \lambda\varphi. \tag{A.3}$$

The boundary condition $Bu = 0$ will be satisfied for a product solution $u(\mathbf{x}, t) = \varphi(\mathbf{x})\psi(t)$ if φ satisfies the boundary condition

$$B\varphi = 0 \quad \text{on } \partial G. \tag{A.4}$$

In this case (A.3) and (A.4) constitute an eigenvalue problem. It may be possible to find a sequence λ_n of eigenvalues and a complete orthonormal set of eigenfunctions $\varphi_n(\mathbf{x})$ which satisfy (A.3) and (A.4). For example, in one dimension, the eigenvalue problem

$$-\varphi''(x) = \lambda\varphi(x), \qquad 0 < x < L,$$

$$\varphi(0) = \varphi(L) = 0,$$

has solutions

$$\varphi_n(x) = \sin(n\pi x/L), \qquad \lambda_n = (n\pi/L)^2.$$

For further discussion of these eigenvalue problems in one dimension, see Chapter 4.

We find a corresponding sequence of solutions $\psi_n(t)$ of (A.2) by substituting the values λ_n in (A.2). If the ODE (A.2) has order m, then ψ_n will depend on

m parameters which must be determined by the initial conditions. Thus for each $n = 1, 2, 3, \ldots$, there will be "building-block" solutions

$$u_n(\mathbf{x}, t) = \varphi_n(\mathbf{x})\psi_n(t). \tag{A.5}$$

Since (A.1) is linear and homogeneous, a sum of solutions of (A.1) is again a solution. To satisfy the initial condition, we seek solutions of (A.1) in the form of an infinite sum

$$u(\mathbf{x}, t) = \sum_{1}^{\infty} u_n(\mathbf{x}, t).$$

Example 1

The heat equation with constant coefficients.

$$u_t - k\Delta u - \gamma u = 0 \qquad \text{in } G,$$

$$u(x, 0) = f(x) \qquad \text{in } G,$$

$$Bu = 0 \qquad \text{on } \partial G.$$

We divide the equation by k, and in (A.1) we take $M = -\Delta$ and $N = (1/k)(\partial_t - \gamma)$. The eigenvalue problem (A.3), (A.4) is

$$-\Delta\varphi = \lambda\varphi \qquad \text{in } G,$$

$$B\varphi = 0 \qquad \text{on } \partial G.$$

We assume that there are eigenfunctions $\varphi_n(\mathbf{x})$ and eigenvalues λ_n. We set $\mu_n = k\lambda_n - \gamma$. Then (A.2) becomes

$$\psi_n'(t) + \mu_n\psi_n(t) = 0$$

with solutions

$$\psi_n(t) = A_n e^{-\mu_n t}.$$

The form of the solutions to the time-dependent problem is

$$u(\mathbf{x}, t) = \sum_{1}^{\infty} A_n e^{-\mu_n t} \varphi_n(\mathbf{x}). \tag{A.6}$$

The coefficients A_n are determined by the initial conditions

$$A_n = \frac{\langle f, \varphi_n \rangle}{\langle \varphi_n, \varphi_n \rangle}.$$ (A.7)

Recall that we are using the scalar product for functions on G,

$$\langle f, g \rangle = \int_G f(\mathbf{x}) g(\mathbf{x}) d\mathbf{x}.$$

Example 2

The heat equation with a source.

$$u_t - k\Delta u - \gamma u = q \qquad \text{in } G,$$ (A.8)

$$u(x, 0) = f(x) \qquad \text{in } G,$$

$$Bu = 0 \qquad \text{on } \partial G.$$

This problem is solved by expanding the desired solution

$$u(\mathbf{x}, t) = \sum_1^\infty u_n(t) \varphi_n(\mathbf{x})$$ (A.9)

and the source

$$q(\mathbf{x}, t) = \sum_1^\infty q_n(t) \varphi_n(\mathbf{x}).$$

We substitute the expansions in the left and right sides of (A.8) to produce the family of ODE's for the coefficient functions $u_n(t)$

$$u_n'(t) + \mu_n u_n(t) = q_n(t),$$

$$u_n(0) = A_n = \frac{\langle f, \varphi_n \rangle}{\langle \varphi_n, \varphi_n \rangle},$$

with solutions

$$u_n(t) = A_n e^{-\mu_n t} + \int_0^t e^{-\mu_n(t-s)} q_n(s) ds.$$

These expressions for $u_n(t)$ are substituted in (A.9) to provide the solution of (A.8).

Example 3

The wave equation.

$$u_{tt} + 2du_t - c^2 \Delta u + ku = 0 \qquad \text{in } G, \qquad (A.10)$$

$$u(x,0) = f(x), \quad u_t(x,0) = g(x) \qquad \text{in } G,$$

$$Bu = 0 \qquad \text{on } \partial G.$$

Here we take $N(t, \partial_t) = (\partial^2/\partial t^2 + 2d\partial/\partial t + k)/c^2$, and again $M = -\Delta$. The eigenvalue problem is the same as that of the heat equation. (A.2) becomes

$$\psi''(t) + 2d\psi'(t) + (k + c^2\lambda)\psi(t) = 0. \qquad (A.11)$$

Now assuming $c^2\lambda_n > d^2 - k$ for all n (all modes are underdamped), the solution of IBVP (A.10) is given by

$$\psi_n(t) = e^{-dt}[A_n \cos(\omega_n t) + B_n \sin(\omega_n t)],$$

with $\omega_n = \sqrt{c^2\lambda_n + k - d^2}$. The coefficients are given by

$$A_n = \frac{\langle f, \varphi_n \rangle}{\langle \varphi_n, \varphi_n \rangle}, \qquad \omega_n B_n = \frac{\langle g, \varphi_n \rangle}{\langle \varphi_n, \varphi_n \rangle} + dA_n. \qquad (A.12)$$

Example 4

The wave equation with a source.

The wave equation with a source is handled in the same way as the heat equation with a source. Substitute $u = \sum u_n(t)\varphi_n(x)$ and $q = \sum q_n(t)\varphi_n(x)$ in

$$u_{tt} + 2du_t - c^2 \Delta u + ku = q \qquad (A.13)$$

to arrive at the family of ODE's

$$u_n''(t) + 2du_n'(t) + (c^2\lambda_n + k)u_n(t) = q_n(t).$$

Assuming zero initial conditions, the solutions are

$$u_n(t) = \frac{1}{\omega_n} \int_0^t e^{-d(t-s)} \sin(\omega_n(t-s))q_n(s)ds. \qquad (A.14)$$

Inhomogeneous boundary conditions

Problems with inhomogenous boundary conditions can be reduced to problems with sources by subtracting off a term to make the boundary conditions homogeneous. We illustrate with a problem for the heat equation.

$$u_t - k\Delta u = 0 \qquad \text{in } G, \tag{A.15}$$

$$u(\mathbf{x}, 0) = f(\mathbf{x}) \qquad \text{in } G,$$

$$u = g \qquad \text{on } \partial G.$$

Let $p(\mathbf{x}, t)$ be any smooth function defined on the cylinder $\bar{G} \times [0, T]$ such that

$$p(\mathbf{x}, t) = g(\mathbf{x}, t) \qquad \text{for } x \in \partial G.$$

Then

$$v(\mathbf{x}, t) = u(\mathbf{x}, t) - p(\mathbf{x}, t)$$

solves an inhomogeneous equation with homogeneous boundary conditions:

$$v_t - k\Delta v = q \equiv -p_t + k\Delta p, \tag{A.16}$$

$$v(\mathbf{x}, 0) = f(\mathbf{x}) - p(\mathbf{x}, 0),$$

$$u = 0 \qquad \text{on } \partial G.$$

Then we apply the techniques used to solve (A.8).

A.2 Separation of variables in steady-state problems

Separation of variables can used to solve some steady-state problems in two and three dimensions in regions such as sectors, disks, or rectangles.

Eigenvalue problems

The eigenvalue problem (A.3) can sometimes be solved by separation of variables. We must be able to separate the variables in the equation in a way which fits with the geometry of the region G. In a domain with radial symmetry, we usually use polar or spherical coordinates. For examples of this procedure, see Sections 8.3 and 8.4. For reference we recall the eigenfunctions for the disk of radius a, for Dirchlet and Neumann boundary conditions, and for the ball of radius a for the Dirichlet boundary condition.

The disk

The disk of radius a is $G = \{x^2 + y^2 < a^2\}$. The eigenvalue problem for the Dirichlet boundary condition is

$$-\Delta\varphi = \lambda\varphi \quad \text{in } G, \qquad \varphi(a, \theta) = 0, \quad 0 \le \theta \le 2\pi.$$

The radially symmetric eigenfunctions are

$$\varphi_{m,0}(r) = J_0(r\sqrt{\lambda_{m,0}}), \qquad m = 1, 2, \ldots,$$

where $\lambda_{m,0} = (\rho_{m,0}/a)^2$, and $\rho_{m,0}$ is the mth zero of the Bessel function $J_0(\rho)$. For $n \ge 1$, there is a double sequence of independent eigenfunctions with angular dependence.

$$\varphi^1_{m,n}(r, \theta) = J_n(r\sqrt{\lambda_{m,n}}) \cos(n\theta)$$

$$\varphi^2_{m,n}(r, \theta) = J_n(r\sqrt{\lambda_{m,n}}) \sin(n\theta).$$

$\lambda_{m,n} = (\rho_{m,n}/a)^2$, and $\rho_{m,n} > 0$ is the mth positive zero of the Bessel function $J_n(\rho)$.

The eigenvalue problem for the Neumann boundary condition is

$$-\Delta\varphi = \mu\varphi \quad \text{in } G, \qquad \partial\varphi/\partial r(a, \theta) = 0, \quad 0 \le \theta \le 2\pi.$$

The radially symmetric eigenfunctions are

$$\varphi_{m,0}(r) = J_0(r\sqrt{\mu_{m,0}}), \qquad m = 0, 1, 2, \ldots,$$

where $\mu_{m,0} = (\sigma_{m,0}/a)^2$ and $\sigma_{m,0}$ is the mth zero of $J_0'(\rho)$. Note here that $\varphi_{0,0} \equiv 1$ is the first radially symmetric eigenfunction. The eigenfunctions with angular dependence are:

$$\varphi^1_{m,n} = J_n(r\sqrt{\mu_{m,n}}) \cos(n\theta)$$

$$\varphi_{m,n}^2 = J_n(r\sqrt{\mu_{m,n}})\sin(n\theta),$$

where $\mu_{m,n} = (\sigma_{m,n}/a)^2$, and $\sigma_{m,n} > 0$ is the mth positive zero of $J_n'(\rho)$.

The ball

The ball of radius a, is $G = \{x^2 + y^2 + z^2 < a^2\}$. We use spherical coordinates

$$x = r\sin\theta\cos\phi$$

$$y = r\sin\theta\sin\phi$$

$$z = r\cos\theta$$

where $r = \sqrt{x^2 + y^2 + z^2}$, θ is the polar angle, $0 \le \theta \le \pi$, and ϕ is the azimuthal angle, $0 \le \phi \le 2\pi$. See Figure 9.1 with $x_0 = 0$. The eigenvalue problem with Dirchlet boundary condition is

$$-\Delta\varphi = \lambda\varphi \quad \text{in } G, \qquad u(a, \theta, \phi) = 0.$$

The eigenfunctions depend on three indices,

$$\varphi(r, \theta, \phi) = \varphi_{l,m,j}(r, \theta, \phi)$$

with $l = 0, 1, 2, \ldots$, $j = 1, 2, 3, \ldots$, and $m = -l, -l+1, \ldots, 0, 1, 2, \ldots, l$. The purely radial eigenfunctions are

$$\varphi_{0,0,j} = \frac{J_{1/2}(r\sqrt{\lambda_{0,j}})}{\sqrt{r}}$$

and the corresponding eigenvalues are $\lambda_{0,j}$, where $a\sqrt{\lambda_{0,j}}$ is the jth positive root of $J_{1/2}(\rho) = 0$.

The eigenfunctions involving angular dependence are

$$\varphi_{l,m,j} = \frac{J_{l+1/2}(r\sqrt{\lambda_{l,j}})}{\sqrt{r}} Y_l^m(\theta, \phi).$$

Here the eigenvalues are $\lambda_{l,j}$ which are determined by the condition that $a\sqrt{\lambda_{l,j}}$ is the jth nonzero root of $J_{l+1/2}(\rho) = 0$. The functions Y_l^m are called the *spherical harmonics*. They play the role in these eigenfunctions that $\sin(n\theta)$ and $\cos(n\theta)$ play in the eigenfunctions for the disk. The spherical harmonics form a complete orthogonal set in the space of piecewise continuous functions on the the unit sphere with the complex scalar product

$$\langle z, w \rangle = \int_0^\pi \int_0^{2\pi} z(\theta, \phi) \bar{w}(\theta, \phi) \sin(\theta) d\phi d\theta.$$

The spherical harmonics are in turn written as products

$$Y_l^m(\theta, \phi) = P_l^m(\cos\theta) e^{im\phi}.$$

In the standard notation that we are following, the ϕ-dependence is expressed with the complex exponential instead of the real sine and cosine. The real and imaginary parts of Y_l^m also yield eigenfunctions with the same eigenvalues.

The functions $P_l^m(s)$ are the *Legendre* functions. They are defined on the interval $-1 \le s \le 1$ and solve the ODE

$$\frac{d}{ds}[(1-s^2)\frac{dp}{ds}] + [l(l+1) - \frac{m^2}{1-s^2}]p = 0.$$

Note that the coefficients in this ODE are singular at $s = \pm 1$. The Legendre functions are the solutions of this ODE which satisfy the boundary condition that p be finite at $s = \pm 1$. The formula for P_l^m is

$$P_l^m(s) = \frac{(-1)^m}{2^l l!}(1-s^2)^{m/2}(\frac{d}{ds})^{l+m}[(s^2-1)^l].$$

Note that for each l, there are $2l + 1$ eigenfunctions for each eigenvalue $\lambda_{l,j}$. This means that the eigenspace of $\lambda_{l,j}$ has dimension $2l + 1$. The eigenfunctions $\varphi_{l,m,j}$ form a complete orthogonal set in the complex scalar product on the space of piecewise continuous functions on G,

$$\langle f, g \rangle = \int \int \int_G f(x, y, z)\bar{g}(x, y, z) \, dxdydz$$

$$= \int_0^a \int_0^\pi \int_0^{2\pi} f(r, \theta, \phi)\bar{g}(r, \theta, \phi)r^2 \sin(\phi) \, d\phi d\theta dr.$$

The *solid harmonic* functions are the building-block solutions of the Laplace equation $\Delta u = 0$ in the sphere of radius ρ. They are

$$r^l Y_l^m(\theta, \phi), \quad l = 0, 1, \ldots, \quad m = -l, -l+1, \ldots, 0, 1, 2, \ldots, l.$$

It can be shown that the solid harmonic functions are polynomials in the Cartesian variables x, y, z. The solution of the Dirichlet problem in the ball of radius ρ

$$\Delta u = 0 \quad \text{in } G, \qquad u = f \quad \text{on } r = \rho$$

is given by

$$u(x, y, z) = u(r, \theta, \phi) = \sum_{l=0}^{\infty} \sum_{m=-l}^{m=l} a_l^m \left(\frac{r}{\rho}\right)^l Y_l^m(\theta, \phi).$$

The coefficients a_l^m are the coefficients in the expansion of the boundary data f in terms of the spherical harmonics

$$f(\theta, \phi) = \sum_{l=0}^{\infty} \sum_{m=-l}^{m=l} a_l^m Y_l^m(\theta, \phi).$$

For a complete discussion of the eigenfunctions of the Laplace operator in the ball, see [Str1] or [Ja].

The Poisson equation

Let G be a bounded, open set in R^2 or R^3, and consider the problem

$$- \Delta u = q \qquad \text{in } G, \tag{A.17}$$

$$Bu = 0 \qquad \text{on } \partial G.$$

Assume that a complete set of eigenfuctions $\varphi_n(\mathbf{x})$ with eigenvalues λ_n is known for the problem

$$-\Delta \varphi = \lambda \varphi \qquad \text{in } G,$$

$$Bu = 0 \qquad \text{on } \partial G.$$

Expand the desired solution u of (A.17) in terms of the eigenfunctions, $u(\mathbf{x}) = \sum u_n \varphi_n(\mathbf{x})$, and do the same for the given function q, $q(\mathbf{x}) = \sum q_n \varphi_n(\mathbf{x})$. Substitute in (A.17) and equate the coefficients of φ_n. We deduce that $u_n = q_n/\lambda_n$. The solution of (A.17) is given by

$$u(\mathbf{x}) = \sum_{1}^{\infty} \left(\frac{q_n}{\lambda_n}\right) \varphi_n(\mathbf{x}). \tag{A.18}$$

Boundary sources

Now consider two problems where the inhomogenous term is on the boundary of the region.

$$\Delta u = 0 \qquad \text{in } G, \qquad\qquad (A.19)$$

$$u = f \qquad \text{on } \partial G,$$

and

$$\Delta u = 0 \qquad \text{in } G \qquad\qquad (A.20)$$

$$\frac{\partial u}{\partial n} = g \qquad \text{on } \partial G.$$

Here $\partial/\partial n$ is the normal derivative at the boundary.

One way to solve (A.19) is to find a smooth function $p(\mathbf{x})$, such that $p = f$ on ∂G. Then $v = u - p$ satisfies a Poisson equation with homogeneous boundary conditions:

$$-\Delta v = q \equiv \Delta p \qquad \text{in } G,$$

$$v = 0 \qquad \text{on } \partial G.$$

This problem can be solved as we solved (A.17).

$$u(\mathbf{x}) = p(\mathbf{x}) + v(\mathbf{x}) = p(\mathbf{x}) + \sum_1^\infty (\frac{q_n}{\lambda_n}) \varphi_n(\mathbf{x}).$$

However, a more direct way is possible in certain geometries. For example, in a rectangle another separation of variables is possible, as discussed in Section 9.3.

Next consider problem (A.20) in the disk G of radius ρ. When G is the disk of radius ρ, the separated building-block solutions (see Section 9.2) are

$$u_n(r, \theta) = [a_n \cos(n\theta) + b_n \sin(n\theta)]r^n,$$

and we seek the solution of (A.20) in the form

$$u(r, \theta) = \frac{a_0}{2} + \sum_1^\infty [a_n \cos(n\theta) + b_n \sin(n\theta)]r^n.$$

To satisfy the boundary conditions we require that

$$\frac{\partial u}{\partial r}(\rho, \theta) = \sum_1^\infty [a_n \cos(n\theta) + b_n \sin(n\theta)]n\rho^n$$

$$= g(\theta) = \frac{A_0}{2} + \sum_1^\infty [A_n \cos(n\theta) + B_n \sin(n\theta)]$$

where A_n and B_n are the Fourier coefficients of g. It is clear that we must choose

$$a_n = \frac{A_n}{n\rho^n}, \qquad b_n = \frac{B_n}{n\rho^n}, \qquad n = 1, 2, \ldots$$

Furthermore g must satisfy the compatibility condition that $A_0 = 0$. The coefficient a_0 is not determined. Thus the solution of (A.20) is

$$u(r, \theta) = \frac{a_0}{2} + \sum_1^\infty (1/n)[A_n \cos(n\theta) + B_n \sin(n\theta)](r/\rho)^n,$$

where a_0 is an arbitrary constant.

Problems in a sector

To illustrate the use of the technique to solve a problem in a sector we consider

$$\Delta u = 0 \quad \text{in } G, \tag{A.21}$$

where $G = \{0 < \theta < \alpha, 0 < r < \rho\}, 0 < \alpha < 2\pi$, with the boundary conditions

$$u(r, 0) = 0 \qquad \text{and} \qquad u_\theta(r, \alpha) = 0, \quad 0 \le r \le \rho, \tag{A.22}$$

$$u(\rho, \theta) = g(\theta), \qquad 0 \le \theta \le \alpha.$$

Separation of variables yields the building-block solutions which satisfy (A.21) and the boundary conditions (A.22)

$$r^{[n\pi/(2\alpha)]} \sin(\frac{n\pi\theta}{2\alpha}), \quad n = 1, 3, 5, \ldots.$$

Consequently the solution is given by

$$u(r, \theta) = \sum_1^\infty A_n \sin(\frac{n\pi\theta}{2\alpha}).$$

The coefficients A_n are determined from the expansion of the boundary data g in terms of the eigenfunctions $\sin(n\pi\theta/(2\alpha))$.

Problems in an annulus

Now let G be the annulus

$$G = \{(r, \theta) : \rho_1 < r < \rho_2; \ 0 \le \theta \le 2\pi\},$$

and let $f_1(\theta)$ and $f_2(\theta)$ be given functions. Consider the boundary value problem in G,

$$\Delta u = 0 \quad \text{in } G,$$

$$u(\rho_1, \theta) = f_1(\theta), \quad u(\rho_2, \theta) = f_2(\theta), \quad 0 \le \theta \le 2\pi.$$

The desired solution is now sought in the form

$$\frac{a_0}{2} + \sum_1^\infty [a_n \cos(n\theta) + b_n \sin(n\theta)] r^n + \sum_1^\infty [c_n \cos(n\theta) + d_n \sin(n\theta)] r^{-n}.$$

The coefficients a_n, b_n, c_n, and d_n will satisfy a 4×4 linear system when the boundary conditions are imposed.

A.3 Fundamental solutions

In this section we list the fundamental solutions to several equations we have discussed.

The heat equation

$$u_t - k\Delta u = 0, \quad \mathbf{x} \in R^n, \ t > 0. \tag{A.23}$$

The fundamental solution is

$$S_n(\mathbf{x}, t) = \frac{1}{(4\pi kt)^{(n/2)}} e^{-\frac{|\mathbf{x}|^2}{4kt}}$$

and solves (A.23) with the IVP

$$S_n(\mathbf{x}, 0) = \delta(\mathbf{x}).$$

The solution to the initial-value problem consisting of (A.23) with the initial condition

$$u(\mathbf{x}, 0) = f(\mathbf{x}) \tag{A.24}$$

is given by the convolution

$$u(\mathbf{x}, t) = \int_{R^n} S_n(\mathbf{x} - \mathbf{y}, t) f(\mathbf{y})d\mathbf{y}. \tag{A.25}$$

If (A.23) is modified to become

$$u_t - k\Delta u + \gamma u = 0,$$

the fundamental solution is

$$e^{-\gamma t} S_n(\mathbf{x}, t),$$

and the solution of the IVP is given by

$$u(\mathbf{x}, t) = \int_{R^n} e^{-\gamma t} S_n(\mathbf{x} - \mathbf{y}, t) f(\mathbf{y})d\mathbf{y}. \tag{A.26}$$

The wave equation

We recall the formulas for the solutions of the IVP's and show that they may also be written as convolutions of a fundamental solution with the initial data.

The IVP in $R^n \times R$ is

$$u_{tt} - c^2 \Delta u = 0, \quad \mathbf{x} \in R^n, \ t \in R, \tag{A.27}$$

$$u(\mathbf{x}, 0) = f(\mathbf{x}), \qquad u_t(\mathbf{x}, 0) = g(x), \quad \mathbf{x} \in R^n.$$

When $n = 1$ the solution is given by d'Alembert's formula

$$u(x, t) = \frac{f(x - ct) + f(x + ct)}{2} + \frac{1}{2c} \int_{x-ct}^{x+ct} g(y)dy. \tag{A.28}$$

When $n = 2$ the solution of the IVP (A.27) is given by

$$u(\mathbf{x}, t) = \frac{1}{2\pi c} \frac{\partial}{\partial t} \int_{|\mathbf{y}-\mathbf{x}| \leq ct} \frac{f(\mathbf{y})d\mathbf{y}}{\sqrt{c^2 t^2 - |\mathbf{x} - \mathbf{y}|^2}} \tag{A.29}$$

$$+ \frac{1}{2\pi c} \int_{|\mathbf{x}-\mathbf{y}| \leq ct} \frac{g(\mathbf{y})d\mathbf{y}}{\sqrt{c^2 t^2 - |\mathbf{x} - \mathbf{y}|^2}}.$$

For $n = 3$, the solution of (A.27) is given by the Kirchoff formula

$$u(\mathbf{x}, t) = \frac{\partial}{\partial t}\left[\frac{1}{4\pi c^2 t}\int_{|\mathbf{x}-\mathbf{y}|=ct} f(\mathbf{y})dS(\mathbf{y})\right] + \frac{1}{4\pi c^2 t}\int_{|\mathbf{x}-\mathbf{y}|=ct} g(\mathbf{y})dS(\mathbf{y}).$$
(A.30)

For $t > 0$ each of these formulas can be expressed as a convolution

$$u(\mathbf{x}, t) = \frac{\partial}{\partial t}\int_{R^n} E_n(\mathbf{x}-\mathbf{y}, t)f(\mathbf{y})d\mathbf{y} + \int_{R^n} E_n(\mathbf{x}-\mathbf{y}, t)g(\mathbf{y})d\mathbf{y}. \qquad (A.31)$$

E_n is a fundamental solution of the wave equation, sometimes called the *Riemann function* for the wave equation. It satisfies

$$(E_n)_{tt} - c^2\Delta E_n = \delta(\mathbf{x}, t),$$

$$E_n(\mathbf{x}, 0) = 0, \qquad (E_n)_t(\mathbf{x}, 0) = \delta(\mathbf{x}).$$

Let $K^+ = \{(\mathbf{x}, t) : t > 0, \ |\mathbf{x}| < ct\}$. K^+ is the *forward light cone* with vertex at the origin.

For $n = 1$ the Riemann function is given by

$$E_1(x, t) = \begin{cases} \frac{1}{2c}, & (x, t) \in K^+ \\ 0, & \text{otherwise} \end{cases}. \qquad (A.32)$$

In particular, E_1 satisfies (A.27) for $t \neq 0$ in the weak sense.

For $n = 2$, the Riemann function E_2 is given by

$$E_2(\mathbf{x}, t) = \begin{cases} \frac{1}{\sqrt{c^2 t^2 - |\mathbf{x}|^2}}, & (\mathbf{x}, t) \in K^+ \\ 0, & \text{otherwise} \end{cases}. \qquad (A.33)$$

For $n = 3$, the Riemann function is not a function in the usual sense, but rather a distribution,

$$E_3(\mathbf{x}, t) = \begin{cases} \delta(|\mathbf{x}| - ct)/(4\pi c|\mathbf{x}|), & t > 0 \\ 0, & t < 0 \end{cases}. \qquad (A.34)$$

There are distribution Riemann functions for the wave equation when $n > 3$ (see [RR]).

The Poisson equation

The fundamental solutions for the Laplace operator $-\Delta u$, are

$$s_2(\mathbf{x}) = \frac{1}{2\pi} \log(\frac{1}{|\mathbf{x}|}) \quad \text{when } n = 2,$$

and

$$s_3(\mathbf{x}) = \frac{1}{4\pi} \frac{1}{|\mathbf{x}|} \quad \text{when } n = 3.$$

The general formula for $n \geq 3$ is

$$s_n(\mathbf{x}) = \frac{1}{(n-2)\Omega_n |\mathbf{x}|^{n-2}}.$$

$\Omega_n = 2\pi^{n/2}/\Gamma(n/2)$ is the area of the unit sphere in R^n and $\Gamma(x)$ is the gamma function. The solution of the Poisson equation

$$-\Delta u = q$$

is given by the convolution

$$u(\mathbf{x}) = \int_{R^n} s_n(\mathbf{x} - \mathbf{y})q(\mathbf{y})d\mathbf{y}.$$

Uniqueness of solutions of the Poisson equation is discussed in Section 9.4.

A.4   The Laplace operator in polar and spherical coordinates

Polar coordinates in R^2 are

$$x = r\cos\theta \qquad y = r\sin\theta.$$

It follows that $r^2 = x^2 + y^2$ and

$$r_x = \frac{x}{r}, \qquad r_y = \frac{y}{r}, \qquad \Delta r = \frac{1}{r}.$$

Furthermore

$$\theta_x = -\frac{\sin \theta}{r}, \qquad \theta_y = \frac{\cos \theta}{r}, \qquad \Delta\theta = 0.$$

Now using the chain rule, the change of variable in the Laplace operator is

$$\Delta u = u_{rr}(r_x^2 + r_y^2) + 2u_{r\theta}(r_x\theta_x + r_y\theta_y) + u_{\theta\theta}(\theta_x^2 + \theta_y^2) + (\Delta r)u_r + (\Delta\theta)u_\theta.$$

Hence the Laplace operator in polar coordinates is given by

$$\Delta u = u_{rr} + \frac{1}{r}u_r + \frac{1}{r^2}u_{\theta\theta}. \tag{A.35}$$

We can introduce polar coordinates in R^2 centered at a point (x_0, y_0) with the change of coordinates $r = \sqrt{(x - x_0)^2 + (y - y_0)^2}$ and

$$x = x_0 + r \cos \theta, \quad y = y_0 + r \sin \theta.$$

The formula for Δu in these coordinates is the same, and is derived in the same way.

Spherical coordinates in R^3 (see Figure 9.1))are

$$x = r \sin \theta \cos \phi, \qquad y = r \sin \theta \sin \phi, \qquad z = r \cos \theta.$$

Here ϕ is the azimuth, $0 \le \phi \le 2\pi$, and θ is the polar angle, $0 \le \theta \le \pi$. Let $\rho = \sqrt{x^2 + y^2}$ be the distance from the z axis. Then we can write

$$\rho = r \sin \theta, \qquad x = \rho \cos \phi, \qquad y = \rho \sin \phi.$$

Using the two-dimensional result (A.35), we see that

$$u_{\rho\rho} + u_{zz} = u_{rr} + \frac{1}{r}u_r + \frac{1}{r^2}u_{\theta\theta},$$

and

$$u_{xx} + u_{yy} = u_{\rho\rho} + \frac{1}{\rho}u_\rho + \frac{1}{\rho^2}u_{\phi\phi}.$$

Adding these two lines and cancelling out the $u_{\rho\rho}$ term from both sides, we obtain

$$u_{xx} + u_{yy} + u_{zz} = u_{rr} + \frac{1}{r}u_r + \frac{1}{\rho}u_\rho + \frac{1}{\rho^2}u_{\phi\phi} + \frac{1}{r^2}u_{\theta\theta}. \tag{A.36}$$

Since $r^2 = \rho^2 + z^2$ it follows that

$$r_\rho = \frac{\rho}{r}, \qquad \theta_\rho = \frac{\cos\theta}{r}, \qquad \phi_\rho = 0.$$

Thus

$$u_\rho = u_r r_\rho + u_\theta \theta_\rho + u_\phi \phi_\rho$$

$$= \frac{\rho}{r} u_r + \frac{\cos\theta}{r} u_\theta.$$

Then we substitute this expression for u_ρ in (A.36) and $\rho = r\sin\theta$ in the denominator of the next to last term of (A.36) to obtain

$$\Delta u = u_{rr} + \frac{2}{r} u_r + \frac{1}{r^2}\left[u_{\theta\theta} + (\cot\theta)u_\theta + \frac{1}{\sin^2\theta} u_{\phi\phi} \right]. \qquad (A.37)$$

Spherical coordinates in R^3 centered at a point $\mathbf{x}_0 = (x_0, y_0, z_0)$ with $r = |\mathbf{x} - \mathbf{x}_0|$ are given by

$$x = x_0 + r\sin\theta\cos\phi, \quad y = y_0 + r\sin\theta\cos\phi, \quad z = z_0 + r\cos\theta.$$

The expression for Δu in these coordinates is also given by (A.37).

Appendix B

Elements of MATLAB

In this appendix we give a brief summary of some of the more important MATLAB features that are used in this text. A very convenient short introduction to MATLAB is available in the MATLAB Primer by Kermit Sigmon. A more complete introduction is given in the MATLAB User's Guide, published by Mathworks. To find every possible command, look at the Reference Guide, also published by Mathworks. Even more convenient is the on-line help. For information about any command, function, or function on functions, enter help *name*. For example, to find how to use the fast Fourier Transform of MATLAB , enter help fft.

In addition there are numerous books published by authors dealing just with the rich subject of MATLAB graphics. Many special application codes using MATLAB have been written and are available at the Mathworks web site: http://www.Mathworks.com.

The basic element of the MATLAB environment is the matrix. Scalars are considered to be 1×1 matrices and an n vector is either a row vector (an $1 \times n$ matrix) or a column vector (an $n \times 1$ matrix).

B.1 Forming vectors and matrices

Matrices can be entered by typing in the elements one at a time. Typing this

```
>> [1 2 3;4 5 6]
```

produces

```
ans =

    1    2    3
    4    5    6
```

Notice that we use a semicolon to separate the rows. Usually we want to give a vector or matrix a name. To assign a matrix value to a variable we proceed as follows: Type this

```
>> A = [1 2 3;3 4 5]
```

to get

```
A =

    1    2    3
    4    5    6
```

Many times we do not want to see the value displayed on the screen, especially if the matrix or vector has thousands of elements. This can be accomplished by adding a semicolon after the defining statement.

```
>> A = [1 2 3;4 5 6];
```

If we want to see the numbers in A, we enter

```
>> A
```

which produces

```
A =

    1    2    3
    4    5    6
```

The transpose of a matrix is formed by the command A'. If the row vector x is defined by

```
>> x = [1 5 4 8 10]
```

then x is turned into a column vector with the command x'. To determine the dimensions of a vector or matrix, use the commands

```
>> size(A)

ans =

    2    3

>> size(x)

ans =
```

```
      1      5
```

```
>> size(x')
```

```
      5      1
```

To view a certain element in a matrix or vector, we specify its location with a command:

```
>> A(1,2)
```

```
ans =
```

```
      2
```

```
>> x(5)
```

```
ans =
```

```
      10
```

In many cases the vectors or matrices are far too large to enter one element at a time. For instance if we want to enter a vector **x** consisting of points (0.1, .2, .3, .4, . . . , 5.5, 6), we can use the command

```
>> x = 0:.1:6 ;
```

This row vector has 61 elements. To create a vector of zeros or of ones of the same length there are commands

```
>> y = ones(size(x));
>> z = zeros(size(x));
```

The same works for matrices

```
>> Z = zeros(size(A));
>> Y = ones(size(A))
```

```
Y =
```

```
      1      1      1
      1      1      1
```

One can also specify a matrix of zeros or ones by giving the dimensions:

```
>> Z = zeros(2,3)
```

The $n \times n$ identity matrix is produced with the command eye(n). There are special commands for entering sparse matrices or diagonal matrices.

B.2   Operations on matrices

Addition and subtraction of matrices are done in the obvious way.

```
>> B = [2 0 -1; 1 2 7];
>> A + B

ans =

    3    2    2
    5    7   13
```

In general, MATLAB can add together only matrices having the same dimension. However, there is one special and very useful exception. If A is a matrix and c is a scalar, then the sum A +c means to add c to every element of A. In particular if x is a vector and t a scalar, then x+t is a vector of the same length with t added to each component.

Multiplication of matrices is done with the symbol * and presumes that the matrices have the proper dimensions.

```
>> C = [3 3; -1 4; 6 2]
>> A*C

ans =

   19   17
   43   44

>> b = [1 5 -2]
>> A*b'

ans  =

    5
   17

>> c = [1 3]
>> c*A

ans =

    3    6    9
   12   15   18
```

A vector or matrix can always be multiplied or divided by a scalar.

```
>> 2 * A

ans =

        2       4       6
        8      10      12

>> A/2

ans =

   0.5000    1.0000    1.5000
   2.0000    2.5000    3.0000
```

B.3 Array operations

Another kind of matrix operation, called array multiplication, is also possible in MATLAB. If A and B are two matrices of the same size with elements $A(i, j)$ and $B(i, j)$, then the command (using .* instead of *)

```
>> C = A.*B
```

produces another matrix C of the same size with elements $C(i, j) = A(i, j)B(i, j)$. For example, using the same 2×3 matrices A and B we defined earlier, we have

```
>> C = A.*B

C =

        2       0      -3
        4      10      42
```

To raise a scalar to a power, say two, we use the command 5^2. If we want the operation to be applied to each element of a matrix, we use .^2. For example, if we want to produce a new matrix whose elements are the square of the elements of the matrix A, we enter

```
>> A.^2

ans =
```

```
   1     4     9
  16    25    36
```

There is also a kind of array division for two matrices of the same size which divides the two matrices element by element.

```
>> D = [1 3 5; -2 4 -1]
>> A./D
```

```
ans =
```

```
    1.0000    0.6667    0.6000
   -2.0000    1.2500   -6.0000
```

B.4 Solution of linear systems

Given an $n \times n$ matrix A and an n column vector b, the linear system $Ax = b$ can be solved in several ways. The simplest way is to use the following method.

```
>> A = [1 2 3; 4 5 6; 6 7 9];
>> b = [ 1 0 1]'
>> x = A\b
```

```
x =
```

```
   -0.0000
   -2.0000
    1.6667
```

The solution to the equation $xA = b$, where x and b are now row vectors, is given by the division command x = A/b. Another way to solve $Ax = b$ is to find the inverse of A and then multiply by b.

```
>> C = inv(A);
>> x = C*b;
```

In the first method MATLAB is using the Gaussian elimination procedure with partial pivoting. In general is it more reliable for numerical work requiring high accuracy.

B.5   MATLAB functions and mfiles

MATLAB has the usual built-in functions, such as $\sin x$, $\cos x$, $\tan x$, $\exp x$, $\log x$, $\sqrt{x}$, etc. These functions can take matrices as arguments, in which case the function is applied to each element of the matrix.

```
>> T = [2 3 pi; 8 pi/2 1];
>> cos(T)

ans =

        0.4161   -0.9900   -1.0000
       -0.1455    0.0000    0.5403

>> sqrt(A)

ans =

        1.0000    1.4142    1.7321
        2.0000    2.2361    2.4495
```

In addition, Bessel functions of various kinds and all orders are available. For example, the Bessel function $J_\alpha(x)$ is called by entering besselj(alpha, x). Another important function is the error function, called simply by entering erf(x). There are also many specialized functions of matrices, such as eig(A) which finds the eigenvalues of a square matrix.

It is possible to build more complicated functions using the extremely flexible MATLAB feature known as an *mfile* . There are two kinds of mfiles, *function* mfiles and *script* mfiles. In this section we discuss the former. Script mfiles will be discussed in the next section.

Suppose we need to compute the values of the function

$$f(x) = x \exp(-\sin(x))/(1 + x^2).$$

Rather than type this full expression out every time, we can create a function mfile, called f.m, so that to evaluate f at $x = 2$, we need only type the command f(2). Here is what the mfile looks like.

```
function y = f(x)
y = x*exp(-sin(x))/(1+x^2);
```

Written this way, the function can take only scalars for x. However, if we write it using the symbols for the array operations, like this:

```
function y = f(x)
y = x.*exp(-sin(x))./(1+x.^2);
```

the function can now be used on vectors and matrices. Notice that in the denominator we are adding the scalar 1 to the vector `x.^2` to produce another vector, which then divides, in array fashion, the factor `x.^2.*exp(-sin(x))`. When a function is written this way, we will say that it is *array-smart*. Array-smart functions are especially important for graphing.

Functions which are defined piecewise may also be constructed in an array-smart fashion. Consider the example

$$f(x) = \begin{cases} x & x < 0 \\ x^2 & 0 \le x < 2 \\ 4 & x \ge 2 \end{cases} .$$

The building blocks for this kind of function are the *characteristic* functions for intervals of the form $(-\infty, a)$ and (a, ∞). An mfile for this kind of function would be

```
function y = c(x)
y = (x < 3);
```

Check that $c(x) = 1$ for $x < 3$ and $c(x) = 0$ for $x \ge 3$. Now we make an mfile for f which is array-smart as follows:

```
function y = f(x)
y1 = x.*(x < 0);
y2 = x.^2.*( (x < 2) - (x < 0) );
y3 = 4*(1 - (x < 2));
y = y1 + y2 + y3;
```

Of course we can also define functions of several variables, such as the fundamental solution of the heat equation in two space dimensions and time.

```
function z = S(x,y,t)
z = exp(-(x.^2 + y.^2)./(4*t) )./sqrt((4*pi*t))
```

The variables used in the mfile to define the function are "dummy" variables. One can use any variable names to call the function. For example, for the function f defined above, we can use the statements

```
s = -2:.1:4;
r = f(s);
```

The first command defines the vector **s** with 61 components, and the second command computes another vector **r** with $r_i = f(s_i)$ for $i = 1, \ldots, 61$.

B.6 Script mfiles and programs

Script mfiles are used to collect a sequence of commands that may be lengthy or tedious to type over and over again. Calling the name of the script mfile tells MATLAB to execute the sequence of commands, which we can call a program. In particular script mfiles can call function mfiles. We give an example in which we implement the Euler method for numerically integrating an ODE initial value problem. Let the problem be

$$y'(x) = f(x, y), \qquad y(0) = a.$$

If we wish to compute an approximate solution up to $x = X$, we divide the interval $[0, X]$ into N subintervals of length $\Delta x = X/N$. We label the points where we want to compute the solution as $x_n = (n - 1)\Delta x, n = 1, \ldots, N + 1$, and we label the computed values y_n. Then Euler's method for computing the y_n is

$$y_n = y_{n-1} + \Delta x f(x_{n-1}, y_{n-1}).$$

Let us take $f(x, y) = x/(1 + y^2)$, $X = 10$, and $N = 100$ so that $\Delta x = .1$. Finally, let us take $a = 1$. First we write a function mfile for $f(x, y)$.

```
function z = f(x,y)
z = x./(1+y.^2)
```

Here is a script file to do this integration, call it `xeuler.m`. We use the name `xeuler.m` for " example Euler" because MATLAB already has a program called `euler.m`.

```
X = 10;
N = 100;
a = 1;
delx = X/N;
x = 0: delx : 10;
y = zeros(1,101);
y(1) = a;
for n = 2:N+1
    y(n) = y(n-1) + delx*f(x(n-1), y(n-1));
end
```

The output of this program is the vector $\mathbf{y} = (y_1, \ldots, y_{N+1})$. We start the indexing with one because MATLAB does not allow an index to start with zero. The last three lines of the file form a *for loop*. Before the loop begins, we set aside the spaces in memory for the vector consisting of $y_1 = a$ and the 100 computed values $y_n, n = 2, \ldots, 101$.

If we wish to make a smaller step size, or change the interval $[0, X]$, or change the initial data, we must edit the file xeuler.m and put in new numbers. This can be tiresome if we want to run this program many times with different data. To make the program more interactive, we make X, N, and a input data to be read at run time by replacing the first three lines with the commands

```
X = input('Enter the value of X   ')
N = input('Enter the value of N   ')
a = input(' Enter the initial value a   ')
```

Now when we run the program, it will pause and wait for the data to be entered. After the program has run, we can do further manipulations on the output, such as plotting a curve through the data points. We will discuss plotting 2-D and 3-D graphs in later sections.

B.7 Vectorizing computations

MATLAB is very fast at making vector and matrix calculations. It is less efficient using for loops and these should be avoided when possible. In the program xeuler, this could not be avoided because the value of each y_n depended on the previous values. In the following simple example, a for loop can be reduced to a single vector operation.

We want to calculate the sum of the first 100 integers. If we choose not to use the formula $S = N(N + 1)/2$, we can do this with the following sequence of commands:

```
s = 0;
for n = 1:100
    s = s+n;
end
s
```

This short program runs very quickly, but we can accomplish the same results without using a for loop as follows:

```
N = 1:100
s = sum(N)
```

The first command is equivalent to N = 1:1:100 which creates the vector $N = (1, 2, 3, \ldots, 100)$. The second command is a special MATLAB feature which sums the components of a vector. When applied to a matrix A it sums the columns of the matrix, with output a row vector. The command sum(sum(A)) yields the sum of all the elements of the matrix. To stress the vectorial aspect of this computation, suppose that we have a function $g(n)$ given by an array-smart mfile g.m, and we want to compute

$$\sum_{n=1}^{n=100} g(n).$$

Again we can do this in two lines:

```
N = 1:100;
s = sum(g(N));
```

For a second more complicated example which involves some interesting manipulations of indices, we consider the implementation of a simple finite difference scheme for the linear first-order PDE (see Chapter 2)

$$u_t(x, t) + cu_x(x, t) = 0, \qquad u(x, 0) = f(x).$$

If we replace u_t by a forward difference and u_x by a backward difference, we get the finite difference scheme

$$u_{j,n+1} = (1 - c/\rho)u_{j,n} + (c/\rho)u_{j-1,n},$$

where $\rho = \Delta x / \Delta t$. In addition to prescribing the initial data $f(x)$, we shall also assume that the boundary condition $u(0, t) = f(0)$ for all $t > 0$, and we shall solve for u in the set $\{(x, t), 0 \leq x \leq 10, t > 0\}$. First we write a script mfile that implements this scheme using one for loop inside another. Call this file fds.m for "finite difference scheme". We assume that the function f is given by an array-smart mfile f.m.

```
c = .75;
delx = .1;
x = 0: delx: 10;
J = 10/delx; % set the number of spatial steps.
delt = .1
r = c*delt/delx;
nsteps = 40; % set the number of time steps.

u = f(x); % initialize the vector u.

for n = 1:nsteps
    v = u;
```

```
    for j = 2:J+1
        u(j) = (1-r)*v(j) +r*v(j-1);
    end
end
```

Notice that we do not have to recompute the first component of the vector u because it is always the same, namely, equal to $f(0)$. The program does not save the values of u at the intermediate time steps. After the program has run, we can view the initial snapshot with the mfile f.m and the snapshot at the end of the run, which is the solution at time $t = nsteps \times \Delta t$.

The inner for loop can be eliminated by exploiting an indexing feature of MATLAB. In the program as written, the index j is a scalar. However, in MATLAB it is possible to have vector indices. For example if k = [2 4 6 7] then u(k) = [u(2), u(4), u(6), u(7)]. Let us take the index j to be the vector $j = (2, 3, \ldots, J+1)$. Using the fact that we can add or subtract a scalar from a vector (section B.2), we see that the vector index $j - 1 = (1, 2, 3, \ldots, J)$. Now we add the statement j = [2:J+1] before the outer loop and replace the inner for loop with one vector statement so that the last six lines become

```
j = [2:J+1]
for n = 1:nsteps;
    v = u;
    u(j) = (1-r)*v(j) + r*v(j-1);
end
```

The program is only slightly more compact, but it runs a bit faster. When the dimensions of the problem are quite large, this difference becomes more significant.

B.8   Function functions

MATLAB uses the term "function functions" for a class of functions that operate on functions rather than on matrices. We mention five of them here. The arguments of these function functions consist of strings which name the function operated on as well as various parameters.

The root finder fzero. Given a function $f(x)$ defined by an mfile f.m, which has a zero in the neighborhood of the point $x = a$, the command fzero('f',a) finds the root with a default tolerance of .001. For example, to find the root of $f(x) = x/2 - \sin x$ which lies near $x = 2$, write a function mfile for f and then enter fzero('f',2). It give the answer 1.8955. The call for fzero is slightly different in MATLAB 5.0. For more information enter help fzero.

The numerical integrators `quad` and `quad8` . These two quadrature routines estimate the value of the integral

$$\int_a^b f(x)dx.$$

`quad` uses an adaptive Simpson's rule, while `quad8` uses an adaptive eight-panel Newton-Cotes rule. The syntax for both of them is the same. Again the function should be defined by a function mfile. However, avoid the use of single letter names like f or h for the functions. For example if the function has the mfile name `ff.m` the command `quad( 'ff', 2,4)` produces an estimate for the integral over the interval $[2, 4]$.

The ordinary differential solvers `ode23` and `ode45`. These two routines numerically solve the initial value problem

$$y'(x) = f(x, y), \qquad y(x_0) = a.$$

`ode23` uses a 2nd/3rd order Runge-Kutta method while `ode45` uses a 4th/5th order Runge-Kutta-Fehlberg method. For example, consider the initial value problem of section B.6, where $f(x, y) = x/(1+y^2)$, $x_0 = 0$, and $a = 1$. First write a function mfile for f, called `f.m`, as we did in Section B.6. Then use these commands with MATLAB4.2 to integrate from $x = 0$ to $x = 5$:

```
>> x0 = 0;
>> xend = 5;
>> a = 1;
>> [x,y] = ode23('f', x0, xend,a);
```

The output consists of the two vectors $\mathbf{x} = (x_0, x_1, \ldots, x_n)$ and $\mathbf{y} = (a, y_1, y_2, \ldots, y_n)$. The numbers y_i are the computed values at the points x_i. The points x_i run from x_0 to x_n but are not usually equally spaced because the method is adaptive. It takes shorter steps when the solution is changing rapidly and longer ones when the solution is changing more slowly.

Both of these routines can be used on systems of ODE's. Enter `help ode23` or `ode45` for more information. In particular, the command to call `ode45` and `ode23` above are correct for MATLAB4.2. However, in MATLAB5.0, the set of ODE solvers has been expanded to include solvers for stiff systems *and* the calling instructions are slightly different. In MATLAB5.0 the last instruction in the program above would be

```
>> [x,y] = ode23('f', [x0,xend], a);
```

Note the addition of the square bracket for the interval of integration.

B.9   Plotting 2-D graphs

MATLAB has an excellent set of graphic tools. In this section we will only touch on some of the most elementary ones. We begin with 2-D graphs. MATLAB takes two vectors $\mathbf{x} = (x_1, \ldots, x_N)$ and $\mathbf{y} = (y_1, \ldots, y_N)$, locates the points (x_j, y_j), and joins them by straight lines. The command is plot(x,y). The vectors $\mathbf{x} = (1, 2, 3, 4, 5)$ and $\mathbf{y} = (2, 3, 1, 5, -1)$ plotted this way produce the picture below (see Figure B.1).

```
>> x = [1 2 3 4 5];
>> y = [2 3 1 5 -1];
>> plot(x,y)
```

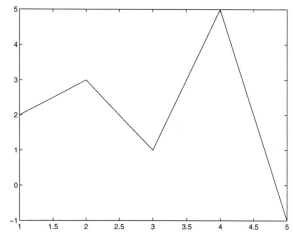

FIGURE B.1
Plot of the vectors $\mathbf{x} = (1, 2, 3, 4, 5)$ *and* $\mathbf{y} = (2, 3, 1, 5, -1)$.

The circle of radius 2, with center at $(1, 3)$ is produced by the commands

```
>> theta = 0:.01:2*pi;
>> x = 1 + 2*cos(theta);
>> y = 3 + 2*sin(theta);
>> plot(x,y)
```

To plot the graph of a function like $\sin x$ on the interval $[0, 2\pi]$, we use the commands

```
>> x = 0:.01:2*pi;
>> y = sin(x);
>> plot(x,y)
```

The last two commands can be combined in one: `plot(x,sin(x))`. If the function f is defined by an array-smart mfile `f.m`, we can also plot it with the command `plot(x,f(x))`.

The output vector **y** from the program `xeuler.m` can be plotted against the vector **x** again with the simple command `plot(x,y)`. This command can be made the last line of the program `xeuler.m` so that the graph is produced automatically.

There are two ways that we can plot several curves on the same graph. Remember that a curve is determined by a pair of vectors **x**, **y** of the same length n. Suppose another pair of vectors **z**, **w** of the same length m where m may differ from n. The first way to plot the two curves on the same graph is with the command

```
>> plot(x,y,z,w)
```

In MATLAB4.2 the first curve will be in yellow, the second in magenta. In MAT-LAB5.0, the colors will be blue and green.

For example, if we want to compare the initial data $f(x)$ with the snapshot of the solution at time $t = nsteps \times \Delta t$ that results from program `fds` of section B.7, we do it with the command

```
>> plot(x,f(x),x,u)
```

This statement could be appended to the file `fds.m` to make the plot automatically.

Two functions f and g given by array-smart mfiles `f.m` and `g.m` can be plotted on $[-1, 4]$ together with $\exp(x)$ by the commands

```
>> x = -1:.1:4;
>> plot(x,f(x),x,g(x),exp(x))
```

The three curves will be in different colors.

The second way to plot several curves on the same graph uses the command `hold on`.

```
>> plot(x,y)
>> hold on
>> plot(z,w)
>> hold off
```

Both curves will now be same color. The three functions $f(x)$, $g(x)$, and $\exp(x)$ are plotted together these commands.

```
>> plot(x,f(x))
>> hold on
>> plot(x,g(x))
>> plot(x,exp(x))
>> hold off
```

This way of plotting curves together can be combined with a for loop. Put the following statements in a script mfile, call it graphs.m.

```
x = 0:.1:20;
plot(x,x.*exp(-x))
hold on
for n = 2:6
    y = x.^n .*exp(-n*x);
    plot(x,y)
end
hold off
```

Now the six curves will be plotted on the same graph by calling the program graphs.

Further 2-D graphing features

The axis command. When we use the command plot(x,y), MATLAB automatically plots the curve on the rectangle $[x_{min}, x_{max}] \times [y_{min}, y_{max}]$. If we wish to change this scale, perhaps to expand a portion of the graph, and instead plot on the rectangle $[a, b] \times [c, d]$, we follow the plot command with axis([a b c d]). You can return the axis scaling to the automatic, default, mode with the command axis('auto').

The zoom command. This is another way to blow up a portion of the graph, using the mouse. Enter the command zoom on. Then move the pointer to the region of the graph you want to enlarge. Click with the left mouse button. This will blow up the size by a factor of two. Clicking again blows it up again by a factor of two. Clicking with the right mouse button has the opposite effect. The command zoom out restores the original figure. zoom off turns off the zoom feature.

The ginput command. This feature allows one to pick off the coordinates of points on a figure using the mouse. The command is [x,y] = ginput(n). Move the pointer to n different points in the figure, and click at each point with the left mouse button. When you have clicked on the last point, the coordinates of the n points are displayed on the screen in column vectors **x** and **y**.

B.10   Plotting 3-D graphs

In this section we will see how to graph functions of two variables. The basic plotting variable for 2-D graphs is the vector. In 3-D graphs, the basic plotting variable is the matrix.

Most often we want to plot a function $f(x, y)$ over the rectangle $a \leq x \leq b, c \leq y \leq d$. First construct a mesh over the rectangle by selecting a stepsize in the x direction, Δx, and a stepsize in the y direction, Δy. Then construct the vectors **x** and **y** with the commands x = a:delx:b and y = c:dely:d. The next command creates two matrices : [X,Y] = meshgrid(x,y). If n is the length of **x** and m is the length of **y**, both X and Y are $m \times n$ matrices. The m rows of X are all equal to the vector **x**, and the n colums of Y are all equal to the vector **y**. A corresponding $m \times n$ matrix of the values of f at the grid points is generated by the command Z = f(X,Y). We assume that $f(x, y)$ is expressed by an array-smart mfile f.m. A three-dimensional wire mesh surface is generated by the command mesh(X,Y,Z) while a faceted surface is generated by the command surf(X,Y,Z). We could, of course, combine two commands into one as mesh(X,Y,f(X,Y)) or surf(X,Y,f(X,Y)).

For an example, write a function mfile for $f(x, y) = \exp(-(x-3)^3 + (y-2)^2)$.

```
function z = f(x,y)
z = exp( ( (x-3).^2 + (y-2).^2) );
```

To graph f over the rectangle $[0, 4] \times [0, 6]$ use the sequence of commands

```
>> x = 0:.1:4;
>> y = 0:.1:6;
>> [X,Y] = meshgrid(x,y);
>> Z = f(X,Y);
>> mesh(X,Y,Z)
```

Follow this with the command surf(X,Y,Z) to see the faceted surface. For another example, try Z = sin(X-Y). The wire mesh surface looks like this (see Figure B.2).

We do not always plot graphs of functions over rectangles. Sometimes the domain of interest for the function is a disk or an ellipse. In this case the matrices X and Y cannot be generated by the simple meshgrid command. To generate a grid for the disk of radius a, centered at the origin, we must introduce polar coordinates. Let Δr and $\Delta \theta$ be the spacing of the mesh in the radial and angular variables. Then we construct the X and Y matrices for polar coordinates with these commands.

```
>> r = 0:delr:a;
>> theta = 0:deltheta:2*pi;
>> X = r'*cos(theta);
>> Y = r'*sin(theta);
```

In the last two lines we take the matrix product of the column vector which is **r** transpose and the row vectors $\cos\theta$ and $\sin\theta$. These operations produce full matrices. To see the mesh of coordinates in the x, y plane use mesh(X,Y,Z), where Z = zeros(size(X)). Now try mesh(X,Y,X+Y) and mesh(X,Y,sin(pi*(X.^2 +Y.^2))). The result is shown in Figure B.3.

If the function f to be graphed is given as a function of polar coordinates $f = f(r, \theta)$, we must construct matrices R and Θ to use as the arguments for f. We want all the rows of R to be the vector **r** and all the columns of Θ to be the vector θ. This is accomplished by taking the matrix products

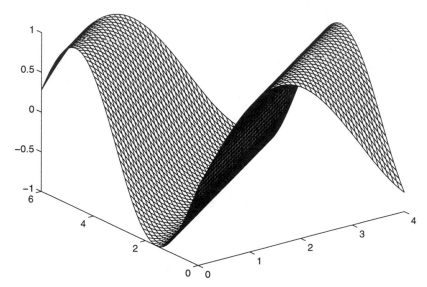

FIGURE B.2
Mesh surface of $z = \sin(x - y)$.

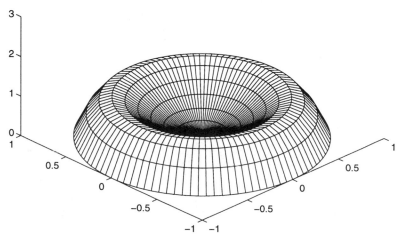

FIGURE B.3
Plot of $f(x, y) = \sin(\pi(x^2 + y^2))$ over the unit disk.

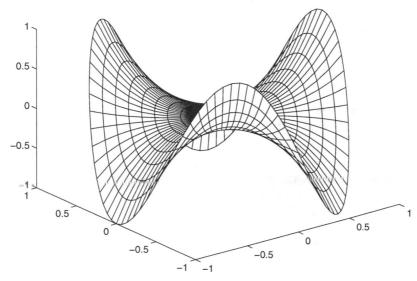

FIGURE B.4
Graph of $f(r, \theta) = r^3 \cos(3\theta)$ over the unit disk.

```
>> R = r'*ones(size(theta));
>> TH = ones(size(r))'*theta;
```

Now we can plot the function $f(r, \theta) = r^3 \cos(3\theta)$ over the disk of radius a with the command `mesh(X,Y,Z)` where `Z = R.^3.*cos(3*TH)`. The result is shown in Figure B.4.

Further 3-D graphing features

The `axis` command has a 3-D analogue. To set the scaling with limits a, b for the x axis, c, d for the y axis, and e, f for the z axis, use the command `axis([a b c d e f])`.

The `view` command. The direction from which a 3-D graph is viewed is specified by two angles: the azimuthal angle, which is the angle of rotation from the negative y axis, and the angle of elevation above the x, y plane. For example the view down the positive x axis toward the origin has azimuthal angle of 90° and elevation of 0°. The default direction for the 3-D graphs has angles $(-40, 30)$. To change the view enter, for example, `view(45,50)`. Try the `view` command with the graph of Figure B.2.

The `pcolor` command. Instead of a 3-D surface plot or mesh plot, the command `pcolor(X,Y,Z)` makes a 2-D plot of the colors used in `surf` or `mesh`. This is especially useful in visualizing heat flow. It can also be combined effectively with the next command.

The contour command. To see the level sets (contour lines or curves) of a function $f(x, y)$, proceed as in the surf or mesh command to make meshgrid matrices and a function value matrix. Then contour(X,Y,Z,n,'k') will divide the interval [min f, max f] into $(n + 1)$ equal subintervals and plot n contour lines. The 'k' means they will be plotted in black. Other colors are possible.

B.11   Movies

Movies are constructed by using a for loop and the getframe command. Here is a sample program to see the vibrations of the lowest mode of the vibrating string on the interval [0, 1]. The solution is the function $u(x, t) = \sin(\pi t) \sin(\pi x)$. We shall make a movie which goes through one full period of the motion, which in this case is $T = 2$. Then by running the movie several times, we will be able to see the periodic vibrations. Let the movie have 20 frames, i.e., one snapshot per time interval $\Delta t = .1$. Each snapshot will be stored as a column of a matrix, call it M, as follows:

```
x = 0:.1:10;
moviein(20);
for j = 1:20
      t = (j-1)*.1
      u = sin(pi*t)*sin(pi*x)
      plot(x,u)
      M(:,j) = getframe;
end
```

The getframe command make a scan of the figure that results from the plot command. The length of columns of M depends on the size of the figure. If you enlarge the figure, the matrix M will be larger, requiring more memory.

Now that the movie is made, we can run it with the command

```
>> movie(M,4,10)
```

The 4 is the number of times the movie will be run, in this case 4 complete periods. The 10 is the number of frames per second. If there is no third argument, the movies runs at the default speed of 12 frames per second.

Appendix C

References

[A] S. Antman, The equations for the large vibrations of strings, *American Mathematical Monthly*, **87**(1980), 359-370.

[AS] M. Abramovitz and I. Stegun, *Handbook of Mathematical functions*, Dover Publications, New York, 1964.

[Bar] R. Bartle and D. Sherbert, *Introduction to Real Analysis*, Wiley, New York, 1982.

[BaS] I. Babuska and B. Szabo, *Finite Element Analysis*, Wiley-Interscience, New York, 1991.

[BD] W. Boyce and R. DiPrima, *Elementary Differential Equations*, sixth edition, Wiley, New York, 1997.

[Be] J. Benedetto, *Harmonic Analysis and Applicatons* CRC Press, Boca Raton, Fl., 1996.

[Bo] F. Bowman, *Introduction to Bessel Functions*, Dover Publications, New York, 1958.

[BH] W. Briggs and V. Henson, *The DFT, An Owners Manual for the Discrete Fourier Transform*, SIAM, Philadelphia, 1995.

[BS] S. Brenner and L. R. Scott, *The Mathematical Theory of Finite Element Methods*, Springer-Verlag, New York, 1994.

[CF] R. Courant and K.O. Friedrichs, *Supersonic Flow and Shock Waves*, Springer-Verlag, New York, 1948.

[CourHilb] R. Courant and D. Hilbert, *Methods of Mathematical Physics*, Vols. 1 and 2, Wiley-Interscience, New York, 1962.

[ChowHale] S. Chow and J.K. Hale, *Methods of Bifurcation Theory*, Springer-Verlag, New York, 1982.

[CL], E. Coddington and N. Levinson, *Theory of Ordinary Differential Equations*, McGraw-Hill, New York, 1955.

[E] L. Edelstein-Keshet, *Mathematical Models in Biology*, Birkhäuser Mathematics Series, McGraw-Hill, New York, 1988.

[EEHJ] K. Eriksson, D.Estep, P. Hansbo, and C. Johnson. *Computational Differential Equations*, Cambridge University Press, New York, 1996.

[Fe] W. Feller, *Probability Theory and Its Applications*, volume 1, Wiley, New York, 1950.

[Fi] S. Fisher, *Complex Variables*, Wadsworth, Brooks/Cole, Belmont, Ca, 1986.

[Fo] G. Folland, *Introduction to Partial Differential Equations*, Princeton University Press, Princeton, N.J., 1976.

[FT] A. French and E. Taylor, *An Introduction to Quantum Physics*, W. W. Norton, New York, 1978.

[GF] I. Gelfand and S. Fomin, *Calculus of Variations*, Prentice Hall, Englewood Cliffs, N.J., 1963.

[GO] D. Gottlieb and S. Orszag, *Numerical Analysis of Spectral Methods*, CBMS-NSF Regional Conference Series in Applied Mathematics, SIAM, Philadelphi, PA, 1977 and 1993.

[HLS] T.Hou, J. Lowengrub, and M. Shelley, Removing stiffness from interfacial flows with surface tension, *Journal of Computational Physics*, **114** (1994), 314-338.

[IM] K. Ingard and P. Morse, *Theoretical Acoustics*, McGraw-Hill, New York, 1968.

[Ja] R. D. Jackson, *Classical Electrodynamics*, Second edition, Wiley, New York, 1975.

[John1] F. John, *Partial Differential Equations*, 4th edition, Springer-Verlag, New York, 1982.

[John2] F. John, *Plane Waves and Spherical Means Applied to Partial Differential Equations*, Interscience, New York, 1955. Republished by Springer-Verlag, New York, 1981.

[Johnson] C. Johnson, *Numerical Solution of Partial Differential Equations by the Finite Element Method*, Cambridge University Press, New York, Cambridge, U.K., 1994

[KC] D. Kincaid and W. Cheney, *Numerical Analysis*, Brooks/Cole Publishing, Pacific Grove, Ca., 1991.

[La] P.D. Lax, *Hyberbolic Systems of Conservation Laws and the Mathematical Theory of Shock Waves*, SIAM Regional Conference Series in Applied Mathematics, 11, 1972.

[Le] R.J. LeVeque, *Numerical Methods for Conservation Laws*, Second edition, Birkhäuser, Boston-Basel, 1992.

[Lo] J. Logan, *An Introduction to Nonlinear Partial Differential Equations*, Wiley-Interscience, New York, 1994.

[M] A. Messiah, Quantum Mechanics, two volumes, Wiley, New York, 1966.

[PW] M. Protter and H. Weinberger, *Maximum Principles in Partial Differential Equations*, Prentice Hall, Englewood Cliffs, N.J., 1967.

[RM] R. D. Richtmeyer and K. W. Morton, *Difference Methods for Initial Value Problems*, Second Edition, Wiley-Interscience, New York, 1967.

[RR] M. Renardy and R. Rogers, *An Introduction to Partial Differential Equations*, Springer-Verlag, New York, 1993.

[Sm] J. Smoller, *Shock Waves and Reaction-Diffusion Equations*, Springer-Verlag, New York, 1983.

[Strichartz] R. Strichartz, *Guide to Distribution Theory and Fourier Transform*, CRC Press, Boca Raton, Fl., 1994.

[St] J. C. Strikwerda, *Finite Difference Schemes and Partial Differential Equations*, Brooks/Cole Publishing, Pacific Grove, Ca., 1989.

[Str1] W. Strauss, *Partial Differential Equations, An Introduction*, Wiley, New York, 1992.

[Str2] W. Strauss, *Nonlinear Wave Equations*, CBMS Regional Conference Series in Mathematics, # 73, American Mathematical Society, 1989.

[T] J. Troutman, *Variational Calculus with Elementary Convexity*, Springer-Verlag, New York, 1983.

[Wh] G.B. Whitham, *Linear and Nonlinear Waves*, Wiley-Interscience, New York, 1974.

[Z] E. C. Zachmanoglou and D. W. Thoe, *Introduction to Partial Differential Equations with Applications*, Dover Publications, New York, 1986.

Appendix D

Solutions to Selected Problems

Chapter 2

Section 2.2

1. $u(x, t) = e^{-t} f(x - ct)$.

2. The ODE satisfied by $v = u(x_0 + ct, t)$ is $v'(t) = t(x_0 + ct)$ with solution

$$v(t) = v(0) + \int_0^t s(x_0 + cs)ds.$$

3. The ODE satisfied by $v = u(x_0 + ct, t)$ is $v' = v^2$. The solution is given in equation (1.12).

4. Do the characteristics cut the line l in a single point?

6. (c) The oscillations become narrower because the characteristics are all tending toward the t axis as t increases.

7. (a) The characteristics are parts of the parabolas $2t = x^2 - x_0^2$.
   (b) $p(x, t)$ is not defined for $2t > x^2$.
   (d) The hump gets narrower because the characteristic speed is greater on the left side of the hump than on the right side.

Section 2.3

2. (a) $f'(x_0) < 0$ means that for some $x_1 > x_0$, we have $f(x_1) < f(x_0)$. Show that the characteristics emanating from $(x_0, 0)$ and $(x_1, 0)$ will cross at some time $t > 0$.

4. Find the time t when the top of the wave (starting at $x = 0$) overtakes the bottom of the wave (starting at $x = L$).

Section 2.4

1. Use the fact that u is constant on characterstics.

2. Linearize about the constant solution $u(x, t) \equiv 1$. The linearized equation is $u_t + u_x = 0$.

Section 2.5

1. (b) Break the verification into three cases: (i) $a < b < t/2$; (ii) $a < t/2 < b$; (iii) $t/2 < a < b$.

2. $t_{**} = 4$, and the step wave moves with speed $3/4$. The R-H condition is satisfied.

4. $t_* = 4/3$, $t_{**} = 2$. The shock speed is $1/2$.

Section 2.6

1. $u_{j,n} = (-1)^j (1 - 2c/\rho)^n$.

2. $c_{max} = \max c(f(x)) = \max(1 + x/(1 + x^2))$.

3. (b) We must restrict the values of f to lie in the interval $[0, \beta/2]$. Since u is constant on characteristics, this implies $0 \le u(x, t) \le \beta/2$. $c_{max} = \max_u F'(u) = \beta$.

4. (a) Plots of cl and shocks begin to differ significantly after the shock has formed at time $t = 8/3$. With $\Delta x = \Delta t = .05$, cl rounds off the step of the shock wave at top and bottom for $t > t_{**} = 4$.
   (b) Plots are now closer, but the same effects are present.

8. (a) $c_{max} = \max f5(x) = 2$, so CFL condition is satisfied if $\Delta t \le \Delta x/2$.
   (b) $\max |f'(x)|$ occurs at $x = .5, 1.5$ with $f'(1.5) = -3 \exp(-1/2)$.
   (c) Shock strength decreases because of rarefaction wave overtaking the shock from behind. Shock speed also decreases according to (2.28).

9. Two shocks and one rarefaction.

10. The width of the transition region is $(n+1)\Delta x$ when $u_{j,n}$ is computed with exact arithmetic. However, when $u_{j,n}$ is computed in the double precision of MATLAB , the transition region reaches a maximum width of $5\Delta x$.

Chapter 3

Section 3.2

3. You cannot apply the maximum principle to either u or v by itself because they do not satisfy the homogeneous heat equation. On the other hand $u - v$ does satisfy (3.4).

Section 3.3

3. (a) Look at the convexity of the graph. Where is $u_{xx} > 0$, and where is $u_{xx} < 0$? (b) If $u(1/2, t) = Ct^\gamma$, then $\log(u(1/2, t)) = \log C + \gamma \log t$, provides two linear equations for $\log C$ and γ when different values of t are substituted. (c) The leading factor in (3.16) is $t^{-1/2}$.

4. (a)
$$u(x, t) = \frac{1}{2}(\alpha + \beta) + \frac{1}{2}(\beta - \alpha) erf(\frac{x}{\sqrt{4kt}}).$$

5. Break up the integral

$$\int_{-\infty}^{\infty} = \int_{-\infty}^{-B} + \int_{-B}^{B} + \int_{B}^{\infty}$$

and show that the middle integral tends to zero as $t \to \infty$ while the other two parts behave like the two integrals of exercise 4.

6. (d) By symmetry $m = 0$, and $p(t) = 2kt$. e) $erf(\sqrt{2})$.

7. (b)

$$e^{-\frac{(y-x)^2}{4kt}} = \left[1 + \frac{(x-m)(y-m)}{2kt} + [-\frac{1}{2kt} + \frac{(x-m)^2}{4(kt)^2}]\frac{(y-m)^2}{2}\right]e^{-\frac{(x-m)^2}{4kt}} \cdots$$

Integrate this expression against $f(y)$. Note that the second term will cancel.

8. For any continuous function $f(x)$,

$$\lim_{h \to 0} \frac{1}{h} \int_{0}^{h} f(x)dx = f(0).$$

10.

$$(v_\varepsilon)_s - k(v_\varepsilon)_{xx} = 0, \quad v_\varepsilon(x, 0) = f(x).$$

Section 3.4

2.

$$\frac{d}{dt}\int_0^\infty u(x,t)dx = k\int_0^\infty u_{xx}(x,t)dx = ku_x(x,t)\Big|_0^\infty.$$

Section 3.3.

6. $v(x,t) = U - u(x,t)$.

7. $v(x,t) = 100(1 - erf(\frac{x}{\sqrt{4kt}}))$.

9. Use the result of exercise 2 (a).

Section 3.5

3. The width $s_r = (\frac{4\varepsilon}{u_l - u_r}) \log 19$.

4. $\varepsilon\varphi'' + \varphi'(c - \varphi) - \gamma\varphi = 0$.

5. (b) $k\varphi'' + c\varphi + \varphi(1 - \varphi) = 0$.

   (c) The eigenvalues of the matrix of the linearized system are $\lambda = \frac{-c \pm \sqrt{c^2 - 4k}}{2k}$

   at $(0, 0)$, and are $\lambda = \frac{-c \pm \sqrt{c^2 + 4k}}{2k}$ at $(1, 0)$.

Chapter 4

Section 4.1

3. When $\lambda < 0$, we write $\lambda = -\mu$ with $\mu > 0$. Equation (4.3) becomes $\varphi'' = \mu\varphi$ with general solution

$$\varphi(x) = A\cosh(x\sqrt{\mu}) + B\sinh(x\sqrt{\mu}).$$

The boundary conditions (4.4) lead to a 2×2 system for A and B with nonzero determinant for any $\mu > 0$.

5. $A_n = -2L(-1)^n/(n\pi) = O(1/n)$. The series does not converge uniformly because $S_n(L) = 0$ for all n, but $f(L) = L$.

6. $A_n = 2[1 - \cos(n\pi/2)]/(n\pi)$

7. $\exp(-\lambda_n kt) \le \exp(-\lambda_1 kt)$ for all $n \ge 1$. Hence

$$\left| \sum A_n e^{-\lambda_n kt} \varphi_n(x) \right| \le \sum |A_n| e^{-\lambda_1 kt}.$$

10. (b) This means dividing the estimate for the error $|u(x, t) - u_1(x, t)|$ by $u_1(L/2, t)$.

12. (a) The flux at the right end of the bar is greater than that at the left end for a while (look at u_x). (b) The equation and boundary conditions are both homogeneous. Hence the only way t appears in the solution is in the expression kt.

13. The steady state $U(x) = x/2 - 2$.

16. If $\gamma > k\lambda_1$ and $A_1 \ne 0$, the solution will grow.

Section 4.3

2. $A_0 = 1$, $A_n = 2 \sin(n\pi/2)/(n\pi)$ for $n \ge 1$.

4.

$$\varphi_n(x) = \cos\left(\frac{n\pi}{2L}\right), \quad A_n = \frac{4L}{n\pi} \sin\left(\frac{n\pi}{2}\right) - \frac{8L}{(n\pi)^2}, \quad n = 1, 3, 5, \ldots$$

(c) Set the index $n = 2m - 1$, so that $m = 1, 2, 3, \ldots$. Then $\lambda_m = [(m - 1/2)\pi/L]^2$ and $\tilde{\lambda}_m = (m\pi/L)^2$.

7. The average value of the initial data is .17725.

9. The first four roots of $\sigma(s) = \tan(s) - Lh/s = 0$ are $s_1 = 1.4289$, $s_2 = 4.3058$, $s_3 = 7.2281$, $s_4 = 10.2003$.

10. (b) When $h < 0$ the equation for the positive eigenvalues is still $\tan(s) = Lh/s$. The first three roots when $L = 10$ and $h = -1$ are $s_1 = 1.7434$, $s_2 = 5.1912$, $s_3 = 8.5621$.
(c) With $s = L\sqrt{\gamma}$, the equation to solve for the negative eigenvalue is $\tanh(s) = -Lh/s$. There is one positive root (remember $h < 0$).

Section 4.4

2. Expand $q(x, t) = \sin(t)r(x) = \sin(t) \sum r_n \varphi_n(x)$. Then the coefficients $u_n(t)$ are given by

$$u_n(t) = r_n \int_0^t e^{-\lambda_n k(t-s)} \sin(s)\,ds.$$

You must evaluate the integral by integrating by parts twice. Finally,

$$u(x, t) = P(x) \sin t - Q(x) \cos t + \text{decaying terms},$$

where P and Q are given by series expansions in the eigenfunctions.

4.   (a) Take $p(x, t) = l(t) + r(t)x$. b) $v(x, t)$ solves

$$v_t - k v_{xx} = -p_t, \quad v(0, t) = v_x(L, t) = 0, \quad v(x, 0) = -p(x, 0).$$

Expand $p(x, t) = \sum p_n(t) \varphi_n(x)$:

$$u(x, t) = p(x, t) + v(x, t)$$

$$= p(x, t) + \sum_1^\infty [-p_n(0) e^{\lambda_n k t} - \int_0^t e^{\lambda_n k(t-s)} p_n'(s) ds] \varphi_n(x).$$

(c) If $p_n'(t) = 0$ for $t \geq T$, then all terms in the sum for v tend to zero as $t \to \infty$. The steady state $U(x) = \lim_{t \to \infty} p(x, t) = \alpha + \beta x$.

5. $\int_0^{10} q(x) dx = 0$ so condition (4.50) is satisfied.

6. $q_0 = (1/10) \int_0^{10} q(x) dx \approx \sqrt{\pi}/10 = .177$. For large t it can be seen that the maximum of the solution grows as $q_0 t = .177 \, t$.

9. $a = 1/6$, $b = -2$, $c = 5/6$, $d = -16/3$. The value of the steady state solution $U(x)$ at $x = 5$ is $-7/6 = -1.16$.

10.   (a) The temperature at $x = 5$ reaches 50° at time $t = 3$.

Chapter 5

Section 5.3

3. $u(x, t)$ is given by the following formulas. Note that in each region the formula is a strict solution of the wave equation. Across the characteristic boundaries, the formulas are continuous and provide a weak solution.

$$u(x, t) = \begin{cases} \frac{1}{2c}[e^{x-ct} - e^{x+ct}], & x < -ct \\ \frac{1}{2c}[e^{x-ct} - e^{-(x+ct)}], & -ct < x < ct \\ \frac{1}{2c}[e^{-(x-ct)} - e^{-(x+ct)}], & x > ct \end{cases}$$

5. $f(x) = u(x, 0) = F(x) + G(x) = \exp[-(x + 5)^2] - 2\exp[-(x - 5)^2]$.
$g(x) = u_t(x, 0) = -F'(x) + G'(x)$
$= 2(x + 5)\exp[-(x + 5)^2] + 4(x - 5)\exp[-(x - 5)^2]$.

7. (b) All except 6.

9. (a) $v(x, 0) = f(x)$, $v_t(x, 0) = u_t(x, 0) + du(x, 0) = g(x) + df(x)$.

14. (a) $c = 1 \pm \sqrt{2}$. (b) $c = 1, 2$. In this case both families of charactertistics lean to the right in the x, t plane.

Section 5.4

3. (a) The reflected wave is upside down and reversed right to left.
(b) The reflected wave is right side up and reversed right to left.

4. $u_x(0, t) = hu(0, t)$, $h > 0$, implies

$$\frac{de(t)}{dt} = -\frac{h}{2}\frac{d}{dt}u^2(0, t),$$

so that $\tilde{e}(t) \equiv e(t) + (h/2)u^2(0, t)$ is the conserved energy.

7. $c_l = c_r(1 - R)(1 + R)$.

8. (b) Reflection from a fixed end. (c) For $c_l \gg 1$, like reflection from a free end with boundary condition $u_x(0, t) = 0$.

11. For $s < 0$, $F(s) = \int_0^s h(-\sigma/c)d\sigma$.

Section 5.5

2. $\omega_n = c(n\pi/L)$ and $\omega_n B_n = 2L[\cos(n\pi) - 1]/(n\pi)^2$ for $n > 0$, and $B_0 = 1$. Only the odd frequencies $\omega_1, \omega_3, \ldots$ are present.

4. $\omega_n B_n = 2U[\cos(3n\pi/4) - \cos(n\pi/4)]/(n\pi) = 4U\cos(3n\pi/4)/(n\pi)$.

5. (a) The "corners" move (horizontally) with speed 1, along the characteristics. For $2.5 < t < 5$, the snapshots have four corners (points where the first derivative has a jump). (b) The period of the motion is $T = 2\pi/\omega_1 = 20$.

6. (c) The frequencies $\omega_n = \sqrt{\lambda_n^2 + k - d^2}$ in the case when all mode are underdamped. All modes are exponentially damped oscillations. (d) When $d > 0$, we do not expect periodic motion because the motion is damping out. When $d = 0$, for the motion to be periodic, all the frequencies present must be integer multiples of some fundamental frequency. However, with $k > 0$, this may not be the case.

7.   (b) The effect of the damping term is to slow down the *moving* part of the string. In all cases, the singularities move (horizontally) with speed 1. They are unaffected by modifying lower order terms in the equation.

11. $\varphi_n(x) = \sin(n\pi x/(2L))$, $\lambda_n = (n\pi/(2L))^2$ for $n = 1, 3, 5, \ldots$. The coefficients $A_n = 8L \sin(n\pi/4)/(n\pi)^2$ for n odd.

12.   (b) For $t > T$,

$$u(x, t) = \frac{1}{\omega_2^2}[\cos(\omega_2(T - t)) - \cos(\omega_2 t)].$$

14.   See exercise 4 of Section 5.4.

Section 5.7

2.   Along a solution trajectory $s \to (\varphi(s), \psi(s))$, the energy quantity (5.80) satisfies

$$\frac{dE}{ds} = \frac{2d\theta\psi^2}{\theta^2 - 1}.$$

This means that for $d > 0$ and $\theta > 1$, the trajectories spiral outward in the (φ, ψ) plane as $s \to \infty$.

3.   (b) Use the Taylor expansion

$$u(x, t) = u(x, 0) + u_t(x, 0) + \frac{1}{2}u_{tt}(x, 0)t^2 + O(t^3)$$

and take into account the fact that u satisfies the PDE.

4.   (b) If $u_{tt} - u_{xx} + u^3 = 0$, then $v = 2u$ satisfies $v_{tt} - v_{xx} + v^3/4 = 0$. (c) The center part of the string is accelerated downward more strongly than in the case $\gamma = 0$. (d) When $\gamma = 0$, the graphs for $\delta = 1$ and $\delta = 2$ are identical except for scale. (e) When $\gamma = 1$, the graphs for $\delta = 1$ and $\delta = 2$ are very different because of part (b).

5.   (b) When $\gamma = 1$, the waves do not emerge intact from their interaction.

6.   (b) Blow up of the solution of the solution occurs around $t = 3.86$. This is not just a problem with the numerical method.

7.   (a) The linear solution lags behind the nonlinear solution because the restoring force is greater in the nonlinear equation.

Chapter 6

Section 6.2

1. In cases (a), (b), and (d), the even, periodic extensions are continous, and piecewise C^1, while the odd, periodic extensions are discontinous either at $x = 0$ or at $x = \pm L$. (c) Both even and odd periodic extensions are continuous and piecewise C^1.

4. $\sigma_N^2 = O(N^{-3})$ for both even and odd extensions.

6. $\sigma_N^2 = O(N^{-1})$.

8. $\sigma_N^2 = O(N^{-1})$.

Section 6.3

3. $\hat{f}(\xi) = O(\xi|^{-1}|)$ as $|\xi| \to \infty$.

4. $\hat{f}(\xi) = O(|\xi|^{-2})$, faster than in exercise 3 because this f is continuous.

5. The Fourier transform of $\delta(x - a)$ is $\exp(-ia\xi)$.

Section 6.5

2. When $N = 2$, $\tilde{c}_0 = 1$, $\tilde{c}_1 = 5$. When $N = 4$, $\tilde{c}_0 = 1$, $\tilde{c}_1 = 3$, $\tilde{c}_2 = 0$, $\tilde{c}_3 = 2$.

3. The period of h_N is $N\pi$. When $N \geq 8$, the period interval of h_N includes $[-4\pi, 4\pi]$. Note that both the functions h_N and $\hat{f}(\xi) = 2\exp(-i\xi)\sin\xi/\xi$ are complex-valued. MATLAB graphs the real part.

4. (b) $h_N(\xi)$ has period $N\pi/8$. We must choose $N \geq 64$ for the period interval of h_N to include the viewing interval $[-4\pi, 4\pi]$. (period of $\exp(i * \xi)$ is $\pi/4$). (c) $N = 64$ gives a very good fit on $[-4\pi, 4\pi]$, better than the fit with $N = 64$ for the function of exercise 3.

Section 6.6

1. (a) $c_0 = \pi$, $c_k = i/k$, $k = \pm 1, \pm 2, \ldots$

2. (a) $c_0 = 2\pi^2/3 \approx 6.5797$, $c_k = -2/k^2$, $k = \pm 1, \pm 2, \ldots$

3. (a) $c_0 = \frac{1}{2\sqrt{\pi}} erf(\pi)$.

5. To evaluate h_N using the mfile h.m takes 826,576 flops while using ftrans takes only 71,811 flops. There may be slight variations.

6.   (a) Define $f(0) = 2/3\pi$. b) $N_* = 100$.

Chapter 7

Section 7.1

2. $g(\pm 2) = 0$ and $g'(\pm 2) = 0$, but g'' is not defined at $x = \pm 2$, so $n = 1$.

4.   The oscillations of the function cancel each other in the integral more and more as t increases.

6.   There will be less cancellation in the integral because the graph does not oscillate as rapidly around the stationary point x_0.

7. $p \approx .525$ and $C \approx 1.25$.

Section 7.2

2.   (a) $\omega = \alpha k - \beta k^3$. (b) The curves of constant phase $x(t)$ satisfy the ODE

$$x'(t) = \omega(k)/k = \alpha - \beta k^2 = (2\alpha + \frac{x}{t})/3.$$

(c) Since the wave number decreases as $\sigma = x/t$ increases, the wave number will decrease as we follow a particular crest.

4.   (a) Since $\omega'(k) = k/\sqrt{m^2 + k^2}$, the range is $[a/\sqrt{m^2 + a^2}, b/\sqrt{m^2 + b^2}]$. b) $\omega'(k) = \alpha - 3\beta k^2$, so the range is $[\alpha - 3\beta b^2, \alpha - 3\beta a^2]$. c) $\omega'(k) = 2\gamma k$, so the range is $[2\gamma a, 2\gamma b]$.   d) Spreads most for KdV, least for Klein-Gordon.

5.   Wavelength $\lambda = 2\pi/2 = \pi$ uniform across the packet when $t = 0$.

6.   (a) Components with $k > 2$ travel faster than those with $k < 2$ because $\omega'(k)$ is increasing. (b) Shorter wavelengths in front; yes. (c) Wavelength $\lambda \approx \pi$, wave number $k \approx 2$.

7.   (a) $\omega'(k) = k/\sqrt{k^2 + 4}$.   b) The local wave number at $x = 70.2$ when $t = 100$ is $k \approx 2$. The phase velocity at $(x, t) = (70.2, 100)$ should be approximately $\omega(k)/k = \sqrt{k^2 + 4}/k = \sqrt{2}$.

Section 7.4

3.   The oscillations away from $x = 0$ become very rapid as $t \to 0$, and the amplitude tends to ∞. As t gets larger, the oscillations broaden and the amplitude tends to zero.

4. A "bulge" in the amplitude when $t = .5$ and $n = 4$, moves off to the right as n increases. When $t = 2$, the Riemann sums have converged on $[0, 20]$ for $n \geq 20$. When $t = .1$, we must take n much larger to get convergence because the integrand oscillates so rapidly.

5. The solution of the Schrödinger equation has many small "kinks" as it converges to the discontinuous initial data, while the solution of the heat equation converges in a smooth manner to the discontinuous data.

Section 7.5

1. The ground state has the smallest probability of being found outside the interval $|x| < 1$. The higher the bound state, the larger the probability of being found outside the interval $|x| < 1$.

2. (a) Four roots on the upper branch (even eigenfunctions) and three roots on the lower branch (odd eigenfunctions).
    (c) $s_1 = 1.4276$, $s_2 = 2.852534$, $s_3 = 4.2711$.

3. (a) As $Q \to \infty$, the roots on the upper branch tend to $k\pi/2$, k odd, while the roots on the lower branch tend to $k\pi/2$, k even. Thus the kth eigenvalue $\lambda_k(Q) \to (k\pi/2)^2$. The kth eigenfunctions are

$$\varphi_k(x, Q) \to \begin{cases} \cos(k\pi x/2), & k \text{ odd} \\ \sin(k\pi x/2), & k \text{ even} \end{cases}.$$

5. As $\hbar \to 0$, $Q = 2ml^2 V/\hbar^2 \to \infty$ and by exercise 3, $\lambda_k(Q) \to (k\pi/2)^2$ as $Q \to \infty$. Hence $E_k \to 0$ as $\hbar \to 0$. The eigenfunctions $\psi_k(x)$ converge to the eigenfunctions for the vibrating string with fixed ends at $x = \pm l$.

Chapter 8

Section 8.1

2. (a) As usual, show that $dQ/dt = 0$.

$$\frac{dQ}{dt} = \int\int u_t \, dxdy = k \int\int \Delta u \, dxdy = k \lim_{\rho \to \infty} \int\int_{D_\rho} \Delta u \, dxdy,$$

where D_ρ is the disk of radius ρ. By the divergence theorem,

$$\int\int_{D_\rho} \Delta u \, dx dy = \int_{C_\rho} \frac{\partial u}{\partial n} \, ds \to 0$$

as $\rho \to \infty$ because of the hypotheses on the decay of u and its derivatives as $\rho \to \infty$.

(b)

$$\frac{dm_x}{dt} = \frac{k}{Q} \int\int x \Delta u \, dx dy$$

and again by the divergence theorem,

$$\int\int_{D_\rho} x \Delta u \, dx dy = \int_{C_\rho} [x \frac{\partial u}{\partial n} - n_1 u] \, ds.$$

(c)

$$\int\int [(x - m_x)^2 + (y - m_y)^2] \Delta u \, dx dy = \int\int r^2 \Delta u \, dx dy$$

and by the divergence theorem,

$$\int\int_{D_\rho} r^2 \Delta u \, dx dy = \int_{C_\rho} [r^2 u_r - 2ru] \, ds + \int\int_{D_\rho} u \Delta(r^2) \, dx dy.$$

4.   (a)

$$\int\int e^{-\frac{(x-a)^2 + (y-b)^2}{4kt}} \, dx dy = \int\int e^{-\frac{x^2+y^2}{4kt}} \, dx dy = 2\pi \int_0^\infty e^{-\frac{r^2}{4kt}} r \, dr.$$

$$\int\int x e^{-\frac{(x-a)^2 + (y-b)^2}{4kt}} \, dx dy = a + \int\int (x - a) e^{-\frac{(x-a)^2 + (y-b)^2}{4kt}} \, dx dy = a.$$

$$\int\int [(x - a)^2 + (y - b)^2] e^{-\frac{(x-a)^2 + (y-b)^2}{4kt}} \, dx dy = 2\pi \int_0^\infty r^3 e^{-\frac{r^2}{4kt}} \, dr$$

$$= (2\pi)(4kt)^2 \int_0^\infty s^3 e^{-s^2} \, ds,$$

which implies that $p(t) \to 0$ as $t \to 0$.
(b) The fraction is $1 - \exp(-4kt)$.

5. The expansion of $(\xi, \eta) \rightarrow S_2(x - \xi, y - \eta, t)$ around $\xi = m_x$, $\eta = m_y$ is

$$S_2(x - \xi, y - \eta, t) = S_2(x - m_x, y - m_y, t) \times$$

$$[1 + \frac{(x - m_x)(\xi - m_x) + (y - m_y)(\eta - m_y)}{2t}$$

$$- \frac{(\xi - m_x)^2 + (\eta - m_y)^2}{4t} + O(t^{-3})].$$

Integrate against $f(\xi, \eta)$ noting that the second term integrates to zero.

6. (b) When $t \approx 7.2$ there is a single maximum at $(4, 6)$.
   (c) $m_x - 2.6$ and $m_y = 3.8$.

9. Let $T(x, y, t) = S_2(x, y, t) - S_2(-x, y, t)$. Then the solution of the initial value problem is

$$u(x, y, t) = \int \int_{\xi > 0} T(x - \xi, y - \eta, t) f(\xi, \eta) \, d\xi d\eta.$$

Section 8.2

1. (a) $u = 0$ on the semicircle, $r = a$, $0 \le \theta \le \pi$, and $u_y = 0$ on the straight segment $y = 0$. (b) All the eigenvalues $\lambda_n > 0$ by (8.15), so

$$u(x, y, t) = \sum_n A_n e^{-\lambda_n k t} \varphi_n(x, y) \rightarrow 0$$

as $t \rightarrow \infty$.

Section 8.3

1. $\varphi_{m,n}(x, y) = \sin(m\pi x/a) \sin((2n-1)\pi/(2b))$, $m, n = 1, 2, 3, \ldots$ $\lambda_{m,n} = (m\pi/a)^2 + ((2n - 1)\pi/(2b))^2$.

2. The level curves are perpendicular to the upper edge because it is insulated. On the insulated edge, the spatial gradient $\nabla_{x,y} u$ is tangent to the boundary, and on the edges where $u = 0$ it is perpendicular to the boundary. The level curves are nearly parallel to the boundary near these edges.

3. (b) With $\mu_n = (n\pi/b)^2$, the problem for the factor $X(x)$ is

$$-X'' + (\lambda - \mu_n)X = 0,$$

with the condition that X remain bounded as $x \to \pm\infty$.
(c) The IVP for u_n is

$$(u_n)_t - k(u_n)_{xx} + \mu_n u_n = 0, \quad u_n(x, 0) = f_n(x).$$

4.

$$\varphi_{m,n}(x, y) = \cos(\frac{m\pi x}{a})\sin(\frac{(2n-1)\pi y}{2b}), \quad \lambda_{m,n} = (\frac{m\pi}{a})^2 + (\frac{(2n-1)\pi}{2b})^2,$$

$m = 0, 1, \ldots \ n = 1, 2, \ldots$.

$$u(x, y, t) = \sum_{m=0}^{\infty}\sum_{n=1}^{\infty} A_{m,n}e^{-(\lambda_{m,n}k+\gamma)t}\varphi_{m,n}(x, y).$$

Section 8.4

1. (a) Because $\theta \to v(r, \theta)$ has period 2π, so does v_θ, and hence $\int_0^{2\pi} v_{\theta\theta}d\theta = 0$.

2. (a) $\rho_1 = 2.40483$, $\rho_2 = 5.52008$, $\rho_3 = 8.65373$.

3. Since the initial data has no θ-dependence, we need only the eigenfunctions $\varphi_{m,0} = J_0(r\rho_m/a)$, so

$$u(r, t) = \sum_{m=1}^{\infty} A_{m,0}e^{-\lambda_{m,0}kt}\varphi_{m,0}(r).$$

4. (a) Suppose U satisfies Bessel's equation of order zero. Differentiate the equation to get a third-order equation in U. The U term can be eliminated using the fact that U satisfies Bessel's equation of order zero, leaving a-second order equation in $V = U'$.

6. Let $\mu_n = n\pi/\beta$. Angular eigenfunctions are $\cos(\mu_n\theta)$, $n = 0, 1, 2, \ldots$, and the radial factors must satisfy

$$r^2 R'' + r R' + (\lambda r^2 - \mu_n^2)R = 0.$$

Thus $R_\lambda(r) = J_{\mu_n}(r\sqrt{\lambda})$ where $J_\mu(\rho)$ must be a Bessel function of non-integer order μ.

Section 8.5

1. (a) Use the same boundary values and approximate the discontinuous initial data $f(x, y)$ by continuous functions

$$f_l(x, y) = (\frac{\pi - x}{\pi})^l \sin(y).$$

The maximum principle may be applied to the solution u_l with this data.

(b)

$$U(x, y) = [\cosh(\pi x) - \frac{\cosh(4\pi)}{\sinh(4\pi)} \sinh(\pi x)] \sin(y).$$

(d) The solution given by the truncated series does not satisfy the initial condition $u(x, y, 0) = 0$ inside G.

4. (b)

$$v(x, y, t) = \sum_{m=1}^{\infty} A_{m,0} e^{-\lambda_{m,0}kt} \varphi_{m,0}(xy),$$

with $A_{m,0} = (a^3/m\pi)[2(-1)^m - 1]$.

(c)

$$w(x, y, t) = \int_0^t e^{-\lambda_{1,1}k(t-s)} \sin(s)ds\, \varphi_{1,1}(x, y)$$

$$= \left[\frac{k\lambda_{1,1} \sin t - \cos t - e^{-\lambda_{1,1}kt}}{1 + (\lambda_{1,1}k)^2} \right] \varphi_{1,1}(x, y)$$

Section 8.6

4. (b)

$$u(r, t) = \frac{1}{2r} \int_{r-t}^{r+t} s e^{-s^2} ds = \frac{1}{4r}[e^{-(r-t)^2} - e^{-(r+t)^2}].$$

7. If $\mathbf{x} \cdot \alpha > 0$, then the sphere (or disk in R^2) with center $(\mathbf{x}, t)$ and radius ct is contained in the half-plane $\mathbf{x} \cdot \alpha > 0$ where the initial data f and g are both zero. Formulas (8.41) in 3D and (8.42) in 2D show that $u(\mathbf{x}, t) = 0$.

Section 8.7

2. Assuming u dies off rapidly along with its derivatives as $|\mathbf{x}| \to \infty$, we make the usual calculation

$$\frac{de}{dt} = \int \int_{x>0} [u_{tt}u_t + c^2 \nabla u \cdot \nabla u_t]\, dxdy$$

$$= - \int_R u_t(0, y, t)u_x(0, y, t)dy.$$

3. (b)

$$e(t) = \pi \int_0^\infty [\tilde{u}_t^2 + c^2\tilde{u}_r^2]rdr$$

$$= 2\pi c^2 \int_0^\infty (\varphi')^2(r - ct)dr + \pi c^2 \int_0^\infty [\frac{\varphi(r - ct)^2}{4r^3} - \frac{\varphi\varphi'(r - ct)}{r^2}]rdr.$$

Section 8.8

2.

$$u(r, t) = \frac{1}{2rc} \int\int_{D(r,t)} \rho q(\rho, s)d\rho ds$$

where $D(r, t)$ is the triangle $\{r - c(t - s) < \rho < r + c(t - s),\ 0 < s < t\}$.

3.   Assuming that $q(\rho, s) = 0$ for $|\rho| > a$, $\int_{-a}^a \rho q(\rho, s)d\rho = 0$ because $\rho q(\rho, s)$ is even. Hence for $r \geq 0$, and for $ct \geq (r + a)$,

$$\int\int_{D(r,t)} \rho q(\rho, s)d\rho ds = \int_{s_{bottom}}^t \int_{r-c(t-s)}^{r+c(t-s)} \rho q(\rho, s)d\rho ds,$$

where $s_{bottom} = t - (r + a)/c$. Now assume that $q(\rho, s) = \exp(i\omega s)q_0(\rho)$. We evaluate the integral in the region $r \geq a$ and $ct \geq r + a \geq 2a$ with the change of variable $\tau = s - s_{bottom}$. In this region,

$$u(r, t) = C\frac{e^{i\omega(t-(r+a)/c)}}{2rc},$$

where

$$C = \int_0^{2a/c} e^{i\omega\tau} \int_{c\tau-a}^a \rho q_0(\rho)d\rho d\tau.$$

Section 8.9

1.   (a) Yes, period $T = 2\pi/(c\sqrt{5})$.   b) No, $\omega_{1,2}$ and $\omega_{3,1}$ are not rational multiples of one another.

2.

$$u(x, y, t) = \left[\frac{1}{\omega_{3,4}} \int_0^t \sin(\omega_{3,4}(t - s)) \sin(\omega s)ds\right]\varphi_{3,4}(x, y)$$

If $\omega \neq \omega_{3,4}$, the solution consists of bounded oscillations. If $\omega = \omega_{3,4}$, there is resonance and

$$u(x, y, t) = \frac{1}{2\omega}[-t\cos(\omega t) + \frac{\sin(\omega t)}{\omega}]\varphi_{3,4}(x, y).$$

3.

$$u(x, y, t) = \sum_{m,n=1}^{\infty} B_{m,n} \sin(\omega_{m,n} t)\varphi_{m,n}(x, y).$$

(b)

$$\omega_{m,n} B_{m,n} = \frac{4}{\pi^2 \varepsilon \sqrt{2}} \left[\frac{2\cos(m(\pi/2 - \varepsilon))}{m} \right] \left[\frac{2\cos(n(\pi/2 - \varepsilon))}{n} \right].$$

(c)

$$\omega_{m,n} B_{m,n} = \frac{4}{\pi^2 \varepsilon \sqrt{2}} \left[\frac{\cos(m(\pi - 2\varepsilon)) - (-1)^m}{m} \right] \left[\frac{\cos(n(pi - 2\varepsilon)) - (-1)^n}{n} \right].$$

For $m = n = 1$, and ε small, use the Taylor expansions of $\cos(\pi/2 - \varepsilon)$ and $\cos(\pi - 2\varepsilon) + 1$ to deduce that in the first case,

$$\omega_{1,1} B_{1,1} \approx \frac{16\varepsilon}{\pi^2 \sqrt{2}},$$

and in the second case,

$$\omega_{1,1} B_{1,1} \approx \frac{16\varepsilon^3}{\pi^2 \sqrt{2}}.$$

When the drum is struck in the corner, the $\omega_{1,1}$ frequency component is smaller than when the drum is struck in the center.

7. (a) $\omega_1 = \omega_{2,4} = 2c\sqrt{5}$ and $\omega_2 = \omega_{6,3} = 3c\sqrt{5}$. (b) ω_1 and ω_2 are both integer multiples of $\omega = c\sqrt{5}$, so the period of motion is $T = 2\pi/\omega = 2\pi/(c\sqrt{5})$. (d) Yes, the line $x = \pi/2$.

Section 8.10

3. The system is
$$\epsilon u_t + v_x = 0$$
$$\mu v_t + u_x = 0$$

with solution

$$u(x,t) = F(x-ct)+G(x+ct) \text{ and } v(x,t) = \sqrt{\frac{\epsilon}{\mu}}[F(x-ct)-G(x+ct)].$$

5.   (a) $u(0,t) = 0$, no restriction on $v(0,t)$, $v(x,t) = -\sqrt{\epsilon/\mu} \cdot G(x+ct)$
     for $t \leq 0$.
     (b) For $t \geq 0$,

$$u(x,t) = G(x+ct)-G(ct-x) \text{ and } v(x,t) = -\sqrt{\epsilon/\mu}[G(x+ct+G(ct-x)].$$

6.   (b) The IBVP for the system is

$$\epsilon u_t + v_x = 0 \quad \mu v_t + u_x = 0, \ 0 < x < L,$$

$$u(0,t) = u(L,t) = 0,$$

$$u(x,0) = f(x), \quad v(x,0) = g(x), \ 0 < x < L.$$

(c) The system of ODE's is

$$\epsilon u_n' - (n\pi/L)v_n = 0, \quad \mu v_n' + (n\pi/L)u_n = 0,$$

with initial conditions

$$u_n(0) = A_n = \frac{2}{L}\int_0^L f(x)\sin(n\pi x/L)dx,$$

$$v_n(0) = \sqrt{\epsilon/\mu}B_n = \frac{2}{L}\int_0^L g(x)\cos(n\pi x/L)dx.$$

Finally, with $\omega_n = c(n\pi/L)$,

$$u(x,t) = \sum_1^\infty [A_n \cos(\omega_n t) + B_n \sin(\omega_n t)]\sin(n\pi x/L).$$

$$v(x,t) = \sqrt{\epsilon/\mu}\left[\frac{B_0}{2} + \sum_1^\infty [B_n \cos(\omega_n t) - A_n \sin(\omega_n t)]\cos(n\pi x/L)\right].$$

Chapter 9

Section 9.1

2.

$$\int\int\int_{B_r} \Delta u dx = \int\int_{S_r} \frac{\partial u}{\partial n} dS(\mathbf{x}) = r^2 \int_0^\pi \int_0^{2\pi} u_r(r, \theta, \phi) \sin\theta d\phi d\theta,$$

whence

$$\frac{1}{4\pi r^2} \int\int\int_{B_r} \Delta u dx = \frac{d}{dr}\left(\frac{1}{4\pi r^2} \int\int_{S_r} u dS\right).$$

3. Because $\theta \to u(r, \theta)$ has period 2π,

$$\frac{1}{2\pi r} \int_{C_r} \Delta u ds = \frac{1}{2\pi r} \int_0^{2\pi} [u_{rr} + \frac{1}{r}u_r + \frac{1}{r^2}u_{\theta\theta}]dr d\theta$$

$$= \frac{1}{2\pi} \int_0^{2\pi} [u_{rr} + \frac{1}{r}u_r](r, \theta)d\theta = (\partial_r^2 + \frac{1}{r}\partial_r)\frac{1}{2\pi} \int_0^{2\pi} u(r, \theta)d\theta.$$

4.

$$v_{\xi\xi} = u_{xx} \cos^2\theta + 2u_{xy} \cos\theta \sin\theta + u_{yy} \sin^2\theta$$

$$v_{\eta\eta} = u_{xx} \sin^2\theta - 2u_{xy} \cos\theta \sin\theta + u_{yy} \cos^2\theta.$$

6. (b)

$$\max_{\bar{D}_\rho} u(x, y) = u(\rho, 0) = \frac{1}{1-\rho}.$$

(c) $u = 0$ everywhere on ∂D_1, except at $(1, 0)$ where u is not defined. The maximum principle cannot be applied to u on $\bar{D}_1$ because u is not continuous on $\bar{D}_1$.

7. (c) Show that $u - v$ is subharmonic with $u - v = 0$ on ∂G. Apply the maximum principle for subharmonic functions.

Section 9.2

1. Keeping in mind that $\rho^2 = a^2 + b^2$, the level curve

$$\frac{\rho^2 - (x^2 + y^2)}{(x-a)^2 + (y-b)^2} = C$$

is the circle with center $(Ca, Cb)/(C + 1)$ and radius $\rho/(C + 1)$.

5.

$$u(r, \theta) = \frac{5}{2} + \frac{r}{\rho}(\frac{1}{2}\sin\theta - \frac{1}{10}\cos\theta) - \frac{1}{2}(\frac{r}{\rho})^2\sin(2\theta) + (\frac{r}{\rho})^4\cos(4\theta).$$

6. $f(\theta) = \cos\theta + 3\sin\theta + \sin(3\theta)$

7. (a) Maximum is attained on unit circle at $\theta = \pi$, and minimum at $\theta = 3\pi/2$.

$$\max_{\bar{D}_1} u = f(\pi) \approx 1, \quad \min_{\bar{D}_1} u = f(3\pi/2) \approx -3.$$

(b) The average of $f(\theta)$ over the unit circle is approximately $-.3158$.

8. Must require that $\int_0^\pi g(\theta)d\theta = 0$ and the solution is only determined up to a constant.

9. $u - v$ is a bounded, harmonic function on all of R^2 or R^3.

Section 9.3

1. We seek u in the form $u = v + w$ where

$$\Delta v = 0, \quad v(0, y) = f(y), \quad v(a, y) = g(y), \quad v(x, 0) = v_y(x, b) = 0.$$

Then

$$v(x, y) = \sum_{n \text{ odd}} [A_n \sinh(\frac{n\pi(a - x)}{2b}) + B_n \sinh(\frac{n\pi x}{2b})] \sin(\frac{n\pi y}{2b}).$$

$$A_n \sinh(\frac{n\pi a}{2b}) = \frac{2}{a}\int_0^a f(y)\sin(\frac{n\pi y}{2b})dy$$

$$B_n \sinh(\frac{n\pi a}{2b}) = \frac{2}{a}\int_0^a g(y)\sin(\frac{n\pi y}{2b})dy.$$

w is found in the form

$$w(x, y) = \sum_{m=1}^\infty [C_m \cosh(\frac{m\pi(b - y)}{a}) + D_m \sinh(\frac{m\pi y}{a})] \sin(\frac{m\pi x}{a}).$$

Section 9.4

1. From (9.27) in the case of two dimensions,

$$\nabla u(\mathbf{x}) = -\frac{1}{2\pi} \int \frac{\mathbf{x} - \mathbf{y}}{|\mathbf{x} - \mathbf{y}|^2} q(\mathbf{y}) \, d\mathbf{y}.$$

For $|\mathbf{x}| > \rho \geq |\mathbf{y}|$, $|\mathbf{x} - \mathbf{y}| \geq |\mathbf{x}| - \rho$, which implies that

$$\left| \frac{\mathbf{x} - \mathbf{y}}{|\mathbf{x} - \mathbf{y}|^2} \right| \leq \frac{1}{|\mathbf{x} - \rho|}.$$

Hence for $|\mathbf{x}| > \rho$,

$$|\nabla u(\mathbf{x})| \leq \frac{1}{2\pi (|\mathbf{x}| - \rho)} \int |q(\mathbf{y})| \, d\mathbf{y}.$$

2. $u(r) = (1 - r^2)/4$.

3.

$$r u_r = -\int_0^r s q(s) \, ds = \begin{cases} -q_0 r^2/2, & 0 < r < \rho \\ -q_0 \rho^2/2, & \rho < r < 1 \end{cases}.$$

$$u(r) = -\int_r^1 u' \, ds = \int_r^1 \frac{1}{s}(su') \, ds$$

$$= \begin{cases} q_0((1/2)(\rho^2 - r^2) - \rho^2 \log(\rho))/2, & r < \rho \\ -\rho^2 \log(\rho), & r > \rho \end{cases}.$$

4.

$$u(r) = \begin{cases} q_0(3\rho^2 - r^2)/6, & 0 < r < \rho \\ q_0 \rho^3/(3r), & r > \rho \end{cases}.$$

6. Greatest deflection occurs when the load is placed at the center $(0, 0)$.

7. By the maximum principle, $\min_{\bar{G}_\varepsilon} \gamma = 0$ because $\gamma = 0$ on ∂G. By the strong form of the maximum principle, $\gamma > 0$ in G_ε.

8. Let $\mathbf{y}$ be in the first quadrant with $\hat{\mathbf{y}} = (-y_1, y_2)$ in the second quadrant. Then the Green's function γ is given by

$$\gamma(\mathbf{x}, \mathbf{y}) = s(\mathbf{x} - \mathbf{y}) - s(\mathbf{x} - \hat{\mathbf{y}}) - s(\mathbf{x} + \hat{\mathbf{y}}) + s(\mathbf{x} + \mathbf{y}).$$

11. (d) If $|y| > a \geq |\eta|$, then $x^2 + (y - \eta)^2 \geq x^2 + (y - a)^2 \geq (r - a)^2$.

12. $u(x, y) = (1/\pi)[\arctan((y + a)/x) - \arctan((y - a)/x)]$.

13. (a) Combine the result of part (d) of execise 11 with the fact that for $x > 0$, $P(x, y, \eta) \leq 1/(\pi x)$.

14. (a)

$$\int_{\partial G} \frac{\partial v}{\partial n_\mathbf{x}} ds = \int_G -\Delta_\mathbf{x} s(\mathbf{x} - \mathbf{y}) d\mathbf{x} = \int_G \delta(\mathbf{x} - \mathbf{y} d\mathbf{x} = 1.$$

(b)

$$\int_{\partial G} \frac{\partial v}{\partial n_\mathbf{x}} ds = \int_G \Delta_\mathbf{x} (s(\mathbf{x} - \mathbf{y}) + \frac{|\mathbf{x} - \mathbf{y}|^2}{2nA}) d\mathbf{x} = 0.$$

(c) In the sense of generalized functions,

$$\Delta_\mathbf{x} N(\mathbf{x}, \mathbf{y}) = -\Delta_\mathbf{x}(s(\mathbf{x} - \mathbf{y}) + \frac{|\mathbf{x} - \mathbf{y}|^2}{2nA} + v(\mathbf{x}, \mathbf{y}))$$

$$= \delta(\mathbf{x} - \mathbf{y}) - \frac{1}{A}$$

15. This choice of N satisfies $N(x, y) = N(y, x)$ for any value of the constant C:

$$N(x, y) = \begin{cases} -y + (x^2 + y^2)/(2L) + C, & 0 < x < y \\ -x + (x^2 + y^2)/(2L) + C, & y < x < L \end{cases}.$$

16. For $\mathbf{y} \in H$, let $\hat{\mathbf{y}} = (-y_1, y_2, y_3)$. Then a choice of N which tends to zero at infinity is given by

$$N(\mathbf{x}, \mathbf{y}) = \frac{1}{4\pi}(\frac{1}{|\mathbf{x} - \mathbf{y}|} + \frac{1}{|\mathbf{x} - \hat{\mathbf{y}}|}).$$

20.

$$v(x, y) = \frac{y}{\pi}\left[\int_{-\infty}^0 \frac{B \, d\xi}{(x - \xi)^2 + y^2} + \int_0^\infty \frac{A \, d\xi}{(x - \xi)^2 + y^2}\right]$$

$$= \frac{A + B}{2} + \frac{A \arctan(x/y) - B \arctan(x/y)}{\pi}.$$

Section 9.5

1. Try $u_n(x) = x^n(1-x)^n$.

3. (a) $(x^2 u_x)_x + (y^2 u_u)_y = 0$. b) $u_{tt} - c^2 u_{xx} = 0$.

5. The Rayleigh quotient for this choice of v gives a value of 6. The first
    zero of $J_0(\rho)$ is $\rho_1 = 2.405$ and the first eigenvalue $\lambda_1 = \rho_1^2 = 5.7832$.
    Choosing $v(x, y) = \cos(.5\pi(1 - (x/a)^2 - (y/b)^2)$ gives a better approx-
    imation.

Chapter 10

Section 10.1

1. (a)

$$
\begin{aligned}
2u_1 - u_2 &= \Delta x^2 f(x_1) \\
-u_1 + 2u_2 - u_3 &= \Delta x^2 f(x_2) \\
\dots &= \dots \\
-u_{J-2} + 2u_{J-1} &= \Delta x^2 f(x_{J-1})
\end{aligned}
$$

(b) $u_j^k = (u_{j-1}^{k-1} + u_{j+1}^{k-1} + \Delta x^2 f(x_j))/2$. A MATLAB program which uses
vector indices is

```
J = input(' enter the number of subintervals   ')
N = input(' enter the number of iterations   ')
delx = 10/J; x = 0: delx : 10;
j = [2:J];
u = zeros(1,J+1);    q = delx^2*f(x);
for n = 1:N
    v = u;
    u(j) = .5*(v(j-1) + v(j+1) + q(j));
end
plot(x,u,x,f(x))
```

2. (b) The maximum error decreases as J^{-2} or Δx^2.

3. (c)

$$
\begin{aligned}
u_1 - u_2 \qquad\qquad &+c\Delta x^2 u_1 &&= \Delta x^2 f(x_1)\\
-u_1 + 2u_2 - u_3 \qquad\qquad &+c\Delta x^2 u_2 &&= \Delta x^2 f(x_2)\\
\cdots \qquad\qquad &\cdots &&\cdots\\
-u_{J-2} + u_{J-1} &+c\Delta x^2 u_{J-1} &&= \Delta x^2 f(x_{J-1}).
\end{aligned}
$$

When $f(x) \equiv 0$ and $c = 0$, the vector of values $u_j = 1$ for all j is a nontrival solution. This corresponds to the eigenfunction of part (a) (when $c = 0$) $\varphi(x) \equiv 1$ with eigenvalue $\lambda = 0$.

(d)

$$
\begin{aligned}
2u_0 - 2u_1 \qquad\qquad &+c\Delta x^2 u_0 &&= \Delta x^2 f(x_0)\\
-u_0 + 2u_1 - u_2 \qquad\qquad &+c\Delta x^2 u_1 &&= \Delta x^2 f(x_1)\\
\cdots \qquad\qquad &\cdots &&\cdots\\
-2u_{J-1} + 2u_J &+c\Delta x^2 u_J &&= \Delta x^2 f(x_J)
\end{aligned}
$$

4. (b) The simplest way to handle the variable coefficient case is to assume that both $a(x, y)$ and $b(x, y)$ are C^1 and write the equation as

$$
-a(x, y)u_{xx} - b(x, y)u_{yy} - a_x(x, y)u_x - b_y(x, y)u_y = q(x, y).
$$

Then approximate u_{xx}, u_{yy}, u_x and u_y by centered differences, making the truncation error $O((\Delta x)^2 + (\Delta y)^2)$. However, this technique requires a knowledge of the derivatives of the coefficients a_x and b_y. We can avoid this by approximating a_x and b_y again by centered differences, yielding a scheme with a truncation error of order $O((\Delta x)^2 + (\Delta y)^2)$.

6. Arranging the unknowns in a vector $\mathbf{u} = (u_{11}, u_{1,2}, u_{2,1}, u_{2,2})$, the matrix for the 4×4 system is

$$
\begin{bmatrix}
1 & -1/4 & -1/4 & 0\\
-1/4 & 1 & 0 & -1/4\\
-1/4 & 0 & 1 & -1/4\\
0 & -1/4 & -1/4 & 1
\end{bmatrix}.
$$

(d) The exact solution of the 4×4 system is

$$
[u_{1,1}, u_{1,2}, u_{2,1}, u_{2,2}] = [7.6234, 7.6234, 2.8298, 2.8298]
$$

Section 10.2

2.   (b) First substitute in v such that $v(0) = v(L) = 0$, and integrate by parts in the left side of the equation to deduce that

$$\int_0^L [u''(x) - q(x)]dx = 0$$

for all such v. This shows that $-u'' = q$. Next substitute a general $v \in C^1[0, L]$ and integrate by parts a second time to show that $u'(0) = u'(L) = 0$.

Section 10.3

1.   (a) The mass matrix

$$B = \begin{bmatrix} 2/3 & 1/6 & 0 & 0 & \dots & 0 \\ 1/6 & 2/3 & 1/6 & 0 & \dots & 0 \\ 0 & 1/6 & 2/3 & 1/6 & \dots & 0 \\ . & . & . & . & . & . \\ . & . & . & . & . & . \\ 0 & 0 & 0 & 0 & 1/6 & 2/3 \end{bmatrix}.$$

Appendix E

List of Computer Programs

Chapter 2

2.3

    `mtc`   Solves the initial value problem for $u_t + uu_x = 0$ using the method of characteristics before shocks form. Also computes linear approximation.

2.5

    `shocks`   Solves the initial value problem for $u_t + uu_x = 0$ using the modified method of characteristics to display development of shocks.

2.6

    `cl`   Solves the initial value problem for $u_t + uu_x = 0$ using upwind difference scheme. Must have nonnegative initial data.

Chapter 3

3.3

    `heat1`   Solves the initial value problem for $u_t - ku_{xx} = 0$ on the whole line by evaluating the convolution. Also computes the Gaussian approximation.

3.4

    `heat2`   Solves the initial value problem for $u_t - ku_{xx} = 0$ on the half-line with Dirichlet or Neumann boundary condition at $x = 0$ by evaluating convolution. Also computes the Gaussian approximation.

Chapter 4

4.2

    `heat3`   Solves initial value problem for $u_t - u_{xx} = 0$ on $0 < x < 10$ with Dirichlet boundary conditions (homogenous and inhomogeneous) using Crank-Nicholson.

4.3

    `heat4`   Solves initial value problem for $u_t - ku_{xx} = q$ on $0 < x < 10$ with Neumann boundary conditions (homogeneous and inhomogeneous) and with sources using Crank-Nicholson.

4.4

    `heat5`   Solves initial value problem for $u_t - ku_{xx} = 0$ on $-10 < x < 10$ with piecewise constant k. u is assigned at $x = \pm 10$.

Chapter 5

5.4

    `film_rfl`   Makes a 40-frame movie of the reflection of a wave from the left end of the half-line with the Dirichlet boundary condition.

    `robin`   Constructs solutions of $u_{tt} - c^2 u_{xx} = 0$ on the half-line $x > 0$ with the Robin boundary condition.

    `trans`   Constructs solutions of $u_{tt} - c^2 u_{xx} = 0$ on whole line with piecewise constant c (transmission problem).

5.5

wave1   Solves initial value problem for $u_{tt} + 2du_t - u_{xx} + ku = 0$ for $0 < x < 10$ with $u = 0$ at $x = 0$ and at $x = 10$. Uses a finite difference scheme.

5.7

wave2   Solves initial value problem for $u_{tt} - u_{xx} + \gamma u^3 = 0$ on $0 < x < 10$, with $u = 0$ at $x = 0$ and $x = 10$. Uses finite difference scheme.

Chapter 6

6.2

fseries   Sums n terms of the Fourier series for several different functions.

6.5

fast   Computes approximate Fourier series coefficients using DFT and fast Fourier transform of MATLAB.

ftrans   Computes an approximate Fourier transform using the DFT and fast Fourier transform of MATLAB.

Chapter 7

7.1

osc   Evaluates oscillatory integral

$$\int_{-2}^{2} f(y)e^{it\phi(t,y)}dy$$

for two choices of ϕ.

7.2

disper   Evaluates

$$u(x, t) = \int U(k) \cos(kx - \omega(k)t)dk$$

for three different dispersion relations coming from linear Klein-Gordon, linearized KdV, and vibrating beam equation.

7.4

unifm   Computes an approximation to the solution of the free Schrödinger equation $iu_t = -u_{xx}$ with initial data a uniform probability density on $[-1, 1]$.

shd   Computes solution of $iu_t = -u_{xx}$, and compares with solution of $u_t = u_{xx}$ on whole line for square pulse initial condition.

7.5

sqwell   (not used in text) Computes eigenvalues and eigenfunctions for $iu_t = -u_{xx} + Vu$ with square well potential V.

Chapter 8

8.1

heat6   Computes exact solution of $u_t - \Delta u = 0$ in R^2 with initial data which is the sum of two Gaussians.

8.5

heat7   Calculates solution of $u_t - \Delta u = 0$ in rectangle with $u = 0$ on three sides, and u given on the fourth side. Sums Fourier series.

Chapter 9

9.1

    `mvp`   Computes files for graphing of a surface together with circle and computes average around the circle.

9.2

    `dirch`   Solves the Dirichlet problem in the unit disk using Fourier series expansion of given boundary data. Fourier coefficients calculated using fast Fourier transform.

9.4

    `deflect`   Solves $-\Delta u = q$ in unit disk with $u = 0$ on boundary by integrating the Green's function against q. q is a cylindrical step function.

    `green`   Graphs the Green's function for the unit disk. User enters the coordinates (a,b) of the singularity.

Chapter 10

10.3

    `kdv`   Finds periodic solutions of the KdV equation using a pseudospectral method.

Index

Acoustic boundary conditions, 185
Acoustic waves, 160
Admissible functions, 414
Aliasing, 243
Ampere's law, 357
Analytic functions, 367
Angular frequency, 191, 194, 264, 331
Annulus, 471
Array operations in MATLAB, 481
Array smart mfiles in MATLAB, 484

Backward difference, 54
Backward Euler method, 445
Balance law of diffusion, 74, 297, 450
Band-limited, 246
Banded matrix, 427
Beam equation, 271
Bessel functions, 317, 465
Bessel's equation, 317
Bessel's inequality, 118
Boundary conditions, 95
    acoustic, 185
    first kind (Dirichlet), 95, 304
    inhomogeneous, 144, 146, 464, 469
    mixed, 305, 391
    second kind (Neumann), 95, 304
    symmetric, 131, 306
    third kind (Robin), 95, 304
Bounded

function, 2
set, 2
Boundstates of Schrödinger operator, 290, 292
Brownian motion, 75
Building-block solutions, 461
    for the heat equation, 112
        in higher dimensions, 305
    for the Laplace equation, 378
    for the wave equation, 193
Burgers' equation, 36, 46

Carrying capacity, 451
Cauchy-Riemann equations, 376
Cauchy-Schwarz inequality, 117, 221
Centered difference, 54, 425
CFL condition, 57, 58, 210
CFL number, 58
Change of variable in multiple integral, 13
Characteristic polynomial, 455
Characteristics, 24, 25, 32, 36, 166, 194
    method of, 25, 26
Charge density, 356
Circular mean, 368
Class C^k, 11
Classification of second order PDE's, 455
    elliptic, 457
    hyperbolic, 457
    parabolic, 457

Closure, 1
Cole-Hopf transformation, 104
Compact, 2
Complete orthogonal set, 119
Complete set of eigenfunctions, 133,
    308, 460
Condensation, 159
Condition at infinity, 394
Conductivity, electrical, 363
Conformal mapping, 375, 413
Conjugate harmonic functions, 376
Conservation Law, 21
Conservation law
    nonlinear, 30
    of cell growth, 65
    of traffic flow, 32
Conservation of energy
    for Maxwell equations, 361
    for the telegraph equation, 174
    for the wave equation, 172, 196
        in higher dimensions, 340, 349
Conservation of mass, 31
    for gas dynamics, 157
    for vibrating string, 161
Conservation of momentum
    for gas dynamics, 158
    for vibrating string, 163
Consistent numerical scheme, 59
Constitutive relation, 31, 163, 357
Continuous spectrum, 287, 290
Control theory, 352
Convergence
    mean square, 119, 223
    of Fourier series, 223
    of functions, 5
    of numerical scheme, 59
    of series of functions, 6
    pointwise, 5, 223
    uniform, 5, 119, 223
Convergence of eigenfunction expan-
        sion
    for the heat equation, 116
    for wave equation, 196

Convolution, 234, 283, 300, 395, 473,
    474
Correspondence principle, 264, 281
Coulomb's law, 356
Crank-Nicolson method, 109, 445
Current density, 356
Curves of constant phase, 268
Cutoff frequency, 246

d'Alembert's formula, 168, 472
de Broglie relation, 277
Delta function, 7, 398
    in higher dimensions, 398
DFT coefficients, 242
DFT coefficients as approximation to
        Fourier coefficients, 243
Dielectric constant, 357
Difference quotients, 54
Diffusion, 73
Direct method of the calculus of vari-
        ations, 417
Dirichlet boundary condition, 95, 108,
    181, 304, 465
Dirichlet problem, 378, 415
    in the annulus, 471
    in the disk, 378, 469
    in the rectangle, 389
Discrete Fourier transform (DFT), 238,
    242
Dispersion relation, 265
    for the Schrödinger equation, 278
Dispersive equations, 264
Displacement field, 356
Divergence free condition, 359
Divergence theorem, 12
Domain of an operator, 131
Domain of dependence
    for wave equation, 170, 210
        in higher dimensions, 336, 344
    of numerical scheme, 58, 210
Domain of influence, 170

Eigenfunction, 113
    approximate, 290

for the disk, 315
for the rectangle, 310
Eigenfunction expansion for periodic
function, 220
Eigenfunction expansion for the heat
equation
in higher dimensions, 308
in one dimension, 114
Eigenfunction expansion for the wave
equation
in higher dimensions, 348
in one dimension, 194
Eigenfunctions, 221
Eigenvalue problem, 112, 194, 460,
464
for the ball, 466
for the disk, 465
in higher dimensions, 305
in quantum mechanics, 288
Eigenvalues of higher multiplicity, 311
Eikonal equation, 20
Einstein hypothesis, 281
Electric field, 356
Electromagetic flux, 362
Electrostatics, 360
Elliptic, 457
Energy
for nonlinear Klein-Gordon equa-
tion, 211
for the wave equation, 171, 196
in higher dimensions, 339, 349
kinetic, 173
potential, 173
Energy density, 173
Energy inequality, 180
Entropy condition, 50
Equation of changing type, 458
Equilibrium state, 414
Error function, 88
Euler formula, 221
Euler method, 54, 485
Euler-Lagrange equation, 415
Even extension, 97
Even function, 96, 182

Explicit numerical scheme, 106, 211

Faraday's law, 356
Fast Fourier transform, 250
Fast Fourier transform, operation count,
252
Fick's law, 75, 450
Finite difference scheme, 55, 487
Finite element method
in one dimension, 433
in two dimensions, 437
Finite energy solution, 173, 332, 342
First order PDE's, 20
Fisher's equation, 451
Flux, 31, 362, 450
For loop, 486
Forward difference, 54
Fourier inversion formula, 232
Fourier matrix, 240, 241
Fourier series
complex, 221
real, 219
Fourier tranform rules, 234
Fourier transform, 231
Fourier's law of cooling, 74, 298
Free Schrödinger equation, 280
Function functions in MATLAB, 488
Function mfiles in MATLAB, 483
Fundamental mode of vibration, 195
Fundamental solution
of free Schrödinger equation, 284
of the heat equation, 84, 238
in higher dimensions, 298, 471
of the Laplace equation, 394,
473
of the wave equation, 473

Galerkin method, 442, 450
Gas dynamics, 157
Gauss-Seidel method, 429
Gaussian, 7, 232
Gaussian approximation, 94
Gaussian elimination, 482
Gear methods, 445

Generalized function, 9
Generating function for Bessel functions, 365
Ghost points for numerical scheme, 108, 432
Gibbs phenomenon, 228, 230
Global error, 59
Green's first identity, 131
Green's first identity in higher dimensions, 306
Green's function
    expansion in eigenfunctions, 413
    for half-plane, 404
    for the ball, 407
    for the disk, 406
    in electrostatics, 403
    in higher dimensions, 401
    in one dimension, 400
    properties of, 402
Green's second identity, 131
Green's second identity in higher dimensions, 307
Ground state, 292
Group lines, 266
Group velocity, 266, 267

Hard acoustic boundary condition, 185
Harmonic function, 367
Harnack inequality
    for the ball, 386
    for the disk, 385
Hat functions, 434
Heat conductivity, 74
Heat equation, 75, 298, 461
    with source, 141
    with source, 462
Heat flux, 74, 297
Heisenberg uncertainty principle, 278
Helmholtz equation, 181, 347, 399
Huyghens' principle, 336
Hyperbolic, 457

IBVP, 96

Implicit numerical scheme, 107, 211, 445
Incident wave, 184, 186
Incoming wave, 332
Initial condition
    for ODE's, 17
    for PDE, 22
Initial value problem
    for ODE's, 17
Initial-value problem
    for first order linear equation, 23
    for heat equation, 85, 237
        in higher dimensions, 300, 471
    for Maxwell equations, 357
    for nonlinear conservation law, 39, 45
    for the free Schrödinger equation, 282
    for traffic flow, 33
    for wave equation, 166
        in higher dimensions, 332
Initial-value problem for wave equation
    in higher dimensions, 472
Insulated boundary condition, 130, 310
Inverse Fourier transform, 232
Isentropic, 158
Iterative methods for solving linear systems, 428
IVP, 23

Jacobi method, 428
Jump conditions, 75, 151

Kinetic energy, 173
Kirchoff's formula, 334, 359, 472
Klein-Gordon equation
    linear, 264
    nonlinear, 211
Korteweg-deVries equation (KdV), nonlinear, 447

Korteweg-deVries equation, linearized, 270

Lanczos procedure for Fourier series, 257
Laplace equation, 367, 467
Laplace operator, 297
   in polar coordinates, 316, 378, 474
   in spherical coordinates, 372, 474
   invariance properties
      under conformal mapping, 375
      under translation and rotation, 374
      with respect to spherical means, 332, 371
Lax equivalence theorem, 59
Legendre functions, 467
Linear functional, 16
Linear operator, 16
Linear PDE, 20
Linear transformation, 15
Linearized equation, 40
Linearized equations
   for vibrating string, 164
   of gas dynamics, 159
Linearized Klein-Gordon equation
   about $u = 0$, 216
   about travelling wave, 213
Liouville's theorem, 384, 398
Local wave number, 267
Local wavelength, 267
Logistics equation, 451
Longitudinal motion of vibrating string, 165
Lowest eigenvalue as minimizer, 418

Mach number, 159
Magnetic field, 356
Magnetic induction, 356
Magnetic permeability, 357
Mass matrix, 443
Maturation velocity, 65
Maximum principle, 77, 122, 300

Maximum principle for harmonic functions, 372, 381
   half-plane, 377
Maxwell equations, 356
Mean square convergence, 119, 197, 223
Mean square error, 225
Mean value property for harmonic functions, 368
Mean value property of harmonic functions, 369, 426
Mesh for numerical scheme, 208
Method of Characteristics, 25
Method of descent, 335
Method of stationary phase, 262
mfiles in MATLAB, 483
Minimal surface problem, 416
Minimax principle for eigenvalues, 421
Minimizer, 414
Minimizing sequence, 415
Mitosis boundary condition, 67, 71
Mixed boundary condition, 305
Modes of vibration
   of drumhead, 349
   of string, 193
Momentum space, 277
Movies in MATLAB, 496

Neumann boundary condition, 95, 108, 184, 198, 304, 465
Neumann function, 411
   expansion in eigenfunctions, 413
   for the half-space, 413
   in one dimension, 412
Neumann problem, 378
   for the disk, 388, 469
Nodal curves of drumhead, 349
Nodes
   in finite element method, 437
   of vibrating string, 195
Nonnegative boundary conditions, 131
Nonuniqueness of weak solution, 48
Norm, 116

Null Theorem, 3
Nyquist frequency, 246

O and o notation, 9
Octave, 195
Odd extension, 97, 182
Odd function, 96, 182
Open set, 1
Order of partial differential equation,
      19
Ordinary differential equations, 17
Orthogonal, 116
Orthogonal linear transformation, 374
Orthogonal set, 116
Oscillatory integral, 259
Outgoing wave, 332
Overtones, 195

Parabolic, 457
Partial derivative, 11
Partial sums, 6
Pathwise-connected set, 372
Perfect conductor, 364
Phase velocity, 265, 268
Piecewise continuous, 119, 224
Plancherel theorem, 233
Plane wave solution of Maxwell equa-
      tions, 359
Plane waves, 331
Plotting 2-D graphs in MATLAB, 490
Plotting 3-D graphs in MATLAB, 492
Plucked string, 195, 197, 215
Pointwise convergence, 5, 223
Poisson equation, 322, 361
      with boundary conditions, 468
      with zero boundary condition,
            399, 417, 433
      without boundary, 394, 473
Poisson formula
      for the disk, 382
Poisson kernel
      for half-plane, 405
      for the disk, 381
Poisson kernel and heat kernel, 384

Polar coordinates in MATLAB, 493
Potential energy, 173
Poynting vector, 362
Principle of Duhamel, 89, 174, 188,
      300, 343
Principle of limiting amplitude, 345
Principle of reciprocity, 403
Principle of superposition, 22, 409
Pseudospectral method, 447
Pythagorean property, 117

Quadratic functional , 414

Rankine-Hugoniot condition, 46, 47,
      52
Rarefaction wave, 49
Rayleigh quotient, 418
Rayleigh-Ritz procedure, 417
Reaction-diffusion equation, 450
Reduced wave equation, 347
Reflected wave, 184, 186
Reflection coefficient, 187
Reflection zone, 183
Resonance, 202
Riemann function, 473
Riemann problem, 45
Riemann sum, 85, 231
Robin boundary condition, 96, 100,
      190, 304, 351

Sampling frequency, 246
Sampling interval, 239, 243, 246
Sampling theorem, 246
Scalar product, 115
      in higher dimensions, 306
      in polar coordinates, 318
Scalar product for complex functions,
      221
Schrödinger equation
      with potential, 281
      without potential, 280
Schrödinger operator , 287
Script mfiles in MATLAB, 483, 485
Sector, 470

Separation of variables, 111, 193, 389, 444, 459
   in higher dimensions, 310
Shock speed, 46
Shock wave, 46, 47
   admissible, 50
Singularity, 170
Sobolev norm, 440
Soft acoustic boundary condition, 185
Solid harmonic functions, 467
Solid mean value property of harmonic functions, 370
Solution
   strict, 41
   weak, 41
SOR (successive overrelaxations), 430
Sound speed of a gas, 159
Source-flux balance, 147
   in higher dimensions, 326, 327
Sparse matrix, 427
Specific heat, 73, 297
Spectral methods, 445
Spectrum of the Schrödinger operator, 287
Speed of light, 358
Spherical coordinates, 371, 466, 475
Spherical harmonics, 466
Spherical mean, 332, 368
Spherical wave, 332
Square well potential, 290
Stability of weak solutions, 48
Stable numerical scheme, 59
Stationary point of phase, 262, 266
Steady state solution, 123, 147, 322, 325, 378
Steady-state solution, 360, 464
Stiff system of ODE's, 445, 489
Stiffness matrix, 436, 439
String
   with fixed ends, 192
   with free ends, 198
Subharmonic, 375
Subspace, 15

Symmetric boundary conditions, 131, 306
Symmetric linear operator, 401
Symmetric sampling range, 246

Taylor expansion, 10, 14, 53
Telegraph equation, 173, 203
Tension, 162
Tent function, 438
Tent functions, 452
   space of, 438
Test function, 42
Traffic flow, 32
Transmission coefficient, 187
Transmission problem, 186
Transmitted wave, 186
Transport equation, 21
Transverse motion of vibrating string, 165
Transverse wave, 360
Travelling wave, 102, 211
Truncated cone, 340
Truncation error, 55, 106, 109, 208

Uncertainty principle, 274
Uniform convergence, 5, 119, 223
Unstable, 49
Unstable numerical scheme, 59
Unstable solution, 213
Upwind numerical scheme, 57, 58

Variation, 415
Variational problems, 414
Vector indices in MATLAB, 488
Vector space, 14
Vector subspace, 15
Vectorizing computations in MATLAB, 486
Velocity of propagation, 24, 30, 36, 265
Velocity potential, 160, 185, 198
Vibrating drumhead, 348
Viscosity, 102
Viscous effect, 102

Wave equation, 463
    with boundary source, 187, 201
    with oscillating source, 345
    with source, 174, 343, 463
Wave function, 276
Wave length, 264
Wave number, 191, 264, 331
Wave packet, 267, 279
Waves
    dispersive, 264
    nonlinear, 30
    plane, 331
    spherical, 332
    transverse electromagnetic, 360
    travelling, 102
Weak solution, 41, 169, 197
    nonuniqueness of, 48
    of conservation law, 44
    of Fisher's equation, 452
    of heat equation, 442
    of variational problem, 415
    unstable, 49
Weierstrass test, 6, 119, 226
Well-posed problem, 20, 49, 86, 169,
      198